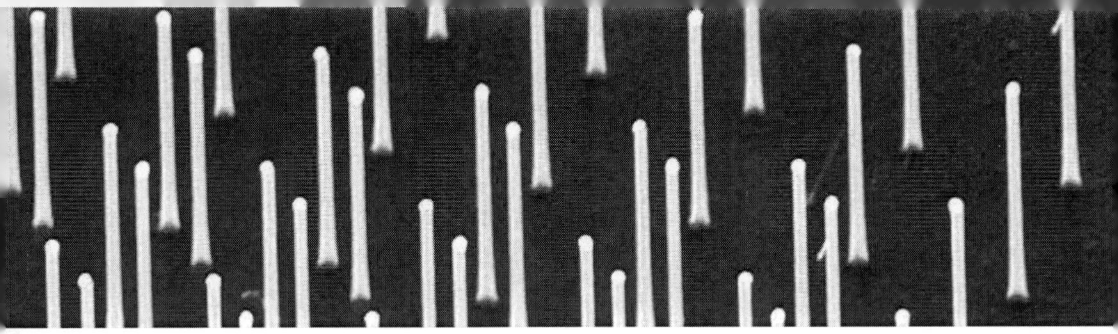

Silicon and Silicide Nanowires

edited by
Yu Huang
King-Ning Tu

Silicon and Silicide Nanowires

Applications, Fabrication, and Properties

Pan Stanford Publishing

Published by

Pan Stanford Publishing Pte. Ltd.
Penthouse Level, Suntec Tower 3
8 Temasek Boulevard
Singapore 038988

Email: editorial@panstanford.com
Web: www.panstanford.com

British Library Cataloguing-in-Publication Data
A catalogue record for this book is available from the British Library.

Silicon and Silicide Nanowires:
Applications, Fabrication, and Properties

ISBN 978-981-4303-46-0 (Hardcover)
ISBN 978-981-4303-47-7 (eBook)

Printed in the USA

Contents

Preface

Nanoscience and nanotechnology are a recent development. At the moment, the research on nanoscale materials science for nanotechnology is ubiquitous. Much has been accomplished in the processing of nanoscale materials, such as growth of silicon nanowires, as well as in the study of physical and device properties of silicon nanowires. Although the research has not reached a stage where nanotechnology can be called mature and mass production of nanodevices can be done, it is time to have a critical review of what has been done and what is the direction and potential of nanotechnology research in the near future. This is the purpose of this book, and it is limited to the processing, properties, and applications of nanowires of silicon and silicides. One of the difficulties to be overcome, for example, is the large-scale integration of nanowires. We can handle a few pieces of nanowires easily, but it is not at all trivial if we have to handle a million of them. Nevertheless, maybe it is realistic to focus first on those applications in which we just need to use a small number of nanowires, such as in biosensor devices.

This review volume has a collection of ten chapters. The first chapter by Kodambaka reviews the *in situ* transmission electron microscopy (TEM) observations of vapor–liquid–solid (VLS) growth of silicon nanowires. The direct observations of dynamic growth phenomena of Si nanowires help identify the key parameters in controlling the morphological and structural evolution of nanowires.

The second chapter by Dayeh and Picraux reviews the growth of coaxial as well as radial direction growth of Ge/Si heterostructured nanowires. The specific aspects of size effect of Ge on Si nanowire growth, defect formation during growth, and the growth mechanism are covered on the basis of thermodynamics, especially the effect of small size on VLS growth and the critical radius for suppressing the growth.

The third chapter by Higgins, Schmitt, and Jin is a review on the synthesis of metal silicide nanowires. Due to the complex multiphase behavior of silicide formation, the synthesis of silicide nanowires via chemical methods is rather complicated. Several synthetic strategies have been developed to overcome this challenge. This chapter highlights the strategies of current approaches, the future synthetic challenges, and a review of the emerging applications.

The fourth chapter by Chen and Wu is to reviews the formation of nano-silicide phases on Si substrates as well as in Si nanowires. The microstructures and the epitaxial interfaces between Si and a large number of silicides of near-noble and transition metals have been studied by high-resolution transmission electron microscopy (HRTEM), and lattice images are presented and reviewed. Controlled growth of nano-heterostructures of silicide/Si/silicide such as NiSi/Si/NiSi and PtSi/Si/PtSi are given.

The fifth chapter by Chou, Lu, and Tu reviews the point-contact reaction–induced epitaxial growth of Ni and Co silicides in Si nanowires. In this chapter, events of repeating homogeneous nucleation have been observed. Historically, on analyzing repeating homogeneous nucleation, it was found that it cannot explain the rapid rate of bulk single-crystal growth during the melting process. Thus, F. C. Frank's model of spiral crystal growth around a screw dislocation was invented. This model was used to observe repeating homogeneous nucleation in nanowires, and it enabled to directly compare theoretical and experimental studies on nucleation.

The sixth chapter by Gusak and Zaporozhers is a review on the interaction between inverse Kirkendall effect and Kirkendall effect in hollow nanoshells and nanotubes. Owing to the two surfaces in a hollow nanoshell and nanotube, a vacancy flux pre-exists in the nanostructure and it induces the inverse Kirkendall effect to affect interdiffusion and, in turn, the stability of the nanoshell and nanotube. The kinetic analysis presented in this chapter is an example of the current active study of kinetic processes in nanoscale materials.

The seventh chapter by Seike and Ohdomari is a review on electrical transport properties of doped Si nanowires under strain. It presents a comprehensive study directed toward a better understanding of electron and hole transport in *n*-type and *p*-type strained Si nanowire metal-oxide-semiconductor field-effect transistors (MOSFETs). It is aimed for applications in nano-complementary metal-oxide-semiconductor (nano-CMOS) circuits and systems, for example, in 32 nm process node, and hopefully to extend it into the 10 nm process node regime. In this nanoscale regime, stress plays the most important role in carrier transport.

Thus, the effect of stress and strain on carrier transport, especially on channel transconductance is discussed.

The eighth chapter by Hu, Cui, Hong, McDonough, and Cui is a review on silicon nanowires and related nanostructures to serve as Li-ion battery anodes. Owing to its highest-known theoretical capacitance (~4200 mAh/g), Si holds great potential as anode material for Li-ion batteries. Yet, its high volume expansion in absorbing Li ions is a reliability concern. Hence, nanosize anode materials, such as Si-based carbon composites, are reviewed in this chapter. The improved electrochemical performance of Si-based anodes for applications is discussed.

The ninth chapter by Qu and Duan is a review on porous silicon nanowires. Due to deep quantum confinement effect and/or complex surface states, the porous Si nanowires show strong photo- and electro-luminescence in visible light range and thus have potential applications in Si-based optoelectronics, bio-imaging, and biosensors. The recent progress on the synthesis, characterization, properties, and pore-formation mechanism are summarized.

The tenth chapter by Lin and Huang reviews nanoscale contact engineering for Si nanowire devices. The one-dimensional nanostructure, such as nanotubes and nanowires, are attractive building blocks for nanoelectronics. Making reliable electrical silicide contacts to individual silicon nanowire devices is one of the key factors that determine the nanodevice performance and reliability. This chapter reviews the detailed synthetic approaches, silicide growth kinetics, electronic properties, and device applications of nanoscale silicide/silicon/silicide heterostructures formed by solid-state reactions.

As editors, we hope that this book would serve as a stepping stone for readers to reach a higher level of scientific understanding and an advanced field of device applications.

Yu Huang
King-Ning Tu
Autumn 2013

Chapter 1

In situ Observations of Vapor–Liquid–Solid Growth of Silicon Nanowires

S. Kodambaka
Department of Materials Science and Engineering, University of California, Los Angeles, CA 90095, USA
kodambaka@ucla.edu

1.1 Introduction

Low-dimensional materials, such as nanowires, have attracted considerable attention over the last decade due to their potential for forming structures with tailored composition, crystallinity, and morphology and hence with tunable electronic, optical, magnetic, mechanical, and other properties. The most widely used approach to grow nanowires involves deposition from vapor phase in the presence of a liquid catalyst and is commonly known as vapor–liquid–solid (VLS) process.[1] The VLS method was first proposed in the early 1960s for the Au-mediated growth of Si whiskers.[2–5] In this process, the metal alloys (e.g., Au–Si eutectic) with the depositing material and forms a liquid alloy which promotes material incorporation at the liquid–solid interface and results in the

Silicon and Silicide Nanowires: Applications, Fabrication, and Properties
Edited by Yu Huang and King-Ning Tu

ISBN 978-981-4303-46-0 (Hardcover), 978-981-4303-47-7 (eBook)
www.panstanford.com

formation of one-dimensional (1-D) pillars or "wires." Variants of VLS method include the so-called vapor–solid–solid (VSS) process, where solid instead of liquid catalyst promotes 1-D growth;[6] self-catalyzed process, where one of the components of the growing material acts as a catalyst;[7] and oxide- and template-assisted processes, where oxygen and patterns aid wire formation.[8] Over the past 50+ years, these methods have been used to grow whiskers and nanowires of a wide variety of elemental metals, semiconductors as well as compound arsenides, borides, carbides, nitrides, oxides, phosphides, selenides, sulphides, and tellurides.[9] An exhaustive list of all the materials grown, methods employed, catalysts and growth parameters used can be found in Ref. 10.

Recently, nanowires of silicon have generated a lot of interest for applications in Li ion batteries and in thermoelectric devices.[11–14] Given the growing demand for sustainable energy, large-scale, low-cost, production of high-efficiency energy storage and harvesting devices is highly desirable. Nanowires can potentially offer great advantage in the fabrication of such devices because they can form single-crystalline, defect-free structures with atomically abrupt interfaces and surfaces at desired locations even on amorphous or lattice mismatched substrates.[15] To-date, Si nanowires have been grown via VLS using Au,[1] Ag,[1] Ga,[16] Au–Ga,[17] and In,[18] and via VSS using Ti,[19] Al,[20] Ni,[21] Cu,[22] and Pd.[23] Over the last decade, remarkable progress has been made in the areas of nanowire synthesis, device fabrication, and characterization. Readers interested in the growth and properties of Si nanowires can find more details in the review by Schmidt and co-workers.[24] Recent attempts to obtain desired nanowire heterostructures have yielded limited success and have proven to be a challenging task.[25–27] Improvements in the performance of nanowire-based devices and the development of large-scale fabrication procedures, however, require a fundamental understanding of and precise control over morphological, compositional, and structural evolution of nanowires.

Pioneering studies by Bootsma and Gassen[28] and by Givargizov[29–31] provided important insights into the VLS growth process. The classic analysis of Givargizov[29] predicted that narrower wires should grow more slowly than wider wires, that is, the growth rate dL/dt increases with wire diameter d, due to the Gibbs–Thomson effect,[32] with a critical diameter below which growth cannot occur. The opposite behavior, that is, dL/dt decreasing with d, is expected when the rate-limiting step is either diffusion through the droplet or accumulation at the catalyst–wire interface via surface diffusion.[33] For the growth of Si nanowires, some experiments[29] agree with Givargizov's prediction while others do not.[34–37] In most of these reports, *ex situ* post-growth measurements of nanowire lengths were used to determine the growth rates. While post-growth characterization is essential for the determination of chemical composition, interface structure, and wire morphologies, inference of the growth mechanisms from such data can be difficult. *In situ* studies offer a potential solution.

In situ experiments are ideally suited to follow the kinetics of nucleation and growth of individual nanowires and to quantitatively determine the underlying mass transport mechanisms and reaction pathways. The advent of sophisticated *in situ* characterization techniques has enabled monitoring of dynamic materials phenomena such as nucleation and growth. A variety of methods, such as scanning and transmission electron microscopies [scanning electron microscopy (SEM) and transmission electron microscopy (TEM)], optical reflectometry, and X-ray diffraction, have recently been used to follow the nucleation and growth of nanowires.[23,38–49]

Of all these techniques, *in situ* microscopy is probably the first method that was used over 40 years ago to observe the growth of nanowires. (Readers are advised to check the references within the book chapter by Givargizov.[10]) *In situ* microscopy permits *direct* determination of nanowire growth, helps minimize the uncertainties in growth-rate measurements, and, since the structures are observed under growth conditions, often reveals surprising and previously unknown aspects of the growth process as we show in the following sections. Moreover, the ability to directly observe the changes as a function of the growth parameters (time, temperature, and/or flux) not only saves enormous amount of time required otherwise but also helps in identification of the key factors influencing the growth. Hence, *in situ* characterization has the potential to promote rapid progress in the development of methods for the fabrication of complex nanowire architectures. Here, we review the recent *in situ* ultra-high vacuum transmission electron microscopy (UHV TEM) observations of the nucleation and growth of Si nanowires via the VLS process at relatively low fluxes. Complementary *in situ* TEM studies of nanowire growth at higher fluxes have been reported elsewhere.[23,47] In the following sections, we will present experimental results focused on the kinetics of nucleation and growth of Si nanowires using Au as the catalyst.

1.2 Experimental

All the experimental results presented in this chapter are obtained using a custom-designed Hitachi H-9000 UHV (base pressure 2×10^{-10} Torr) TEM with a LaB_6 filament and operated at 300 kV. The microscope is part of a multi-chamber system, located at the IBM T. J. Watson Research Center, and is equipped with a thermal evaporator and gas-dosing facilities for physical and chemical vapor deposition (PVD and CVD), respectively (see Fig. 1.1).[50]

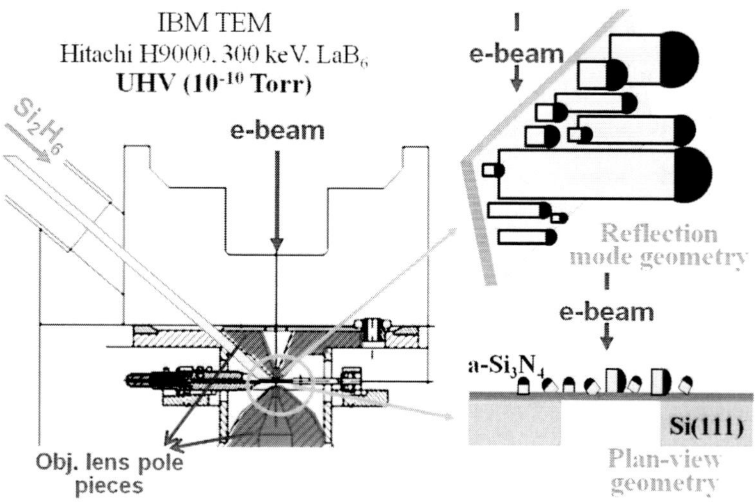

Figure 1.1 (Left) Schematic of the custom-designed sample column, part of the IBM UHV TEM. (Right) The top and bottom images show reflection mode and plan-view geometries of the samples used for nanowire growth and nucleation experiments, respectively.

For CVD, precursor gases, such as disilane (Si_2H_6) for Si deposition, are introduced into the sample region through UHV leak valves which facilitate precise control of the total pressure in the system. In this particular UHV-TEM, the maximum attainable pressures are limited to $\sim 10^{-4}$ Torr during the operation of the TEM. (In other systems, however, it is possible to vary the pressure over wider range, e.g.,

1 Torr and higher, with the aid of differential pumping or by using a high-pressure cell.[51,52]) Since the permissible pressures are lower than that typically used for nanowire growth in a CVD reactor, the nanowire nucleation and growth rates are also lower, which is ideal for real-time monitoring of the growth phenomena. Moreover, low doses of oxygen or water vapor, the common contaminants in a CVD reactor, can be controllably introduced via the leak valves and their effect investigated.

In order to carry out the *in situ* experiments, a specially designed double-tilt holder fitted with electrical leads is used to resistively heat the samples. For the nanowire nucleation and growth experiments, two different sample geometries as shown in Fig. 1.1 are used. Plan-view samples work best for observation of nanowire nucleation and early-stage growth processes while reflection-mode geometry is best suited for the observation of the growth of nanowires.[53] Chemically-thinned polished Si(111) wafers with ~80 nm–thick amorphous silicon nitride (a-SiN_x) films are used as plan-view samples. The a-SiN_x membranes are thermally stable, chemically inert, relatively easy to clean, and hence are ideal for the nucleation experiments. These samples are cleaned in the UHV preparation chamber, attached to the microscope by slow heating to ~600 °C, and held until the base pressure is below 5×10^{-10} Torr. For the nanowire growth experiments, rectangular pieces (4 mm × 1.5 mm × 0.5 mm) of Si(111) are cut from a Si(111) wafer, cleaned chemically by dipping in HF solution to remove the native oxide, and mounted such that one of the polished surfaces is parallel to the electron beam as shown in Fig. 1.1. The samples are cleaned in the preparation chamber by a series of heat treatments: Samples are first heated slowly to ~600 °C and held for a few hours until the base pressure in the system is below 5×10^{-10} Torr. Then, the samples are heated rapidly to ~1250 °C and held for 30–60 s after which they are cooled to ~600 °C. This procedure yields oxide-free Si surface, desirable for accurate description of the growth phenomena. For both the nucleation and growth experiments, ~2 nm–thick Au films are evaporated onto the samples in UHV. (In case of Au–Al alloys, a bilayer film composed of Au and Al was prepared by sequential evaporation with an atomic ratio of 2:1.) The samples are then transferred without breaking vacuum to the microscope and heated to the desired temperature (450 ~ 650 °C) in presence of Si_2H_6 gas at pressures between 1×10^{-8} and 1×10^{-5} Torr. TEM images are acquired at video rate (30 frames per second) using bright- or dark-field imaging conditions and digitized for further analysis. In order to check for and minimize the electron-beam irradiation effects, nucleation and growth kinetics data were collected with continuous as well as intermittent electron beams. In all of our studies, the electron beam does not affect the kinetics. Substrate temperatures are measured before and after the experiments using an infrared pyrometer. For Si samples, the surface can be cleaned again by heating to 1250 °C so that a series of growth experiments can be carried out on the same sample at one area. Each of the nucleation experiments, however, was carried out on new a-SiN_x samples.

1.3 Silicon Nanowire Nucleation Kinetics

All of the nucleation experiments are carried out using Au-covered a-SiN_x plan-view samples.

Upon heating the samples in presence of disilane to temperatures above, the Au–Si eutectic temperature (~363 °C),[54] the solid Au crystallites alloys with the deposited Si. With increasing deposition time, Si concentration in the alloy reaches the eutectic alloy composition, at which point Au–Si alloy will exist as a liquid. Further deposition will lead to Si supersaturation in the liquid alloy followed by nucleation of solid Si (red dot in Fig. 1.2).

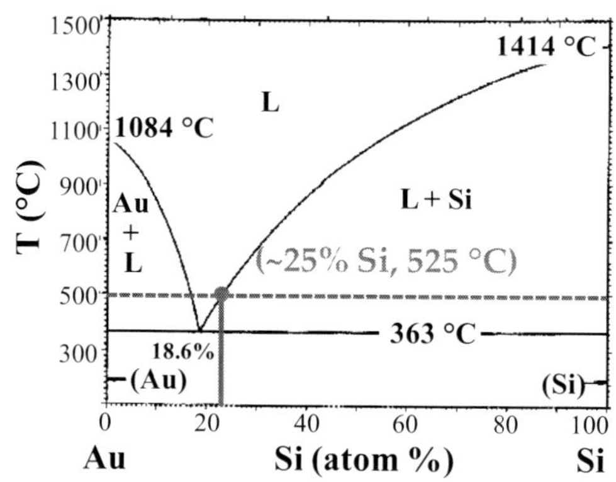

Figure 1.2 Au–Si binary phase diagram, adapted from Ref. 54. The dotted blue line shows Au–Si alloy composition versus Si concentration at 525 °C. The red dot indicates the point after which solid Si will nucleate from the Au–Si liquid alloy.

A schematic and the *in situ* TEM data of this nucleation and early-stage growth processes are shown in Fig. 1.3. After the nucleation event, deposited Si is incorporated at the liquid–solid interface and results in the wire growth. A similar behavior has also been observed in relatively higher pressure growth experiments.[23]

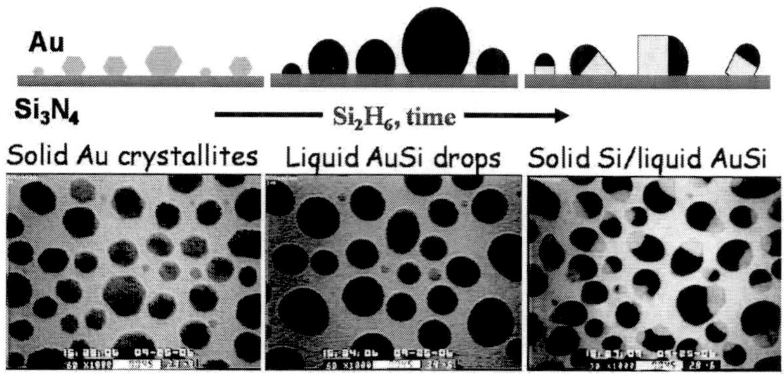

Figure 1.3 (Top) Schematic of Si nanowire nucleation process. Solid Au transforms to liquid Au–Si alloy followed by Si nucleation during disilane deposition onto solid Au crystallites. (Bottom) Bright-field TEM images acquired during Si deposition using 8×10^{-7} Torr disilane at 525 °C. The left, center, and right images show solid Au crystallites, Au–Si eutectic droplets, and a mixture of solid Si (lighter grey) and liquid Au–Si alloy (darker grey), respectively.

The benefit of *in situ* observations is that one can quantify the observed phenomena using the time-resolved data acquired over a range of pressures and temperatures. In this particular case, the kinetics of nucleation can be analyzed by following the early stages of growth.[55] While the exact amount of Si required for liquid alloy formation and for subsequent nucleation depends on the growth flux, time, and temperature, in all cases, nucleation of Si was observed at the droplet edges. This behavior is attributed to the minimization of nucleation barrier at the triple-phase lines, i.e., solid–liquid–vapor boundaries, which exist at the edges of the droplets.[56]

From the TEM movies, times corresponding to the nucleation events are identified and the sizes of nucleus measured. The upper panel in Fig. 1.4 shows typical TEM images of the nucleation process. The higher magnification image sequence in the middle panel highlights one such event in one of the droplets. From a series of such

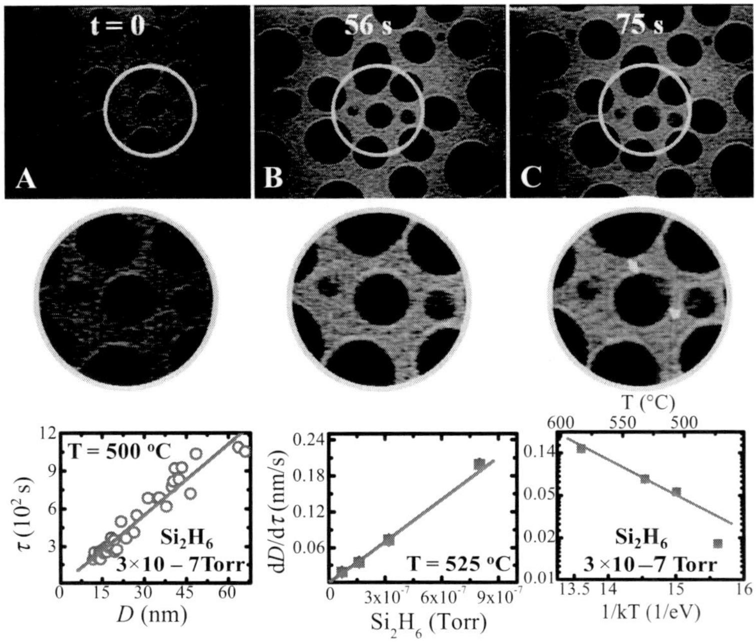

Figure 1.4 (Upper panel) A series of TEM images showing a single nucleation event during disilane deposition (8×10^{-7} Torr) at T = 525 °C. (Middle panel) Higher magnification images from the region highlighted by a circle in the upper panel. (Bottom panel) Plots of (A) incubation time τ versus Au crystallite size D, (B) $dD/d\tau$ versus disilane pressure, and (C) $dD/d\tau$ vs. T.

images acquired for several Au crystallites as a function of deposition time, the total time τ taken to nucleate Si at each crystallite of size D is determined. We refer to this time as the incubation time. A typical plot of τ versus D is shown in Fig. 1.4A for a given disilane pressure and growth temperature. We note that τ increases linearly with D, that is, the incubation time required for nucleation of Si increases linearly with the initial Au crystallite size at a constant rate $dD/d\tau$. Similar behavior is observed in all the experiments carried over a range of temperatures and pressures. From the τ versus D data obtained from these experiments, $dD/d\tau$ was measured. Figures 4B and 4C show plots of $dD/d\tau$ versus disilane pressure at a constant temperature T and $dD/d\tau$ versus T at a constant disilane pressure, respectively. Note the linear relationship between $dD/d\tau$ and disilane pressure. And, the $dD/d\tau$ versus T is indicative of a thermally activated process. Based upon this data, the rate-limiting step for the observed process can be determined. There are three basic steps: (*i*) deposition of Si from the vapor onto the Au crystallite, (*ii*) diffusion through the Au (or Au–Si alloy), and (*iii*) precipitation of Si from the liquid. Assuming that the Si deposition from the vapor is an irreversible process, reasonable at these relatively low temperatures, the rate-limiting step for Si nucleation is determined as the thermally activated, dissociative adsorption of disilane. This result is consistent with the Si nanowire growth kinetics discussed in the later sections.[37]

In addition, the above-presented *in situ* TEM data can be used to measure the critical supersaturation required for nucleation. For this purpose, the nucleus size $r(t)$ is measured as a function of deposition time t. Figure 1.5 shows a representative plot of r versus t. The data is extracted from a measurement sequence such as the one shown in Fig. 1.4. Note the rapid increase in the nucleus size at initial times followed by a slower growth at later times. The point of crossover, labeled C in the plot, is a measure of the critical supersaturation in the droplet.

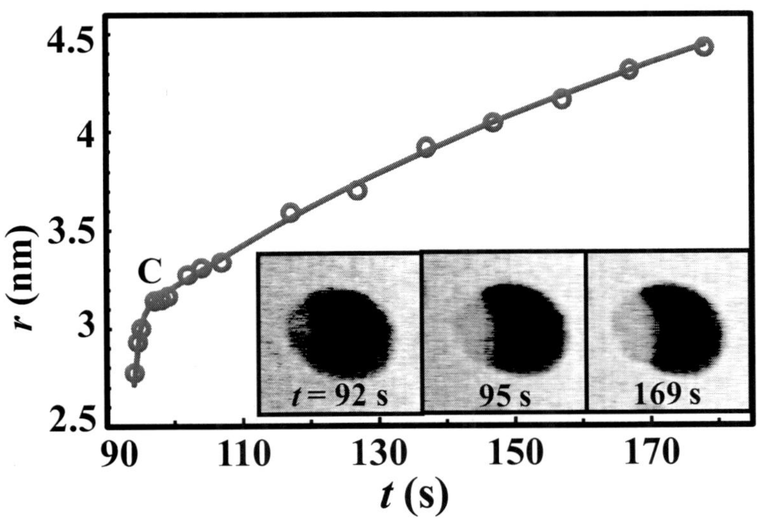

Figure 1.5 Plot of Si nucleus size r as a function of time t during deposition. Disilane pressure = 1.5×10^{-6} Torr) and T = 585 °C. The inset shows a series of TEM images of growing Si along with the droplet. The point of crossover in the data is indicated by the alphabet C. The solid curve is fit to the data obtained using the model described in Ref. 55.

This crossover is observed in all of the experiments and the exact shape of the curve $r(t)$ depends on the growth conditions.

Qualitatively, the observed process can be understood as follows. There are three basic steps involved in the process of nucleating solid silicon. (1) The preferential dissociation of disilane at the Au (or Au–Si alloy) surface, (2) formation of an Au–Si alloy formation which eventually transforms to a liquid alloy with increasing concentration of Si, and (3) precipitation of solid Si from the Au–Si alloy liquid after further deposition. This is because as time progresses, Si concentration increases in the liquid alloy. From the binary phase diagram, solid Si should precipitate out when the Si concentration exceeds the thermodynamic equilibrium composition of the alloy at that temperature, for example, the red dot in Fig. 1.2 at T = 525 °C. However, the presence of a barrier for nucleation would lead to supersaturation of the liquid. At some critical supersaturation, this barrier, vanishes nucleation will occur, and the nucleus grows rapidly at the expense of excess Si in the droplet. Subsequent growth is relatively slower and is controlled by the growth flux.

In order to quantitatively describe the nucleation kinetics, a simple kinetic model has been developed based upon the assumptions that all the liquid droplets are independent from one another, all solid and liquid phases maintain a constant shape during the entire process, and that the Gibbs–Thomson effect is negligible at the observed length scales. The model predicts that the rate of growth of the Si nucleus is proportional to the Si supersaturation in the droplet and is given by the following relation,[55] $\frac{dr}{dt} = k_{\mathrm{LS}}\left[\alpha\left(c_n - c_0\right) + k_{\mathrm{VL}}\frac{P}{R}\left(t - t_n\right) - \left(\frac{r}{R}\right)^3\right]$,

where R is the Au–Si alloy droplet size at the time of nucleation t_n, and r is the size of the growing nucleus that satisfies the condition $r = 0$ at $t = t_n$. In the above equation, the term $(c_n - c_0)$ is a measure of the supersaturation at t_n. The second term (P/R) accounts for the increase in Si concentration c due to dissociation of the disilane gas while the third term (r^3/R^3) corresponds to the loss of Si from the liquid droplet due to the capture by the nucleus. α is a dimensionless parameter that depends on the shape of the particle. The terms k_{VL} and k_{LS} are geometry-corrected vapor–liquid and liquid–solid rate constants, respectively. The solid curve in Fig. 1.5 is fit to the experimental data using this equation. Clearly, the simple kinetic model explains the observed phenomena. Moreover, from similar fits to the data obtained at different temperatures and pressures, the critical supersaturation of Si in the Au–Si liquid alloys has been measured. For details, readers are encouraged to refer to Ref. 55 and the associated supplementary material.

In summary, the phenomenon of nucleation at the nanoscale has been observed *in situ* in a TEM. Simple kinetic models have been developed to quantitatively explain the data. The experimental and theoretical approaches implemented and/or developed are general and applicable to other material systems. The development of nanotechnology relies on precise placement and control over the nanoscale structure and composition of each device. And, since nucleation is essential to the synthesis of any material, there is a need to understand the kinetic factors underlying nucleation at the nanoscale. Hence, we expect that similar studies will be carried to investigate the kinetics of nucleation in other materials.

1.4 Silicon Nanowire Growth Kinetics

In this section, we will review the recent literature on *in situ* observations of the VLS growth of Si nanowires in a UHV TEM. All the experiments are carried out on clean, Au-covered Si(111) substrates mounted in the reflection-mode geometry. The samples are then heated to temperatures above the Au–Si eutectic temperature (~ 363°C) to form liquid Au–Si alloy droplets. Upon the introduction of disilane gas, Si is deposited preferentially under the droplet-substrate interface and results in the growth of nanowires. As the nanowires grow away from the substrate, they are observed in bright or dark field in transmission mode.

In situ observations provide a wealth of information regarding the nanowire growth orientation, structure, crystallinity, and shapes of the wire and the droplet. This is best illustrated in Fig. 1.6, which shows a series of TEM images acquired *in situ* during the growth of Si nanowires. The characteristics of VLS growth are readily visible in the image—1-D morphology with a smoothly curved feature at the tip, characteristic of a liquid droplet, which in this case is the Au–Si alloy. From the images, we notice that the liquid–solid interface between the nanowires and the catalyst droplets is planar. Most of the nanowires in this experiment, carried out in "clean" environment grow along ⟨111⟩. (In the presence of small amounts of oxygen, however, Si nanowires grow along ⟨110⟩.[57]) From the TEM images in

Fig. 1.6 and electron diffraction patterns (not shown), we find that the Si nanowires grown in the UHV TEM are single crystals. A few of the wires show structural defects, such as stacking faults and twins, highlighted in Fig. 1.6A.

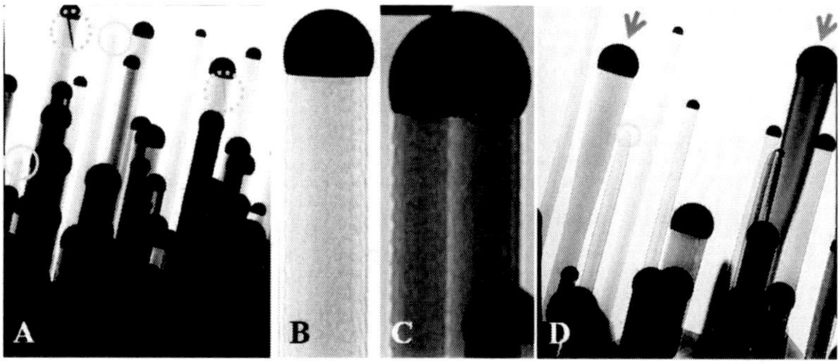

Figure 1.6 Collage of typical *in situ* bright-field TEM images acquired from Si(111) samples during the growth of Si nanowires via VLS using disilane as the Si source and Au as the catalyst. The images show single-crystalline, mostly ⟨111⟩-oriented, Si nanowires with liquid Au–Si alloy droplets (darker contrast smoothly curved features) at the tips. Solid circles indicate the absence of Au–Si droplets on a few wires (see A and D). Dotted circles highlight structural defects that are occasionally found in the wires. (B and C) Higher magnification images of surface facets and droplet shapes. (B) Under appropriate imaging conditions, surface structure of the wires can be better visualized. The arrows in D show tapered nanowires with increasing diameter.

In Au–Si system, all the Si nanowires oriented along ⟨111⟩ are bound by sawtooth facetted surfaces.[58] These surface facets are clearly visible in the TEM (see Fig. 1.6B,C). Moreover, using the appropriate imaging conditions, one can resolve the apparent 3D structure of the nanowires (see Fig. 1.6C). The shapes of the droplets can be measured as a function of deposition parameters and wire diameter. These data were found to be essential in understanding the mechanisms controlling the Au-catalyzed growth of Ge nanowires.[59] Note that the images also show nanowires with a range of diameters. This is because of the uncontrolled dewetting of the Au film during annealing and the variation in droplet size via Ostwald ripening. While this is undesirable for the large-scale fabrication of nanowire devices, we take advantage of this variation in diameters to quantify the growth kinetics.

Another interesting and probably surprising aspect that is observed in the images is the variation in the nanowire diameters along their lengths. As seen in Figs. 1.6A,D, a few of the wires are tapered with wider bases and narrower tips, while others are narrower at the bases and wider at the tips. Such morphologies are often attributed either to direct non-catalyzed deposition on the sidewalls,[28] or to increased accumulation of the material at the tips.[7] *In situ* observations, discussed below, reveal that tapering can also be due to Ostwald ripening of the catalyst droplets during growth.[60]

Real-time observations provide insights into the morphological evolution of nanowires, as shown in Fig. 1.7.

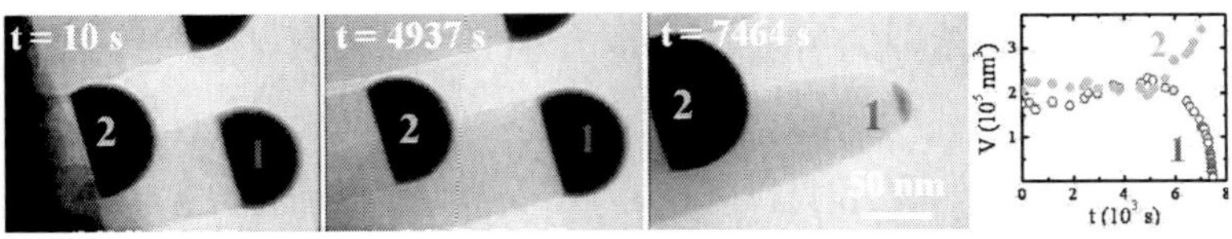

Figure 1.7 *In situ* TEM images acquired during the growth of Si nanowires at T = 655 °C using disilane pressure of 1 × 10^{-6} Torr. The time t = 0 is arbitrarily defined as the time at which the first image in this measurement sequence was acquired. The plot shows time-dependent change in volumes of droplets on two wires labeled 1 and 2. Adapted from Refs. 60 and 61.

The images show morphologies of two Si nanowires along with their droplets obtained at various times during growth. The associated plot shows droplet volume V as a function of deposition time t. Note that volume of the droplet on the wire labeled 1 decreases, while that of wire 2 increases. This behavior is characteristic of Ostwald ripening, a curvature-driven phenomenon. In the Au–Si system, the coarsening/decay of Au–Si alloy droplets occurs as a result of Au diffusion to/away from the droplets.[60,61]

In addition to direct visualization of nanowire growth, *in situ* TEM allows quantitative description of the kinetics from the measurements of the rates of growth of nanowires as a function of diameter and deposition parameters such as time, flux, and temperature. In our experiments, we find that the axial growth rate, dL/dt, can be controllably varied over three orders of magnitude.

From the TEM images such as Fig. 1.6A, acquired as a function of deposition time, dL/dt data were extracted for several wires with different diameters d. Fig. 1.8 shows plots of dL/dt as a function of d, P, and T. For the Si nanowires grown using disilane with Au as the catalyst, under our growth conditions, dL/dt is practically invariant with d, increases linearly with P, and increase exponentially with T. The measured activation energy of 0.53 ± 0.02 eV for the VLS growth of Si wires is nearly one-fourth of the activation energy associated with the same process on a non-catalyzed Si surface, indicative of the catalytic effect of the droplets.[62]

The experimental data shown in Fig. 1.8 can be understood as follows. During VLS growth of nanowires, the first step is the adsorption followed by dissociation of the gas precursor. This can occur at the catalyst surface, on the sidewalls, and/or on the substrate. In order to form a 1-D wire, the deposited adatoms will then have to diffuse to the wire–droplet interface. Finally, the material is incorporated at the interface. For Au–Si, the rate-limiting step for Si wire growth is the dissociative chemisorption of disilane at the Audi alloy droplet.[28]

In conclusion, *in situ* TEM revealed new and surprising aspects of the VLS process of silicon nanowire growth. The images provided direct evidence for the presence of a liquid catalyst, characteristic of the VLS process. From the wire morphologies and time-dependent observations, important insights into the growth dynamics and the droplet stability were obtained. These studies are useful and probably essential for developing large-scale fabrication of precisely controlled nanostructures.

1.5 Summary and Outlook

In this book chapter, an overview of *in situ* electron microscopy studies of silicon nanowire nucleation and growth via vapor–liquid–solid process was presented. The combination of real-space and real-time observation of phenomena such as the nucleation and growth kinetics enables accurate determination of the underlying reaction and diffusion mechanisms.

One of the unique advantages of an UHV TEM is the ability to controllably vary the composition of the environment during *in situ* experiments. This capability can be used to investigate the influence

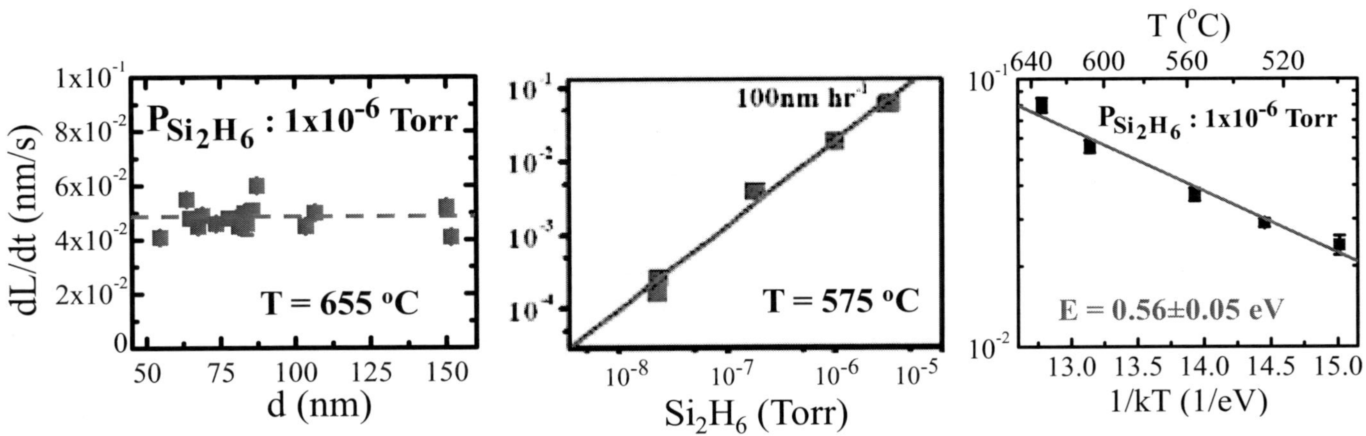

Figure 1.8 (a) Plot of axial growth rate dL/dt as a function of Si nanowire diameter d. The data is obtained from Si nanowires grown at $T = 655$ °C using disilane pressure of 1×10^{-6} Torr. (b) Log–log plot of dL/dt versus pressure at 500 °C. (c) Arrhenius plot of dL/dt versus T for Si nanowires grown using 1×10^{-6} Torr disilane. The dotted and solid lines in all the plots are least-squares fits to the data. Adapted from Ref. 37.

of typical dopants, such as boron and phosphorous, contaminants such as oxygen and water vapor, common to most conventional CVD reactors, and alloying elements such as germanium on the growth structure, composition, and morphology of nanostructures. *In situ* TEM observations (not discussed here) also show that the presence of oxygen during Si nanowire growth leads to a change in growth orientation and significantly influences the morphology. Typically, narrower and longer ⟨110⟩-oriented wires are obtained with the introduction of oxygen either in the vapor phase or on the substrate. The Au–Si alloy droplet is retained at the tip of the nanowires as they do not coarsen/decay during growth. As a result, Si nanowires grow longer with minimal tapering. *In situ* microscopy studies indicate that oxygen prevents (or minimizes) Au migration on the Si nanowire surfaces and helps retain the catalyst droplets at the nanowire tips. In the Au–Ge system, Ge nanowires were observed to grow via VLS *below* the bulk Au–Ge eutectic temperature. *In situ* TEM measurements of the Au–Ge droplet shapes showed that stabilization of the liquid phase required high Ge concentrations in the droplet.[59] More recent studies revealed the importance of the catalyst composition on nanowire heterostructure growth. The use of Au-Al alloys instead of pure Au was found to be better for the growth of atomically abrupt interfaces in Si–Ge nanowire heterostructures. *In situ* TEM studies of the effect of Al on nanowire growth kinetics was vital to this achievement.[63]

Overall, *in situ* observations provide unique information, allowing the effects of growth parameters to be quantified for individual wires and allowing observation of the structures present during growth, rather than after cooling and exposure to air. The technique has thus far provided valuable insights into the physical processes controlling the morphological, structural, and compositional evolution of Si, Ge, and Si/Ge heterostructure nanowires. While the use of low-pressures can be perceived as an advantage for such studies, the conclusions may not be directly applicable to experiments carried out in conventional high-pressure systems. Some of the recent literature does indicate that the growth kinetics are different even while using the same precursors and catalysts.[47] However, the experimental and theoretical approaches implemented here are general and can be applied to investigate nucleation and growth kinetics in other similar materials systems. In addition to TEM, the use of complementary *in situ* techniques, such as SEM, can be useful in obtaining a holistic understanding of the growth dynamics. For example, SEM helps determine the 3D shapes of catalysts and wires and wire growth orientations during growth while TEM can identify structure defects in the wires, orientation relationships between solid catalyst and wire, and can visualize the growth dynamics with near-atomic resolution. The combination of multiple characterization techniques are likely to yield better insight into the factors influencing the growth morphologies, determine accurately the chemical reaction pathways and mass transport mechanisms, and extract activation energies for the underlying processes. The recent growing demand for high-efficiency energy storage and harvesting devices will require revolutionary breakthroughs in the development of materials and device architecture. We expect that *in situ* studies will be essential in addressing these challenging tasks.

Acknowledgments

All of the results presented in this book chapter were obtained at the IBM T. J. Watson Research Center in Yorktown Heights, NY. I am grateful to F. M. Ross, J. Tersoff, M. C. Reuter, J. B. Hannon, R. M. Tromp, B. J. Kim, C.-Y. Wen, E. A. Stach, A. W. Ellis, and K. B. Reuter for help with the experiments, interpretations, and for providing access to the data.

References

1. Wagner, R.S. and Ellis, W.C., Vapor-liquid-solid mechanism of single crystal growth, *Appl. Phys. Lett.*, 4, 89–90 (1964).
2. Wagner, R.S., Ellis, W.C., Arnold, S.M. and Jackson, K.A., Study of filamentary growth of silicon crystals from vapor, *J. Appl. Phys.*, 35, 2993–3000 (1964).
3. Wagner, R.S. and Doherty, C.J., Controlled vapor-liquid-solid growth of silicon crystals, *J. Electrochem. Soc.*, 113, 1300–1305 (1966).
4. Wagner, R.S., Defects in silicon crystals grown by VLS technique, *J. Appl. Phys.*, 38, 1554–1560 (1967).
5. Wagner, R.S., *Crystal Growth*, Pergamon Press, Oxford & New York, 45–119 (1967).
6. Lensch-Falk, J.L., Hemesath, E.R., Perea, D.E. and Lauhon, L.J., Alternative catalysts for VSS growth of silicon and germanium nanowires, *J. Mater. Chem.*, 19, 849–857 (2009).

7. Woo, R., Gao, L., Goel, N., Hudait, M., Wang, K.L., Kodambaka, S., and Hicks, R.F., Kinetic control of self-catalyzed indium phosphide nanowires, nanocones, and nanopillars, *Nano Lett.*, 2009, 9, 2207.
8. Mohan, P., Motohisa, J. and Fukui, T., Controlled growth of highly uniform, axial/radial direction-defined, individually addressable InP nanowire arrays, *Nanotechnology*, 2005, 16, 2903.
9. Givargizov, E.I., Growth of whiskers by VLS mechanism, *Curr. Top. Mater. Sci.*, 1, 82–145 (1978).
10. Givargizov, E.I., *Highly Anisotropic Crystals*, Springer-Verlag, New York, 70–229 (1987).
11. Tian, B., Zheng, X., Kempa, T.J., Fang, Y., Yu, N., Yu, G., Huang, J. and Lieber, C.M., Coaxial silicon nanowires as solar cells and nanoelectronic power sources, *Nature*, 449, 885–890 (2007).
12. Hochbaum, A.I., Chen, R., Delgado, R.D., Liang, W., Garnett, E.C., Najarian, M., Majumdar, A. and Yang, P., Enhanced thermoelectric performance of rough silicon nanowires, *Nature*, 451, 163–167 (2008).
13. Boukai, A.I., Bunimovich, Y., Tahir-Kheli, J., Yu, J.-K., Goddard III, W.A. and Heath, J.R., Silicon nanowires as efficient thermoelectric materials, *Nature* 451, 168–171 (2008).
14. Chan, C.K., Peng, H., Liu, G., Mcilwrath, K., Zhang, X.F., Huggins, R.A., and Cui, Y., High-performance lithium battery anodes using silicon nanowires, *Nat. Nanotechnol.*, 3, 31–35 (2008).
15. Eymery, J., François Rieutord, F., Favre-Nicolin, V., Robach, O., Niquet, Y.-M., Fröberg, L., Mårtensson, T., and Samuelson, L., Strain and shape of epitaxial InAs/InP nanowire superlattice measured by grazing incidence X-ray techniques, *Nano Lett.*, 7, 2596–2601 (2007).
16. Sharma, S. and Sunkara, M.K., Direct synthesis of single-crystalline silicon nanowires using molten gallium and silane plasma, *Nanotechnology*, 15, 130–134 (2004).
17. Lugstein, A., Steinmair, M., Hyun, Y.J., Bertagnolli, E., and Pongratz, P., Ga/Au alloy catalyst for single crystal silicon-nanowire epitaxy, *Appl. Phys. Lett.*, 90, 023109 (2007).
18. Iacopi, F., Richard, O., Eichhammer, Y., Bender, H., Vereecken, P.M., De Gendt, S. and Heyns, M., Size-dependent characteristics of indium-seeded Si nanowire growth, *Electrochem. Solid-State Lett.*, 11, K98–K100 (2008).
19. Kamins, T.I., Williams, R.S., Chen, Y., Chang, Y.-L. and Chang, Y.A., Chemical vapor deposition of Si nanowires nucleated by $TiSi_2$ islands on Si, *Appl. Phys. Lett.*, 76, 562–564 (2000).

20. Wang, Y., Schmidt, V., Senz, S. and Gösele, U., Epitaxial growth of silicon nanowires using an aluminium catalyst, *Nature Nanotechnol.*, 1, 186–189 (2006).

21. Hsu, J.-F. and Huang, B.R., The growth of silicon nanowires by electroless plating technique of Ni catalysts on silicon substrate, *Thin Solid Films*, 514, 20–24 (2006).

22. Yao, Y. and Fan, S., Si nanowires synthesized with Cu catalyst, *Mater. Lett.*, 61, 177–181 (2007).

23. Hofmann, S., Sharma, R., Wirth, C.T., Cervantes-Sodi, F., Ducati, C., Kasama, T., Dunn-Borkowski, R.E., Drucker, J., Bennett, P. and Robertson, J., Ledge-flow-controlled catalyst interface dynamics during Si nanowire growth, *Nature Mater.*, 7, 372–375 (2008).

24. Schmidt, V., Wittemann, J.V., Senz, S. and Gösele, U., Silicon nanowires: A review on aspects of their growth and their electrical properties, *Adv. Mater.*, 21, 2681–2702 (2009).

25. Dick, K.A., Kodambaka, S., Reuter, M.C., Deppert, K., Samuelson, L., Seifert, W., Wallenberg, L.R. and Ross, F.M. The morphology of axial and branched nanowire heterostructures, *Nano Lett.*, 7, 1817–1822 (2007).

26. Mouchet, C., Latu-Romain, L., Cayron, C., Rouviere, E., Cell, C. and Simonato, J.-P. Growth of one-dimensional Si/SiGe heterostructures by thermal CVD, *Nanotechnology*, 19, 335603 (2008).

27. Caroff, P., Wagner, J.B., Dick, K.A., Nilsson, H.A., Jeppsson, M., Deppert, K., Samuelson, L., Reine Wallenberg, L., and Wernersson, L.-E., High-quality InAs/InSb nanowire heterostructures grown by metal–organic vapor-phase epitaxy, 4, 878–882 (2008).

28. Bootsma, G.A. and Gassen, H.J., Quantitative study on growth of silicon whiskers from silane and germanium whiskers from germane, *J. Crystal Growth*, 10, 223–234 (1971).

29. Givargizov, E.I. and Sheftal, N.N., Morphology of silicon whiskers grown by the VLS-technique, *J. Crystal Growth*, 9, 326–329 (1971).

30. Givargizov, E.I. and Chernov, A.A., Rate of whisker growth by the vapor-liquid-solid mechanism and the role of surface energy, *Sov. Phys. Crystallogr.*, 18, 89–92 (1973).

31. Givargizov, E.I., Fundamental aspects of VLS growth, *J. Crystal Growth*, 31, 20–30 (1975).

32. Herring, C., *The Physics of Powder Metallurgy*, McGraw-Hill, New York, 143–144 (1951).

33. Sears, G.W., A growth mechanism for mercury whiskers, *Acta Metallurgica*, 3, 361–369 (1955).
34. Weyher, J., Some notes on the growth kinetics and morphology of VLS Silicon crystals grown with platinum and gold as liquid-forming agents, *J. Crystal Growth*, 43, 235–244 (1978).
35. Liu, J.L., Cai, S.J., Jin, G.L. and Wang, K.L., Wirelike growth of Si on an Au/Si(111) substrate by gas source molecular beam epitaxy, *Electrochem. Solid-State Lett.*, 1, 188–190 (1998).
36. Schubert, L., Werner, P., Zakharov, N.D., Gerth, G., Kolb, F.M., Long, L., Gösele, U. and Tan, T.Y., Silicon nanowhiskers grown on <111> Si substrates by molecular-beam epitaxy, *Appl. Phys. Lett.*, 84, 4968–4970 (2004).
37. Kodambaka, K., Tersoff, J., Reuter, M.C. and Ross, F.M., Diameter-independent kinetics in the vapor-liquid-solid growth of Si nanowires, *Phys. Rev. Lett.*, 96, 096105 (2006).
38. Wu, Y. and Yang, P., Direct observation of vapor-liquid-solid nanowire growth, *J. Am. Chem. Soc.*, 123, 3165 (2001).
39. Stach, E. A., Pauzauskie, P., Kuykendall, T., Goldberger, J. and Yang, P., Watching GaN nanowires grow, *Nano Lett.*, 3, 867–869 (2003).
40. Zhou, G. W., Yang, J. C., Xu, F. T., Barnard, J. A. and Zhang, Z. Quantitative V-L-S growth model and experiments of Fe catalyzed Si nanowire formation, *Mat. Res. Soc. Symp. Proc.*, 737, F6.3.1 (2003).
41. Ozaki, N., Ohno, Y., Kikkawa, J. and Takeda, S., Growth of silicon nanowires on H-terminated Si {111} surface templates studied by transmission electron microscopy, *J. Electron Microsc.*, 54 (Supplement 1), i25–i29 (2005).
42. Clement, T., Ingole, S., Ketharanathan, S., Drucker, J., and Picraux, S. T., In situ studies of semiconductor nanowire growth using optical reflectometry, *Appl. Phys. Lett.*, 89, 163125 (2006).
43. Kalache, B., Cabarrocos, P. R. I., Moral, A. F. I., *Jpn. J. Appl. Phys.*, 45, L190 (2006).
44. Kirkham, M., Wang, X., Wang, Z. L. and Snyder, R. L., Solid Au nanoparticles as a catalyst for growing aligned ZnO nanowires: A new understanding of the vapour–liquid–solid process, *Nanotechnology*, 18, 365304 (2007).
45. Zhang, J. M., Zhang, X. Z., Chen, L., Xu, J., You, L. P., Ye, H. Q. and Yu, D. P., In situ study of the growth of ZnO nanosheets using environmental scanning electron microscope, *Appl. Phys. Lett.*, 90, 233104 (2007).

46. Taheri, M. L., Reed, B. W., LaGrange, T. B. and Browning, N. D., In situ laser synthesis of Si nanowires in the dynamic TEM, 4, 2187–2190 (2008).

47. Madras, P., Dailey, E. and Drucker, J., Kinetically induced kinking of vapor-liquid-solid grown epitaxial Si nanowires, *Nano Lett.*, 9, 3826–3830 (2009).

48. Sun, Y. H., Gao, J. Y., Zhu, R., Xu, J., Chen, L., Zhang, J. M., Zhao, Q., and Yu, D. P., In situ observation of ZnO nanowire growth on zinc film in environmental scanning electron microscope, *J. Chem. Phys.*, 132, 124705 (2010).

49. Kirkham, M., Wang, Z. L. and Snyder, R. L., Tracking the catalyzed growth process of nanowires by in situ x-ray diffraction, *J. Appl. Phys.*, 108, 014304 (2010).

50. Hammar, M., LeGoues, F., Tersoff, J., Reuter, M.C., and Tromp, R.M., *Surf. Sci.*, 349, 129 (1995).

51. Robertson, I.M., Teter, D., *Microsc. Res. Tech.*, 42, 260 (1998).

52. Gai, P.L., Sharma, R. and Ross, F.M., *MRS Bulletin*, 33, 107 (2008).

53. Ross, F.M., *AIP Conference Proceedings*, 916, 363 (2007).

54. Okamoto, H. and Massalski, T.B., *Binary Alloy Phase Diagrams I*, ASM International, Ohio, 263–264 (1986).

55. Kim, B. J., Tersoff, J., Kodambaka, S., Reuter, M. C., Stach, E. A. and Ross, F. M., Kinetics of individual nucleation events observed in nanoscale vapor-liquid-solid growth, *Science*, 322, 1070–1073 (2008).

56. Sear, R. P., *J. Phys.: Condens. Matter*, 19, 033101 (2007).

57. Kodambaka, S., Hannon, J.B., Tromp, R.M. and Ross, F.M., , 6, 1292 (2006).

58. Ross, F.M., Tersoff, J. and Reuter, M.C., Sawtooth faceting in silicon nanowires, *Phys. Rev. Lett.*, 95, 146104 (2005).

59. Kodambaka, S., Tersoff, J., Reuter, M.C. and Ross, F.M., Germanium nanowire growth below the eutectic temperature, *Science*, 316, 729–732 (2007).

60. Hannon, J.B., Kodambaka, S., Ross, F.M. and Tromp, R.M., The influence of the surface migration of gold on the growth of silicon nanowires, *Nature*, 440, 69–71 (2006).

61. Kodambaka, S., Tersoff, J., and Ross, F. M., Growth kinetics of Si and Ge nanowires, in *Proceedings of SPIE*, Vol. 7224, 72240C (2009). doi: 10.1117/12.810672.

62. Bramblett, T.R., Lu, Q., Hasan, M.A., Jo, S.K. and Greene, J.E., Si(001)2x1 gas-source molecular-beam epitaxy from Si2H6—growth-kinetics and boron doping, *J. Appl. Phys.*, 76, 1884–1888 (1994).
63. Wen, C.-Y., Reuter, M.C., Bruley, J., Tersoff, J., Kodambaka, S., Stach, E.A. and Ross, F.M., *Science*, 326, 1247 (2009).

Chapter 2

Growth of Germanium, Silicon, and Ge–Si Heterostructured Nanowires

Shadi A. Dayeh and S. Thomas Picraux
Center for Integrated Nanotechnologies, Los Alamos National Laboratory, Los Alamos, NM 87545, USA
picraux@lanl.gov

2.1 Introduction

Anisotropic growth of one-dimensional (1-D) crystals has been the subject of research from as early as the 1950s.[1,2] Such studies intensified with the discovery of the vapor–liquid–solid (VLS) mechanism for silicon wires in 1964[3] and gained global interest in nanoscale science and engineering since the 1990s.[4,5] The 1-D structure of nanometer scale wires, or nanowires(NWs), serves as a platform for understanding basic materials science[6] and surface science behavior[7] as well as electron,[8] photon,[9] phonon,[10] and ion[11] transport phenomena at reduced dimensions. Numerous applications in electronics,[12] photonics,[13] thermoelectrics,[14] and energy storage[15] have been postulated and are being pursued. To achieve full control over the properties of NW structures, uncover

Silicon and Silicide Nanowires: Applications, Fabrication, and Properties
Edited by Yu Huang and King-Ning Tu

ISBN 978-981-4303-46-0 (Hardcover), 978-981-4303-47-7 (eBook)
www.panstanford.com

the underlying mechanisms that control such properties, and exploit their promise for technological applications, a greater understanding of growth mechanisms, structure, and properties is needed. Even further opportunities to achieve new properties may be anticipated by combining different materials into heterostructured NWs.

Since the general features of semiconductor NW synthesis and growth mechanisms have been discussed in review articles in the literature,[12,13,16,17] we concentrate in this chapter on the specific aspects of size effects in Ge and Si NW growth, defect formation during growth, and the understanding of growth mechanisms to synthesize heterostructured Si and Ge combinations in axial and radial directions. We have two main goals in this chapter. The first goal is to provide new perspective on the VLS mechanism and concrete experimental evidence on size effects in semiconductor NW synthesis, as well as to resolve relevant controversies about this effect. The second is to discuss the VLS growth of heterostructures with examples of the growth mechanisms and challenges, and the resultant NW morphologies. For background we first provide an overview of the VLS mechanism and its application to Ge and Si NWs. We then discuss the effects of small size on VLS growth and introduce the critical radius for suppression of the VLS growth mechanism. The underlying thermodynamic mechanism giving rise to these effects is illustrated through the influence of temperature, pressure, and surface energies on VLS growth at small size. We then discuss semiconductor NW heterostructure growth, beginning with ledge nucleation and fault propagation during VLS growth and followed by a broader discussion of Ge–Si axial and radial heterostructure growth. The novel aspects of such heterostructures are highlighted by an example of an energy band-edge-engineered device that is uniquely achieved by VLS heterostructure growth.

2.2 The VLS Growth Mechanism

Growth of metal-catalyzed semiconductor NWs using chemical vapor deposition techniques is generally accepted to follow the VLS growth mechanism[3] which uses a metallic nanoparticle, typically Au, that forms a low-temperature liquid eutectic with the desired growth material (Fig. 2.1a). Gas precursors containing reactant material are introduced and catalytically decompose with the incorporation of the growth species through the nanoparticle surface forming a liquid droplet near the eutectic temperature. This liquid droplet continues to adsorb decomposing solute atoms from the vapor, leading to a supersaturated state (Fig. 2.1b), at which point crystallization of the semiconductor occurs at the liquid–solid interface resulting in 1-D NW growth (Fig. 2.1c).

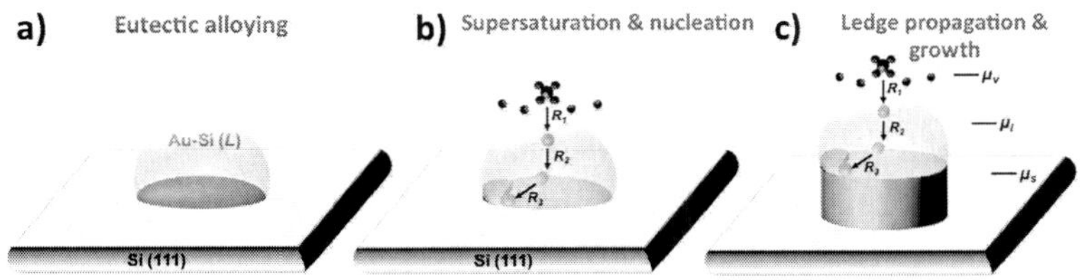

Figure 2.1 Schematic illustration of the VLS process for Si NWs. (a) A heated Si (111) substrate with a Au nanoparticle atop forms an eutectic molten nanoparticle. (b) Introduction of gas precursors (SiH_4 or $SiCl_4$) decompose at the particle surface and Si atoms diffuse into the nanoparticle which reaches a supersaturated state leading to the onset of ledge nucleation. (c) Continuation of ledge nucleation and propagation at the liquid–solid interface leads the growth of a Si NW with a diameter set by the starting Au nanoparticle diameter.

The most fascinating and distinguishing aspects of the VLS process arise from the use of the liquid metal–semiconductor alloy catalyst. Some of these remarkable aspects include (i) growth of semiconductor crystals at temperatures much lower than those used in thin film growth,[18] (ii) high diffusivity of reactant materials through or at the surface of the eutectic particle,[19] and (iii) the formation of a single nucleus, typically at the triple-phase (VLS) boundary,[20] followed by (iv) rapid ledge propagation during atomic layer growth.[21] These particular aspects are what lead to the relatively high NW-growth rates mediated by the eutectic particle, and the formation of crystals with well-predicted crystallographic structure.

In the case of growth of the isoelectronic Ge–Si materials system, a number of basic thermodynamic and kinetic considerations take effect. The metal nanoparticle—semiconductor system Au–Ge or Au–Si must reach a liquid condition prior to ledge formation in VLS NW growth. Considering the bulk phase diagram shown in

Fig. 2.2a, the eutectic temperature for Au–Ge and Au–Si is ~360 °C. In nanoscale alloys, we show below that the eutectic temperature can be significantly lower than in bulk materials. Moreover, *in situ* electron microscopy studies of nucleation of a liquid phase for a Au–Ge nanoparticle system has been shown to occur at much lower temperatures than in the bulk Au–Ge system.[22] We denote in Fig. 2.2a an approximate temperature and composition range over which growth of Ge and Si NWs can occur.[18] The precise temperatures for growth experiments are inevitably pressure dependent since the decomposition efficiency of the widely used precursors GeH_4 and SiH_4 for Ge and Si NW growth, respectively, is pressure dependent. GeH_4 is fully decomposed at a temperature of 280 °C,[23] and SiH_4 at a temperature of ~600 °C.[24] This therefore dictates that growth of Si NWs is generally accomplished at temperatures higher than the eutectic temperature of 360 °C as shown in Fig. 2.2a whereas Ge NW growth extends to temperatures below the eutectic point.

Aside from the eutectic formation, the Au–Ge or Au–Si system must reach a supersaturated state with a difference in the chemical potentials (Fig. 2.1c) expressed as

$$\Delta\mu = \mu_v - \mu_s \cong \mu_l - \mu_s = kT\ln(P_{in} / P_{0(NW)}), \tag{2.1}$$

where μ_v, μ_l, and μ_s are the chemical potentials of Ge or Si in the vapor, liquid, and solid phases respectively, k is Boltzmann's constant, T is the temperature in K, P_{in} is the input partial pressure of Ge or Si adatoms, $P_{0(NW)}$ is the equilibrium partial pressure of Ge or Si in Au. $P_{0(NW)}$ can be expressed as[25]

$$P_{0(NW)} = P_\infty \left(e^{\frac{4\Omega\alpha_{vl}}{dkT}} \right) \cong P_\infty \left(e^{\frac{4\Omega\alpha_{vs}}{dkT}} \right), \tag{2.2}$$

where P_∞ is the partial pressure of Ge or Si in infinite Au medium, Ω is the atomic volume of liquid Ge or Si in Au, α_{vl} is the surface energy density of the eutectic particle, α_{vs} is the surface energy density of Ge or Si sidewall facets under typical NW growth temperatures, and d is the NW diameter. Under typical growth conditions, say for Ge, $\alpha_{vl} \approx \alpha_{vs}$ ($\alpha_{vl} \sim 0.78$ J/m^2 and 0.8 J/m^2 for temperatures of 404 °C and 470 °C, respectively, which is similar to the (110) Ge surface energy density of 0.88 J/m^2).[26] This difference in α_{vl}, α_{vs} is typically neglected in the literature.[27] Using Henrys law ($P_i = k_H C_i$), where k_H is a constant, together with Eqs. 2.1 and 2.2, we can express the

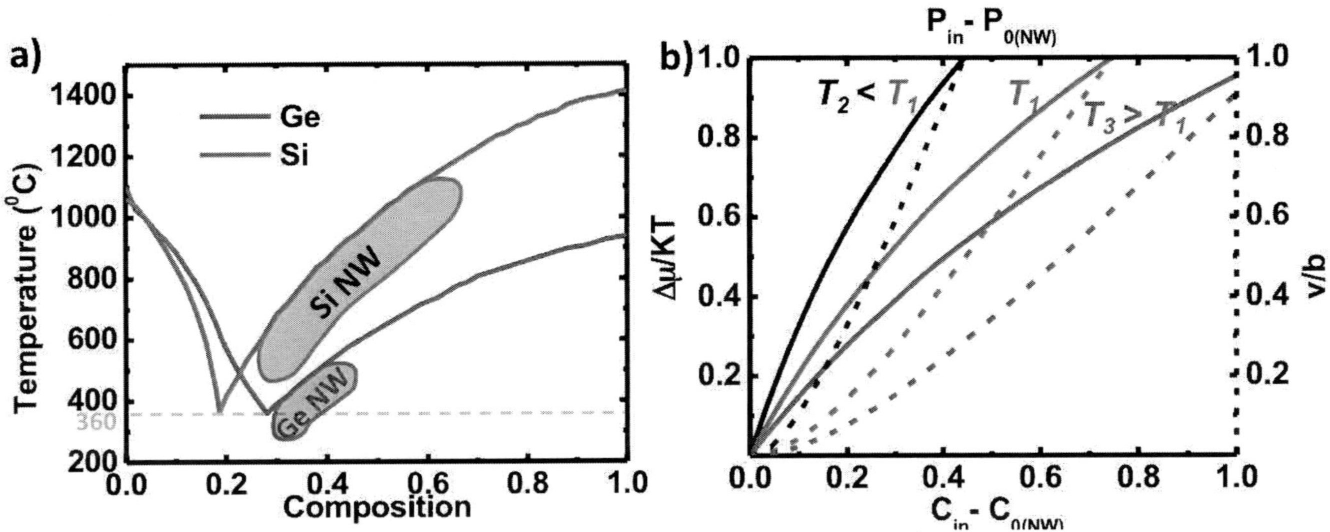

Figure 2.2 (a) Bulk phase diagrams of Ge and Si with highlighted regions of Ge and Si NW growth under commonly used CVD growth conditions. (b) Plot of the normalized supersaturation, $\Delta\mu/kT$, and the normalized growth rate, v/b, as a function of the input concentration gradient, C_{in}–$C_{0(\text{NW})}$, or partial pressure gradient, P_{in}–$P_{0(\text{NW})}$, of input precursors at the vapor–liquid interface to those in the liquid eutectic at equilibrium [liquidus line in (a)].

concentration gradient of Ge between the surface of the eutectic alloy (vapor–liquid), C_{in}, and the equilibrium concentration of Ge at the NW growth surface (liquid–solid) at, $C_{0(NW)}$, as

$$C_{in} - C_{0(NW)} = C_{\infty}\left(e^{\frac{4\Omega\alpha_{vs}}{dkT}}\right)\left(e^{\frac{\Delta\mu}{kT}} - 1\right). \tag{2.3}$$

Equation 2.3 portrays the important influence of thermodynamic and kinetic effects on NW growth behavior, and is plotted in Fig. 2.2b. For a constant diameter and at a constant temperature T_1 an increase in the input precursor partial pressure (equivalent to an increase in input concentration gradient) leads to an increase in the effective supersaturation, $\Delta\mu/kT$, and therefore an increase in the growth rate. In Fig. 2.2b, the normalized growth velocity, v/b, is plotted to exclude additional kinetic effects such as precursor decomposition or reaction rate at the liquid–solid interface, as will be discussed in the temperature section. As the temperature decreases toward the eutectic temperature (i.e., moving left on the liquidus line), the equilibrium concentration of Ge or Si in Au decreases as shown in Fig. 2.2a. Therefore, for the same input partial pressure, the supersaturation and consequently the normalized growth rate must increase as the temperature is increased as depicted for the case of $T_2 < T_1$ in Fig. 2.2b. On the other hand, as the temperature is increased further away from the eutectic point (i.e., moving right on the liquidus line), the equilibrium concentration of Ge or Si in Au increases and the supersaturation and consequently the normalized growth rate must decrease as shown for the case of $T_3 > T_1$ in Fig. 2.2b. This unified picture of the VLS growth precisely depicts our experimental observations on the dependence of Ge NW growth rate on temperature and pressure, as discussed in the forthcoming sections.

To expand further on the VLS framework introduced above, we put the analysis in the context of the three processes commonly debated as rate-limiting steps in the VLS growth. These processes, depicted in Fig. 2.1c, include (1) incorporation rate at the vapor–liquid interface by precursor decomposition, R_1, (2) diffusion rate through the Au particle, R_2, and (3) crystallization rate at the liquid–solid interface, R_3, limited by nucleation of step ledges.[28] R_2 is not limiting but resembles a unique aspect of the VLS growth compared with the vapor–solid–solid (VSS) growth. It is the fast reactant (Ge or Si) diffusion in the liquid growth seed that enables fast growth rates with excellent morphological control and high aspect ratios in VLS grown NWs compared with VSS grown NWs whose axial elongation rate is not fast enough compared with its sidewall deposition growth rate for comparable high aspect ratio NW growth.[28,29]

Considering the case of Ge NW growth, R_1 resembles an equilibrium reaction of GeH_4 decomposition at the vapor–liquid interface and R_3 resembles a nucleation reaction at the liquid–solid interface. In steady state, these two rates are equal such that the amount of Ge consumed by R_3 at the liquid–solid interface is equal to the amount of Ge supplied by R_1 from the vapor–liquid interface (both are area normalized), and the growth proceeds at the liquid–solid interface according to a chemical potential difference $\Delta\mu_{vs}$ expressed in Eq. 2.1. If the NW diameter decreases, at otherwise constant temperature and pressure, μ_s increases and $\Delta\mu_{vs}$ decreases (see Eq. 2.4) leading to lower growth rates. In this case, R_3 decreases and consequently, R_1 decreases by a reduction of the rate constant of GeH_4 decomposition and incorporation at the vapor–liquid interface. For a constant diameter and temperature, if the GeH_4 partial pressure increases, R_1 increases and consequently R_3 increases leading to a higher growth rate, as we will discuss in the *Pressure Effects* section. From this, we conclude that during steady state NW growth, it is misleading to isolate either R_1 or R_3 as the rate-limiting step, as these are by definition equal during steady state growth. For a constant diameter and pressure, if the temperature increases, $\Delta\mu_{vs}$ decreases but the temperature-dependent reaction constants at the vapor–liquid interface increases R_1 and similarly at the liquid–solid interface increases R_3 leading to higher growth rates with temperature.

Since elemental group IV NWs typically utilize a single growth precursor, validation of the VLS concepts discussed above eliminates complexities and misinterpretations which can arise in the growth of other compound semiconductor NWs. The concepts themselves however are generally applicable to any material system undergoing the VLS growth mechanism provided all rate-limiting factors are taken into account. For example, in the binary III–V compounds, differences of decomposition temperatures[30] and the different adatom mean free paths[31] need to be accounted for. In almost every NW materials system, a kinetically limited growth behavior, that is, increase in growth velocity with temperature, is observed.[32,33] For the special case of III–V semiconductor NWs, the solubility of group III materials is generally orders of magnitude higher than that of the group V species.[34] The eutectic growth nanoparticle comprises therefore of Au-group III composition. Since the mean free path of group III is dependent on temperature and the relative molar fraction of group V to group III, diffusion of group III to the nanoparticle can be limited by the complex interplay between these two growth precursors.[31] As such, a high growth temperature can increase the amount of decomposed group V precursor and reduce the surface mobility of In to the Au or Au–In nanoparticle leading to suppression or complete cessation of III–V NW growth at higher temperatures.[30] We state therefore that pulling the growth conditions largely away from those that produce uniform NW morphologies, for any semiconductor NW material system, can lead to significant deviation from the growth model discussed above and consequent misinterpretation of experimental data. Care must be exercised when applying the VLS concept to any materials system that is more complex than the monatomic Ge or Si systems, as well as under the presence of contaminants or low pressures and/or temperatures, including for Ge and Si NW materials growth.

2.3 Size Effects in Nanowire Growth

The diameter and consequently the electronic and optical properties of NWs are dependent on the size of the metal nanoparticle. Substituting Eq. 2.1 into Eq. 2.2, we get

$$\frac{\Delta\mu}{kT} = \frac{\Delta\mu_0}{kT} - \frac{1}{kT}\frac{4\alpha_{\mathrm{vs}}}{d} \, . \tag{2.4}$$

where $\Delta\mu_0$ is the supersaturation of the semiconductor material for an infinite medium (i.e., the bulk limit of a planar surface for the catalytic growth particle). Equation 2.4 has a number of consequences for size-dependent semiconductor NW growth. First, as the diameter of the NW decreases the magnitude of the right hand term increases, and this corresponds to a reduction in supersaturation and consequently a reduction in growth rate. Such a thermodynamic size-dependent effect is usually referred to as the Gibbs–Thomson effect. A continued decrease of NW diameter to a few nanometers eliminates the supersaturation condition in the growing nanoparticle such that NW growth ceases below a minimal critical diameter, d_c, which from Eq. 2.4 is given by

$$d_c = 4\Omega\alpha_{vs}/\Delta\mu_0. \quad (2.5)$$

Second, NWs grown in different orientations, and therefore with different facets and α_{vs} are expected to result in different d_c. Third, different NW materials have distinguishable $\Omega\alpha_{vs}$ product, and therefore different d_c. Fourth, tuning of $\Delta\mu_0$ with temperature and pressure can change d_c, however at the expense of compromised NW morphology as we shall discuss later. These basic concepts generally apply to all semiconductor NW growth systems.

Since germanium oxide can be easily desorbed from Ge surfaces at $T \geq 300$ °C, which is close to the Ge–Au eutectic temperature and higher than the temperatures at which GeH_4 input precursor decomposes, it is therefore straightforward to conclude that high-quality epitaxial [111] Ge NW growth on Ge(111) surfaces is more readily feasible at ~300 °C than Si. Figure 2.3a shows a plot the temperature profile typically used for the growth of epitaxial [111] Ge NWs on Ge (111) surfaces as demonstrated in Fig. 2.3b. A first high-temperature step is used to desorb germanium oxide and form an eutectic Au–Ge alloy prior to introducing GeH_4 to initiate the Ge NW growth. Once GeH_4 is introduced to the chamber for few tens of seconds, epitaxial growth of the base of the NW is established. The temperature is then quickly reduced to elongate the NW at temperatures that can exclude Ge deposition on the NW sidewalls by vapor–solid growth, and therefore minimize tapering in the NW diameter.

The clean growth procedure of Ge NWs can therefore be used for establishing the size effects in semiconductor NW growth,[35] without the complications of competing reactions in III–V NW growth, or the effects of contaminants or nucleation-deficient conditions in Si NWs that lead to contradictory results.[36] Equation 2.4 predicts that smaller diameter NWs have reduced supersaturations and lower growth rates. This is indeed observed in Ge VLS NW growth as illustrated in Fig. 2.4a, where equally spaced Au metal dots of different sizes have been patterned with electron beam lithography and were subject to Ge NW growth according to the conditions described in Fig. 2.3a. To attain a deeper insight into the governing process at which this drop in growth rate occurs, we consider the variation of the partial pressure or liquidus line with diameter. The diameter-dependent vapor pressure of solute in a Au nanoparticle, $P_{NW}(d)$, can be expressed as a function of its vapor pressure in bulk Au, P_∞, as follows:[25]

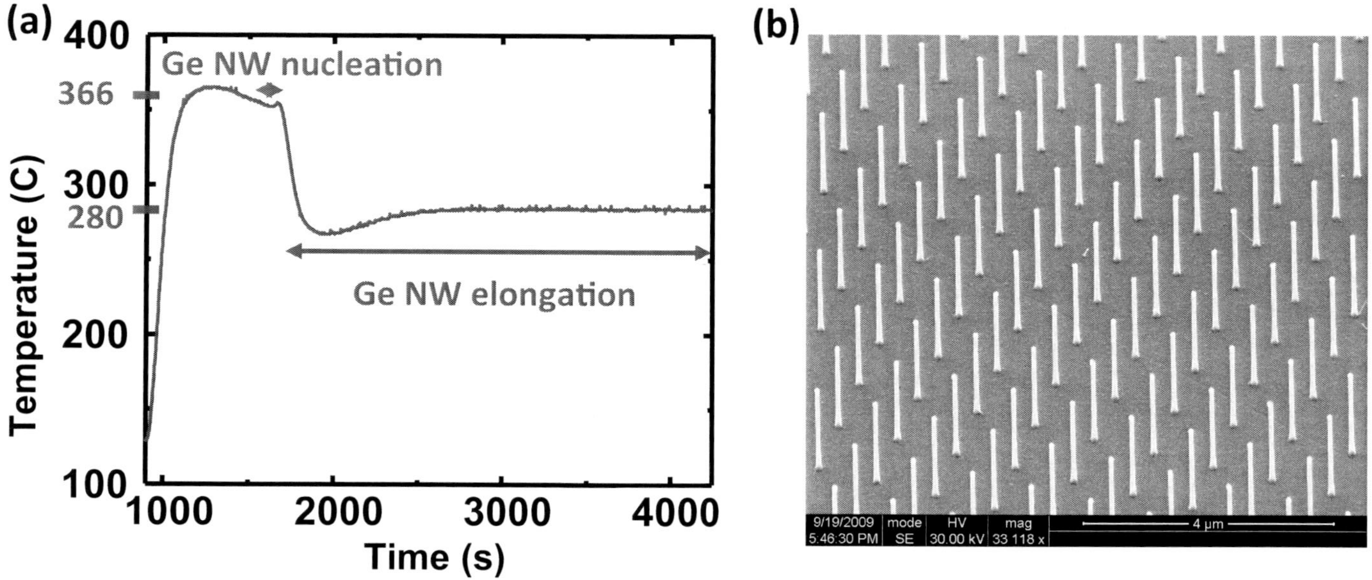

Figure 2.3 (a) Temperature profile for growing epitaxial [111] oriented Ge NWs on a Ge (111) surface as shown in the SEM image of panel (b).

$$P_{0(\text{NW})}(d) = P_{\infty} exp(4\Omega\alpha_{\text{lv}} / dkT) \cong P_{\infty} exp(4\Omega\alpha_{\text{vs}} / dkT) \qquad (2.6)$$

The term κ is added to Eq. 2.6 to take into account surface energy density changes, and consequently curvature changes of the Au–Ge alloy atop the Ge NW with temperature.[35] We therefore can express the equilibrium Ge concentration in the Au–Ge alloy, $C_{0(\text{NW})}$, as

$$C_{0(\text{NW})} = C_{\infty} \exp(4\kappa\alpha_{\text{vs}} / dkT), \qquad (2.7)$$

which indicates an $exp(1/d)$ increase in the equilibrium concentration of Ge in Au from that of the bulk. Equation 2.7 gives the measured equilibrium concentration of Ge in Au at different temperatures when measured over NWs shown in the array of Fig. 2.4a. It now becomes intuitive to deduce the size-dependent effects in the growth of semiconductor NWs from Fig. 2.4b. Under a constant input partial pressure, that is, constant input Ge concentration at the surface of the Au–Ge eutectic droplet, the concentration gradient with respect to the equilibrium concentration of the Au–Ge eutectic decreases as their diameter decreases, as depicted in Fig. 2.4b. As such, the supersaturation at smaller diameters decreases, and therefore the growth rate decreases at smaller diameters. This can be further deduced from combining Eqs. 2.1 and 2.7 to obtain

$$\Delta\mu / kT = \Delta\mu_0 / kT - (1/\kappa)\ln(C_{0(\text{NW})} / C_{\infty}). \qquad (2.8)$$

Here, at constant growth conditions or supersaturations ($\Delta\mu_0$), a reduction in NW diameter means an increased $C_{0(\text{NW})}$ and therefore a reduced effective supersaturation as deduced from Eq. 2.8 which in turns lead to a reduced growth rate. The dependence of the NW growth velocity (or rate) on supersaturation has been debated extensively, with the generally accepted detailed work of Dubrovskii and Sibirev[37] indicating that such dependence has a dominant quadratic character, validating the original empirical assumptions of Givargizov and Chernov on such dependence.[38] We take the NW growth velocity, $v = dL/dt$, to depend quadratically on the supersaturation in the present case: $v = b(\Delta\mu / kT)^2$, where b is a kinetic coefficient of crystallization that is independent of supersaturation. Thus, Eq. 2.4 can be rearranged as,

$$\sqrt{v} = \sqrt{b}(\Delta\mu_0 / kT) - \sqrt{b}(4\alpha_{\text{vs}} / kT)(1/d). \qquad (2.9)$$

It becomes evident from Eq. 2.9 that in order to examine the dependence of the growth velocity on supersaturation isolated from any other kinetic process, one needs to normalize the growth rate by the kinetic coefficient of crystallization, b, as plotted in Fig. 2.2b.

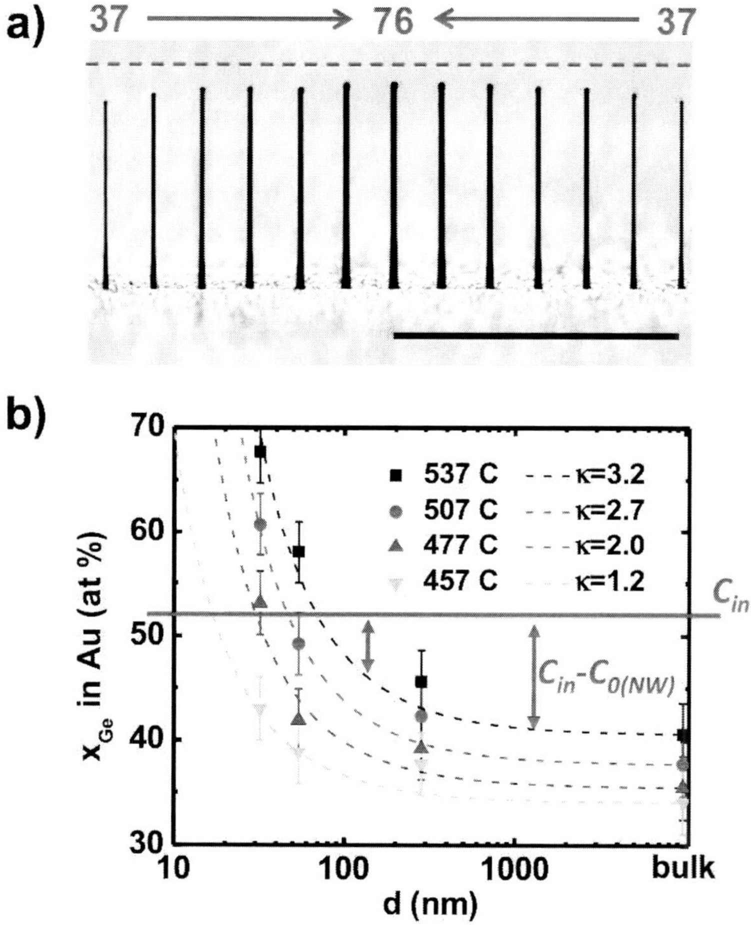

Figure 2.4 (a) SEM image of a linear array of Ge NWs grown from lithographically patterned Au discs. The dashed line highlights the systematic changes in NW length at different diameters. Scale bar is 3 μm. (b) Ge atomic % concentration in Au as a function of diameter at different temperatures. Symbols are experimental data points; dashed lines are fits according to Eq. 2.7. Reprinted with permission from the American Chemical Society, copyright 2010, Ref. [35].

We now discuss the experimental data for Ge VLS NW growth in the context of the size-dependent formalism presented above. Fig. 2.5a shows the NW length as function of diameter for different growth times. Figure 2.5b shows a plot of $v^{1/2}$ as function of $1/d$ for the data

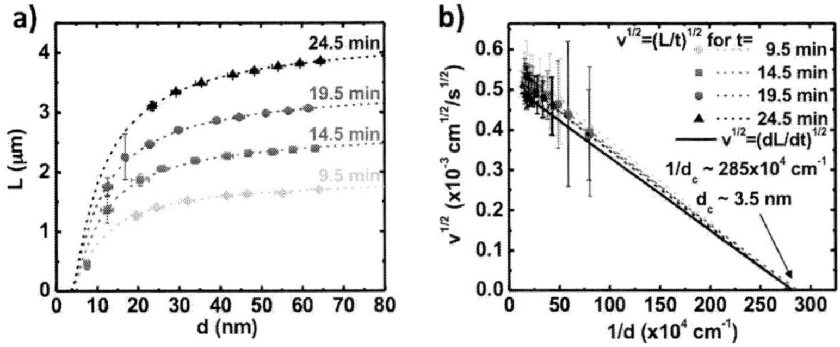

Figure 2.5 (a) Plot of Ge NW length as function of diameter with a P[GeH_4] = 0.6 Torr for different growth times at the low-temperature step [t(276 °C) = 5, 10, 15, and 20 min] resulting in a linear increase of NW length with time for all diameters. (b) Plot of the square root of the growth velocity for the data shown in (a) calculated using $v = L/t$ and using $v = dL/dt$ showing convergence of all data to the same cut-off diameter of ~3.55 nm. Reprinted with permission from the American Chemical Society, copyright 2010, Ref. [35].

of Fig. 2.5a according to the assumption of $v = L/t$ with L and t being the total NW length and growth times, respectively (an assumption that is valid only for long growth times), as well as considering $v = dL/dt$. By fitting the data in Fig. 2.5b with Eq. 2.9, the slope of the fitted lines determines the kinetic coefficient b, and the intercept with the $1/d$ axis determines d_c, from which $\Delta\mu_0$ can be estimated using Eq. 2.5. All fitted lines in Fig. 2.5b lead to a d_c of 3.5 nm, justifying the growth velocity form $v = L/t$ for the present conditions, which becomes valid at sufficiently long growth times. Also integrating Eq. 2.9 is seen to result in an excellent fit to the experimental data of Fig. 2.5a. These experimental results are in agreement with many previous assessments of size-dependent growth rates for Si NWs[39–42] and in contradiction to a few others.[36,43,44] Inconsistencies in size-dependences have been observed when the growth was performed under low SiH_4 or Si_2H_6 partial pressures due to Au loss by diffusion from the Au seed at the NW tips, or due to the presence of oxide masks and contaminants on the growth substrate surface.[36,44] The data presented in Fig. 2.5 clearly shows that a reduction of the growth rate with smaller NW diameters is valid, particularly at optimal growth conditions that lead to controlled NW morphology and orientation such as those NWs shown in Fig. 2.3b. It is worth noting that due to their differences in surface energies, atomic volumes, and supersaturations, Si NWs should grow to about 60% smaller diameter than Ge,[41] and changes in growth orientation, that is, in the exposed facets, will influence the minimum diameter.[35]

2.4 Temperature Effects on Nanowire Growth

The most substantial effect on efficient NW nucleation and growth, as well as on the resulting morphology, composition, and doping profiles, arises from temperature effects. For uniform nucleation at the liquid–solid interface and epitaxial NW growth, the growth temperature must be close to or above the eutectic temperature. In some cases, this temperature is sufficiently large for strong decomposition of precursors such as for GeH_4, which leads to concurrent Ge deposition on the NW sidewalls as the NW continues to grow axially. Therefore, the temperature profile for Ge NW growth typically follows that of Fig. 2.3a for successful epitaxial growth with little to no radial coating. The dependence of the growth rate on temperature for VLS NW growth is fairly complex and involves such factors as the relative change of material decomposition, reaction rates at the growth eutectic–NW interface, diffusivity of reactants into, or at the surface of the growth eutectic, diffusion lengths on the substrate and NW sidewalls, etc. Even with the detailed work of Dubrovskii and Sibirev, such complex temperature dependences can lead to incorrect conclusions, particularly when dealing with the temperature dependence of the critical diameter, as we will discuss below.[37]

Experiments typically indicate that a higher growth temperature leads to a more efficient NW nucleation and enhanced growth rates. If that is the case, the question arises as to whether the growth of smaller diameter NWs at higher temperatures becomes more feasible. To answer this question, we must remind the reader that the equilibrium concentration of the semiconductor material in Au increases as the temperature increases. For a constant input precursor partial pressure, this translates to a reduction in the supersaturation as the temperature increases. A reduction in supersaturation causes an increase in d_c according to Eq. 2.5, $d_c = 4\Omega\alpha_{vs}/\Delta\mu_0$, which means that smaller diameter NWs are less likely to grow at higher temperatures. To simplify this argument, let's consider the GeH_4 precursor, which is decomposed at the lowest possible growth temperature (~280 °C). Therefore, increasing the temperature above 280 °C does not in principle contribute to extra input Ge concentration at the vapor–liquid interface. It is the increase in the kinetic reaction rates, dictated by the kinetic coefficient b in Eq. 2.9 that leads to an increase in the growth rate with temperature. The supersaturation on the other hand should decrease with temperature. This result can also be deduced from Eq. 2.8 where $C_{0(\mathrm{NW})}$ increases faster with temperature than C_∞. Neglecting this latter dependence of $C_{0(\mathrm{NW})}$ on temperature is what led Dubrovskii and Sibirev to reach a conclusion that the d_c decreases as T increases, which is opposite to the above intuitive argument and what we find in experiments.

The observed temperature dependence of the growth velocity and its impact on the critical size dependence is illustrated in Fig. 2.6a. As $v^{1/2}$ increases with temperature, the intercept with the $1/d$ axis decreases indicating larger d_c, which is plotted as a function of temperature in Fig. 2.8a. To infer the dependence of growth rate on temperature, while excluding other kinetic effects, we plot the growth rate normalized to the kinetic coefficient, v/b, in Fig. 2.6b. A reduced supersaturation with temperature leads to a lower normalized growth rate and larger cut-off diameters as demonstrated in Fig. 2.6b. The detailed dependence of b, $\Delta\mu_0/KT$, and d_c on temperature as extracted from the experimental data of Fig. 2.6 is listed in Table 2.1, which is in excellent agreement with the formalism presented here.

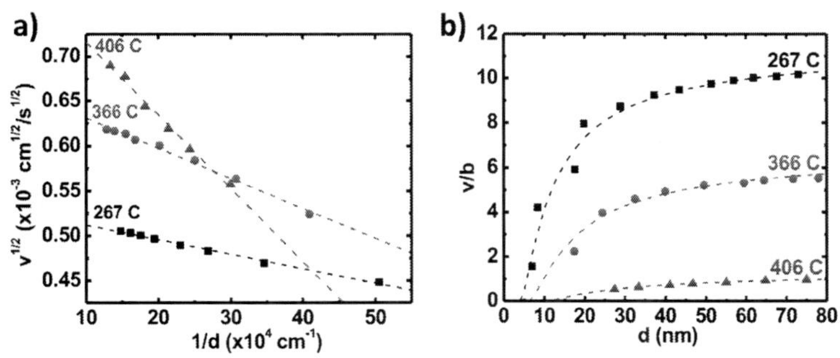

Figure 2.6 (a) Plot of $v^{1/2}$ as a function of $1/d$ at different temperatures of 276 °C, 366 °C, and 406 °C showing different slopes and intercepts. (b) Plot of the growth velocity normalized by the kinetic coefficient b for the data shown in (a). Dashed lines are fits according to Eq. 2.9. Reprinted with permission from the American Chemical Society, copyright 2010, Ref. [35].

Table 2.1 Calculated kinetic coefficients, supersaturation values, and critical diameters based on measurements for Ge NWs grown at 276 °C, 366 °C, 406 °C, and 440 °C

T (°C)	b (10^{-8} cm/s)	$\Delta\mu_0/kT$	d_c (nm)
276	3.0	3.1	3.4
366	20.5	1.5	6.0
406	80.5	0.9	9.7
440	86.9	0.7	11.6

2.5 Pressure Effects on Nanowire Growth

Similar to temperature, the growth precursor partial pressure effects on NW nucleation, growth, and morphology are substantial. For instance, a straight forward conclusion from Eq. 2.1 indicates that larger supersaturations and growth rates are expected as P_{in} increases. However, as P_{in} increases, some precursors, especially III–V ones, display enhanced decomposition efficiency, an enhanced radial coating, and increased layer stacking in faulty positions during NW growth,[45] with the resulting planar defects typically perpendicular to the growth direction in III–V NWs, and parallel to the growth direction in Si and Ge NWs.[46]

At or near optimal pressure settings for NW growth, the growth behavior is well-described with the formalism presented above. The key features of such dependence can be seen in Fig. 2.7. First, higher P_i leads to higher growth rates, consistent with Eqs. 2.1 and 2.9 where the latter provides the good fit to the data as shown in Fig. 2.7. Second, higher P_i results in lower cut-off diameters as expected from Eq. 2.5, due to an increase in $\Delta\mu_0$. Third, one can see directly from Eq. 2.9 that an increase in P_i will shift $v^{1/2}$ positively (first term dependent on $\Delta\mu_0$) but the slope of $v^{1/2}(1/d)$ remains unchanged (second term in Eq. 2.9), which is also evident in Fig. 2.7b. It is worth noting that a pressure increase also leads to enhanced radial growth, which can compromise the ability to grow smaller NW diameter wires at higher P_i.[35]

We can now summarize the effects of temperature and pressure, and their influence on the critical diameter, d_c, for VLS NW growth as demonstrated in Fig. 2.8. A temperature increase leads to enhanced kinetics and growth velocities, however with a reduced supersaturation that increases the cut-off diameter for growth

(Fig. 2.8a). A pressure increase causes a larger supersaturation and growth rates and with larger supersaturations, a decrease in d_c is obtained (Fig. 2.8b). These combined effects of temperature and pressure (Figs. 2.6–2.8), as presented here for Ge NW growth, demonstrate clearly key phenomena of the VLS growth mechanism.

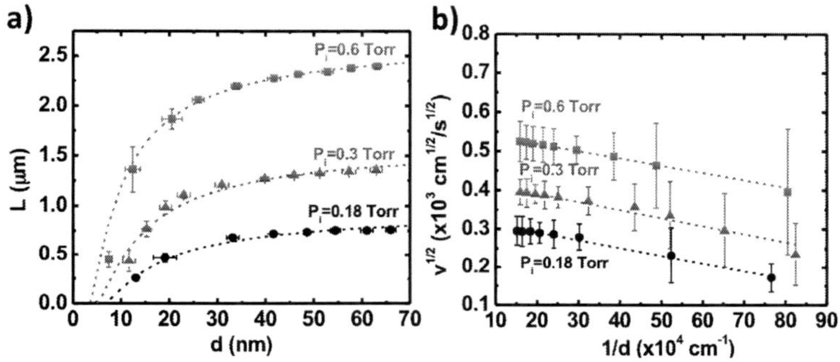

Figure 2.7 (a) Plot of the NW length as a function of diameter for different GeH_4 partial pressures. (b) Plot of $v^{1/2}$ as a function of $1/d$ at different GeH_4 partial pressures for data of (a) showing the progressive reduction in the growth velocity and decrease in critical diameter with increasing pressure. Reprinted with permission from the American Chemical Society, copyright 2010, Ref. [35].

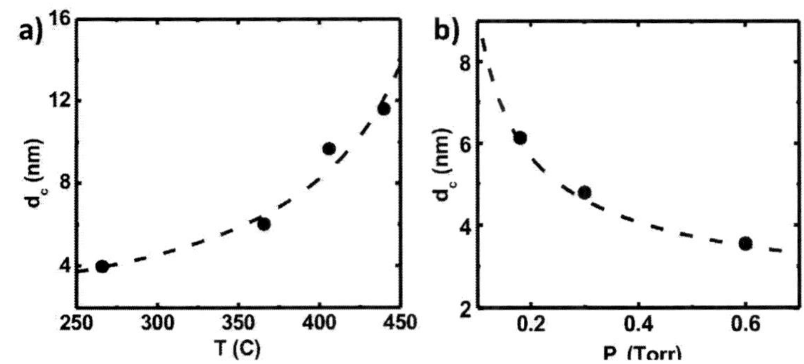

Figure 2.8 (a) Plot of the critical diameter as function of temperature extracted from Fig. 2.6a, illustrating higher cut-off diameters with higher temperatures. The dashed line is calculated according to Eq. 2.5. (b) Plot of the critical diameter as a function of input GeH_4 partial pressure extracted from Fig. 2.7a and showing logarithmic dependence and lower cut-off diameters with higher partial pressures. The dashed line is calculated according to Eq. 2.5, where $\Delta\mu_0$ has been obtained by a fit to the experimental data as function of $\ln|P_i|$.

2.6 Dopant Precursor Influence on Nanowire Growth

Technological applications of semiconductor NWs require controlled doped interfaces, and the introduction of dopant precursors can have a direct influence on the NW growth rate and morphology. N-type and p-type doping is commonly achieved with the introduction of PH_3 and B_2H_6 reactant gases during NW growth. Fig. 2.9a shows a plot of the pressure normalized growth rate, $v/P_i = L/tP_i$, as a function of $1/d$ for Ge NW growth, which is slightly lowered in the presence of B_2H_6 at small diameters. Introduction of PH_3 into the growth chamber shows that v/P_i is lower than that of undoped NWs for $d \geq 20$ nm (progressively changing from a ~1% reduction at $d = 20$ nm to ~6.5% at $d = 80$ nm), while higher v/P_i are obtained for $d < 20$ nm (again changing from a ~2% increase at $d = 15$ nm to ~22% at $d = 7.5$ nm). Since the normalized supersaturation to partial pressure ratio $\Delta\mu_0/kT/|ln(P_i)|$ is lowered when dopants are introduced (see Table 2.2), this interesting behavior with PH_3 is attributed to a change in surface energy, α_{vs}. Indeed, as one can see in Table 2.2, the $\sqrt{b\alpha_{vs}}$ product remains essentially the same for the cases of GeH_4 and GeH_4/B_2H_6, and lower for the GeH_4/PH_3 case, suggesting a lower α_{vs}, in the latter case after phosphine exposure and resulting in a lower d_c as observed experimentally (Fig. 2.9b and Table 2.2) as deduced from Eq. 2.5. This observation is in general agreement with incorporation of high concentrations of P into the NW sidewalls.[47]

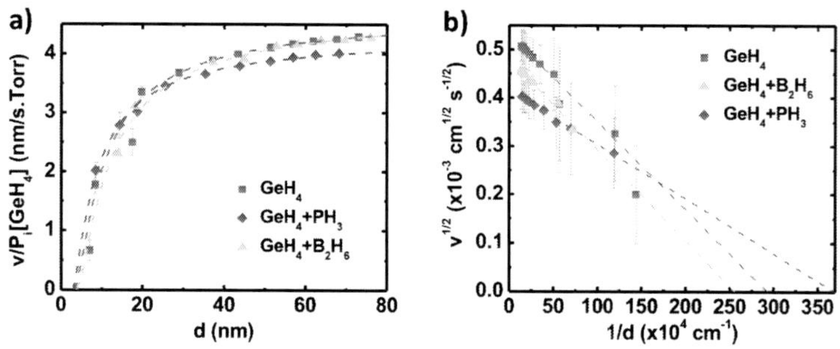

Figure 2.9 (a) Plot of the growth rate normalized by the GeH_4 partial pressure showing a reduction in the growth rate when PH_3 is present during growth. (b) Plot of the square root of the growth rate as a function of inverse diameter illustrating slope and d_c change in the presence of PH_3. Reprinted with permission from the American Chemical Society, copyright 2010, Ref. [35].

Table 2.2 The calculated critical diameters, pressure-normalized supersaturation values and kinetic coefficients based on measured results for Ge NWs grown with and without dopants at 276 °C using Eqs. 2.5 and 2.9

Precursor	d_c (nm)	$\Delta\mu_0/kT/ln(|P_i|)$	$b^{1/2}\alpha_{vs}$ $(10^{-8}\ J/cm^{5/2}/s^{1/2})$
GeH_4	3.4	6.0	1.5
$GeH_4+B_2H_6$	3.9	4.1	1.6
GeH_4+PH_3	2.8	3.7	0.98

Dopant incorporation through the liquid eutectic into the VLS grown semiconductor NW is still a subject of research. Atom probe tomography measurements on Si NWs have shown that the competitive radial deposition from vapor phase growth on the sidewalls of Si NWs (vapor–solid growth) resulted in substantially more efficient dopant incorporation compared with incorporation through the liquid growth eutectic.[47] The degree of doping by such competitive processes depends on the growth conditions, as discussed before. While such processes occur during NW growth, gradients in the doping densities occur in both the radial and axial directions (Fig. 2.10a). Another method for controlled dopant incorporation

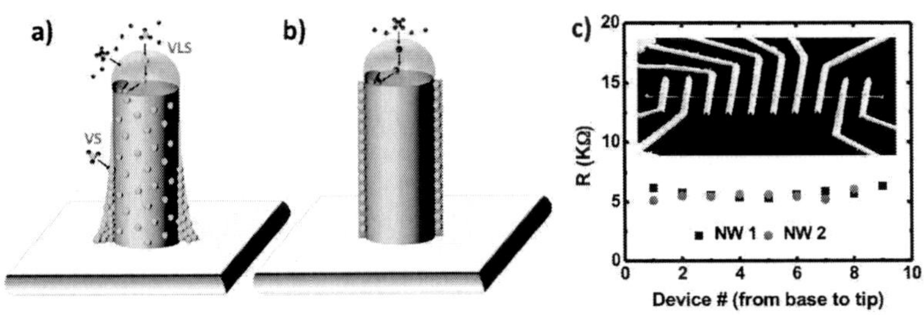

Figure 2.10 (a) Cartoon illustrating competitive incorporation of dopants from the vapor (VS growth) or through the liquid growth eutectic (VLS growth). (b) Doped shells formed on undoped core NWs in a two-step process can result in better axial uniformity of dopants. (c) Plot of the 2-terminal resistance across the entire length (~6 μm) of an i-core/n-shell Ge NW doped using the approach described in (b), showing uniform resistance throughout for two of such devices. Inset is an SEM image of the device labeled NW1.

in semiconductor NWs is to grow an undoped semiconductor NW and properly alter the growth conditions to prevent axial elongation and induce radial deposition of a conformal doped shell (Fig. 2.10b). Electrical measurements on NWs grown using the latter doping method (i-core/doped-shell) resulted in uniform NW resistance, that is, uniform effective doping, through the entire NW length, as shown in Fig. 2.10c.

2.7 Defects during VLS Growth of Semiconductor Nanowires

Growth according to the VLS mechanism proceeds in a layer-by-layer process with ledge nucleation occurring near the triple-phase interface.[20,48] Because of their small size nucleation and propagation of defects in semiconductor NWs can be dramatically different from that observed in thin film growth, where due to the much larger growth area many defect nucleation sites coexist. At higher supersaturations and growth rates in NWs, the probability of stacking adatoms at faulted locations becomes higher. Wulff constructions have shown that a NW growing in the [111] orientation will have six hexagonal {110} type facets with three inclined {111} planes triangular facets at their intersections (Fig. 2.11). Molecular dynamics (MD) simulations have shown that nucleation occurs at the three triangular facets, $(1\bar{1}1)$ labeled as F_1, $(11\bar{1})$ as F_2, and $(\bar{1}11)$ as F_3, which results in elongation along the NW growth direction in layer-by-layer growth.[49] It is found that in the case of a Si NW, a four-atom cluster on these inclined {111} facets can lead to uniform NW diameter growth. MD simulations have also shown that a dimer formed in a faulted position on F_1, F_2, and F_3 can lead to the nucleation of a SF in the NW. Ledge propagation then occurs from these {111} facets at an inclined angle to the NW growth direction. By considering different possible nucleation sites for each new layer after SF nucleation (Fig. 2.11e), MD simulations have shown that cluster nucleation is energetically favorable by a factor of ~2 (Fig. 2.11f) at the triple-phase interface with the SF. This condition imposes the presence of a single axial defect per NW segment length as we shall see in the experimental results discussed below.

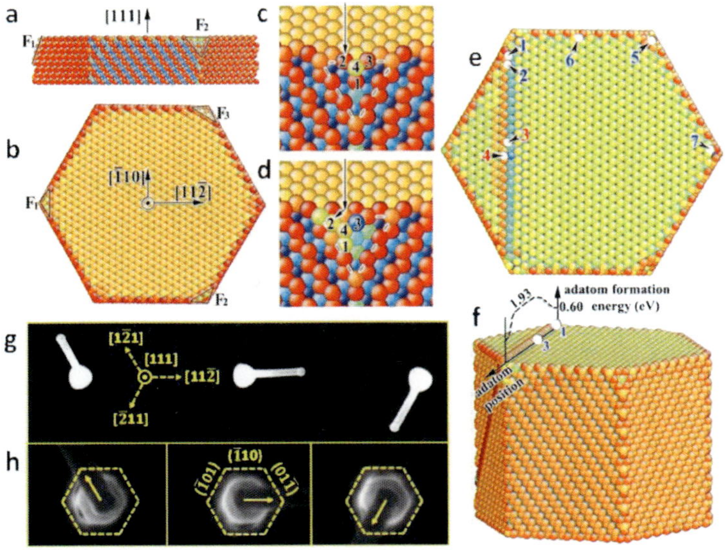

Figure 2.11 (a,b) Side and top view of a [111]-oriented Si NW showing 6 {110} facets with three {111} inclined triangular facets at their intersections where stacking fault (SF) nucleation occurs. (c,d) Side view taken from MD simulations showing nucleation on the (111) F_1 facet in proper (c) or faulted (d) positions. The black arrows are inserted to guide the eye for a four-atom cluster shifting to the faulted position. (e,f) Top (cross-section) and perspective side view of a NW with a SF on the $(1\bar{1}1)$ plane as used for MD simulations of nucleation energy barriers at points labeled 1 to 7 in (e). The calculated nucleation energy is superimposed on the structure in (f) and the lowest energy is found at the {110}/SF interface (i.e., SF/triple-phase interface during NW growth). (g,h) High-contrast (g) and low-contrast magnified top view (h) of Ge–Si heterostructure NW grown from lithographically patterned Au dots showing the three possible kink directions (projected onto the top (111) plane) in agreement with MD simulations. Nucleation of the Si segment appears clearly to emerge at two {110} facet interfaces (dashed hexagons in h) corresponding to F_1, F_2, or F_3 in (b). Reprinted with permission from the American Chemical Society, copyright 2011, Ref. [49].

Axial Ge–Si heterostructure NWs provide a natural marker to decisively assess the formation of defects during semiconductor NW growth. Often, an axial Si segment grown on a Ge segment displays a systematic kink from the [111] growth orientation. We utilize this kink to prove that NW nucleation does occur at the {111} facets at the interface between two {110} facets. When Ge–Si heterostructure NW growth is performed from lithographically patterned Au dots on (111) Ge surfaces, kinking was typically observed to occur in the three well-defined <112> orientations as projected in the top-view SEM image of Fig. 2.11g. Low-contrast SEM images have revealed that the kink in the Si segment originates at the interface between two {110} facets, as labeled in Fig. 2.11h, consistent with the MD simulation results of nucleation at one of the three {111} facets, labeled F_1, F_2, and F_3 in Fig. 2.11b.

To further resolve defect nucleation in NWs, high-resolution TEM was performed on such axial Ge–Si heterostructures as shown in Fig. 2.12. Detailed analysis of the microstructure of the Si segments, as shown in Fig. 2.12 reveals the presence of a single fault per NW segment that nucleates from the same NW facet (Fig. 2.11b). First, a SF nucleates and propagates in the [112] orientation indicative of a SF on a $(1\bar{1}1)$ growth surface. During layer-by-layer growth, nucleation at the triple-phase interface is pinned at the intersection of the SF with the top liquid–NW interface. As a result, access to the low-energy {111} facet is prohibited, thereby preventing nucleation of any additional faults until the SF is terminated at the opposite side of the NW. This is in complete agreement with the results of Fig. 2.11f and the discussion above. Such a SF requires a NW segment length of $d/\tan(19.5°)$, where d is the NW diameter and 19.5° is the angle between the [112] and [111] orientations, and for the case of Fig. 2.12 is ~85 nm for d = 30 nm, in agreement with experimental observations. Once the SF terminates on the opposite side of the NW, access to the low-energy {111} facet is again permitted as nucleation of the next atomic layer occurs. From Fig. 2.12b, we observe that a twin boundary (TB) is immediately nucleated at the same low-energy $(1\bar{1}1)$ facet at which the preceding SF nucleated, where adatom stacking in two consecutive fault positions leads to TB formation. HRTEM confirms that this process occurs within one atomic bilayer of the termination of the SF with the nucleation switching back to the low-energy $(1\bar{1}1)$ facet. Once formed, the ledge nucleation occurs at the TB/{110} facet interface and propagates on the (111) growth surface (Fig. 2.12b,c). The NW diameter consequently increases just before the kink until the increasing line tension of the stretched liquid Au surface at the triple-phase interface forces the NW growth direction to switch to the [112] orientation, resulting in a 19.5° kink with respect to the initial [111] growth axis. Continuum modeling for other cases of NW growth has also noted that perturbations in the line tension of the growth seed may change the NW growth direction.[50] These striking *ex situ* observations of defect nucleation during NW growth establish new features of the VLS mechanism for layer-by-layer NW growth and are supported by the MD simulations discussed earlier.

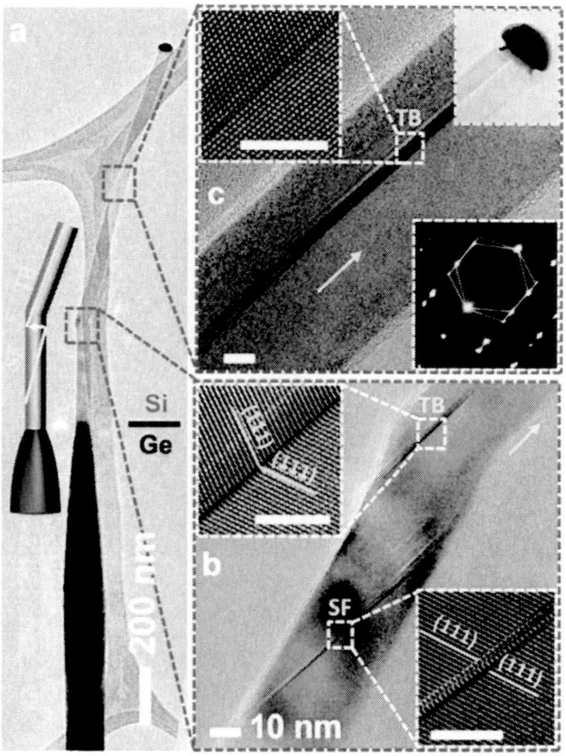

Figure 2.12 (a) TEM image (right side) and cartoon (left side) of a kinked Ge (dark)–Si (bright) axial NW heterostructure. (b) HRTEM image showing nucleation of an intrinsic SF followed by a Σ3 (111) coherent TB from the same NW facet. Both SF and TB nucleate at $(1\bar{1}1)$ facets and propagate along the [112] direction as described in Fig. 2.11. (c) Once formed, the TB pins the nucleation for each new atomic layer at the triple-phase interface and continues to propagate during NW growth allowing no further access to the low-energy (111) facets and thus no additional fault nucleation. The unlabeled scale bars are 5 nm. Arrows in (b) and (c) indicate a [112] NW growth orientation after NW kinking. Reprinted with permission from the American Chemical Society, copyright 2011, Ref. [49].

With such detailed understanding of defect nucleation in semiconductor NWs on an atomic level, we are now in a position to discuss the more generic behavior of defect nucleation and Au diffusion on semiconductor NWs with "global" pressure and temperature variations. Figure 2.13 shows what we refer to as a *global Si NW growth morphology–structure diagram*. At high temperatures and pressures, Au diffusion and axial twin faults can coexist as shown in Fig. 2.13 (green datum point). If the temperature is kept constant and the pressure is reduced, the growth rate is lowered and fault nucleation is lowered, resulting in single crystal NWs (blue data points in Fig. 2.13). However, Au diffusion is relatively severe under these conditions (black spots in Fig. 2.13 are Au nanoparticles). On the other hand, a high SiH_4 partial pressure but a reduced temperature can result in less Au diffusion but faults and twin boundaries persist due to the high growth rates (black datum point in Fig. 2.13). A low SiH_4 partial pressure and low temperature result in optimal Si NW morphology with single crystal structure and no Au

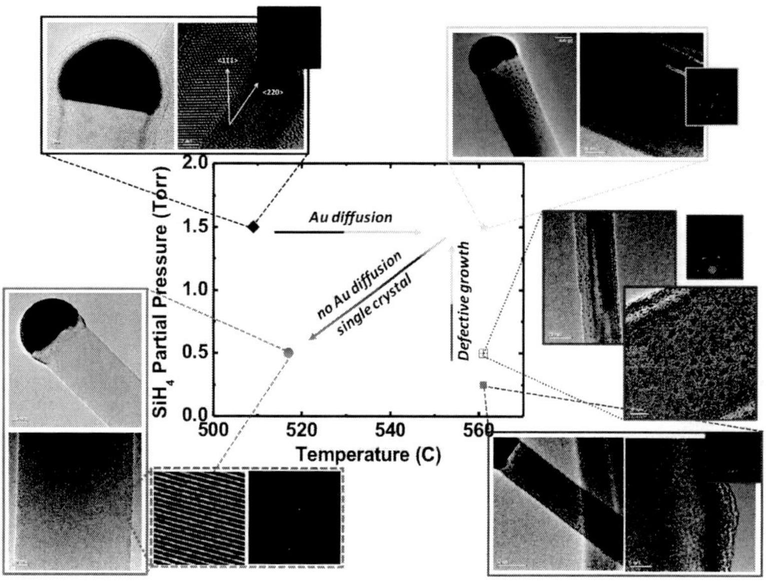

Figure 2.13 *A global Si NW growth morphology–structure diagram* illustrating changes in NW crystal structure, Au diffusion from the growth eutectic droplet, and NW morphology as function of temperature and pressure.

diffusion. However, at such growth conditions, nucleation efficiency of Si NW growth can dramatically decrease and a compromise has to be dealt with regarding crystal structure and morphology compared with nucleation efficiency.

2.8 Ge Core/Si Shell Heterostructured Nanowires

It has been noted early on that the formation of high-quality epitaxial core/shell materials[51] composed of a semiconducting NW surrounded by one or more single crystal shells can offer opportunities for new control of charge transport in nanostructures.[52] Realizing such heterostructures and assessing their resulting performance is challenging due to the diffusion of the Au growth-mediating seed during Si shell deposition and the typically amorphous nature of the deposited Si shell. The detrimental Au diffusion at higher temperatures, such as for the Ge core/Si shell heterostructures (Fig. 2.14e–h) has been addressed by either introducing O_2 during growth[53] or etching of the Au seed *ex situ* prior to growing Si shells,[54] both of which introduce contaminants that adversely affect the NW properties.

Since Si surface facets have a higher surface energy than Ge,[26] Au diffusion on the Si surface is expected to be less facile than on the Ge surface. Thus, the introduction of a thin Si layer beneath the Au growth seed is anticipated to create a barrier to the Au diffusion down the Ge NW sidewalls. This Au–NW interface energy difference for the Ge and Si NW surface facets should help to stabilize the Au seed at the NW tip to higher temperatures and at lower pressures. Deposition of such an interface layer must be carried out at a low temperature to prevent the onset of Au diffusion along the Ge NW surface prior to formation of the blocking layer. The catalytic effect of the Au nanoparticle was employed to locally decompose SiH_4 to form a thin Si layer below the Au particle at the Ge NW growth temperature (276 °C). When the temperature was ramped to 410 °C, no Au diffusion occurred as depicted in Fig. 2.14i–l, in dramatic contrast to the situation where no Si interfacial barrier layer was grown (Fig. 2.14e–h).

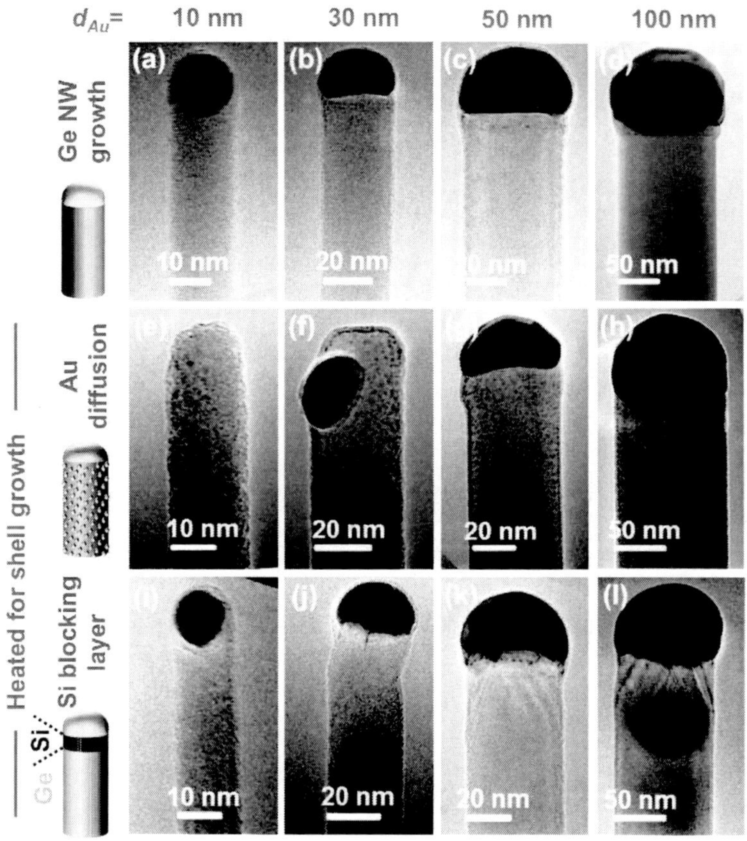

Figure 2.14 TEM images and morphology of different diameter Ge NWs demonstrating inhibition of Au diffusion. (a–d) TEM images of Ge NWs grown at 276 °C from 10 nm, 30 nm, 50 nm, and 100 nm Au nanoparticles, respectively with no Au diffusion. TEM images of Ge NWs grown at 276 °C and subject to a temperature ramp-up to 410 °C; Au diffusion is evident throughout all diameters with total loss of Au nanoparticle for the smallest diameter in (e). (i–l) same as in (e–h), however, with a Si-blocking layer where Au diffusion is blocked and the nanoparticles volume is conserved. Reprinted with permission from the American Institute of Physics, copyright 2011, Ref. [55].

The presence of a Si interfacial layer to block Au diffusion has a dramatic effect on the subsequent growth of Ge and Si shells for heterostructured NWs. Such growth of high-quality shells allows the realization of radial doped structures which can spatially alter and confine carriers, leading to improved performance of device structures such as enhanced on-currents or reduced surface scattering in NW field-effect transistors (FETs). Figure 2.15a–c shows a sequence of TEM images of Ge–Si core/shell NWs grown from 10 nm, 30 nm, and 50 nm diameter Au colloids without a Si barrier layer and without maintaining SiH_4 partial pressure during temperature ramp-up, which results in Au diffusion and a rough Si shell morphology. With a low-temperature Si interfacial barrier layer and a SiH_4 partial pressure of 125 mTorr during temperature ramp up, Au diffusion on the Ge NW sidewalls was avoided and single crystal Ge–Si core/shell NWs were grown in a single growth run as demonstrated in Fig. 2.15d–f. The Si shell thickness was chosen to be ~3 nm, which is within the coherent critical thickness limit for all Ge core diameters and hence ensures high crystalline quality of the Si shells without misfit dislocations.

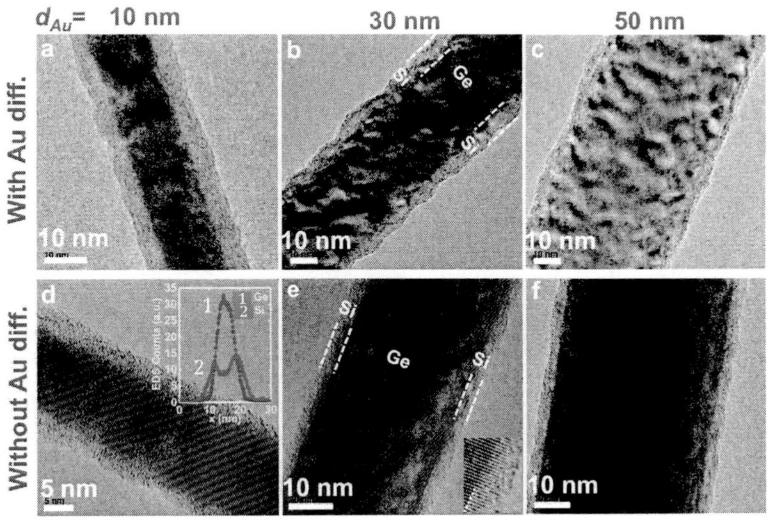

Figure 2.15 (a–c) TEM images of Ge–Si core/shell NWs grown at 276/500 °C from 10 nm, 30 nm, and 50 nm Au colloids, respectively, *without* a low-temperature Si barrier layer leading to 5–7 nm thick Si shells and rough NW surface morphologies. (d–f), TEM images of Ge–Si core/shell NWs grown at 276/500 °C from 10 nm, 30 nm, and 50 nm Au colloids, respectively, with a low-temperature Si barrier layer and a SiH_4 partial pressure of 0.25 Torr during temperature ramp (10 min) and Si shell deposition (7 min) leading to a 3 nm thick single crystal Si shell. Reprinted with permission from the American Institute of Physics, copyright 2011, Ref. [56].

The new understanding of the interplay between interface energies and Au diffusion on NW sidewalls for morphology control over Ge–Si core/shell heterostructures enable a single step growth process for high-quality growth of single crystal heterostructures. This provides a route for such synthesis without having to expose the NWs to *ex situ* chemical treatments that adversely affect their morphology,[54] or to *ex situ* thermal treatments that may induce thermally driven contaminants into the Si shell and Ge core and diffuse their otherwise abrupt interface.[51] With such a growth approach, a 2× improvement in experimental hole mobility, transconductance, and on-currents was demonstrated for heterostructures with smooth surface morphologies compared with those with rough surface morphologies and record on-currents for p-type FET devices were achieved.[55,56]

2.9 Unique Opportunities for Bandgap Engineering in Semiconductor Nanowires

The most compelling aspect of the VLS growth approach is the potential to modulate doping and composition along the length of the NW during its layer-by-layer growth,[57,58] thereby enabling energy band-edge engineering along the NW axis for additional control over charge transport. This control is in contrast to planar device growth where composition modulation in the charge transport direction (in-plane) is typically not possible during material growth which proceeds in the orthogonal (out-of-plane) direction.[59] This aspect is particularly attractive for heterostructured tunneling FET devices to simultaneously achieve steep turn on transistor characteristics while dramatically reducing the leakage current in the off state.[60,61]

Ge–Si heterostructured NWs not only offer two different charge injection and collection barriers at the metal semiconductor interfaces, but also add the important ability to accommodate band-offsets and built-in electric fields in the conduction or valence bands, separately at the Ge and Si sides of the NW, by proper selection of metal–semiconductor barrier heights with the metal contacts. The axial compositional asymmetry and band-offset accommodation, if chosen properly, can assist in more efficient sweeping of injected carriers from one terminal to the other and minimize or eliminate carrier leakage in the opposite direction. For the device demonstrated in Fig. 2.16,[62] a p+ doped Ge NW shell on i-Ge core, followed by an

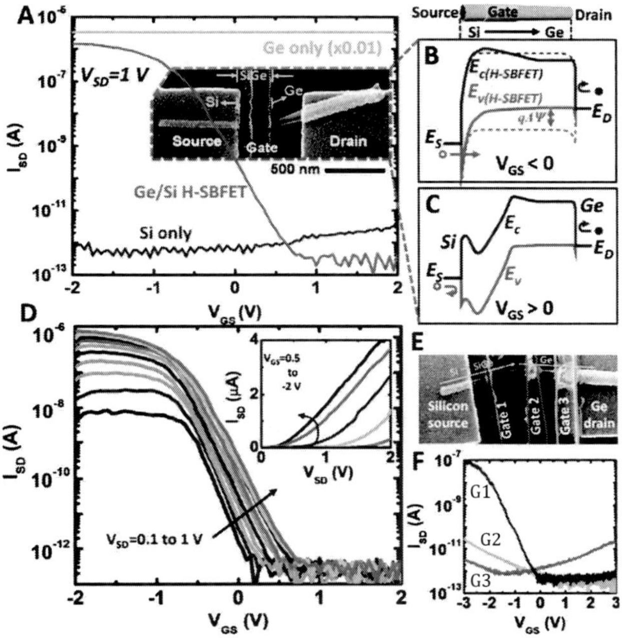

Figure 2.16 (a) Transfer curves of a p$^+$Ge–Si NW heterostructure S-TFET showing 10^7 I_{on}/I_{off} ratio. Devices fabricated in the same process on reference p$^+$Ge segments showed no gate modulation and those on reference Si segments showed ambipolar behavior over a larger V_{GS} range. (b,c) Simulated energy band-edge diagrams, extracted from the center of the NW channel, illustrating device operation where V_D is applied to the Si side (left, labeled as source) using the same gate configuration as in (a). A pure Si channel [dashed band-edge diagram in (b)] lifts up the potential by $\Delta\Psi$ in the channel therefore reducing the probability of hole emission into the channel and eliminating the built-in field in the valence band, only attained with the heterostructure. (d) Transfer curves at several V_{DS} biases showing no source-drain leakage and no ambipolar transport. Threshold-voltage shifts toward positive voltage are due to the lowering of the Schottky-barrier height at higher V_{DS} biases. Inset shows output characteristics of the same device. (e) SEM image of a multiple-gate device on a single Ge–Si axial NW heterostructure. (f) Transfer curves with V_D applied to the Si side of the device and gate voltage applied to gate 1 (G1), gate 2 (G2), and gate 3 (G3). Current modulation as function of V_{GS} is strongest near the Si side of the device (G1) and is weakest on the Ge side of the device (G1 and G2). Reprinted with permission from the American Institute of Physics, copyright 2011, Ref. [62].

undoped axial Si segment was grown and heterostructure Schottky barrier FETs (H-SBFETs) were fabricated. As shown in Fig. 2.16a, the highly doped Ge NWs show high currents with no current modulation over a large gate bias range, indicating no channel and no barrier modulation for the Ge segments. Low currents and no gate bias dependence were also observed for Si NWs within the gate voltage range shown in Fig. 2.16a. In contrast, a Si–Ge heterostructure NW shows seven orders of magnitude current modulation for the same source-drain bias voltage (V_{SD} = 1 V) as shown in Fig. 2.16a.

The high on-current observed in the axial Ge–Si heterostructure compared with that of homogeneous Si devices results from accommodation of the Ge to Si band offset being mostly in the valence band at the Si side of the channel, which provides an additional field that can assist hole thermoionic emission at the Si source (see band diagrams in Fig. 2.16b,c). This interpretation of transport behavior is also supported with another device with multiple gates placed along the channel of the Ge–Si axial NW heterostructure where transconductance is maximal for the gate placed on the Si segment of the device near the Ni source contact (Fig. 2.16e,f). This enhanced combined performance of high on-currents, high I_{on}/I_{off} ratios, and elimination of ambipolar behavior for this heterostructured S-TFET has not been reported in previous TFET devices. Such an axial band-edge-engineered device is made possible through the composition modulation achieved during the VLS growth process.

2.10 Conclusions

In this chapter we review and provide new perspectives on the VLS growth mechanism in semiconductor NWs along with detailed experimental evidence of emergent size effects in the growth and heterostructuring of Ge/Si NWs. Our analysis of size effects in the synthesis of Ge NWs provides the first direct and quantitative validation of a thermodynamic limit on the achievable NW diameter in VLS growth by combining observations of the reduced growth rate and enhanced equilibrium Ge solubility in the liquid Au–Ge growth seeds. The unified framework for understanding these nanoscale size effects—based on the Gibbs–Thomson effect—provides a basis for analyzing the cutoff diameter and other size-dependent growth phenomena at small diameters for a wide variety of semiconductor materials that form NWs by the VLS process. Aside from conventional growth parameters, such as temperature and precursor pressure, surface energies are a key factor in synthesis processes at the nanoscale, as demonstrated in the section on the influence of dopants. By tracking defect nucleation and propagation as a fingerprint of the layer-by-layer growth process in heterostructured NWs, a more complete understanding of the VLS growth mechanism, fault nucleation and NW kinking is enabled. We have also highlighted the influence of surface energies on controlling NW morphology, particularly for the case of Ge/Si core/shell NWs where we demonstrate how this effect can be used to block Au sidewall diffusion to produce optimal heterostructured NWs in a single growth step. These combined processes provide an improved and unified understanding of the VLS growth mechanism. We also demonstrate how to utilize these methods to achieve devices uniquely grown by the VLS method for Ge–Si axial heterostructures that provide enhanced performance compared with their homogenous counterparts. Such key concepts in NW growth and devices are envisioned to bring maturity to the field of semiconductor NWs and the technological applications they promise.

Acknowledgments

This research was funded in part by the Laboratory Directed Research and Development Program at Los Alamos National Laboratory and performed, in part, at the Center for Integrated Nanotechnologies, a U.S. Department of Energy, Office of Basic Energy Sciences user facility at Los Alamos National Laboratory (Contract DE-AC52-06NA25396).

References

1 G. W. Sears, A Growth Mechanism for Mercury Whiskers. *Acta Metal.* **3**, 361 (1953).

2 E. S. Greiner, J. A. Gutowski, and W. C. Ellis. Preparation of Silicon Ribbons. *J. Appl. Phys.* **32**, 2489 (1961).

3 R. S. Wagner, and W. C. Ellis, Vapor-Liquid-Solid Mechanism of Single Crystal Growth. *Appl. Phys. Lett.* **4**, 89 (1964).

4 K. Haraguchi, T. Katsyama, K. Hiruma, and K. Ogawa. GaAs p-n Junction Formed in Quantum Wire Crystals. *Appl. Phys. Lett.* **60**, 745 (1992).

5 A. M. Morales, and C. M. Lieber. A Laser Ablation Method for the Synthesis of Crystalline Semiconductor Nanowires. *Science* **279**, 208 (1998).

6 J. Johansson *et al.* Structural Properties of <111>B-oriented III–V Nanowires. *Nature Materials* **5**, 574 (2006).

7 R. Rosario *et al.* Lotus Effect Amplifies Light-Induced Contact Angle Switching. *J. Phys. Chem. B* **108**, 12640 (2004).

8 C. Thelander *et al.* Electron Transport in InAs Nanowires and Heterostructure Nanowire Devices. *Solid State Commun.* **131**, 573 (2004).

9 J. C. Johnson, H. Yan, P. Yang, and R. J. Saykally. Optical Cavity Effects in ZnO Nanowire Lasers and Waveguides. *J. Phys. Chem. B* **107**, 8816 (2003).

10 P. Martin, Z. Aksamija, E. Pop, and U. Ravaioli. Impact of Phonon-Surface Roughness Scattering on Thermal Conductivity of Thin Si Nanowires. *Phys. Rev. Lett.* **102**, 125503 (2009).

11 J. Y. Huang *et al.* In Situ Observation of the Electrochemical Lithiation of a Single SnO_2 Nanowire Electrode. *Science* **330**, 1515 (2010).

12 C. Thelander *et al.* Nanowire-based One-dimensional Electronics. *Mater. Today* **9**, 28 (2006).

13 P. J. Pauzauskie, and P. Yang. Nanowire Photonics. *Mater. Today* **9**, 36 (2006).

14 A. I. Boukai *et al.* Silicon Nanowires as Efficient Thermoelectric Materials. *Nature* **451**, 168 (2008).

15 C. K. Chan *et al.* High Performance Lithium-Battery Anodes Using Silicon Nanowires. *Nature Nano.* **3**, 31 (2008).

16 V. Schmidt, J. V. Wittemann, S. Senz, and U. Gosele. Silicon Nanowires: A Review on Aspects of their Growth and their Electrical Properties. *Adv. Mat.* **21**, 2681 (2009).

17 S. A. Dayeh. Electron Transport in Indium Arsenide Nanowires. *Semicond. Sci. Technol.* **25**, 024004 (2010).

18 S. T. Picraux, S. A. Dayeh, P. Manandhar, D. E. Perea, and S. G. Choi. Silicon and Germanium Nanowires: Growth, Properties, and Integration. *JOM.* **62**, 35 (2010).

19 H. Wang, and G. S. Fischman. Role of Liquid Droplet Surface Diffusion in the Vapor-Liquid-Solid Whisker Growth Mechanism. *J. Appl. Phys.* **76**, 1557 (1994).

20 F. Glas, J.-C. Harmand, and G. Patriarche. Why Does Wurtzite Form in Nanowires of III-V Zinc Blende Semiconductors. *Phys. Rev. Lett.* **99**, 146101 (2007).

21 C.-Y. Wen *et al.* Formation of Compositionally Abrupt Axial Heterojunctions in Silicon-Germanium Nanowires. *Science* **27,** 1247 (2009).

22 A. D. Gamalski, J. Tersoff, R. Sharmas, C. Ducati, and S. Hofmann. Formation of Metastable Liquid Catalyst During Subeutectic Growth of Germanium Nanowires. *Nano Lett.* **10**, 2972 (2010).

23 V. N. Smirnov, Germane Decomposition: Kinetic and Thermochemical Data. *Kinet. Catal.* **48**, 1608 (2007).

24 B. S. Meyerson, and J. M. Jasinski, Silane Pyrolysis Rates for the Modeling of Chemical Vapor Deposition. *J. App. Phys.* **61**, 785 (1987).

25 L. E. Fröberg, W. Seifert, and J. Johansson. Diameter Dependent Growth Rate of InAs Nanowires. *Phys. Rev. B* **76**, 153401 (2007).

26 Q. Jiang, H. M. Lu, and M. Zhao. Modeling of Surface Energies of Elemental Crystals. *J. Phys.: Condens. Matter.* **16**, 521 (2004).

27 T. Y. Tan, N. Li, and U. Gosele. Is There a Thermodynamic Size Limit of Nanowires Grown by the Vapor-Liquid-Solid Process. *Appl. Phys. Lett.*, **183**, 1199 (2007).

28 C.-Y. Wen, M. C. Reuter, J. Bruley, J. Tersoff, S. Kodambaka, E. A. Stach, F. M. Ross, Formation of Compositionally Abrupt Axial Heterojunctions in Silicon-Germanium Nanowires. *Science* **326**, 1247 (2009).

29 S. Kodambaka, J. Tersoff, M. C. Reuter, and F. M. Ross. Germanium Nanowire Growth below the Eutectic Temperature. *Science* **316**, 729 (2007).

30 S. A. Dayeh, E. T. Yu, and D. Wang. III-V Nanowire Growth Mechanism: V/III Ratio and Temperature Effects. *Nano Lett.* **7**, 2486 (2007).

31 S. A. Dayeh, E. T. Yu, and D. Wang. Surface Diffusion and Substrate-Nanowire Adatom Exchange in InAs Nanowire Growth. *Nano Lett.* **9**, 1967 (2009).

32 P. Aella, S. Ingole, W. T. Petuskey, and S. T. Picraux. Influence of Plasma Stimulation on Si Nanowire Nucleation and Orientation Dependence. *Adv. Mat.* **19**, 2603 (2007).

33 S. A. Dayeh, C. Soci, X. Bao, and D. Wang. Advances in the Synthesis of InAs and GaAs Nanowires for Electronic Applications. *Nano Today* **4**, 347 (2009).

34 K. A. Dick *et al.* A New Understanding of Au-assisted Growth of III-V Nanowires. *Adv. Func. Mat.* **15**, 1603 (2005).

35 S. A. Dayeh, and S. T. Picraux. Direct Observation of Nanoscale Size Effects in Ge Semiconductor Nanowire Growth. *Nano Lett.* **10**, 4032 (2010).

36 H. Schmid, M. T. Bjork, J. Knoch, H. Riel, and W. Riess. Patterned Epitaxial Vapor-Liquid-Solid Growth of Silicon Nanowires on Si(111) Using Silane. *J. App. Phys.* **103**, 024304 (2008).

37 Dubrovskii, V. G., and Sibirev N. V. Growth Rate of a Crystal Facet of arbitrary Size and Growth Kinetics of Vertical Nanowires. *Phys. Rev. E* **70**, 031604 (2004).

38 Givargizov, E. I., and Chernov A. A. Rate of Whisker Growth by the Vapor-Liquid-Crystal Mechanism and the Role of Surface Energy. *Kristalografiya* **18**, 147 (1973).

39 Givargizov, E. I. Fundamental Aspects of VLS Growth. *J. Cryst. Growth* **31**, 20 (1975).

40 Wu, Y., Fan, R., Yang, P. Block-by-Block Growth of Single-Crystalline Si/SiGe Superlattice Nanowires. *Nano Lett.* **2**, 83 (2002).

41 X. Zhang, K.-K. Lew, P. Nimatoori, J. M. Redwing, and E. C. Dickey. Diameter-Dependent Composition of Vapor-Liquid-Solid Grown $Si_{1-x}Ge_x$ Nanowires. *Nano Lett.* **7**, 3241 (2007).

42 T. E. Clark, P. Nimmatoori, K.-K. Lew, L. Pan, J. M. Redwing and E. C. Dickey, E. C. Diameter Dependent Growth Rate and Interfacial Abruptness in Vapor-Liquid-Solid $Si/Si_{1-x}/Ge_x$ Heterostructure Nanowires. *Nano Lett.* **8**, 1246 (2008).

43 J. Weyher. Some Notes on the Growth Kinetics and Morphology of VLS Silicon Crystals Grown with Platinum and Gold as Liquid-Forming Agents. *J. Cryst. Growth* **43**, 235 (1978).

44 S. Kodambaka, J. Tersoff, M. C. Reuter, and F. M. Ross. Diameter-Independent Kinetics in the Vapor-Liquid-Solid Growth of Si Nanowires. *Phys. Rev. Lett.* **96**, 096105 (2006).

45 R. E. Algra *et al.* Twinning Superlattices in Indium Phosphide Nanowires. *Nature* **456**, 369 (2008).

46 F. M. Davidson, D. C. Lee, D. D. Fanfair, and B. A. Korgel. Lamellar Twinning in Semiconductor Nanowires. *J. Phys. Chem. C* **111**, 2929 (2007).

47 D. E. Perea *et al.* Direct Measurement of Dopant Distribution in an Individual Vapor-Liquid-Solid Nanowire. *Nat. Nanotechnol.* **4**, 315 (2009).

48 S. H. Oh, M. F. Chisholm, Y. Kauffmann, W. D. Kaplan, W. Luo, M. Rühle, and C. Scheu. Oscillatory Mass Transport in Vapor-Liquid-Solid Growth of Sapphire Nanowires. *Science* **330**, 489 (2010).

49 S. A. Dayeh, J. Wang, N. Li, J. Y. Huang, A. V. Gin, and S. T. Picraux. Growth, Defect Formation and Morphology Control of Germanium-Silicon Semiconductor Nanowire Heterostructures, *Nano Letters* **11**, 4200, (2011).

50 K. W. Schwarz, and J. Tersoff. From Droplets to Nanowires: Dynamics of Vapor-Liquid-Solid Growth. *Phys. Rev. Lett.* **102**, 206101 (2009).

51 L. J. Lauhon, M. S. Gudiksen, D. Wang, and C. M. Lieber. Epitaxial Core-shell and Core-multishell Nanowire Heterostructures. *Nature* **420**, 57 (2002).

52 G. Liang, J. Xiang, N. Kharche, G. Klimeck, C. M. Lieber, and M. Lundstrom. Performance Analysis of a Ge/Si Core/Shell Nanowire Field Effect Transistor. *Nano Lett.* **7**, 642 (2007).

53 S. Kodambaka, J. B. Hannon, R. M. Tromp, and F. M. Ross. Control of Si Nanowire Growth by Oxygen. *Nano Lett.* **6**, 1292 (2006).

54 I. A. Goldthrope, A. F. Marshall, and P. C. McIntyre. Inhibiting Strain-Induced Surface Roughening: Dislocation-Free Ge/Si and Ge/SiGe Core—Shell Nanowires. *Nano Lett.* **9**, 3715 (2009).

55 S. A. Dayeh, N. H. Mack, J. Y. Huang, and S. T. Picraux. Advanced Core/Multi-shell Germanium/Silicon Nanowire Heterostructures: The Au Diffusion Bottleneck. *Appl. Phys. Lett.* **99**, 023102 (2011).

56 S. A. Dayeh, A. V. Gin, and S. T. Picraux. Advanced Core/Multi-shell Germanium/Silicon Nanowire Heterostructures: Transport and Morphology. *Appl. Phys. Lett.* **98**, 163112 (2011).

57 M. S. Gudiksen, L. J. Lauhon, J. Wang, D. C. Smith, and C. M. Lieber. Growth of Nanowire Superlattice Structures for Nanoscale Photonics and Electronics. *Nature* **415**, 617 (2002).

58 M. T. Bjork *et al.* One-dimensional Steeplechase for Electrons Realized. *Nano Lett.* **2**, 87 (2002).

59 M. L. Lee, E. A. Fitzgerald, M. T. Bulsara, M. T. Currie, and A. Lochtefeld. Strained Si, SiGe, and Ge Channels for High-Mobility Metal-Oxide-

Semiconductor Field-Effect Transistors. *J. Appl. Phys.* **97**, 011101 (2005).

60 O. M. Nayfeh *et al.* Design of Tunneling Field-Effect Transistors using Strained-Silicon/Strained-Ge Type-II Staggered Heterojunctions. *IEEE Elect. Dev. Lett.* **29**, 1074 (2008).

61 L. Wang, E. Yu, Y. Taur, and P. Asbeck. Design of Tunneling Field-Effect Transistors Based on Staggered Heterojunctions for Ultralow-Power Applications. *IEEE Elect. Dev. Lett.* **31**, 431 (2010).

62. S. A. Dayeh, R. M. Dickerson, and S. Tom Picraux. Axial Bandgap Engineering in Germanium-Silicon Heterostructured Nanowires. *Appl. Phys. Lett.* **99**, 113105 (2011).

Chapter 3

Transition Metal Silicide Nanowires: Synthetic Methods and Applications

Jeremy M. Higgins, Andrew L. Schmitt, and Song Jin
University of Wisconsin–Madison, 1101 University Avenue, Madison, WI 53706, USA
jin@chem.wisc.edu

3.1 Introduction

Transition metal silicides comprise a large family of refractory materials that have found success in many applications, including complementary metal-oxide-semiconductor (CMOS) devices, thin-film coatings, bulk structural components, electrical heating elements, photovoltaics (PVs), and thermoelectrics. One-dimensional (1-D) silicide nanomaterials, such as nanowires (NWs), are currently being investigated for potential performance improvements in these and new applications. However, synthesis of silicide NWs via chemical methods is more complicated than in many other classes of nanomaterials due to the complex phase behavior between metals and Si and the numerous stoichiometries and structures of their resulting compounds. Several synthetic strategies have

Silicon and Silicide Nanowires: Applications, Fabrication, and Properties
Edited by Yu Huang and King-Ning Tu

ISBN 978-981-4303-46-0 (Hardcover), 978-981-4303-47-7 (eBook)
www.panstanford.com

been developed to overcome this challenge, as evidenced by the increasing literature reports of silicide NWs. In this chapter we will review the synthesis of thin film and bulk silicide materials, highlight the strategies currently used to synthesize silicide NWs, present current and future synthetic challenges, and review the applications emerging from current silicide NWs.

Transition metal silicides are members of a general class of solid state compounds commonly referred to as intermetallics. Intermetallic compounds are composed of at least two metallic elements, potentially including poor metals (e.g., Al, Ga, In, Sn, Pb) and metalloids (e.g., Si, Ge, As, Sb), often have complicated crystal structures and are increasingly finding applications in a variety of fields. Examples include the use of Heussler alloys in spintronics,[1] Nowotny chimney ladder phases in thermoelectrics,[2] and superalloys in structural materials.[3] Structure-property relationships in such compounds are often quite interesting and strongly dependent on doping. These refractory, intermetallic metal silicides have diverse physical properties. NiSi, $CoSi_2$, and $TiSi_2$[4,5] are metallic materials currently used as ohmic contact,[6] interconnect, and gate contacts[7] in CMOS microelectronic transistors. Si-based optoelectronics[8] has been enabled using semiconducting silicides[9] to produce LEDs[10] and IR detectors.[11] The narrow bandgap semiconducting silicides, in particular $CrSi_2$, β-$FeSi_2$, $MnSi_{1.8}$, and $ReSi_{1.8}$, have been targeted and used for robust, stable, and inexpensive thermoelectric materials[9,12] and have shown promise for PV[13,14] applications.

Beyond their numerous technological applications, metal silicides have fascinated physicists for over 70 years, as they continue to yield surprises at the frontiers of theoretical and experimental condensed matter physics. For example, the metal monosilicides that share the *B20* structure type (MSi, M = Fe, Co, Mn) are a group of highly correlated electron materials with some very unusual properties.[15,16] FeSi, the only known transition metal Kondo insulator,[15] is a heavy-electron compound that exhibits Kondo lattice behavior at room temperature but has a small bandgap, insulating ground state. MnSi, once thought of as a classical itinerant ferromagnet, has recently been discovered to exhibit non-Fermi liquid behavior without a quantum critical transition.[17,18] The *B20* monosilicides and their alloys, $Fe_{1-x-y}Co_xMn_ySi$ ($0 < x, y < 1$) also display a myriad of magnetic behaviors including unusual helical magnetic ordering,[19–21] and the exotic Skyrmion magnetic phases.[22,23] Furthermore, $Fe_xCo_{1-x}Si$

alloys were recently discovered to be magnetic semiconductors,[24,25] bringing about the exciting prospect of CMOS-compatible Si-based spintronics,[26] a growing field that seeks to exploit the electronic spin degree of freedom in lieu of or in addition to the electron charge in electronic and photonic devices.[27]

The reduction of a material's size to nanoscale dimensions has often resulted in new physical phenomena and interesting applications.[28,29] Among the various nanoscale morphologies, 1-D structures, such as NW materials,[29–34] have enjoyed prominent attention and success. The structure, diameter, length, chemical composition, and doping/electronic properties of single crystalline NWs can now be precisely controlled using a variety of bottom-up synthetic methods, including the nanoparticle catalyzed vapor–liquid–solid (VLS)[35] growth process, which has arisen as the most common method for 1-D nanostructure synthesis of materials having relatively simple phase behavior. The numerous 1-D nanomaterials synthesized to date exhibit diverse device behavior and have enabled the successful demonstration of many functional devices[36] following the bottom-up paradigm of nanotechnology.[37]

Synthetic efforts thus far have concentrated on producing NWs of semiconductor channel materials for use in CMOS devices, but metallic contact and gate materials are equally important for continued scaling of semiconductor technology.[38] NWs of CMOS-compatible metallic silicides having low resistivity and suitable work functions, such as the nickel silicides,[7,39] would serve as superior interconnect and gate contacts in nanoelectronic architectures. High-quality, single-crystal semiconducting NWs, such as the half-metallic $Fe_{1-x}Co_xSi$ alloys, will lay the nanomaterials foundation for the exploration of many technologically relevant fields, including Si-based spintronic nanodevices. Reducing the size and dimensionality of strongly correlated electron systems, such as the *B20* metal monosilicides and their alloys, could lead to interesting new physical phenomena not yet considered by theoretical physicists, since it is known that electron correlations are intimately influenced by length scales and symmetry.[15,16] Semiconducting silicides, such as $MnSi_{1.8}$[40,41] and $ReSi_{1.8}$[42,43] are thermoelectric materials with reported figure of merit (*ZT*) values up to 0.7 and 0.8 in the bulk, respectively. These structures, along with Ru_2Si_3 and their alloys,[44] belong to the "Nowotny chimney ladder" phases, having one lattice constant as large as tens of nanometers.[45–47] It has been predicted[48] and experimentally confirmed[49–52] that dimensional reduction of semiconductors enhances *ZT* relative to the bulk values due to surface phonon scattering and/or quantum confinement.

Due to their interesting properties and potential applications, our research group initiated a research program in 2005 to develop free-standing silicide NWs, resulting in the successful synthesis of FeSi,[53,54] CoSi,[55] $Fe_{1-x}Co_xSi$,[56] MnSi,[57] $MnSi_{1.8}$,[58] $CrSi_2$,[59] Ni_2Si,[60] and Ni_3Si.[61] The rational synthesis of NWs from the vapor phase has two fundamental challenges:[29,31,33] (1) delivery of source materials, and (2) anisotropic 1-D crystal growth. The rapid development of semiconducting NWs of materials that exhibit simple phase behavior with low melting eutectics (Fig. 3.1a), such as group IV elements and normal valence II-VI and III-V compounds, have been greatly served by the VLS growth scheme,[29] but a similar mechanistic scheme for the synthesis of free-standing transition metal silicide NWs is currently lacking. The general and rational chemical synthesis of silicide nanomaterials is further complicated by the multiple stoichiometries and complex phase behavior exhibited by many silicide compounds.[62] For instance, there are seven known manganese silicide intermetallic compounds, namely Mn_6Si, Mn_4Si, Mn_3Si, Mn_5Si_2, Mn_5Si_3, MnSi, and $MnSi_{1.8}$, all of which have a stable polymorph at room temperature (Fig. 3.1b).[63]

This complicated phase behavior and the current lack of mechanistic understanding make careful control over synthesized NW phases a difficult prospect. Despite these difficulties, we and other researchers have developed several techniques for the synthesis of free-standing silicide NWs. They can be conveniently categorized into four groups: silicidation of Si NWs, decomposition of Si sources on metal thin films, reaction of metal sources with Si substrates, and deposition of both metal and Si via chemical vapor deposition (CVD) or chemical vapor transport (CVT). Two review articles including various aspects of the growth and nucleation of metal silicide NWs have recently been published.[64,65] However, it is clear that little of the accumulated knowledge obtained overcoming similar difficulties in the synthesis of bulk and thin-film transition metal silicides[66] has been integrated into these NW approaches. Such lessons can greatly serve the growth and development of this field, and in this chapter we will endeavor to highlight instances where obvious parallels exist.

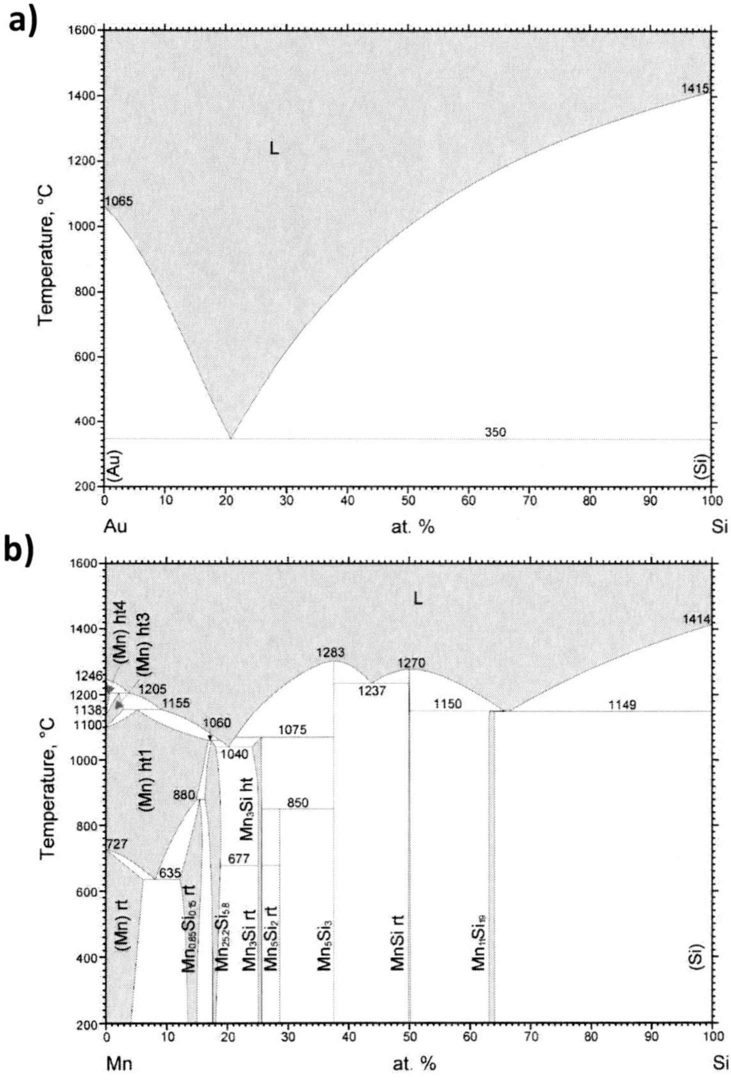

Figure 3.1 Binary phase diagram between (a) Au–Si[67] and (b) Mn–Si.[63]

To illustrate the progress made thus far in the production of transition metal silicide NWs, we have assembled a table of all known silicides stable at room temperature (Table 3.1). Specifically, of the nine 3*d* transition metals that form room temperature stable silicide phases, NW morphologies have been reported for seven (boldfaced in Table 3.1). There are examples of solvothermal silicide nanorods[68]

Table 3.1 Periodic table of the transition metal silicide phases[67,72–109]

IA	IIA	IIIB	IVB	VB	VIB	VIIB	VIII	VIII	VIII	IB	IIB
H_4Si											
$Li_{22}Si_5$											
$Li_{13}Si_4$											
Li_7Si_3											
$Li_{12}Si_7$											
LiSi											
NaSi	Mg_2Si										
$NaSi_2$											
Na_8Si_{46}											
Na_xSi_{136}											
KSi	Ca_2Si	*Sc_5Si_3*	Ti_3Si	V_3Si	Cr_3Si	M_6Si	Fe_3Si	**(Co_3Si)**	**Ni_3Si**	Cu_5Si	
$K_{12}Si_{17}$	Ca_5Si_3	Sc_5Si_4	**Ti_5Si_3**	**V_5Si_3**	Cr_5Si_3	Mn_4Si	**(Fe_5Si_3)**	***Co_2Si***	**$Ni_{31}Si_{12}$**	$Cu_{15}Si_4$	
K_8Si_{46}	CaSi	ScSi	**Ti_5Si_4**	*VSi_2*	CrSi	Mn_3Si	***FeSi***	**CoSi**	***Ni_2Si***	Cu_3Si	
K_7Si_{46}	Ca_3Si_4	Sc_3Si_5	**TiSi**		***$CrSi_2$***	Mn_5Si_2	**$FeSi_2$**	$CoSi_2$	**Ni_3Si_2**		
	$CaSi_2$		***$TiSi_2$***			*Mn_5Si_3*			**NiSi**		
						MnSi			**$NiSi_2$**		
						$MnSi_{1.8}$					

IA	IIA	IIIB	IVB	VB	VIB	VIIB	VIII	VIII	VIII	IB	IIB
RbSi	Sr_2Si	*Y_5Si_3*	Zr_3Si	Nb_5Si_3	Mo_3Si	Tc_4Si	Ru_4Si_3	Rh_2Si	$Pd_{21}Si_5$		
$Rb_{12}Si_{17}$	Sr_5Si_3	Y_5Si_4	Zr_2Si	*$NbSi_2$*	Mo_5Si_3	Tc_3Si	RuSi	Rh_5Si_3	$Pd_{14}Si_3$		
Rb_5Si_{46}	SrSi	YSi	Zr_3Si_2		*$MoSi_2$*	Tc_5Si_3	*Ru_2Si_3*	Rh_3Si_2	Pd_3Si		
	$SrSi_2$	Y_2Si_3	*Zr_5Si_4*			TcSi	$RuSi_2$	*RhSi*	*Pd_2Si*		
		Y_3Si_5	ZrSi			Tc_4Si_7		Rh_4Si_5	$Pd_{39}Si_{20}$		
			$ZrSi_2$					Rh_3Si_4	$Pd_{19}Si_{10}$		
CsSi	Ba_2Si	La_5Si_3	Hf_2Si	Ta_3Si	W_5Si_3	Re_2Si	OsSi	Ir_3Si	$Pt_{25}Si_7$		
Cs_7Si_{136}	Ba_5Si_3	La_3Si_2	*Hf_3Si_2*	Ta_2Si	*WSi_2*	*$ReSi_{1.8}$*	*Os_2Si_3*	Ir_3Si_2	Pt_3Si		
	BaSi	La_5Si_4	Hf_5Si_4	Ta_5Si_3			$OsSi_2$	*IrSi*	Pt_5Si_2		
	Ba_3Si_4	*LaSi*	HfSi	***$TaSi_2$***				Ir_4Si_5	$Pt_{17}Si_8$		
	$BaSi_2$	$LaSi_2$	$HfSi_2$					Ir_3Si_4	*Pt_2Si*		
								Ir_3Si_5	Pt_6Si_5		
								$IrSi_3$	**PtSi**		

Note: Boldfaced phases have been observed in nanowires, italicized phases are predicted 1st phase rule phases, and metastable nanowire phases are in parenthesis.

and self-assembled epitaxial silicide NWs,[69–71] mostly of rare earth silicides on vicinal Si surfaces, but these silicide nanomaterials are distinctly different from free-standing silicide NWs and are not the subject of this review. Herein, we review transition metal silicide bulk and thin-film diffusion couples, the rational chemical synthesis of free-standing transition metal silicide NWs, and their current and potential future applications.

3.2 Formation of Bulk and Thin-Film Metal Silicides in Diffusion Couples

The formation of bulk and thin-film transition metal silicides has been extensively studied over the past 40 years. Synthesis by various classical bulk crystal methods including CVT[110] and melt or flux growth[111] have been used to obtain large silicide crystals for study, but silicides widely used in technological applications are often formed by the direct temperature-induced interfacial reaction of a transition metal with Si, a process known as reactive phase formation.[4,112,113] This synthetic method proceeds via interdiffusion of metal and Si atoms to a growth interface, nucleation of new silicide phases, and subsequent new phase growth. When reactive phase formation occurs between two solid reactants, the systems are commonly referred to as diffusion couples (Fig. 3.2). Diffusion couples are an excellent conceptual tool for illustrating basic solid state reaction processes, including the general kinetic and thermodynamic considerations important in bulk and thin-film silicide formation.

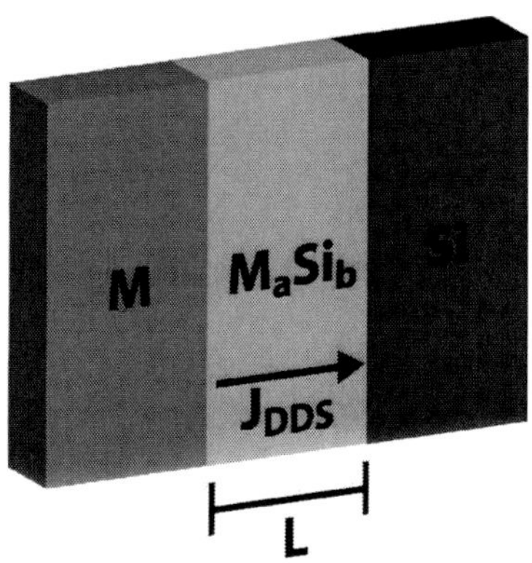

Figure 3.2 Schematic drawing of a metal–Si diffusion couple.

3.2.1 Basic Description

The diffusion couple concept is an excellent paradigm for discussing the reaction between two nominally immiscible phases, be they solid, vapor, or liquid;[114] however, one most commonly encounters diffusion couples in the reactions between two solid materials. In these cases, there have been two classically described varieties: (1) bulk diffusion couples and (2) thin-film diffusion couples.[113,115] A bulk diffusion couple is the reaction between two semi-infinite (on the molecular level, i.e., greater than 10 μm thick) sources. In bulk diffusion couples, most, if not all, of the stable equilibrium phases will be observed. Phases tend to nucleate sequentially, and often multiple phases will be observed to grow simultaneously. These characteristics have lead to the use of bulk diffusion couples in the development of phase diagrams. In the case of a thin-film diffusion couple with a finite supply of at least one element or compound source, that is, a layer less than 100 nm thick, new phases are observed to form sequentially but generally at the expense of the other phases. Further, some equilibrium phases have never been seen in thin-film diffusion couples.

A number of growth models[113] have been proposed for transition metal silicide diffusion couples, but the basic physical picture exhibited in each of these reactive phase formation models is succinctly captured in a three-step process: (1) diffusion of mobile species to the reactive interface between starting materials, (2) reaction of mobile species with immobile species to nucleate and grow a new crystalline phase, and (3) further diffusion of mobile species to grow and nucleate other silicide phases. The actual sequence of the silicides produced depends greatly on the relative quantities of starting materials, the kinetics associated with these processes, and the experimental conditions. The important considerations of this process will be discussed step by step below, and important differences for NW systems will be mentioned as appropriate.

3.2.2 Diffusion, Thermodynamics, and Nucleation in Silicide Reactive Phase Formation

3.2.2.1 Diffusion and the dominant diffusing species

Most silicide kinetic growth models make the assumption that one of the atoms (i.e., either the metal species or Si) is much more mobile than the other. The identity of this mobile species, called the dominant diffusing species (DDS),[116] depends strongly on the nature of the phases present in the system. Initially, the metal and Si interfaces are in direct contact. These species will mutually diffuse into each other, forming a solid solution at the interface. However, once a silicide layer has formed, the most mobile species in the newly formed layer is the DDS.

The mobility of constituent atomic species in an intermetallic compound, such as a silicide, generally follows the Cu_3Au ordered rule.[117] According to this rule, the majority constituent of an intermetallic compound can quickly diffuse on its own sublattice via vacancy mechanism without creating significant disorder. In fact, the rate of this diffusion ought to be similar to self-diffusion within the pure element. Silicides tend to follow this rule fairly well, with the DDS of metal-rich silicides generally being the metal atom, the DDS of Si-rich silicides generally being Si, and in the case of monosilicides, Si tends to be the DDS.[118] It should be noted, though, that there are quite a few notable exceptions to this rule (e.g., $NiSi_2$ and $CoSi_2$).[116]

We have already generally discussed the concept of diffusion within silicides, but a sense of scale is useful. Diffusion follows the Arrenhius relation,

$$D = D_0\, e^{(-Q/RT)}$$

with D being the diffusion constant, D_0 the maximum diffusion coefficient sometimes called the frequency factor, Q the activation energy of the process, and RT the thermal energy. Further the isotropic diffusion length in an n-dimensional system is given by the formula,

$$L_D = \sqrt{2nDt}$$

with L_D being the diffusion length, n being the dimensionality of the system (i.e., 1, 2, or 3), and t being an arbitrary time step. Self-diffusion constants of metals are usually on the order of 10^{-8} cm^2/s near the melting temperature of the solids, but in Si, D is closer to 10^{-12} cm^2/s.[116] We should note that the diffusion coefficients in Si-rich silicides tend to be somewhat higher than that in Si due to the more metallic nature of the bonding in silicides. These numbers are intended to give a sense of the scale over which one must consider diffusion in these compounds, that is, D can vary over six orders of magnitude, giving rise to a three orders of magnitude difference in average diffusion length!

Additional factors need to be considered in NW systems. Volume diffusion, the mass transport mechanism we have discussed above, occurs as the random movement of atoms into lattice defect sites under the overarching influence of a chemical potential gradient. This is appropriate to describe the diffusion behavior in thin-film and bulk diffusion couples when surfaces or grain boundaries do not play a major role. Nanomaterials, on the other hand, have high surface area–to-volume ratios so the effects of surface diffusion and/ or grain boundary diffusion may become quite important, especially at low temperatures. Grain boundary diffusion has been considered for a small number of transition metal silicide compounds; the measured diffusion coefficients for the DDS were 5–6 orders of magnitude higher than corresponding DDS volume diffusion coefficients, a potential increase in diffusion length of three orders of magnitude.[116]

3.2.2.2 Thermodynamics of silicide reactions in binary diffusion couples

Now that we have some concept of how elements move within a diffusion couple, one might ponder the thermodynamics of silicide formation. In reactive phase formation, the first silicide will be formed by the reaction of a transition metal and Si. The free energy change for the reaction is given by the Gibbs free energy of formation for the silicide, that is

$$m\mathrm{M(s)} + n\mathrm{Si(s)} \rightarrow \mathrm{M}_m\mathrm{Si}_n\mathrm{(s)}$$

$$G_{\mathrm{rxn}} = G^{f}_{\mathrm{silicide}}$$

Alternatively, a silicide can be formed by the reaction of either a transition metal or Si with a silicide already present, giving

$$p\mathrm{M(s)} + q\mathrm{Si(s)} + \mathrm{M}_m\mathrm{Si}_n\mathrm{(s)} \rightarrow \mathrm{M}_{m+p}\mathrm{Si}_{n+q}\mathrm{(s)}$$

$$\Delta G_{\mathrm{rxn}} = \Delta G^{f}_{\mathrm{silicide,\ products}} - \Delta G^{f}_{\mathrm{silicide,\ reactants}}$$

At relatively low temperatures, one can make the assumption that $\Delta G^{f}_{\mathrm{silicide}} = \Delta H^{f}_{\mathrm{silicide}}$; these heats of formation, $\Delta H^{f}_{\mathrm{silicide}}$, have been experimentally measured for many silicides.[119] Within a given M–Si system these values, which are often given in units of energy per mole of atoms to make them comparable, vary within a factor of 2–4. It seems to be a general theme of binary diffusion couples, and many other natural systems, that thermodynamics ultimately determines what can happen, but knowledge of the kinetics of nucleation and growth is absolutely necessary if one is to know what will happen.

3.2.2.3 Basics of nucleation

The nucleation of a new phase is governed by the total free energy of the nucleation process, namely the energy gain associated with creation of new surface and the energy loss associated with the reaction,[113]

$$\Delta G_{\text{nuc}}(r) = br^2\sigma - ar^3\Delta G_{\text{rxn,V}}$$

where a and b are geometric factors, r the radius of the nucleus formed, σ the surface energy of the newly formed surface, and $\Delta G_{\text{rxn,V}}$ is the free energy change of the nucleation reaction per unit volume, often written instead as $\Delta\mu/v_{\text{m}}$, the supersaturation over the molar volume. Differentiation of $\Delta G_{\text{nuc}}(r)$ reveals a maximum occurring at the critical nucleus radius, r^*:

$$r^* = \frac{2b\sigma}{3a\Delta G_{\text{rxn,V}}}$$

Substituting r^* into $\Delta G_{\text{nuc}}(r)$ yields the maximum value of the nucleation free energy, that is, the activation barrier for nucleation

$$\Delta G^* = \frac{4b^3\sigma^3}{27a^2\Delta G^2_{\text{rxn,V}}}$$

This short derivation reveals that the activation energy for the nucleation of a new phase is proportional to the energy loss caused by the creation of a new surface and inversely proportional to the free energy gain of the chemical reaction.

3.2.3 Kinetics of Silicide Layer Growth

A general approach to the growth kinetics of transition metal silicide diffusion couples should take into consideration diffusion, thermodynamics, and nucleation, all of which we have briefly covered. In the course of the following discussion we will use these concepts to discuss three important theoretical growth models that describe different types of silicide reactive phase formation: (1) the linear-parabolic approach,[112,114] (2) the physicochemical approach,[120] and (3) interfacial reaction barrier approach[115] These three models lead to a number of important experimentally observed growth phenomena—sequential nucleation of silicide phases, a critical thickness requirement for new silicide nucleation, and the potential importance of different growth-limiting mechanisms. Experimental observation of diffusion couples over the last few decades has revealed three different kinetic growth modes: nucleation-controlled, diffusion-controlled, and reaction-rate-controlled kinetics.

3.2.3.1 Nucleation-controlled kinetics

We have shown that the activation energy for the nucleation of a compound is inversely proportional to the square of the free energy of the silicide formation reaction per unit volume and that the free energy of formation for a given metal silicide can be approximated with its experimentally measured enthalpy of formation. These values are related to each other by the molar volume. In the instance where a compound is formed from its component elements, its free energy of formation is relatively large but its activation energy is low. Therefore, it is unlikely for the initial phase formed in a diffusion couple to be nucleation controlled. This is, in fact, observed experimentally.[114]

Subsequent reactions in the diffusion couple occur as the silicide phase reacts with transition metal or Si to form a new transition metal silicide. The free energy change of this reaction can be approximated as the difference in the heats of formation of the two silicide phases. In the case where the two silicides have similar enthalpies of formation, the free energy change of the reaction is small. The activation energy for nucleation would, therefore, be quite high and can potentially dominate the growth of such a phase.[114]

Nucleation-controlled growth is observed in the later stages of Ni–Si thin-film couples made by evaporating a thin film of nickel onto a thick Si substrate.[4] NiSi can be easily formed between 250 and 500 °C, giving a final state appropriate for use in microelectronics contacts and gates, but above 700 °C the entire film will spontaneously change to $NiSi_2$. Such an instantaneous reaction indicates that the reaction was not limited by the supply of reactants or the reaction itself, but rather by the nucleation.[114]

3.2.3.2 Diffusion-controlled kinetics

Diffusion-controlled kinetics are easily understood if one follows the derivation of the parabolic and linear-parabolic growth models developed by D'Heurle and coworkers.[112] This approach begins with the Nernst–Einstein equation,

$$J_{\text{DDS}} = c_{\text{DDS}} \times m_{\text{DDS}} \times \frac{d\mu_{\text{DDS}}}{dL}$$

with J_{DDS} being the flux of DDS, c_{DDS} the concentration of DDS, m_{DDS} the mobility of DDS, also equal to D_{DDS}/kT, and $d\mu_{\text{DDS}}/dL$ the chemical potential gradient across the growing layer. If D_{DDS} is independent of position in the growing layer, integration gives

$$J_{\text{DDS}} = c_{\text{DDS}} \times \frac{D_{\text{DDS}}}{kT} \times \frac{\Delta G_{\text{DDS}}}{L}$$

where ΔG_{DDS} is the free energy required to move one mole of DDS atoms across the growing layer (in practice, the free energy of the silicide formation reaction is often used instead). If one then notes that c_{DDS} is $1/v_{\text{DDS}}$ (with v_{DDS} being the volume of the compound per DDS atom) and $J_{\text{DDS}} \times v_{\text{DDS}}$ is dL/dt, substitution reveals

$$\frac{dL}{dt} = \frac{D_{\text{DDS}}}{kT} \times \frac{\Delta G_{\text{DDS}}}{L}$$

Integration of this equation with respect to L gives us

$$L = Kt^{1/2}$$

with, $K = [2\ D_{\text{DDS}} \times (\Delta G_{\text{DDS}}/kT)]^{1/2}$ the classic parabolic growth law with L proportional to $t^{1/2}$. The parabolic growth law works well to describe the growth of many silicide phases, but experiments often show faster growth and unusual behavior in the early stages. Further, inspection of the above equations shows that very thin films ought to grow infinitely fast, which is clearly a physical impossibility. To correct this apparent issue, D'Heurle and coworkers introduced the constant K', sometimes called the *chemical constant*, into the growth law as follows

$$\frac{dL}{dt} = \frac{D_{\text{DDS}}}{kT} \times \frac{\Delta G_{\text{DDS}}}{L + K'}$$

Integration of the new rate law results in the linear–parabolic growth law

$$L^2 + 2K'L = K\,(t + t_0)$$

where t_0 is an integration constant that accounts for the possibility of some growing phase already present at $t = 0$. At the limit of large L the parabolic growth law is recovered, but for small L, one obtains the linear growth law

$$L = \frac{K}{2K'}(t + t_0)$$

The linear–parabolic growth law does a better job of fitting experimental data for systems which show unusual early growth behavior, though it can be said that this early stage behavior is rarely well fit to a linear rate law. This theory has been expanded to include the behavior of two different silicide layers growing simultaneously,[112] revealing that phases ought to nucleate sequentially and often only after a critical thickness of one phase has been reached.

For an example of diffusion-controlled kinetics, we can turn once again to the Ni–Si thin-film binary diffusion couple. The initial phase formed between Ni and Si at relatively low temperature (less than 500 °C) is Ni_2Si, which grows to consume the entire supply of Ni. Once all the Ni has reacted, NiSi nucleates and consumes all of the Ni_2Si. The growth of these phases are well described by diffusion-controlled kinetics.[114]

3.2.3.3 Reaction rate–controlled kinetics

Reaction rate–controlled kinetics may be the most important mechanism for controlling silicide NWs synthesis since most of the silicide NWs we discuss below are synthesized via vapor-phase chemical reactions. However, this mechanism is less commonly discussed in the experimental literature on diffusion couples than nucleation-controlled or diffusion-controlled kinetics. The theoretical works by Dybkov (the physicochemical approach),[120] Gosele and Tu (the interfacial barrier approach),[115] and even d'Heurle (the diffusion based approach described above) contain elements of reaction rate–controlled kinetics. All three models predict an initial period where the growth of the new silicide is controlled by the reaction of the DDS with immobile species, followed by a diffusion-limited growth phase. We consider two related ways in which reactions might limit phase formation, both of which seem to arise from the considerations of reactant supply.

Once the DDS species reaches the reactive interface of a nucleated phase, the non-mobile species must be released from its parent interface before reaction can occur. The release of this atom and the subsequent bond formation with the DDS species sometimes requires significant atomic rearrangement. If this rearrangement has a high activation energy then it might be the rate-limiting step. In this case, the release of the immobile species and associated bond rearrangement, that is, the reaction, would be the limiting process in silicide growth, at least initially. The experimental behavior which led d'Heurle to include the "chemical parameter" in the linear–parabolic rate equation derived above is likely due to issues associated with both nucleation of the new phase and reaction rate controlled kinetics.

As mentioned at the beginning of this section, the term "diffusion couple" is a general term which does not require both reacting phases to be solid. If a vapor or liquid molecular species is involved in reactive phase formation, it is possible for the atomic rearrangement associated with the decomposition of the molecular precursor to be the rate-limiting step in the formation process. This has actually been observed; for example, the oxidation of Si exhibits reaction-controlled kinetics in the early portion of the process and diffusion controlled kinetics once the oxide film is thicker.[114]

The growth of silicide NWs is often accomplished by the reaction of a gaseous source with a solid substrate or the delivery of both Si and metal vapor sources to a substrate. The use of such chemical and physical processes for the delivery of precursor material to NW growth substrates introduces many potential processes and reactions, which might potentially limit the growth of silicide NWs. Diffusion-limited growth may arise if a growth substrate is the limiting source of material, chemical transport of vapor-phase reactants may lead to reaction rate–limited growth, and the strict requirement of 1-D nucleation and growth certainly may make NW growth a nucleation-limited process. Researchers working on vapor-phase silicide NW growth appear to be cognizant of some of these points, but we are not aware of a detailed chemical kinetic analysis of the synthesis of any silicide NW phase to date. Even though silicide NW growth currently appears mysterious, the lessons learned about the growth of silicide films should be useful, and we should expect that the physical laws that govern these silicides likely govern the growth of transition metal silicide NWs as well.

3.2.3.4 Bulk versus thin-film diffusion couples

A cursory glance at the experimental diffusion couple literature would lead one to believe that bulk and thin-film diffusion couples are manifestations of completely different phenomena, but the theoretical work done to date on the nucleation and growth of transition metal silicide phases do not make the distinction between these two different varieties of diffusion couple, that is, they are merely manifestations of the same phenomenon observed in different contexts.[113,115] First, while both types of diffusion couples show sequential nucleation of silicide phases, bulk diffusion couples often have multiple phases growing simultaneously, while in thin-film diffusion couples, new phases tend to grow at the expense of phases already present (Fig. 3.3). The lack of simultaneous phase growth in thin-film couples has been explained as the inability of the growing silicide layer to reach the critical thickness required for the nucleation of a new phase, an effect predicted by all three of the well-developed kinetic models, before consuming one of the elemental sources. In this case, nucleation and growth of the new phase *requires* the shrinkage of the previous phase, since it now serves as one of the material sources (Fig. 3.3). In contrast, in a bulk diffusion couple, neither the metal nor Si sources are used up before a growing silicide layer reaches the critical thickness required to nucleate the next phase; since both elemental sources are still available, both of the silicide phases can continue to grow (Fig. 3.3). Similar arguments could be made for phases one might expect to nucleate after the second phase, but which are not observed.

Another important difference between thin-film and bulk diffusion couples is the number of observed phases. Most, if not all, of the equilibrium phases known can be found in bulk diffusion couples, but regardless of starting conditions, some equilibrium bulk phases have never been observed in thin-film diffusion couples. This has been explained as a general issue with phase nucleation, but this may manifest in a number of ways. The three different theoretical models all arrive at a critical thickness requirement of one phase in order to nucleate a subsequent phase, but if this critical thickness is never reached (as in some thin-film diffusion couples) then one might expect certain phases would never nucleate. Further, most bulk diffusion couples are carried out at high temperatures where silicides whose growth is limited by nucleation can still grow.[114] Thin-film diffusion couples are generally studied under relatively low temperatures, and so, compounds which are nucleation-limited do not have the opportunity to begin growing. In these cases, these differences arise from the particular situations in which these different diffusion couple types are observed. In fact, these variations are the origin of the different technological advantages of bulk and thin-film diffusion couples. We expect that similar differences in phase behavior will be present in NW silicide systems and may limit technological applications of some NW silicide materials while potentially enabling the production of some NW phases not accessible as thin films.

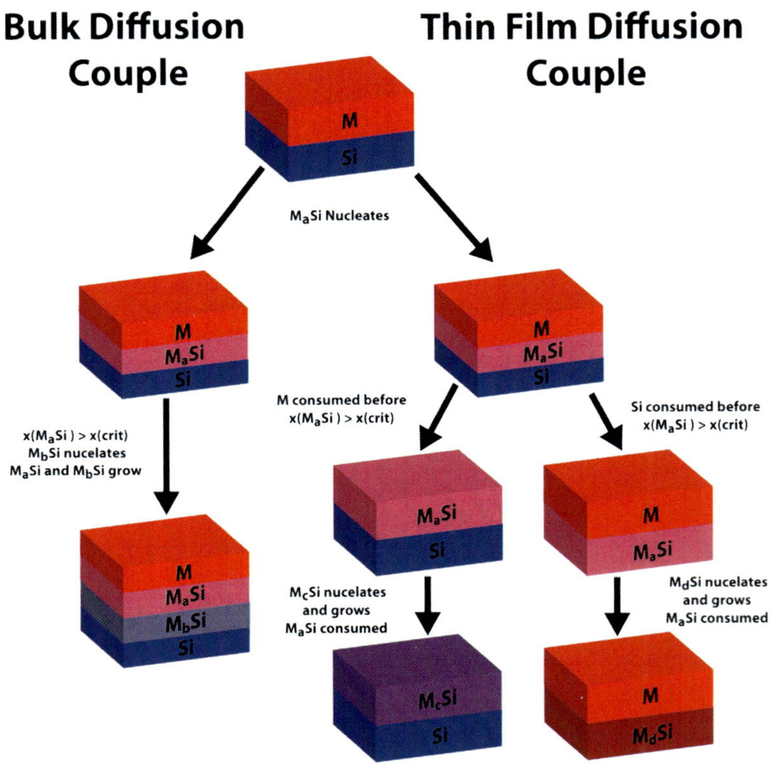

Figure 3.3 Three different possibilities for formation of a second silicide phase under different supply conditions for bulk and thin-film diffusion couples. The metal (M, red) reacts with silicon (Si, blue) producing various metal silicides. Metal-rich phases are redder in color, while Si-rich phases appear bluer. Depending on the particular metal–Si system M_bSi may be identical to M_cSi or M_dSi. Additionally, if M_aSi is the richest phase in either Si or M for the binary system then M_cSi or M_dSi will not exist, respectively.

3.2.4 Phase Formation

Our discussion to this point has focused primarily on how silicides grow, but in both bulk and thin-film diffusion couples phases nucleate sequentially and in a particular order.[112,113] While much work has gone into the details of the phase formation sequence, it is clear that we still do not have a good enough understanding to accurately predict even the first silicide phase formed in a given binary couple with complete accuracy, let alone the entire formation sequence. This has become a rather contentious topic, with a number different models vying for acceptance. In the words of d'Heurle, "A lot of ill-spent ink has been poured on the question of what is the first phase to form."[112] This topic has evolved considerably since its initial conception in order to account for new phenomena, but we will only mention the two simplest models: (1) the Walser-Bene rule[121] and (2) the effective heat of formation rule.[119]

3.2.4.1 Walser–Bene first phase rule

The Walser–Bene rule,[121] sometimes called the first phase rule, was developed in the mid-1970s in response to the increasing interest of the electronics community for low temperature growth of silicides. The quantity of diffusion couple data had increased rapidly, most of which was focused on transition-metal silicide couples. According to the original statement of this rule, "the first compound nucleated in planar binary reaction couples is the most stable congruently melting compound adjacent to the lowest-temperature eutectic on the bulk equilibrium phase diagram."[121]

This phase formation model was based on a number of assumptions. First, the sputtering and evaporation techniques commonly used to prepare the interface between metal and Si should form an amorphous "metallic glass" at the interface. The composition of this glassy layer should be similar to the lowest temperature eutectic of the binary phase diagram, where metallic glasses under supercooled conditions are known to be most stable. Interestingly, this preparation-induced amorphous layer has been observed using modern transmission electron microscopy (TEM) methods.[122,123] Second, as the annealing temperature is increased, the glassy layer will grow thicker via diffusion until it becomes metastable. Nucleation and growth of a crystalline phase will subsequently proceed. Third, the nucleation of congruently melting phases is expected to be favored over non-congruently melting phases because, in general, they exhibit smaller changes in short-range order when undergoing a transformation from a liquid-like to a crystalline atomic arrangement and thereby have a smaller activation energy for reaction. Finally, if the concentration fluctuations that drive nucleation arise due to diffusion, one would expect that initial phase formed will be the congruently melting compound closest in concentration to the lowest melting point eutectic. If more than one congruently melting compound exists near the lowest melting point eutectic on the phase diagram then the more stable compound, indicated by the higher melting point, will form.

There are a number of criticisms one could make about this model. First, it is primarily based on a consideration of the phase diagram with the addition of a little kinetic common sense, and so is inherently empirical. Further, it implies that for a given binary diffusion couple only one silicide may nucleate first, but evidence suggests that the situation may be quite a bit more complicated.[124] Perhaps most importantly this model provides no insight regarding the second, third, or subsequent silicides formed. Despite these shortcomings, this rule has been wildly successful and capable of successfully predicting the first phase formed in a large number of silicide systems, so long as the results of older studies are used. Predicted first phases based on the Walser–Bene first phase rule are italicized in Table 3.1. This rule has been generalized to metal–metal binary systems with notable success by removing the congruently melting restriction.[125] There are, however, more modern models which have attempted to solve some of the problems associated with the Walser–Bene first phase rule.[126–129]

3.2.4.2 Effective heat of formation approach

The second model we will discuss is the effective heat of formation model developed by Pretorius.[119] The basic idea of this model is to take the concentration of available elements into consideration as well as the thermodynamics of formation. It postulates that elements mix and interact at the growth interface with an effective concentration potentially quite different from the physical concentration present at the interface. To account for this, the model makes use of an effective heat of formation calculated per mole of available species during compound formation. For a silicide $M_{1-x}Si_x$ formed at an interface with effective concentration $M_{1-x^*}Si_{x^*}$ with

$x^* < x$, Si will be the limiting element. The effective heat of formation, $\Delta H'$, is given by the formula

$$\Delta H' = \Delta H_{\mathrm{f}} \times \left(\frac{\text{available concentration limiting element}}{\text{compound concentration limiting element}} \right)$$

$$= \Delta H_{\mathrm{f}} \times \left(\frac{x^*}{x} \right)$$

where ΔH_{f} is the standard heat of formation for the silicide.

Calculation of the available concentration-limiting element is impossible, so the authors argue that mixing at the interface should result in an effective concentration similar to that of the liquidus minimum, the idea being the activation energy for solid-state interdiffusion is directly proportional to the melting temperature of the solid. Since this activation energy determines the mobility of the atoms, the concentration at the liquidus minimum would be the end result of effective mixing. Like Walser and Bene, Pretorius and coworkers argue that congruently melting phases are more likely to form in silicide systems due to the decreased activation energy for their nucleation. After all considerations, the effective heat of formation rule states, "the first compound phase to from during metal–Si interaction is the congruent phase with the most negative effective heat of formation ($\Delta H'$) at the concentration of the liquidus minimum of the binary system".[119]

Like the Walser–Bene first phase rule, this model has been criticized for being based primarily on thermodynamics with little explicit attention paid to the kinetics of the formation process, but despite this, it has been at least as successful as the Walser–Bene rule. Further, it explicitly accounts for the possibility that more than one phase might be the "first phase" and even allows for the potential of metastable or amorphous "first phases", although ΔH_{f} of these phases have not yet been well characterized.

3.2.5 Modern Developments

As new tools have been developed, we have learned more about the phase formation process in bulk and thin-film diffusion couples. Initially it was thought that within a given diffusion couple only one phase could form first, but more recent experiments have shown that the identity of the first phase formed may depend strongly on the conditions of the experiment and nature of the sample.[119] Further, only equilibrium phases were once considered for the nucleation of a new phase, but in fact metastable and even amorphous silicide phases are now known to be relatively common as the first nucleated phase.[112,113,123] Finally, investigation of the first crystalline phase nucleated from an amorphous phase has begun; sub-nanometer crystallites of up to five different phases have been identified to nucleate from within the amorphous phase observed in the Ti–Si system using advanced TEM image analysis.[123,130] The phase formation and kinetic models discussed here do a reasonable job of predicting the nucleation and growth of crystalline phases and often do not dismiss the possibility of metastable or amorphous formation from occurring. However, data on these newly recognized phases and information about the earliest steps of the formation processes are still unavailable making application of the models to the entire process difficult. Newly available experimental techniques have expanded (or perhaps magnified) our view of the nucleation and growth occurring during the reactive phase formation of binary transition metal silicide diffusion couples, bringing many answers but also many new questions.

The results accumulated in the last few decades on the diffusion, kinetics, thermodynamics, and phase formation of transition metal silicides in diffusion couples have revealed a wealth of knowledge about these systems and have lead directly to many technological innovations. These lessons should be carefully considered when discussing the synthesis of silicide NWs. While most researchers interested in silicide NW synthesis would agree on its importance, few attempts have been made to connect the growth of silicide NWs to the important lessons learned from diffusion couple studies in a comprehensive way, and perhaps because of this, the current understanding of silicide NW nucleation, one-dimensional growth, and phase selection remains limited. In the next section, we will review the current methods used to synthesize silicide NWs, with an eye toward areas where we might make these connections.

3.3 Silicide Nanowire Growth Techniques

In the following section we will review the methods developed in the last few years to synthesize free-standing silicide NWs, highlight observations potentially important to our understanding of the silicide growth mechanism(s), and summarize the connections between various methods. Broadly, silicide NW synthesis methods can be grouped into two different categories: (1) silicidation of Si NW templates and (2) single-step vapor-phase reactions. In either case, researchers must overcome the two main hurdles for synthesis of silicide NWs—delivery of metal and Si sources and anisotropic 1-D growth of silicide.

3.3.1 Silicidation of Silicon Nanowires

Lieber and coworkers first reported transition metal silicide NWs produced via silicidation of Si NWs (Fig. 3.4a).[131] A Si NW backbone, formed using a Au nanoparticle-mediated VLS process, was covered with a thermally evaporated Ni thin film. Subsequent annealing in forming gas produced single-crystalline NiSi NWs via solid state reaction. Furthermore, epitaxial heterojunctions of NiSi and Si were demonstrated allowing the integration of high performance NW transistors beyond the lithographic limit. The main challenges in silicide NW formation—material delivery and 1-D anisotropic growth—were overcome independently using this synthetic method. This work has been pushed much further in the Ni–Si system[132–134] and easily extended to other metals, such as Fe to form FeSi, α-$FeSi_2$, and β-$FeSi_2$ NWs;[135] Pt resulting in PtSi NWs;[136,137] and Mn to produce MnSi NWs.[138] However, the NWs produced in this way are not always single-crystalline.

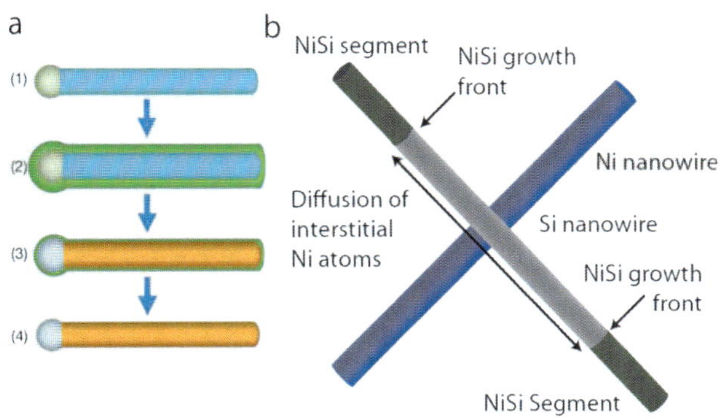

Figure 3.4 Preparation of NiSi NWs by silicidation. (a1) Si NWs (blue) are synthesized via VLS and (a2) coated with Ni metal (green), (a3) reacted at 550 °C to form NiSi NWs (brown). (a4) Extra Ni metal is etched away leaving single crystalline NiSi NWs.[131] (b) Preparation of NiSi NWs by point contact silicidation to form NiSi–Si NW heterostructures.[64]

Other researchers have focused on the *in situ* observation of the silicidation process using TEM.[139–141] In these experiments, VLS grown Si NWs are drop-cast with transition metal NWs, Ni[139] or Co,[140] onto TEM grids in high enough concentration to guarantee direct contact between the two different nanostructures (Fig. 3.4b). For the case of Ni, which is an extremely fast diffusing species in Si (0.6 μm/s at 400 °C),[142] Ni atoms "dissolve" into the Si NW at the rapid rate of 1 μm/min. This results in a supersaturated Ni–Si solid solution until NiSi crystals nucleate at the NW ends. The Si NW is then converted to NiSi one atomic plane at a time until the entire wire becomes fully silicided. Furthermore, highly strained (12%) single-crystalline Si segments can be formed between growing NiSi crystals. In the case of *in situ* silicidation of Si NWs by Co, the phase first nucleated is highly temperature dependent. At low temperature (700 °C), CoSi nucleated, while higher temperature (800 °C) led to $CoSi_2$ nucleation. These phases each showed different silicide/Si interfacial behavior and radically different axial growth rates.

Enough data has now been accumulated on the silicidation of Si NWs to begin a discussion of the NW diffusion couple kinetics and growth mechanisms, which are likely closely related to classical diffusion couples discussed in the previous section. First, regardless of the identity of the phase grown, the silicide in Si NWs grows linearly as a function of time, though the actual growth rate seems to depend significantly on the radius of the Si NW ($L \sim 1/r^2$)[132] and silicide phase grown.[140] *In situ* experiments and post-growth TEM analyses have revealed that silicide phases form with either rough or epitaxially distinct interfaces.[140] *In situ* analysis of the growth rate of CoSi showed it to depend strongly on the dissolution rate of Co and, hence, is likely limited by the interfacial reaction of Co with the Si NWs.[140] On the other hand, the epitaxially matched silicides, NiSi and $CoSi_2$, have been observed to be limited by the homogeneous nucleation of new atomic planes.[141] It is not currently clear whether this is a physical manifestation of an interfacial reaction barrier or a form of diffusion limitation. This result is interesting since one would generally expect that diffusion controlled growth, either volume or surface diffusion, ought to give rise to parabolic silicidation kinetics, but a recent theoretical examination of the problem reveals that if diffusion along the Si/silicide interface is rate limiting, then one should expect a linear growth rate with a $1/r^2$ or $1/r^3$ dependence, depending on the nature of the Si/silicide interface diffusion.[143] The picture emerging from these growth studies emphasizes the importance of the Si/silicide interface in the silicidation of Si NWs.

The stoichiometry of silicided NWs can likely be controlled by carefully managing the quantity of deposited metal and the conditions of the silicidation reaction. However, certain phases will probably be unobtainable through this technique due to the kinetics and thermodynamics of reactive phase formation, as was described for thin-film diffusion couples above. Ni–Si is the best studied NW diffusion couple system thus far. To date, the phases nucleated first have been strongly dependent on processing conditions and the crystallographic nature of the initial Si NW. The phases observed, namely Ni_2Si,[132,133] NiSi,[131,134] and $NiSi_2$,[133] are similar to those observed in thin-film diffusion couples, except that the Ni-rich phase is θ–Ni_2Si, a metastable phase in the bulk.[133] The crystallographic orientation of the initial Si NW has also been shown to play a role in the identity of the first nucleated phase.[133] The appearance of this metastable phase and the dependence of the first nucleated phase identity on Si NW axial crystallographic direction have both been explained by favorable nucleation conditions associated with the silicide–Si epitaxial relationships.

It is now clear that the simple empirical Walser–Bene first phase rule is not suitable for predicting the phases formed in silicidation of Si NWs. In the careful *in situ* TEM studies, NiSi, CoSi, and $CoSi_2$ were all observed to form first, depending on the experimental conditions, although Ni_2Si and Co_2Si are the predicted first phases. In the case of deposited metal diffusing into Si NWs, FeSi is the observed and predicted first phase; Pt_2Si is the predicted phase, while PtSi is actually observed; and three different Ni–Si phases have been observed to form first, depending on experimental conditions, while Ni_2Si is predicted.

Although silicidation of Si NWs has thus far proven to be a very low throughput synthetic method for the production of transition metal silicide NWs, it has been shown to be an excellent way to directly observe the NW silicidation process and may be important in understanding the growth mechanisms employed in other silicide NW synthesis schemes. Further, direct silicidation of Si NWs is an excellent method to produce silicide/Si heterostructures, thereby enabling the straightforward integration of silicide NWs in a variety of NW devices.

3.3.2 Delivery of Silicon to Metal Films

One way to accomplish material delivery for binary systems without using NW templates is to deliver either the Si or the metal and use the other as a substrate. This approach has the benefit of a simple experimental setup, since few gas species are used. Synthetic strategies have emerged for the decomposition of SiH_4 on metal thin films, similar to the reports detailing the CVD synthesis of Si NWs,[144–146] or through using vapor-phase elemental Si through evaporation or sputtering to yield free-standing silicide NWs.[142,147–153] Interestingly, to date, this strategy has been reported primarily for NWs of nickel silicides: $NiSi_2$,[151] NiSi,[146,148,149,152] Ni_2Si,[142,151,152] Ni_3Si_2,[150] $Ni_{31}Si_{12}$,[153] and Ni_3Si[147] (i.e., the room temperature stable nickel silicide phases). Ti_5Si_4 NWs are the only example of successful application of this technique outside the Ni–Si system.[154] Little significant work has been done since the initial 2004 report[151] to uncover the mechanism behind this particular growth method. A possible reason that this strategy has been reported primarily for nickel silicides is the extremely fast diffusivity of Ni in Si.[142] Many transition metals can be used to catalyze Si NW growth using either the VLS mechanism[62] or vapor–solid–solid (VSS) mechanism.[145] Ni itself is actually one of the earliest reported VLS catalysts for Si NW growth.[155] If higher temperatures and a practically limitless metal source (i.e., a metal foil or thick film) are used, nickel silicide NWs are observed.

The growth of Ni_2Si NWs on a Ni thin film (Fig. 3.5) is one of the most thoroughly analyzed silicide NW syntheses.[142] Generally, SiH_4 decomposes onto the Ni surface leading to the growth and elongation of Ni_2Si nanocrystals. As these nanocrystals lengthen their diameters decrease, eventually resulting in the formation and propagation of thin, untapered Ni_2Si NWs. As the nanocrystals grow, the Ni available near the nanocrystal tip decreases, due to the decreasing available surface area for Ni diffusion, until the Ni incorporation equals the rate of SiH_4 decomposition (or Si addition) giving rise to NWs of a constant diameter. Kang and coworkers have also showed that Ni_xO layers can be used to help control the initial stages of 1-D growth by serving as a diffusion barrier for Ni.[152] These results highlight the importance of Ni diffusion from the surface for the growth of these NWs.

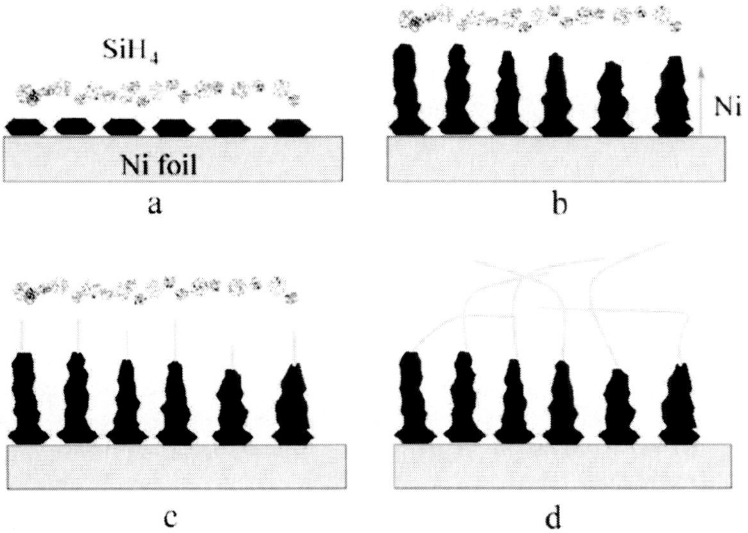

Figure 3.5 Preparation of Ni_2Si NWs using SiH_4 gas precursor and Ni foil. (a) Ni diffusion is faster than silane decomposition, creating nanoparticles on the surface of the metal film. (b) As more Si decomposes, Ni metal from the substrate has to diffuse further into the crystalline material, creating an elongated structure. (c) The rate of metal diffusion and silane decomposition are comparable and single crystalline NWs form. (d) NWs with constant diameter are propagated during growth.[142]

Perhaps the most interesting recent development involves the careful X-ray photoelectron spectroscopy analysis of Ni surfaces before and after various treatments to analyze the conditions required for NiSi NW nucleation and growth.[146] Heating of Ni thin films evaporated onto a Ni diffusion barrier (sputtered TaN to prevent reaction with the underlying Si substrate) in the presence of H_2 without SiH_4 reveals the formation of globular agglomerates on the surface of the substrate believed to be Ni_2O_3, which has a low melting temperature (~600 °C). Further analysis suggests that NiSi NW nucleation and growth involves quasi-liquid Ni–Si–O particles present on the substrate surface, perhaps akin to a VLS process, but more in depth structural studies of the surface are necessary to clarify the details of this proposed growth mechanism.

In spite of its simplicity, primarily Ni silicides have been produced using this technique, serving as a strong indicator of the role of diffusion in this process. The formation of Ti_5Si_4 NWs was observed using a variant of this method,[154] but high Ti foil substrate temperatures (900 °C) and a high-temperature Si source (Si powder) were required. The high temperature was likely required to activate Ti diffusion, and the resultant NWs were quite short, even after long reaction times.

3.3.3 Reactions of Transition Metal Sources with Silicon Substrates

The delivery of transition metal species to Si substrates has also been demonstrated as a viable synthetic pathway to silicide NWs. While a variety of simple organometallic complexes are commercially available as metal sources,[156] researchers have preferred pure metal vapors or anhydrous metal halides as the sources for free-standing metal silicide NWs.

3.3.3.1 Metal vapor

The reaction of metal vapor with a Si substrate to produce NWs has been shown for $TaSi_2$[157,158] and $TiSi_2$.[159] Chueh and coworkers initially reported the synthesis of short (0.2–2 μm) $TaSi_2$ NWs by annealing Si substrates with $FeSi_2$ thin films or nanodots on their surface in a Ta atmosphere.[157] The authors believe that Si atoms segregate from the $FeSi_2$ thin film or nanodots, forming a nanoparticulate Si base, as shown in TEM analysis. Ta atoms adsorb onto the surface of the $FeSi_2$ thin films or nanodots and react with Si diffusing through the substrate to form $TaSi_2$ nanocrystals. They argue that the use of high temperatures and low pressures favors certain crystal facets, leading to the observed 1-D growth. The use of different silicide films was shown to increase the aspect ratio of generated NWs, providing further evidence of the importance of Si diffusion through the silicide thin film.[158] In all cases, the metal impurity from the starting silicide thin film (or nanodots) is found in the final product, accounting for as much as 10% of the total elemental content. Other work reported by the same group provides insight: annealing similar $FeSi_2$ substrates in the absence of Ta vapor but under higher temperature conditions produces Si NWs.[160] This result is not surprising, considering the well-known use of Fe and its metal silicides as VLS or VSS catalysts for the growth of Si NWs; in fact, Fe was one of the first VLS catalysts reported.[144,155] Metal silicide particles likely served as both the 1-D catalyst and favored diffusion path for Si from the substrate. If these results are considered in conjunction with the Si NW silicidation studies described above, it seems probable that the $TaSi_2$ NWs observed were formed by the silicidation of Si NWs as they formed or shortly thereafter. This method of silicide NW growth requires high temperatures (950 °C), long reaction times (16 h), and yields impure products, but it is the only method known to produce $TaSi_2$ NWs.

3.3.3.2 Metal halides

The application of anhydrous metal halides as the transition metal source for reaction with Si substrates has proven to be a general and popular method to grow silicide NWs. Ouyang and coworkers reported this technique as a facile way to produce FeSi NWs using $FeCl_3$.[161] Varadwaj and Seo used a two-zone furnace to give better control of the metal halide vapor pressure.[162] Typically, a halide precursor is vaporized in the upstream zone of a furnace while the substrate is held at a higher temperature during the course of the reaction. Interestingly, by using substrates other than Si, metal-rich phases can be favored (Si powder or a Si substrate is still required for NW growth).[162] Using this approach, NWs of Fe_5Si_3,[162] FeSi,[161,163] Co_3Si,[164] Co_2Si,[164–166] CoSi,[163–166] Ti_5Si_3,[167] TiSi,[168] $TiSi_2$,[169] $CrSi_2$,[170,171] $Fe_{1-x}Co_xSi$,[172] MnSi,[57,173] and Mn_4Si_7[174] have been synthesized. Although no convincing mechanism has been reported, the so-called vapor-solid (VS) mechanism is usually invoked, based largely on the delicate dependence of halide supersaturation on the resultant silicide morphology and phase. However, this mechanism does not provide an explanation for the growth and nucleation of 1-D NWs. At lower pressures, halides become more volatile and metal-rich phases are preferred. Metal chlorides are well known to react with Si to produce $SiCl_4$, MSi_x, and/or Cl_2. When such metal halide sources are used, $SiCl_4$ vapor likely reacts to form elemental Si or a silicide, releasing more Cl_2 into the vapor phase. $SiCl_4$ is widely used to grow Si NWs at such high temperatures.

One possible mechanistic explanation, usually discounted by researchers, is the growth of Si NWs catalyzed by *in situ* grown metal silicide nanoparticles followed by the eventual, or even nearly simultaneous, silicidation of these NWs, carried out by reaction with metal deposited on NW surfaces or diffusing from the substrate. The lack of a catalyst particle tip does not rule out catalyst particles as being responsible for 1-D growth, since it is possible for such particles to be surface bound or eventually incorporated into the wire itself, as is seen, for example, in the case of Pb-catalyzed Pb chalcogenide NWs.[175] Indeed, the synthesis of TiSi NWs[168] reinforces that this idea ought to be pursued further. In this case, $TiCl_4$ is passed over Ti powder, producing reactive Ti subchlorides. This Ti precursor reacts with a Si substrate to produce a $TiSi_2$ thin film from which TiSi NWs grow. Careful TEM analysis of the NW-substrate surface has revealed the presence of a Si deficient amorphous $TiSi_{2-x}$ layer, similar to that observed in thin-film diffusion couples, on the surface of the growth substrate containing nanocrystalline TiSi particles. The authors believe these nanoparticles serve as the seeds for NW phase selection and growth, but have not yet shown a direct correlation between the presence of these seeds and the NWs themselves.

Thus far, the delivery of transition metals to a Si surface via a metal halide has proven to be a very versatile and general method to produce silicide NWs of both slow and intermediate diffusing metal species, but the clear importance of halide supersaturation on the resultant NW phases and morphologies reveals the importance of the decomposition of the halide species. Such chemistry points to reaction rate control for the growth of silicide NWs, but the lack of knowledge about the nature of these reaction processes (often guessed at in publications) currently limits the mechanistic understanding of this growth method.

3.3.4 Simultaneous Metal and Silicon Delivery

Simultaneous delivery of Si and transition metal source material to a substrate to grow metal silicide NWs has been achieved using two different techniques: (1) CVT and (2) CVD.

3.3.4.1 Chemical vapor transport

CVT, a classical crystal growth method, takes advantage of the reversible thermodynamic reaction between two chemical species, a source material and a transport agent, to produce gas-phase precursors *in situ*. The intermediates provide the elements needed to synthesize the desired phase, as shown in Fig. 3.6. This technique has been widely used to grow single crystals of many transition metal silicides.[176] However, NW growth may be obtained by altering the process for kinetically favored growth of 1-D morphologies.

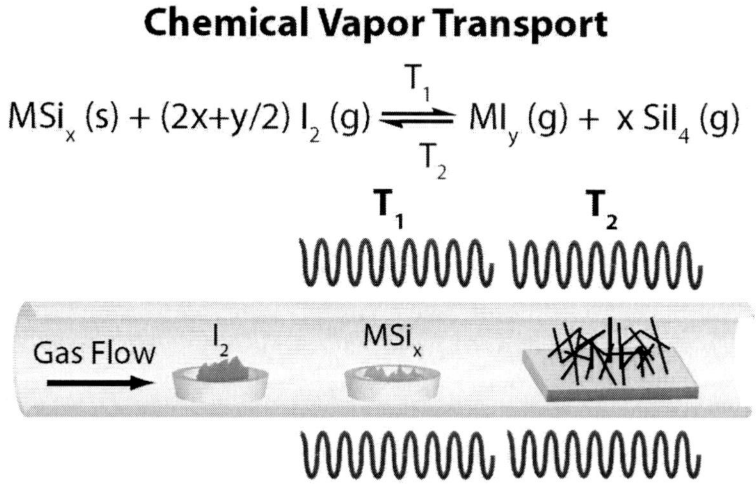

Figure 3.6 Synthesis of metal silicide NWs using CVT. Silicides undergo reversible reaction with I_2 at T_1 to form *in situ* gas-phase products. At T_2 gas phase components undergo the reverse reaction to form silicide NWs.[64]

We have demonstrated the synthesis of four distinct single-crystalline silicide NW phases via continuous-flow CVT using I_2 as the transport agent.[54,59–61] In each case, growth occurred on Si/SiO_2 substrates, sometimes requiring a dilute metal salt solution of $Ni(NO_3)_2$ on the surface. Hexagonally faceted $CrSi_2$ NWs with lengths up to a few hundred microns were synthesized using powdered $CrSi_2$ as the source material.[59] NWs of the δ-Ni_2Si[60] and β-Ni_3Si[61] phases were also synthesized using Ni_2Si and $TiSi_2$ as source materials, respectively. For both $CrSi_2$ and δ-Ni_2Si, the product stoichiometry was identical to the stoichiometry present in the source material. Interestingly, the β-Ni_3Si phase was obtained when $TiSi_2$ was used as a source material; it appears that the Ni in this case was supplied from the $Ni(NO_3)_2$ solution, with the Si originating from the generated SiI_4. Elemental analysis, using energy dispersive spectroscopy (EDS), of the β-Ni_3Si NWs revealed that no Ti was incorporated into the NWs. Finally, the synthesis of FeSi NWs and hyperbranched NWs was recently demonstrated using $FeSi_2$ powder as the starting material.[54] These examples clearly illustrate that the stoichiometry of the starting precursors is not the only factor determining the stoichiometry and phase of the NW product. Careful TEM of microtomed cross-sectional samples of FeSi NWs revealed merohedral twinning of these NWs perhaps providing a clue to their growth mechanism. Impurities from the transport gas, namely I, were not detected by EDS for any of the above phases.

Although the exact growth mechanism has not been determined for CVT-grown silicide NWs, the lack of a visible catalyst tip has been used to rule out VLS-type growth and typically vapor–solid or metal-assisted NW growth is cited. Although there are no confirmed reports of VLS catalyzed transition metal silicide NWs, the potential for *in situ* silicidation of VLS-catalyzed Si NWs remains. The chemical reactions these systems leverage to produce vapor-phase Si and transition metal sources are relatively complicated compared to the delivery of only one material source. An in-depth understanding of the kinetics of these reactions will be extremely beneficial to understanding how to fully control and utilize the CVT silicide NW synthetic approach.

3.3.4.2 Chemical vapor deposition

3.3.4.2.1 *CVD using single-source precursors*

The use of single-source precursors (SSPs), inorganic complexes containing the transition metal and Si atoms necessary for silicide formation, has been shown to be one of the most successful and versatile transition metal silicide NW synthesis techniques to date. SSPs have been used in metal organic chemical vapor deposition (MOCVD) to obtain transition metal silicide thin films.[156] Compared with conventional multisource MOCVD, the SSP approach allows simpler and safer experimental setups due to the elimination of highly hazardous liquid precursors (typically $Fe(CO)_5$ for Fe and $SiCl_4$ for Si), easier and more precise control over stoichiometry, and higher quality material growth;[156,177] for example, FeSi and CoSi thin films have been deposited by the pyrolysis of *cis*-$Fe(CO)_4(SiCl_3)_2$[178] and $Co(CO)_4SiCl_3$.[179] The selection of simple inorganic molecules containing both metal and Si that vaporize and decompose through well-known pathways allows precise control of the product stoichiometry and enables the reproducible growth of NWs (Fig. 3.7).

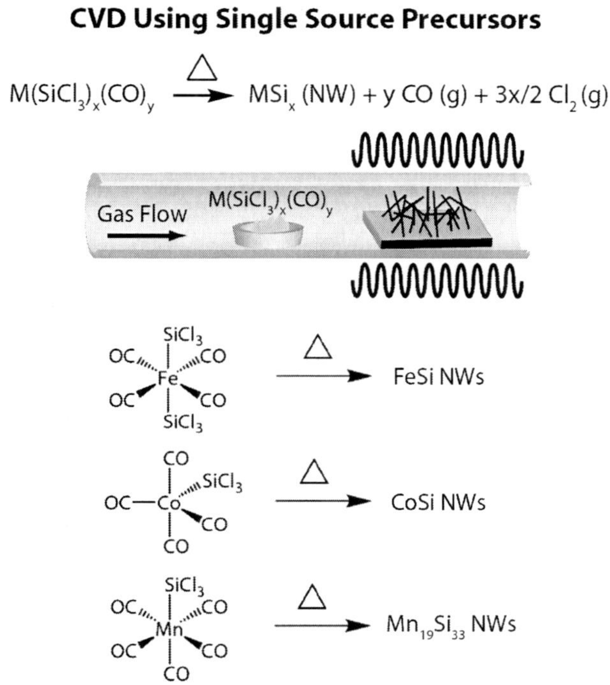

Figure 3.7 Synthesis of metal silicide NWs using CVD of SSPs. Each SSP pyrolyzes to deliver metal silicide material to the growth substrate.[64]

We first demonstrated this approach by using *trans*-$Fe(CO)_4(SiCl_3)_2$ (the *cis* isomer is converted to the *trans* isomer at higher temperatures) to synthesize FeSi NWs.[53] This compound can be readily synthesized by reacting $Fe_3(CO)_{12}$ with excess $SiHCl_3$ in an inert atmosphere at 120 °C for 2 days, and it readily decomposes into FeSi in the gas phase at elevated temperatures (Fig. 3.7). High-quality FeSi NWs can be synthesized using the CVD deposition of this precursor at 200 Torr and 750 °C when Si substrates are prepared with a thin (1–2 nm) Si oxide layer.[53] The dependence of this oxide layer was shown through patterning experiments,[180] which also serves as a simple method to define the locations of silicide NW growth without the use of catalysts, but the exact role this oxide layer plays in the NW nucleation is still under study. To date, this general approach (Fig. 3.7) has been extended to CoSi NWs[55] using

$Co(CO)_4SiCl_3$ and $MnSi_{2-x}$ ($MnSi_{1.8}$) NWs[58] using $Mn(CO)_5SiCl_3$. *Trans*-$Fe(CO)_4(SiCl_3)_2$ and $Co(CO)_4SiCl_3$ SSP mixtures can themselves form a liquid two-component SSP via melting point depression. This solution can be conveniently evaporated for vapor-phase delivery resulting in the synthesis of $Fe_{1-x}Co_xSi$ alloy NWs.[56] Such NWs are particularly interesting for spintronic applications.

Interestingly, the Fe silicide and Mn silicide phases formed do not share their stoichiometry with that of the starting precursor. In the case of FeSi NW formation, the precursor is known to undergo a reductive elimination to $FeSiCl_2$ upon pyrolysis[178] which may give rise to the resultant NW stoichiometry, but the stoichiometry of the $MnSi_{2-x}$ NWs produced from $Mn(CO)_5SiCl_3$ cannot be similarly rationalized as the pyrolysis behavior of the precursor is not fully understood. The higher manganese silicides produced using this precursor are an incredibly complex set of incommensurate structures based on elemental sublattices of Mn atoms which form "chimneys" within which a spiraling sublattice of Si atoms lies.[45] These so-called Nowotny chimney ladder phases can have unit cells (under commensurate approximations) of up to tens of nanometers.[181] The NWs were determined to be $Mn_{19}Si_{33}$ by elaborate electron diffraction experiments. Such structures, having both complex crystal structures and enhanced nanomorphology, are likely efficient phonon scatters potentially interesting for their thermoelectric performance.[182]

Importantly, the decomposition pathways of the SSPs discussed here are not all well characterized. Depending on whether pyrolysis of the SSP occurs in the gas phase or at the growth substrate surface, the transition metal and Si may be delivered to the growth substrate either in a stoichiometry defined by the precursor or a stoichiometry which depends more importantly on the gas phase pyrolytic pathway. The observed difference between SSP and NW stoichiometry in the FeSi and $Mn_{19}Si_{33}$ cases seems to point out that the delivery of source materials by these SSPs are dominated by the decomposition of the SSP while still in the gas phase. Since recent studies[146,168] have highlighted the potential role of surface layers in NW nucleation, similar to the surface oxide layer required in some SSP syntheses, and since it appears as though vapor-phase Si is likely present in the reactor, produced via direct decomposition of the SSP or reaction of Cl_2 with the Si substrate, we should not rule out the possibility that some sort of catalyst produced *in situ* gives rise to Si NWs which are then converted into silicide NWs by silicidation reactions.

Preliminary experiments into the mechanism for this process reveal that the Si incorporated into the NW products may originate from either the substrate or vapor-phase sources, but the transition metal species are incorporated only when introduced from the vapor phase. Controlled production of alloys and doped materials by substitution of either the metal or Si will make available nanomaterials with improved properties or new functionality, for example, $Fe_{1-x}Co_xSi$ alloys, for spintronics and doped $MnSi_{2-x}$ for thermoelectrics. Such controlled doping studies may also allow access to silicide heterostructures, enabling new device functionality and perhaps highlighting often ignored aspects of the growth mechanism, for example, do silicide NWs grow from the tip or the base. The information available so far reveals that a better understanding of the gas phase and surface chemistries taking place in these systems will be required to enable full control over SSP-based silicide NW synthesis.

Although there are many more transition metals (e.g., Ru, Re, and Rh) capable of making analogous molecules which are both volatile and contain metal–Si bonds, many of the early transition metals do not. Such transition metal silicide NW phases of interest are likely accessible only through alternative source material delivery techniques. As new developments occur in metal–silyl complex chemistry, new SSPs could be developed that allow other stoichiometries and new combinations of transition metals and Si to be delivered via gas phase pyrolysis for NW growth.

3.3.4.2.2 *Two-source CVD*

CVD of Si and metal gas phase sources is the classical way in which to prepare silicide thin films for electronic applications. Typically, silane and metal carbonyls are used at low temperatures (250–450 °C) to produce thin films with controlled stoichiometries. Zhou *et al.*, were able to adapt a similar approach to form $TiSi_2$ nanostructures.[183] In these experiments $TiCl_4$ and SiH_4 were coflowed with H_2 gas to a Si substrate at 675 °C to form orthorhombic $TiSi_2$ NWs (C54) and 2-D nanonets (C49). The nanonets are composed of hyperbranched $TiSi_2$ NWs restricted along the [100] and [001] growth directions of the C49 orthorhombic structure. The same authors also showed that by varying the Si:Ti ratio (intermediate between the conditions optimized for the C54 and C49 structures), nanonets can intersect at 120° angles to form 3-D nanonets.[184] This likely occurs through a hexagonal C40 $TiSi_2$ coating along the axis of intersection.

Another likely example of two-source CVD has recently been shown in the reaction of VCl_3 with Si powder, giving rise to the growth of V_5Si_3 NWs and nanotubes on V foils. In this case, VCl_3 disproportionates into VCl_2, a solid, and VCl_4, a gas, when heated. The VCl_4 is then transported to the Si powder source likely giving rise to vaporous $SiCl_4$, or more reactive sub-chlorides, to be transported to the V foil growth substrate. The two different reactants, one purposefully produced *in situ*, then react with each other (and likely the V foil substrate), yielding the observed nanostructures. Interestingly, the authors observed the growth of partially filled and empty nanotubes, a hallmark of dislocation driven NW growth.[185] It is known that metal foils often contain many defects, such as dislocations, which are not as common on the single-crystal Si substrates most commonly used as silicide NW growth substrates.

So far, the two-source CVD approach has only been successfully applied to the synthesis of Ti and V silicide NWs, two refractory metals known to favor Si diffusion. The inherent strength of the dual source approach is the potential for independent control over Si and metal vapor pressures, but this strength can only be fully utilized when the nature of the reactions of the precursors with the substrate and with each other are fully understood.

3.3.5 Vapor-Phase Technique Comparison

Many observations and conclusions can be drawn when summarizing the growth techniques used to produce silicide NWs (Fig. 3.8). Diffusion is clearly an important parameter for silicide NW growth, considering that there have been more reports of silicide NW growth from the Ni–Si binary phase system than any other. This is likely due to the exceptionally fast diffusion behavior of Ni in Si. By simply decomposing SiH_4 on Ni metal films using a wide variety of conditions (Table 3.2), different parts of the phase diagram can be accessed, resulting in the successful NW synthesis of every known Ni silicide phase. Additionally, CVT has been used to give ultra-long Ni silicides wires (up to 3 mm), with long reaction times (2 h) and higher temperatures (1000 °C).[60]

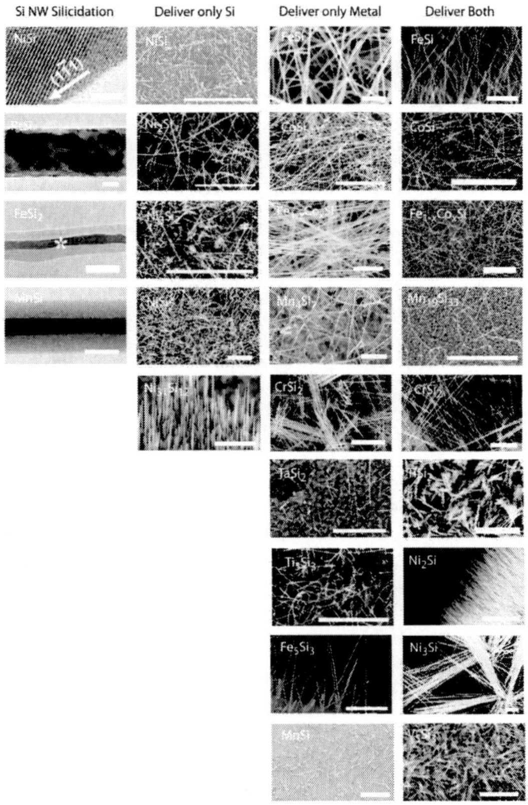

Figure 3.8 Morphologies of transition metal silicide NWs synthesized using four different growth paradigms.

Si NW silicidation: Si NW silicidation yields NiSi,[139] PtSi,[137] $FeSi_2$,[135] and MnSi[138] NWs; 1st column, TEM images (scale bars 5, 20, 100 and 100 nm, respectively).

Deliver only Si: Delivery of SiH_4 to metal substrates results in NiSi,[149] Ni_2Si,[142] Ni_3Si_2,[150] $NiSi_2$,[188] and $Ni_{31}Si_{12}$[153] NWs; 2nd column, SEM images (scale bars are all 5 μm).

Deliver only metal: Delivery of metal vapors and metal halides to Si substrates yield FeSi,[161] CoSi,[165] $Fe_{1-x}Co_xSi$,[172] Mn_4Si_7,[174] $CrSi_2$,[170] $TaSi_2$,[158] Ti_5Si_3,[167] Fe_5Si_3,[162] and MnSi[57] NWs; 3rd column, SEM images (FeSi and Mn_4Si_7 scale bars are 1 μm, MnSi scale bar is 25 μm, and the others are 5 μm).

Deliver both: Both metal and Si delivered to growth substrates using either CVT or CVD results in FeSi,[53] CoSi,[55] $Fe_{1-x}Co_xSi$,[56] $Mn_{19}Si_{33}$,[58] $CrSi_2$,[59] $TiSi_2$,[183] Ni_2Si,[60] Ni_3Si,[61] and V_5Si_3[187] NWs, fourth column SEM images (all scale bars are 5 μm except Ni_2Si, which is 200 μm).

Table 3.2 Comparison of silicide nanowire growth techniques

Scheme	TM source	Si source	Temp. range	Time	Pressure	Lengths	Phases
Si NW Silicidation	TM evaporation	Si NWs	550–700 °C	3–4 h	760 Torr	10–20 μm	$MnSi$,[138] $FeSi$,[135] $FeSi_2$,[135] $CoSi$,[140] $CoSi_2$,[140] Ni_2Si,[132,133] $NiSi$,[131,134,139,141] $NiSi_2$,[133] $PtSi$[136,137]
Si Delivery to Metal	TM films or foils	SiH_4	400–650 °C	10–60 min	10–160 Torr	2–20 μm	$NiSi_2$,[151] $NiSi$,[146,148,149,152] Ni_2Si,[142,151,152] Ni_3Si_2,[150] $Ni_{31}Si_{12}$,[153] and Ni_3Si,[147] Ti_5Si_4[154]
TM Vapor Delivery	TM vapor	Si substrate	850–950 °C	4–16 h	1×10^{-6} Torr	1–10 μm	$TaSi_2$,[157,158] $TiSi_2$[159]
TM Halide Delivery	TM halides	Si					

Scheme	TM source	Si source	Temp. range	Time	Pressure	Lengths	Phases
		substrate	900–1100 °C	0.5–2 h	500–760 Torr	1–20 μm	Fe_5Si_3,[162] FeSi,[161,163] Co_3Si,[164] Co_2Si,[164–166] CoSi,[163–166] Ti_5Si_3,[167] TiSi,[168] $TiSi_2$,[169] $CrSi_2$,[170,171] $Fe_{1-x}Co_xSi$,[172] MnSi,[57,173] Mn_4Si_7[174]
CVT	TM silicides		900–1000 °C	2–3 h	760 Torr	10 μm to 5 mm	$CrSi_2$,[59] FeSi,[54] Ni_2Si,[60] Ni_3Si[61]
2-Source							
CVD	TM halides	SiH_4					
		Si Powder	650–950 °C	15 min	5–760 Torr	2–5 μm	TiSi,[186] $TiSi_2$[183,184], $V5Si_3$[187]
SSP							
CVD	$M(CO)_x(SiCl_3)_y$		650–750 °C	30 min	100–200 Torr	10–40 μm	FeSi,[53,180] CoSi,[55] $Fe_{1-x}Co_xSi$,[56] $Mn_{19}Si_{33}$[58]

Early transition metal silicide NWs, sometimes called refractory metal silicides, are probably the most difficult to make from the perspective of metal diffusion. Refractory metals, such as Ti, V, Nb, Ta, and others, are slow-diffusing species relative to Si and form silicides with extremely high melting points. Ti_5Si_3,[167] Ti_5Si_4,[154] TiSi,[168,186] $TiSi_2$,[159,169,183,184] V_5Si_3,[187] and $TaSi_2$[157,158] have been reported. $TaSi_2$ and $TiSi_2$ were synthesized using the reactions of metal vapor with silicide thin films or bare Si substrates, respectively. $TiSi_2$, TiSi, and Ti_5Si_3 NWs were synthesized using the reaction of $TiCl_4$, or subchlorides, with a Si substrate, but interestingly, Ti_5Si_4 NWs were synthesized by reacting SiH_4 with Ti foils. TiSi, $TiSi_2$, and V_5Si_3 NWs were recently synthesized using a dual-source CVD approach. These synthetic procedures all required high temperatures, and in the case of $TaSi_2$ extremely long reaction times (up to 16 h) yield short (1–2 μm) NWs. The lack of available SSPs and slow diffusion behavior of the early transition metal species implies that successful synthesis of silicide NWs of early transition metals require vapor-phase transition metal sources and/or extremely high temperatures in order to activate refractory metal diffusion.

Silicides of the refractory Group VIB metals, namely Cr, Mo, and W, are an interesting intermediate case. So far, the only silicide phase synthesized among this group has been $CrSi_2$; NWs of this phase have been produced independently by both CVT and metal halide transport.[59,170,171] The mechanism for growth in these two processes is likely similar. Metal halide transport involves evaporation of $CrCl_2$ and production of $SiCl_4$, from reaction with the Si substrate. By delivering $CrSi_2$ via I_2 transport agent, CVT produces CrI_2 and SiI_4. These are borne out in the nearly identical morphologies observed in the two cases (Fig. 3.8). It remains to be seen whether either of these techniques (or others,) can be used to make W or Mo silicide NWs, but because of the generally slow diffusion behavior of these metals, synthesizing such NWs will be difficult through delivery of a Si source to a metal substrate, except at relatively high temperatures. For these metals, there are no $M(CO)_x(SiCl_3)_y$ analogues because the M-Si bond in such structures is quite weak, but precursors stabilized by cyclopentadienyl or π-arene ligands are known, and delivery of such SSPs or independent Si and TM sources via CVD may lead to successful NW synthesis.

The late transition metal (except Groups IB and IIB) silicides have been explored using metal halide transport,[57,161–174] CVT,[54] and CVD of SSPs.[53,55,56,58,180] In the case of SSPs, the metal-to-Si ratio delivered to the substrate may be built into the molecular structure as a means of gaining more control over the reaction. In metal halide transport similar results can be obtained, albeit at higher temperatures of approximately 900–1100 °C (Table 3.2). Varadwaj and Seo were able to show that by eliminating substrate diffusion as a Si source and instead relying on gaseous $SiCl_4$ products, metal-rich phases can be favored.[162] This makes it possible to choose between two phases by changing the substrate.

One could argue that the direct growth of free-standing NWs is either inherently diffusion-limited or reaction-limited, rather than nucleation limited, since anisotropic NW growth is the strict requirement for the existence of a NW morphology. However, the NW nucleation conditions may be the important factors giving rise to this 1-D growth. The current lack of control over silicide NW nucleation and the limited information gathered regarding the initial stages of this process make silicide NW nucleation one of the most interesting aspects of silicide NW growth studies. The VS mechanism, which implies a reaction-limited growth, is often posited for silicides NWs, but it provides an unsatisfactory explanation because the VS mechanism provides little insight into the conditions of nucleation giving rise to anisotropic growth. The nanoparticle- and dislocation-driven mechanisms certainly provide explanations, but thus far there have been no clear examples of dislocations or nanoparticles driving transition metal silicide NW nucleation and growth. Tantalizing hints of quasi-liquid[146] and/or amorphous seed layers[168] have recently arisen, and the observation of twin planes[54] and hollow NWs[187] provide new directions for research. Based on the reaction parameters discussed for each process, sequential or simultaneous growth and silicidation of Si NWs is still plausible. Until recently little attention has been paid to the interface between the NW and the growth substrate, leaving nanoparticle-mediated 1-D nucleation from the base an open possibility.

Today we are capable of synthesizing an array of different transition metal silicide NWs, but the mechanism(s) for the direct anisotropic 1-D growth of silicide NWs remains unclear. The clear connection with diffusion behaviors in Si, the dependence on various oxides and surface structures, and other observations imply that metal and Si diffusion and precursor reactions are extremely important factors in controlling the growth of silicide NWs, in excellent agreement with the lessons gleaned from silicide diffusion couples. However, in-depth studies of NW growth kinetics are made difficult by the current lack of nucleation control, that is, if new NWs nucleate throughout the entire reaction, it is difficult to gauge the relationship of the average NW length with time and thus gain insight about the nature of growth from kinetics. While it would certainly prove useful to identify the gas species and the limiting reactions in the complicated reaction sequences often guessed at in publications, such experiments are time-consuming and involve complicated experimental setups. Regardless it is clear that some simple aspects of silicide NW growth may be elucidated by asking simple questions and careful experimentation: for example, (1) do silicide NWs grow from the tip or the base? (2) what does the interface between a silicide NW and its growth substrate look like? and (3) what is the role of surface layers in silicide NW growth? Answering these questions and others like them will greatly enhance our understanding of the nucleation and growth of silicide NWs.

3.4 Applications of Silicide Nanowires

Transition metal silicides, with their diverse physical properties, have been used in a large number of applications. High melting points, excellent chemical stability, and compatibility with Si processing technology have made them good candidate materials for use in electronics,[4,5] thermoelectrics,[2] and PVs.[9] Metal silicide NWs also possess these favorable properties, and their size may allow their use in potentially improved nanoscale versions of these devices and in new roles, such as field emission sources and spintronics. In the following section, we briefly review the applications and current device concepts utilizing silicide NWs.

3.4.1 Nanoelectronics

Because of the ubiquity of metal silicides as interconnects and contact electrodes in modern microelectronics,[7] silicide NWs are expected to serve a similar role in nanoelectronic devices. Their low resistivities and compatibility with Si make them excellent candidates to continue in this role. Notably, ultralong Ni_2Si NWs with an extremely low resistivity of 21 μΩ-cm are capable of supporting remarkably high failure current densities of greater than 108 A/cm^2 (Fig. 3.9a).[60] The difficulty of selective nanostructure placement for devices serves as a strong barrier against the use of randomly oriented free-standing NWs. A number of groups have found a solution through the direct silicidation of Si NWs, creating intruded Si contacts.[131,134,189] In this process metal contacts lithographically applied to Si NWs are used as the metal source to controllably create MSi/Si/MSi heterostructures (Fig. 3.9b). Careful control of this phenomenon has been applied in the successful fabrication of NiSi NW contacted Silicon Field Effect Transistors (Si FETs) with arbitrary channel length.[134] Indeed, such a strategy may be used as an exploration of new types of devices, since quantum confined[189] and highly strained[139] Si segments have also been demonstrated.

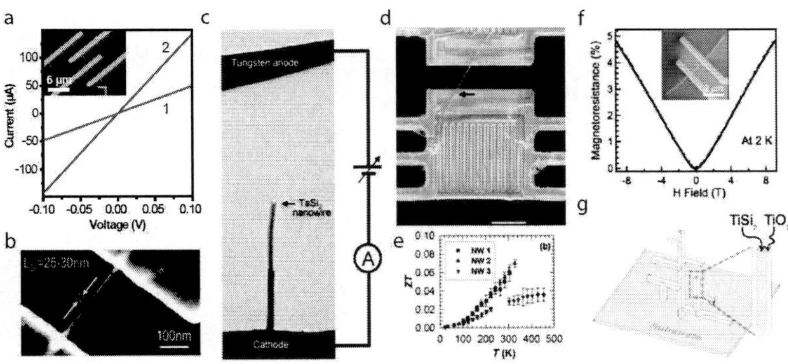

Figure 3.9 Applications of transition metal silicide NWs. (a) Two-probe (1) and four-probe (2) electrical transport for single Ni_2Si NW.[60] (b) Intruded contact NiSi–Si–NiSi quantum dot device.[189] (c) Single $TaSi_2$ NW field emission device.[190] (d) $CrSi_2$ NW on suspended measurement device to investigate thermoelectric properties, notice open central region available for TEM characterization of NW.[191] (e) ZT as a function of temperature for three different $CrSi_2$ NWs.[191] (f) Large positive magnetoresistance of $Fe_{1-x}Co_xSi$ NW; inset is an example measurement device.[56] (g) $TiSi_2$ nanonet—TiO_2 heterostructure for photoelectrocatalytic (PEC) devices.[192]

3.4.2 Nanoscale Field Emitters

Another commonly discussed application for metal silicide NWs is their use as nanoscale field emitters (Fig. 3.9c). Field emission occurs when an electric field between a metallic surface and a ground electrode allows electrons to overcome the work function potential that binds them on the surface of the emitter. The field emission current follows the Fowler–Nordheim relation,

$$J = \frac{A\beta^2 E^2}{\Phi} \times e^{\left(-B\Phi^{3/2}/\beta E\right)}$$

where J is the current density, E the applied electrical field, Φ the work function, β the field enhancement factor, and A and B are constants 1.56×10^{-10} (AV^{-2} eV) and 6.83×10^{3} ($eV^{-3/2}\mu m^{-1}$), respectively.[193] Field emission devices are useful in wide-ranging applications such as flat-panel displays, microwave-generation devices, and high-powered vacuum microelectronic devices. Since the late 1990s there has been much interest in developing 1-D nanomaterials for field emission due to the enhancement of electric field at high curvature tips. Thus, as the geometric enhancement factor β increases, the voltage required for emission decreases. Because transition metal silicide NWs have low resistivities and work functions, high melting points, high aspect ratios, and well-known compatibility with Si-based microelectronics, they have been considered good candidates for future field emission devices. The field emission of as-grown, free-standing transition metal silicide NWs of a variety of phases have been reported including Ni_2Si,[142] NiSi,[147] $TaSi_2$,[157,158] $TiSi_2$,[159] Ti_5Si_3.[167] Authors regularly report turn-on fields (for a current density of 10 $\mu A/cm^2$) in the range of 3–4 V/μm and field enhancement factors as high as 4280.[142] Studies on single NW field emitters have also begun in hopes of advancing current theoretical understanding of the field emission in these nanoscale structures.[190]

3.4.3 Spintronics

As the size of transistors approaches the quantum regime, many researchers have begun to consider alternative computing paradigms. Spintronics is an approach which seeks to use the electron's spin degree of freedom in lieu of or in addition to the electron's charge for computing applications.[27] Long-spin relaxation times and well-developed processing technology make Si an ideal candidate material for devices such as spin field effect transistors.[194] Such technologies are still in their infancy as the difficulties associated with spin-polarized and pure-spin current injection and detection are still being investigated. The conductance mismatch problem[195] prohibits the use of conventional ferromagnetic metals as the contacts in a spin-FET due to greatly enhanced interfacial spin scattering, but novel contact materials, such as half-metals and ferromagnetic semiconductors, would overcome this barrier.[27]

NWs of the $Fe_{1-x}Co_xSi$ alloys, which contain both half-metallic and ferromagnetic semiconducting members in the bulk phase,[24,196] have been synthesized independently by two different groups, one using SSPs with $0 < x < 0.44$,[56] and one using metal halides with $0 < x < 0.1$.[172] In both preparations, the NWs have been shown to retain bulk-like magnetic and electrical characteristics (Fig. 3.9f). Although large magneto-resistances of ~10% have been reported, there have not been measurements of the spin polarization (P_s) in this material. We are currently working on using point contact Andreev reflection to precisely determine the value of P_s in this system. More recently, MnSi NWs, a highly polarized helimagnetic material,[197] have been incorporated into a MnSi/Si/MnSi intruded silicide NW device.[138] Using such a device, the injection of a spin-polarized current via a tunneling mechanism and detection of this spin-polarized current were demonstrated in Si NWs.

3.4.4 Thermoelectrics

The aim of thermoelectric research is the efficient interconversion of thermal and electrical energy using solid state devices. Metal silicides have long been considered a promising group of materials for thermoelectric applications, especially power generation at elevated temperature. The efficiency of a thermoelectric material is judged by the unitless figure of merit, ZT,

$$ZT = \frac{S^2\sigma}{\kappa_L \kappa_e} T$$

with S the Seebeck coefficient, σ the electrical conductivity, κ_L the thermal conductivity due to the lattice, κ_e the thermal conductivity due to electrons, and T the average measurement temperature. Recent research materials development for such applications follow the "electron crystal phonon glass" paradigm,[182] producing materials which strongly scatter phonons but allow efficient electrical transport. The two major strategies for accomplishing this involve using materials with complex crystal structures or nanoscale morphology, both of which lead to relatively small lattice thermal conductivity. NWs of $CrSi_2$[59,170] and the Nowotny chimney ladder phase $MnSi_{2-x}$,[58] which are excellent bulk thermoelectric materials, have been successfully synthesized. Further, specific thermoelectric properties of $CrSi_2$ NWs have been investigated using a specialized technique allowing the measurement of all thermoelectric properties on a single wire and structural characterization via TEM, which revealed increased phonon-surface scattering with decreasing NW diameter at low temperatures (Fig. 3.9d,e).[191] Additionally, phonon confinement will occur more readily in nanostructured materials which, with enhanced phonon-surface scattering and complex crystal structures, will result in dramatic reduction of thermal conductivity. If other properties (i.e., σ and S) are maintained for single crystal NCL NWs, their ZT may be enhanced beyond the already high bulk values of 0.7–0.8.[9,12,41–43] Future work in this area will likely focus on synthesizing new thermoelectric NWs such as Re_4Si_7, β-$FeSi_2$, and Ru_2Si_3, incorporating dopants into transition metal silicide NWs in order to optimize their thermoelectric properties, and thermal property investigation of silicides with complex crystal structures.[48,198]

3.4.5 Solar Energy Conversion

In addition to thermoelectrics, silicide NWs are promising for other renewable energy technologies, such as PV and PEC devices. Transition metal silicides made of inexpensive and abundant elements, like β-$FeSi_2$ (Si and Fe are the 2nd and 4th most abundant elements in the earth's crust, respectively), have appropriate bandgaps (0.8 eV for β-$FeSi_2$) and high light absorption coefficients[13] potentially useful in PV devices. To date there have been no reports of TMSi-based NW PV devices. This is an interesting area for future exploration. Lin and coworkers have made use of $TiSi_2$ nanonets coated with a TiO_2 layer in a PEC application.[192] In this device, $TiSi_2$ nanonets, a planar construct of orthogonal NW branches, act as a highly conductive scaffold and charge collection structure while a semiconducting TiO_2 layer acts as the active photoabsorption layer for water splitting (Fig. 3.9g). Such device configuration takes advantage of the high surface area of nanostructures to produce devices with much enhanced photocurrent compared to planar device geometry. The authors observed a peak incident photon-to-electron conversion efficiency of 16.7% using monochromatic 330 nm light. Our group has also synthesized highly branched and hierarchical nanostructures of other silicides, such as FeSi[54] and $MnSi_{2-x}$. Such complex nanostructures could provide advantages to solar energy harvesting and conversion due to enhanced absorption and short electron-hole diffusion paths.[175,199]

3.5 Conclusion

We have reviewed the synthesis of bulk and thin-film transition metal silicide diffusion couples and the current strategies for synthesizing free-standing transition metal silicide NWs. Thus far, 22 silicide NW phases across nine binary metal–Si systems have been synthesized. Even with increasing interest in silicide NW applications, many synthetic challenges remain. The mechanism for silicide NW formation is not completely understood, but the application of lessons learned from classical diffusion couples and the intriguing observations made on successfully synthesized NW systems provide many lines of inquiry. A number of interesting possibilities have arisen including the silicidation of initially synthesized Si NWs, VLS-like synthesis beginning from quasi-liquid oxide or amorphous silicide materials formed on the growth substrate, and dislocation driven NW synthesis. These hypotheses should aid in designing experiments to determine the exact mechanism for previously reported syntheses and should lead to the synthesis of new phases. Further, it is clear that our understanding of the nucleation and growth of silicide NWs is in its infancy; such studies are quite important since they may provide mechanistic insights due to the differences in nucleation, diffusion, and reaction rate–limited processes. NWs of many more interesting silicides for use in nanoelectronics, field emission, spintronics, thermoelectrics, and solar energy conversion remain to be synthesized, and new yet unexplored applications remain to be discovered.

References

1. Felser, C.; Fecher, G. H.; Balke, B. Spintronics: A challenge for materials science and solid-state chemistry. *Angew. Chem., Int. Ed.,* 2007, ***46***, 668–699.
2. Vining, C. In *CRC Handbook of Thermoelectrics*; Rowe, D. M., Ed.; CRC Press: Boca Raton, FL, 1995, p. 277–285.
3. Williams, J. C.; Starke, E. A. Progress in structural materials for aerospace systems. *Acta Mater.*, 2003, ***51***, 5775–5799.
4. Zhang, S. L.; Ostling, M. Metal silicides in CMOS technology: Past, present, and future trends. *Crit. Rev. Solid State Mat. Sci.,* 2003, ***28***, 1–129.
5. Chen, L. J. *Silicide Technology for Integrated Circuits*; The Institution of Electrical Engineers: London, UK, 2004.
6. Zhang, S. L.; Smith, U. Self-aligned silicides for ohmic contacts in complementary metal-oxide-semiconductor technology: $TiSi_2$, $CoSi_2$ and NiSi. *J. Vac. Sci. Technol., A*, 2004, ***22***, 1361–1370.
7. Maszara, W. P. Fully silicided metal gates for high-performance CMOS technology: A review. *J. Electrochem. Soc.*, 2005, ***152***, G550–G555.
8. Derrien, J.; Chevrier, J.; Lethanh, V.; Mahan, J. E. Semiconducting silicide silicon heterostructures—Growth, properties and applications. *Appl. Surf. Sci.*, 1992, ***56***(8), 382–393.
9. Borisenko, V. E. *Semiconducting Silicides*; Springer: Berlin, 2000; Vol. 39.
10. Leong, D.; Harry, M.; Reeson, K. J.; Homewood, K. P. A silicon/iron-disilicide light-emitting diode operating at a wavelength of 1.5 μm. *Nature*, 1997, ***387***, 686–688.
11. Bost, M. C.; Mahan, J. E. An investigation of the optical-constants and band-gap of chromium disilicide. *J. Appl. Phys.*, 1988, ***63***, 839–844.
12. Rowe, D. M. *CRC Handbook of Thermoelectrics*; CRC Press: Boca Raton, 1994.
13. Senthilarasu, S.; Sathyamoorthy, R.; Lalitha, S. Synthesis and characterization of β-$FeSi_2$ grown by thermal annealing of Fe/Si bilayers for photovoltaic applications. *Sol. Energy Mater. Sol. Cells*, 2004, ***82***, 299–305.
14. Fukuzawa, Y.; Ootsuka, T.; Otogawa, N.; Abe, H.; Nakayama, Y.; Makita, Y. Characterization of β-$FeSi_2$ films as a novel solar cell semiconductor. *P. Soc. Photo-Opt. Inst.*, 2006, ***6197***, 61970N/61971-61970N/61910.

15. Aeppli, G.; DiTusa, J. F. Undoped and doped FeSi or how to make a heavy fermion metal with three of the most common elements. *Mater. Sci. Eng., B*, 1999, ***63***, 119–124.

16. Riseborough, P. S. Heavy fermion semiconductors. *Adv. Phys.*, 2000, ***49***, 257–320.

17. Pfleiderer, C.; Julian, S. R.; Lonzarich, G. G. Non-fermi-liquid nature of the normal state of itinerant-electron ferromagnets. *Nature*, 2001, ***414***, 427–430.

18. Pfleiderer, C.; Reznik, D.; Pintschovius, L.; von Lohneysen, H.; Garst, M.; Rosch, A. Partial order in the non-fermi-liquid phase of MnSi. *Nature*, 2004, ***427***, 227–231.

19. Ishikawa, Y.; Tajima, K.; Bloch, D.; Roth, M. Helical spin structure in manganese silicide MnSi. *Solid State Commun.*, 1976, ***19***, 525–528.

20. Uchida, M.; Onose, Y.; Matsui, Y.; Tokura, Y. Real-space observation of helical spin order. *Science*, 2006, ***311***, 359–361.

21. Manyala, N.; DiTusa, J. F.; Aeppli, G.; Ramirez, A. P. Doping a semiconductor to create an unconventional metal. *Nature,* 2008, ***454***, 976–980.

22. Neubauer, A.; Pfleiderer, C.; Binz, B.; Rosch, A.; Ritz, R.; Niklowitz, P. G.; Boni, P. Topological hall effect in the a phase of MnSi. *Phys. Rev. Lett.,* 2009, ***102***, 186602/186601–186602/186604.

23. Muehlbauer, S.; Binz, B.; Jonietz, F.; Pfleiderer, C.; Rosch, A.; Neubauer, A.; Georgii, R.; Boeni, P. Skyrmion lattice in a chiral magnet. *Science*, 2009, ***323***, 915–919.

24. Manyala, N.; Sidis, Y.; Ditusa, J. F.; Aeppli, G.; Young, D. P.; Fisk, Z. Large anomalous hall effect in a silicon-based magnetic semiconductor. *Nat. Mater.*, 2004, ***3***, 255–262.

25. Manyala, N.; Sidis, Y.; DiTusa, J. F.; Aeppli, G.; Young, D. P.; Fisk, Z. Magnetoresistance from quantum interference effects in ferromagnets. *Nature*, 2000, ***404***, 581–584.

26. Pearton, S. Magnetic semiconductors—Silicon-based spintronics. *Nat. Mater.*, 2004, ***3***, 203–204.

27. Zutic, I.; Fabian, J.; Das Sarma, S. Spintronics: Fundamentals and applications. *Rev. Mod. Phys.*, 2004, ***76***, 323–410.

28. Alivisatos, A. P. Semiconductor clusters, nanocrystals, and quantum dots. *Science*, 1996, ***271***, 933–937.

29. Xia, Y.; Yang, P.; Sun, Y.; Wu, Y.; Mayers, B.; Gates, B.; Yin, Y.; Kim, F.; Yan, H. One-dimensional nanostructures: Synthesis, characterization, and applications. *Adv. Mater.* , 2003, ***15***, 353–389.

30. Yang, P. The chemistry and physics of semiconductor nanowires. *MRS Bull.*, 2005, ***30***, 85–91.

31. Samuelson, L.; Thelander, C.; Bjoerk, M. T.; Borgstroem, M.; Deppert, K.; Dick, K. A.; Hansen, A. E.; Martensson, T.; Panev, N.; Persson, A. I.; Seifert, W.; Skoeld, N.; Larsson, M. W.; Wallenberg, L. R. Semiconductor nanowires for 0d and 1d physics and applications. *Physica E*, 2004, ***25***, 313–318.

32. Wang, Z. L. Functional oxide nanobelts: Materials, properties and potential applications in nanosystems and biotechnology. *Annu. Rev. Phys. Chem.*, 2004, ***55***, 159–196.

33. Wang, F.; Dong, A.; Sun, J.; Tang, R.; Yu, H.; Buhro, W. E. Solution-liquid-solid growth of semiconductor nanowires. *Inorg. Chem.*, 2006, ***45***, 7511–7521.

34. Rao, C. N. R.; Deepak, F. L.; Gundiah, G.; Govindaraj, A. Inorganic nanowires. *Prog. Solid State Ch.*, 2003, ***31***, 5–147.

35. Wagner, R. S.; Ellis, W. C. Vapor-liquid-solid mechanism of single crystal growth. *Appl. Phys. Lett.*, 1964, ***4***, 89–90.

36. Li, Y.; Qian, F.; Xiang, J.; Lieber, C. M. Nanowire electronic and optoelectronic devices. *Mater. Today*, 2006, ***9***, 18–27.

37. Lieber, C. M. Nanoscale Science and Technology: Building a Big Future From Small Things. *MRS Bull.*, 2003, ***28***, 486–491.

38. Leong, M.; Narayanan, V.; Singh, D.; Topol, A.; Chan, V.; Ren, Z. Transistor scaling with novel materials. *Mater. Today*, 2006, ***9***, 26–31.

39. Kittl, J. A.; Lauwers, A.; Pawlak, M. A.; van Dal, M. J. H.; Veloso, A.; Anil, K. G.; Pourtois, G.; Demeurisse, C.; Schram, T.; Brijs, B.; de Potter, M.; Vrancken, C.; Maex, K. Ni fully silicided gates for 45 nm CMOS applications. *Microelectron. Eng.*, 2005, ***82***, 441–448.

40. Aoyama, I.; Fedorov, M. I.; Zaitsev, V. K.; Solomkin, F. Y.; Eremin, I. S.; Samunin, A. Y.; Mukoujima, M.; Sano, S.; Tsuji, T. Effects of Ge doping on micromorphology of MnSi in $MnSi_{\sim 1.7}$ and on their thermoelectric transport properties. *Jpn. J. Appl. Phys.*, 2005, ***44***, 8562–8570.

41. Zaitsev, V. K. In *CRC Handbook of Thermoelectrics*; Rowe, D. M., Ed.; CRC Press: Boca Raton, FL, 1995, p. 299–309.

42. Inui, H. Rhenium silicide as a new class of thermoelectric material. *Mater. Res. Soc. Symp. Proc.*, 2006, ***886***, 219–229.

43. Terada, E.; Oh, M.-W.; Wee, D.-M.; Inui, H. Crystal structure and thermoelectric properties of Al-containing Re silicides. *Mater. Res. Soc. Symp. Proc.*, 2005, ***842***, 425–430.

44. Ishida, A.; Okamoto, N. L.; Kishida, K.; Tanaka, K.; Inui, H. Crystal structures and thermoelectric properties of $Ru_{1-x}Re_xSi_y$ chimney-ladder compounds. *Mater. Res. Soc. Symp. Proc.*, 2007, ***980***, 235–240.

45. Nowotny, H. In *Chemistry of Extended Defects in Non-Metallic Solids*; Eyring, L., O'Keeffe, M., Eds.; North-Holland Publishing Company: Amsterdam, 1970, p. 223–237.

46. Fredrickson, D. C.; Lee, S.; Hoffmann, R.; Lin, J. The Nowotny chimney ladder phases: Following the c_{pseudo} clue toward an explanation of the 14 electron rule. *Inorg. Chem.*, 2004, ***43***, 6151–6158.

47. Rohrer, F. E.; Lind, H.; Eriksson, L.; Larsson, A. K.; Lidin, S. On the question of commensurability—the Nowotny chimney-ladder structures revisited. *Z. Kristallogr.*, 2000, ***215***, 650–660.

48. Hicks, L. D.; Dresselhaus, M. S. Thermoelectric figure of merit of a one-dimensional conductor. *Phys. Rev. B*, 1993, ***47***, 16631–16634.

49. Boukai, A. I.; Bunimovich, Y.; Tahir-Kheli, J.; Yu, J.-K.; Goddard, W. A., III; Heath, J. R. Silicon nanowires as efficient thermoelectric materials. *Nature*, 2008, ***451***, 168–171.

50. Hochbaum, A. I.; Chen, R.; Delgado, R. D.; Liang, W.; Garnett, E. C.; Najarian, M.; Majumdar, A.; Yang, P. Enhanced thermoelectric performance of rough silicon nanowires. *Nature*, 2008, ***451***, 163–167.

51. Dresselhaus, M. S.; Chen, G.; Tang, M. Y.; Yang, R.; Lee, H.; Wang, D.; Ren, Z.; Fleurial, J.-P.; Gogna, P. New directions for low-dimensional thermoelectric materials. *Adv. Mater.*, 2007, ***19***, 1043–1053.

52. Poudel, B.; Hao, Q.; Ma, Y.; Lan, Y.; Minnich, A.; Yu, B.; Yan, X.; Wang, D.; Muto, A.; Vashaee, D.; Chen, X.; Liu, J.; Dresselhaus, M. S.; Chen, G.; Ren, Z. High-thermoelectric performance of nanostructured bismuth antimony telluride bulk alloys. *Science*, 2008, ***320***, 634–638.

53. Schmitt, A. L.; Bierman, M. J.; Schmeisser, D.; Himpsel, F. J.; Jin, S. Synthesis and properties of single-crystal FeSi nanowires. *Nano Lett.*, 2006, ***6***, 1617–1621.

54. Szczech, J. R.; Jin, S. Epitaxially-hyperbranched fesi nanowires exhibiting merohedral twinning. *J. Mater. Chem.*, 2010, ***20***, 1375–1382.

55. Schmitt, A. L.; Zhu, L.; Schmeier, D.; Himpsel, F. J.; Jin, S. Metallic single-crystal CoSi nanowires via chemical vapor deposition of single-source precursor. *J. Phys. Chem. B*, 2006, ***110***, 18142–18146.

56. Schmitt, A. L.; Higgins, J. M.; Jin, S. Chemical synthesis and magnetotransport of magnetic semiconducting $Fe_{1-x}Co_xSi$ alloy nanowires. *Nano Lett.*, 2008, ***8***, 810–815.

57. Higgins, J. M.; Ding, R.; De Grave, J. P.; Jin, S. Signature of helimagnetic ordering in single-crystal MnSi nanowires. *Nano Lett.*, 2010, ***10***, 1605–1610.

58. Higgins, J. M.; Schmitt, A. L.; Guzei, I. A.; Jin, S. Higher manganese silicide nanowires of Nowotny chimney ladder phase. *J. Am. Chem. Soc.*, 2008, ***130***, 16086–16094.

59. Szczech, J. R.; Schmitt, A. L.; Bierman, M. J.; Jin, S. Single-crystal semiconducting chromium disilicide nanowires synthesized via chemical vapor transport. *Chem. Mater.*, 2007, ***19***, 3238–3243.

60. Song, Y.; Schmitt, A. L.; Jin, S. Ultralong single-crystal metallic Ni_2Si nanowires with low resistivity. *Nano Lett.*, 2007, ***7***, 965–969.

61. Song, Y.; Jin, S. Synthesis and properties of single-crystal β_3-Ni_3Si nanowires. *Appl. Phys. Lett.*, 2007, ***90***(17), 173122.

62. Okamoto, H. *Desk handbook : Phase Diagrams for Binary Alloys*; ASM International: Materials Park, OH, 2000.

63. Gokhale, A. B.; Abbaschian, R. The Mn-Si (manganese-silicon) system. *Bull. Alloy Phase Diagrams*, 1990, ***11***, 468–480.

64. Schmitt, A. L.; Higgins, J. M.; Szczech, J. R.; Jin, S. Synthesis and applications of metal silicide nanowires. *J. Mater. Chem.*, 2010, ***20***, 223–235.

65. Kang, K.; Kim, C.-J.; Jo, M.-H. Unconventional roles of metal catalysts in chemical-vapor syntheses of single-crystalline nanowires. *J. Appl. Phys.*, 2009, ***105***(12), 122407.

66. Anselmi-Tamburini, U.; Milanese, C.; Maglia, F.; Buscaglia, V. Transition metal silicides. Growth in metal-silicon couples and diffusion. *Adv. Sci. Technol.*, 2003, ***37***, 61–72.

67. Hassam, S.; Agren, J.; Gauneescard, M.; Bros, J. P. The Ag-Au-Si system—Experimental and calculated phase diagram. *Metall. Trans. A*, 1990, ***21***, 1877–1884.

68. Ma, J.; Gu, Y.; Shi, L.; Chen, L.; Yang, Z.; Qian, Y. Synthesis and oxidation behavior of chromium silicide (Cr_3Si) nanorods. *J. Alloys Compd.*, 2004, ***375***, 249–252.

69. Chen, Y.; Ohlberg, D. A. A.; Medeiros-Ribeiro, G.; Chang, Y. A.; Williams, R. S. Self-assembled growth of epitaxial erbium disilicide nanowires on silicon (001). *Appl. Phys. Lett.*, 2000, ***76***, 4004–4006.

70. Ding, T.; Song, J.; Cai, Q. Rare-earth metal silicide nanostructures on Si(001) surface. *Int. J. Mod. Phys. B*, 2007, ***21***, 1799–1815.

71. He, Z.; Smith, D. J.; Bennett, P. A. Endotaxial silicide nanowires. *Phys. Rev. Lett.*, 2004, ***93***(25), 256102.

72. Okamoto, H. The Si-Zr (silicon-zirconium) system. *Bull. Alloy Phase Diagrams*, 1990, ***11***, 513–519.

73. Olesinski, R. W.; Abbaschian, G. The Si-Zn (silicon-zinc system). *Bull. Alloy Phase Diagrams*, 1985, ***6***, 545–548.

74. Olesinski, R. W.; Gokhale, A. B.; Abbaschian, G. The Ag-Si (silver-silicon) system. *Bull. Alloy Phase Diagrams*, 1989, ***10***, 635–640.

75. Miettinen, J. Thermodynamic description of the Cu-Ni-Si system in the copper-rich corner above 700 °C. *CALPHAD: Comput. Coupling Phase Diagrams Thermochem.*, 2005, ***29***, 212–221.

76. Zhang, C.; Du, Y.; Xiong, W.; Xu, H. H.; Nash, P.; Ouyang, Y. F.; Hu, R. X. Thermodynamic modeling of the v-si system supported by key experiments. *CALPHAD: Comput. Coupling Phase Diagrams Thermochem.*, 2008, ***32***, 320–325.

77. Darby, J. B.; Downy, J. W.; Norton, L. J. Intermediate phases in the techetium-aluminum and technetium-silicon systems. *J. Less Common Met.*, 1965, ***8***, 15–19.

78. Gokhale, A. B.; Abbaschian, G. The Mo-Si (molybdenum-silicon) system. *J. Phase Equil.*, 1991, ***12***, 493–498.

79. Guminski, C. The Hg-Si system (mercury-silicon). *J. Phase Equil.*, 2001, ***22***, 682–683.

80. Lindholm, M. A Thermodyanamic description of the Fe-Cr-Si system with emphasis on the equilibria of the sigma (σ) phase. *J. Phase Equil.*, 1997, ***18***, 432–440.

81. Okamoto, H. Si-Y (Silicon-Yttrium). *J. Phase Equil.*, 1991, ***12***, 397–399.

82. Okamoto, H. The Mn-Si (manganese-silicon system). *J. Phase Equil.*, 1991, ***12***, 505–507.

83. Okamoto, H. Comment on Sc-Si (scandium-silicon). *J. Phase Equil.*, 1995, ***16***, 477.

84. Okamoto, H. Pt-Si (platinum-silicon). *J. Phase Equil.*, 1997, ***18***, 226–226.

85. Okamoto, H. Cr-Si (chromium-silicon). *J. Phase Equil.*, 2001, ***22***, 593–593.

86. Okamoto, H. Cu-Si (copper-silicon). *J. Phase Equil.*, 2002, ***23***, 281–282.

87. Sangster, J.; Pelton, A. D. The Na-Si (sodium-silicon) system. *J. Phase Equil.*, 1992, ***13***, 67–69.

88. Schlesinger, M. The Rh-Si (Rhodium-Silicon System). *J. Phase Equil.*, 1992, ***13***, 54–59.

89. Zhao, J. C.; Bewlay, B. P.; Jackson, M. R.; Chen, Q. Hf-Si binary phase diagram determination and thermodynamic modeling. *J. Phase Equil.*, 2000, ***21***, 40–45.

90. Guo, Z. N.; Yuan, W. X.; Sun, Y.; Cai, Z. F.; Qiao, Z. Y. Thermodynamic assessment of the Si-Ta and Si-W systems. *J. Phase Equilib. Diffus.*, 2009, ***30***, 564–570.

91. Okamoto, H. Nb-Si (niobium-silicon). *J. Phase Equilib. Diffus.*, 2005, ***26***, 649–649.

92. Okamoto, H. Re-Si (rhenium-silicon). *J. Phase Equilib. Diffus.*, 2006, ***27***, 432–432.

93. Okamoto, H. Ru-Si (ruthenium-silicon). *J. Phase Equilib. Diffus.*, 2006, ***27***, 203–203.

94. Okamoto, H. Si-Sr (silicon-strontium). *J. Phase Equilib. Diffus.*, 2006, ***27***, 204–204.

95. Okamoto, H. La-Si (lanthanum-silicon). *J. Phase Equilib. Diffus.*, 2007, ***28***, 585–585.

96. Okamoto, H. Ir-Si (iridium-silicon). *J. Phase Equilib. Diffus.*, 2007, ***28***, 495–495.

97. Okamoto, H. Mg-Si (magnesium-silicon). *J. Phase Equilib. Diffus.*, 2007, ***28***, 229–230.

98. Okamoto, H. Pd-Si (palladium-silicon). *J. Phase Equilib. Diffus.*, 2007, ***28***, 231–232.

99. Okamoto, H. Ca-Si (calcium-silicon). *J. Phase Equilib. Diffus.*, 2007, ***28***, 404–404.

100. Okamoto, H. Os-Si (osmium-silicon). *J. Phase Equilib. Diffus.*, 2007, ***28***, 410–410.

101. Okamoto, H. Ba-Si (barium-silicon). *J. Phase Equilib. Diffus.*, 2008, ***29***, 464–464.

102. Okamoto, H. Co-Si (cobalt-silicon). *J. Phase Equilib. Diffus.*, 2008, ***29***, 295–295.

103. Okamoto, H. Li-Si (lithium-silicon). *J. Phase Equilib. Diffus.*, 2009, ***30***, 118–119.

104. Okamoto, H. Be-Si (beryllium-silicon). *J. Phase Equilib. Diffus.*, 2009, ***30***, 115–115.

105. Sangster, J. Cs-Si (cesium-silicon) system. *J. Phase Equilib. Diffus.*, 2006, ***27***, 188–189.

106. Sangster, J. K-Si (potassium-silicon) System. *J. Phase Equilib. Diffus.*, 2006, ***27***, 190–191.

107. Sangster, J. Rb-Si (rubidium-silicon) System. *J. Phase Equilib. Diffus.*, 2006, ***27***, 192–193.

108. Odin, I. N.; Ivanov, V. A. Liquidus surface of the Cd-Si-Te system. *Russ. J. Inorg. Chem.*, 1991, ***36***, 1039–1042.

109. Seifert, H. J.; Lukas, H. L.; Petzow, G. Thermodynamic optimization of the Ti-Si system. *Z. Metallk.*, 1996, ***87***, 2–13.

110. Gruehn, R.; Glaum, R. New results of chemical transport as a method for the preparation and thermochemical investigation of solids. *Angew. Chem., Int. Ed.*, 2000, ***39***, 692–716.

111. Mason, K. N. Growth and characterization of transiton-metal silicides. *Prog. Cryst. Growth Charact. Mater.*, 1979, ***2***, 269–307.

112. d'Heurle, F. M.; Lavoie, C.; Gas, P.; Philibert, J. In *Diffusion Processes in Advanced Technological Materials*; Gupta, D., Ed.; William Andrew, Inc.: New York, NY, USA, 2005, p. 283–332.

113. Laurila, T.; Molarius, J. Reactive phase formation in thin film metal/metal and metal/silicon diffusion couples. *Crit. Rev. Solid State Mater. Sci.*, 2003, ***28***, 185–230.

114. d'Heurle, F. M. In *Silicides: Fundamentals and Applications -16th Course of the International School of Solid State Physics*; Miglio, L., d'Heurle, F. M., Eds.; World Scientific: Erice, Italy, 1999, p. 169–186.

115. Gosele, U.; Tu, K. N. Growth-kinetics of planar binary diffusion couples—Thin-film case versus bulk cases. *J. Appl. Phys.*, 1982, ***53***, 3252–3260.

116. Gas, P.; d'Heurle, F. M. In *Silicides: Fundamentals and Applications -16th Course of the International School of Solid State Physics*; Miglio, L., d'Heurle, F. M., Eds.; World Scientific: Erice, Italy, 1999, p. 34–51.

117. d'Heurle, F. M.; Gas, P.; Lavoie, C.; Philibert, J. Diffusion in intermetallic compounds: The ordered Cu_3Au rule, its history. *Z. Metallkd.*, 2004, ***95***, 852–859.

118. Gas, P.; d'Heurle, F. M. In *Properties of Metal Silicides*; Maex, K., Van Rossum, M., Eds.; Institution of Engineering and Technology, 1995, p. 279–292.

119. Theron, C. C.; Ndwandwe, O. M.; Lombaard, J. C.; Pretorius, R. First phase formation at interfaces: Comparison between Walser-Bene and effective heat of formation model. *Materials Chemistry and Physics*, 1996, ***46***, 238–247.

120. Dybkov, V. I. Reaction diffusion in binary solid-solid, solid-liquid and solid-gas systems: Common and distinctive features. *Defect Diffus. Forum*, 2001, ***194–199***, 1503–1522.

121. Walser, R. M.; Bene, R. W. First phase nucleation in silicon-transition-metal planar interfaces. *Appl. Phys. Lett.*, 1976, ***28***, 624–625.

122. Sinclair, R. Study of interfacial reaction in metal-silicon and related systems by high-resolution electron-microscopy and thermodynamics analysis. *Mater. Trans.*, 1990, ***31***, 628–635.

123. Chen, L. J. Solid state amorphization in metal/Si systems. *Mater. Sci. Eng., R*, 2000, ***29***, 115–152.

124. Chen, L. J. Metal silicides: An integral part of microelectronics. *J. Miner. Met. Mater. Soc.*, 2005, ***57***, 24–30.

125. Bene, R. W. 1st nucleation rule for solid-state nucleation in metal-metal thin-film systems. *Appl. Phys. Lett.*, 1982, ***41***, 529–531.

126. Zhang, L.; Ivey, D. G. Kinetic plots for predicting silicide formation sequence. *Thin Solid Films*, 1994, ***245***, 234–248.

127. Zhang, L.; Ivey, D. G. Criterion for silicide formation in transition metal-silicion diffusion couples. *Can. Metall. Q.*, 1995, ***34***, 51–71.

128. Gosele, U.; Tu, K. N. Critical thickness of amorphous phase formation in binary diffusion couples. *J. Appl. Phys.*, 1989, ***66***, 2619–2626.

129. Bene, R. W. A Kinetic-model for solid-state silicide nucleation. *J. Appl. Phys.*, 1987, ***61***, 1826–1833.

130. Yang, T. H.; Cheng, S. L.; Chen, L. J. Autocorrelation function analysis of phase formation in the initial stage of interfacial reactions of multilayered titanium-silicon thin films. *Thin Solid Films*, 2004, *469–470*, 513–517.

131. Wu, Y.; Xiang, J.; Yang, C.; Lu, W.; Lieber, C. M. Single-crystal metallic nanowires and metal/semiconductor nanowire heterostructures. *Nature*, 2004, ***430***, 61–65.

132. Appenzeller, J.; Knoch, J.; Tutuc, E.; Reuter, M.; Guha, S. Dual-gate silicon nanowire transistors with nickel silicide contacts. *Tech. Digest*, 2006, 302–305.

133. Dellas, N. S.; Liu, B. Z.; Eichfeld, S. M.; Eichfeld, C. M.; Mayer, T. S.; Mohney, S. E. Orientation dependence of nickel silicide formation in contacts to silicon nanowires. *J. Appl. Phys.*, 2009, ***105***, 094309/094301-094309/094307.

134. Weber, W. M.; Geelhaar, L.; Graham, A. P.; Unger, E.; Duesberg, G. S.; Liebau, M.; Pamler, W.; Cheze, C.; Riechert, H.; Lugli, P.; Kreupl, F. Silicon-nanowire transistors with intruded nickel-silicide contacts. *Nano Lett.*, 2006, ***6***, 2660–2666.

135. Yamamoto, K.; Kohno, H.; Takeda, S.; Ichikawa, S. Fabrication of iron silicide nanowires from nanowire templates. *Appl. Phys. Lett.*, 2006, ***89***, 083107/083101-083107/083103.

136. Lin, Y.-C.; Lu, K.-C.; Wu, W.-W.; Bai, J.; Chen, L. J.; Tu, K. N.; Huang, Y. Single crystalline PtSi Nanowires, PtSi/Si/PtSi nanowire heterostructures, and nanodevices. *Nano Lett.*, 2008, ***8***, 913–918.

137. Liu, B.; Wang, Y.; Dilts, S.; Mayer, T. S.; Mohney, S. E. Silicidation of silicon nanowires by platinum. *Nano Lett.*, 2007, ***7***, 818–824.

138. Lin, Y.-C.; Chen, Y.; Shailos, A.; Huang, Y. Detection of spin polarized carrier in silicon nanowire with single crystal MnSi as magnetic contacts. *Nano Lett.*, 2010, ***10***, 2281–2287.

139. Lu, K.-C.; Wu, W.-W.; Wu, H.-W.; Tanner, C. M.; Chang, J. P.; Chen, L. J.; Tu, K. N. In situ control of atomic-scale Si layer with huge strain in the nanoheterostructure NiSi/Si/NiSi through point contact reaction. *Nano Lett.*, 2007, ***7***, 2389–2394.

140. Chou, Y.-C.; Wu, W.-W.; Cheng, S.-L.; Yoo, B.-Y.; Myung, N.; Chen, L. J.; Tu, K. N. In-situ TEM observation of repeating events of nucleation in epitaxial growth of nano $CoSi_2$ in nanowires of Si. *Nano Lett.*, 2008, ***8***, 2194–2199.

141. Chou, Y.-C.; Wu, W.-W.; Chen, L.-J.; Tu, K.-N. Homogeneous nucleation of epitaxial $CoSi_2$ and NiSi in Si nanowires. *Nano Lett.*, 2009, ***9***, 2337–2342.

142. Liu, Z.; Zhang, H.; Wang, L.; Yang, D. Controlling the growth and field emission properties of silicide nanowire arrays by direct silicification of Ni foil. *Nanotechnology*, 2008, ***19***, 375602/375601-375602/375604.

143. Katsman, A.; Yaish, Y.; Rabkin, E.; Beregovsky, M. Surface diffusion controlled formation of nickel silicides in silicon nanowires. *Journal of Electronic Materials*, 2010, ***39***, 365–370.

144. Morales, A. M.; Lieber, C. M. A laser ablation method for the synthesis of crystalline semiconductor nanowires. *Science*, 1998, ***279***, 208–211.

145. Lensch-Falk, J. L.; Hemesath, E. R.; Perea, D. E.; Lauhon, L. J. Alternative catalysts for VSS growth of silicon and germanium nanowires. *J. Mater. Chem.*, 2009, ***19***, 849–857.

146. Sun, Z.-Q.; Whang, S.-J.; Yang, W.-F.; Lee, S.-J. Synthesis of nickel mono-silicide nanowire by chemical vapor deposition on nickel film: Role of surface nickel oxides. *Jpn. J. Appl. Phys.*, 2009, ***48***, 04C138-131 - 104C138-138.

147. Kim, J.; Lee, E.-S.; Han, C.-S.; Kang, Y.; Kim, D.; Anderson, W. A. Observation of Ni silicide formations and field emission properties of Ni silicide nanowires. *Microelectron. Eng.*, 2008, ***85***, 1709–1712.

148. Kim, J.; Anderson, W. A. Spontaneous nickel monosilicide nanowire formation by metal induced growth. *Thin Solid Films*, 2005, ***483***, 60–65.

149. Kim, C.-J. K., Kibum; Woo, Yun Sung; Ryu, Kyung-Guk; Moon, Heesung; Kim, Jae-Myung; Zang, Dong-Sik; Jo, Moon-Ho. Spontaneous chemical vapor growth of NiSi nanowires and their metallic properties. *Adv. Mater.*, 2007, ***19***, 3637–3642.

150. Kim, J.; Shin, D. H.; Lee, E.-S.; Han, C.-S.; Park, Y. C. Electrical characteristics of single and doubly connected Ni silicide nanowire grown by plasma-enhanced chemical vapor deposition. *Appl. Phys. Lett.*, 2007, ***90***, 253103/253101-253103/253103.

151. Yan, X. Q.; Yuan, H. J.; Wang, J. X.; Liu, D. F.; Zhou, Z. P.; Gao, Y.; Song, L.; Liu, L. F.; Zhou, W. Y.; Wang, G.; Xie, S. S. Synthesis and characterization of a large amount of branched Ni_2Si nanowires. *Appl. Phys. A: Mater. Sci. Process.*, 2004, ***79***, 1853–1856.

152. Kang, K.; Kim, S.-K.; Kim, C.-J.; Jo, M.-H. The role of NiO_x overlayers on spontaneous growth of $NiSi_x$ nanowires from Ni seed layers. *Nano Lett.*, 2008, ***8***, 431–436.

153. Lee, C.-Y.; Lu, M.-P.; Liao, K.-F.; Wu, W.-W.; Chen, L.-J. Vertically well-aligned epitaxial $Ni_{31}Si_{12}$ nanowire arrays with excellent field emission properties. *Appl. Phys. Lett.*, 2008, ***93***, 113109/113101-113109/113103.

154. Chang, C.-M.; Chang, Y.-C.; Lee, C.-Y.; Yeh, P.-H.; Lee, W.-F.; Chen, L.-J. Ti_5Si_4 Nanobats with excellent field emission properties. *J. Phys. Chem. C*, 2009, ***113***, 9153–9156.

155. Wagner, R. S.; Ellis, W. C. The vapor-liquid-solid mechanism of crystal growth and its application to silicon. *Trans. Am. Inst. Min., Metall. Pet. Eng.*, 1965, ***233***, 1053–1064.

156. Gleizes, A. N. MOCVD of chalcogenides, pnictides, and heterometallic compounds from single-source molecule precursors. *Chem. Vap. Deposition*, 2000, ***6***, 155–173.

157. Chueh, Y. L.; Chou, L. J.; Cheng, S. L.; Chen, L. J.; Tsai, C. J.; Hsu, C. M.; Kung, S. C. Synthesis and characterization of metallic $TaSi_2$ Nanowires. *Appl. Phys. Lett.* 2005, ***87***, 223113/223111-223113/223113.

158. Chueh, Y.-L.; Ko, M.-T.; Chou, L.-J.; Chen, L.-J.; Wu, C.-S.; Chen, C.-D. $TaSi_2$ nanowires: A potential field emitter and interconnect. *Nano Lett.*, 2006, ***6***, 1637–1644.

159. Xiang, B.; Wang, Q. X.; Wang, Z.; Zhang, X. Z.; Liu, L. Q.; Xu, J.; Yu, D. P. Synthesis and field emission properties of $TiSi_2$ nanowires. *Appl. Phys. Lett.*, 2005, ***86***, 243103/243101-243103/243103.

160. Chueh, Y. L.; Chou, L. J.; Cheng, S. L.; He, J. H.; Wu, W. W.; Chen, L. J. Synthesis of taperlike Si nanowires with strong field emission. *Appl. Phys. Lett.*, 2005, ***86***, 133112/133111-133112/133113.

161. Ouyang, L.; Thrall, E. S.; Deshmukh, M. M.; Park, H. Vapor-phase synthesis and characterization of ε-FeSi nanowires. *Adv. Mater.*, 2006, ***18***, 1437–1440.

162. Varadwaj, K. S. K.; Seo, K.; In, J.; Mohanty, P.; Park, J.; Kim, B. Phase-controlled growth of metastable Fe_5Si_3 nanowires by a vapor transport method. *J. Amer. Chem. Soc.*, 2007, ***129***, 8594–8599.

163. Zhang, H.-X.; Ge, J.-P.; Wang, J.; Li, Y.-D. Atmospheric pressure chemical vapor deposition synthesis of sulfides, oxides, silicides and metal nanowires with metal chloride precursors. *Nanotechnology*, 2006, ***17***, S253–S261.

164. Seo, K.; Lee, S.; Yoon, H.; In, J.; Varadwaj, K. S. K.; Jo, Y.; Jung, M.-H.; Kim, J.; Kim, B. Composition-tuned Co_nSi nanowires: Location-selective simultaneous growth along temperature gradient. *ACS Nano*, 2009, ***3***, 1145–1150.

165. Seo, K.; Varadwaj, K. S. K.; Mohanty, P.; Lee, S.; Jo, Y.; Jung, M.-H.; Kim, J.; Kim, B. Magnetic properties of single-crystalline CoSi nanowires. *Nano Lett.*, 2007, ***7***, 1240–1245.

166. Tsai, C.-I.; Yeh, P.-H.; Wang, C.-Y.; Wu, H.-W.; Chen, U.-S.; Lu, M.-Y.; Wu, W.-W.; Chen, L.-J.; Wang, Z.-L. Cobalt silicide nanostructures: Synthesis, electron transport, and field emission properties. *Crystal Growth & Design*, 2009, ***9***, 4514–4518.

167. Lin, H.-K.; Tzeng, Y.-F.; Wang, C.-H.; Tai, N.-H.; Lin, I. N.; Lee, C.-Y.; Chiu, H.-T. Ti_5Si_3 nanowire and its field emission property. *Chem. Mater.*, 2008, ***20***, 2429–2431.

168. Lin, H.-K.; Cheng, H.-A.; Lee, C.-Y.; Chiu, H.-T. Chemical vapor deposition of TiSi nanowires on C54 $TiSi_2$ thin film: An amorphous titanium silicide interlayer assisted nanowire growth. *Chem. Mater.*, 2009, ***21***, 5388–5396.

169. Chang, C.-M.; Chang, Y.-C.; Chung, Y.-A.; Lee, C.-Y.; Chen, L.-J. Synthesis and properties of the low resistivity $TiSi_2$ nanowires grown with TiF_4 precursor. *J. Phys. Chem. C*, 2009, ***113***, 17720–17723.

170. Seo, K.; Varadwaj, K. S. K.; Cha, D.; In, J.; Kim, J.; Park, J.; Kim, B. Synthesis and electrical properties of single crystalline $CrSi_2$ nanowires. *J. Phys. Chem. C*, 2007, ***111***, 9072–9076.

171. Yu, L.; Ma, Y.; Zhu, J.; Feng, H.; Wu, Q.; Lu, Y.; Lin, W.; Sang, H.; Hu, Z. An in-situ chloride-generated route to various one-dimensional nanostructures of chromium-based compounds. *J. Phys. Chem. C*, 2008, ***112***, 5865–5868.

172. In, J.; Varadwaj, K. S. K.; Seo, K.; Lee, S.; Jo, Y.; Jung, M.-H.; Kim, J.; Kim, B. Single-crystalline ferromagnetic $Fe_{1-x}Co_xSi$ nanowires. *J. Phys. Chem. C*, 2008, ***112***, 4748–4752.

173. Seo, K.; Yoon, H.; Ryu, S.-W.; Lee, S.; Jo, Y.; Jung, M.-H.; Kim, J.; Choi, Y.-K.; Kim, B. Itinerant helimagnetic single-crystalline MnSi nanowires. *ACS Nano*, 2010, ***4***, 2569–2576.

174. Ham, M.-H.; Lee, J.-W.; Moon, K.-J.; Choi, J.-H.; Myoung, J.-M. Single-crystalline ferromagnetic Mn_4Si_7 nanowires. *J. Phys. Chem. C*, 2009, ***113***, 8143–8146.

175. Bierman, M. J.; Lau, Y. K. A.; Jin, S. Hyperbranched PbS and PbSe nanowires and the effect of hydrogen gas on their synthesis. *Nano Lett.*, 2007, ***7***, 2907–2912.

176. Nickl, J. J.; Koukouss, J. D. Transport reactions and monocrystal growth of silicides of Group-IVa and Group-VIa metals. *J. Less-Common Met.*, 1971, ***23***, 73.

177. Barrelet, C. J.; Wu, Y.; Bell, D. C.; Lieber, C. M. Synthesis of CdS and ZnS nanowires using single-source molecular precursors. *J. Amer. Chem. Soc.*, 2003, ***125***, 11498–11499.

178. Aylett, B. J.; Colquhoun, H. M. Chemical vapor-deposition of transition-metal silicides by pyrolysis of silyl transition-metal carbonyl-compounds. *J. Chem. Soc. Dalton*, 1977, 2058–2061.

179. Novak, I.; Huang, W.; Luo, L.; Huang, H. H.; Ang, H. G.; Zybill, C. E. UPS study of compounds with metal-silicon bonds: $M(CO)_nSiCl_3$ (M = Co, Mn; n = 4, 5) and $Fe(CO)_4(SiCl_3)_2$. *Organometallics*, 1997, ***16***, 1567–1572.

180. Schmitt, A. L.; Jin, S. Selective patterned growth of silicide nanowires without the use of metal catalysts. *Chem. Mater.*, 2007, ***19***, 126–128.

181. Villars, P.; Calvert, L. D. *Pearson's Handbook of Crystallographic Data for Intermetallic Phases*; 2nd ed.; ASTM International: Newbury, OH, 1991.

182. Snyder, G. J.; Toberer, E. S. Complex thermoelectric materials. *Nat. Mater.*, 2008, ***7***, 105–114.

183. Zhou, S.; Liu, X.; Lin, Y.; Wang, D. Spontaneous growth of highly conductive two-dimensional single-crystalline $TiSi_2$ nanonets. *Angew. Chem., Int. Ed.*, 2008, ***47***, 7681–7684.

184. Zhou, S.; Liu, X.; Lin, Y.; Wang, D. Rational synthesis and structural characterizations of complex $TiSi_2$ nanostructures. *Chem. Mater.*, 2009, ***21***, 1023–1027.

185. Morin, S. A.; Bierman, M. J.; Tong, J.; Jin, S. Mechanism and kinetics of spontaneous nanotube growth driven by screw dislocations. *Science*, 2010, ***328***, 476–480.

186. Du, J.; Du, P.; Hao, P.; Huang, Y.; Ren, Z.; Han, G.; Weng, W.; Zhao, G. Growth Mechanism of TiSi nanopins on Ti_5Si_3 by atmospheric pressure chemical vapor deposition. *J. Phys. Chem. C*, 2007, ***111***, 10814–10817.

187. In, J.; Seo, K.; Lee, S.; Yoon, H.; Park, J.; Lee, G.; Kim, B. Morphology-tuned synthesis of single-crystalline V_5Si_3 nanotubes and nanowires. *J. Phys. Chem. C*, 2009, ***113***, 12996–13001.

188. Lee, C.-Y.; Lu, M.-P.; Liao, K.-F.; Lee, W.-F.; Huang, C.-T.; Chen, S.-Y.; Chen, L.-J. Free-standing single-crystal $NiSi_2$ nanowires with excellent electrical transport and field emission properties. *J. Phys. Chem. C*, 2009, ***113***, 2286–2289.

189. Zwanenburg, F. A.; van Rijmenam, C. E. W. M.; Fang, Y.; Lieber, C. M.; Kouwenhoven, L. P. Spin States of the first four holes in a silicon nanowire quantum dot. *Nano Lett.*, 2009, **9**, 1071–1079.

190. Kim Joong, J.; Shindo, D.; Murakami, Y.; Xia, W.; Chou, L.-J.; Chueh, Y.-L. Direct observation of field emission in a single $TaSi_2$ nanowire. *Nano Lett.*, 2007, **7**, 2243–2247.

191. Zhou, F.; Szczech, J.; Pettes, M. T.; Moore, A. L.; Jin, S.; Shi, L. Determination of transport properties in chromium disilicide nanowires via combined thermoelectric and structural characterizations. *Nano Lett.*, 2007, **7**, 1649–1654.

192. Lin, Y.; Zhou, S.; Liu, X.; Sheehan, S.; Wang, D. $TiO_2/TiSi_2$ Heterostructures for high-efficiency photoelectrochemical H_2O splitting. *J. Am. Chem. Soc.*, 2009, ***131***, 2772–2773.

193. Xu, N. S.; Huq, S. E. Novel cold cathode materials and applications. *Mater. Sci. Eng. R*, 2005, *R48*, 47–189.

194. Datta, S.; Das, B. Electronic analog of the electro-optic modulator. *Appl. Phys. Lett.*, 1990, ***56***, 665–667.

195. Schmidt, G.; Ferrand, D.; Molenkamp, L. W.; Filip, A. T.; van Wees, B. J. Fundamental obstacle for electrical spin injection from a ferromagnetic metal into a diffusive semiconductor. *Phys. Rev. B*, 2000, ***62***, R4790–R4793.

196. Guevara, J.; Vildosola, V.; Milano, J.; Llois, A. M. Half-metallic character and electronic properties of inverse magnetoresistant $Fe_{1-x}Co_xSi$ Alloys. *Phys. Rev. B*, 2004, ***69***, 184422.

197. Pfleiderer, C.; Neubauer, A.; Muhlbauer, S.; Jonietz, F.; Janoschek, M.; Legl, S.; Ritz, R.; Munzer, W.; Franz, C.; Niklowitz, P. G.; Keller, T.; Georgii, R.; Boni, P.; Binz, B.; Rosch, A. Quantum order in the chiral magnet MnSi. *J. Phys.-Condes. Matter*, 2009, ***21***, 164215.

198. Hicks, L. D.; Dresselhaus, M. S. Effect of quantum-well structures on the thermoelectric figure of merit. *Phys. Rev. B*, 1993, ***47***, 12727–12731.

199. Bierman, M. J.; Jin, S. potential applications of hierarchical branching nanowires in solar energy conversion. *Energy and Environmental Science*, 2009, ***2***, 1050–1059.

Chapter 4

Metal Silicide Nanowires: Growth and Properties

L. J. Chen[a] and W. W. Wu[b]
[a]*Department of Materials Science and Engineering, National Tsing Hua University, Hsinchu 300, Taiwan, China*
[b]*Department of Materials Science and Engineering, National Chiao Tung University, Hsinchu 300, Taiwan, China*
ljchen@mx.nthu.edu.tw

4.1 Introduction

Metal silicides have been widely used in microelectronic devices as ohmic contacts, Schottky barriers, gate electrodes, and interconnect between transistors in very-large-scale-integration (VLSI) technology.[1–8] A large quantity of literature has been published on the science and technology of thin film silicides.[9–26] The control in nucleation and growth of thin film silicides on Si substrates has been one of the important issues for improving the semiconductor device performance.[27–31]

As the trend of device miniaturization continues, nanoscale metal silicides have attracted increasing interest and attention. Thus, there

Silicon and Silicide Nanowires: Applications, Fabrication, and Properties
Edited by Yu Huang and King-Ning Tu

ISBN 978-981-4303-46-0 (Hardcover), 978-981-4303-47-7 (eBook)
www.panstanford.com

have been extensive investigations on silicide nanodots and silicide nanowires (NWs); the latter were demonstrated to play an important role in nanodevices.[32–59] The synthesis, growth, and properties of silicide NWs are expected to be important in developing the future Si-based semiconductor devices.

The silicide NWs possess the following desirable characteristics for device applications:

1. Diversity in growth techniques: There are a variety of techniques to grow silicide NWs.[60]
2. Richness in physical properties: Metal silicides exhibit a wide range of electrical, magnetic, field-emission, and optical properties. As a result, they may be applied in spintronics, thermoelectrics, nanoelectronics, solar energy conversion, and biosensors.[32,33,60]
3. Compatibility with silicon device processing: Processings for silicide NWs are often compatible with existing silicon device processing technology.
4. Ease in the fabrication of semiconductor/metal heterostructures: The heterostructures are useful for the applications as NW-based transistors and forming narrow gaps between metals.

In this chapter, we review the growth and properties of a number of metal silicide NWs with particular emphasis on those that are potentially useful in future nanoelectronics devices.

4.2 Epitaxial Growth of Silicide Nanowires on Si Substrate

As the fabrication of devices scaled down to the nanometer range, self-assembly has been demonstrated to be advantageous over traditional top-down fabrication methods. Self-assembled epitaxial silicide NWs are potentially useful as interconnectors for future devices, due to their small dimension, metal-like resistivity, single-crystallinity with low-defect density, and self-assembly nature.[42,61–68]

Nickel (Ni) silicide is one of the most extensively studied systems due to the potential for IC applications. Among all Ni silicides, $NiSi_2$ can easily appear in epitaxial form on Si substrates since it possess the same cubic structure and the lattice mismatch between $NiSi_2$ and Si is only 0.4%. The Fe/Si silicide system is another popular system for investigation. There are two well-defined $FeSi_2$ phases: one is metallic α-$FeSi_2$ and another is semiconducting β-$FeSi_2$. In this section, formation of self-assembled epitaxial $NiSi_2$ and α-$FeSi_2$ NWs by the reactive-deposition epitaxy (RDE) method will be described. In addition, a concept of nitride-mediated epitaxy (NME) which can significantly promote growth of epitaxial silicide NWs will be presented.

4.2.1 Epitaxial $NiSi_2$ Nanowires

Figure 4.1 shows a Scanning Electron Microscope (SEM) image of self-assembled epitaxial $NiSi_2$ NWs on (001)Si substrates by the RDE method at 730 °C.[36] The RDE procedure was performed in an ultrahigh vacuum electron-beam evaporation chamber with a base pressure better than 3×10^{-10} Torr. Prior to the loading, substrates were cleaned by a standard Radio Corporation of America procedure followed by dipping in a dilute HF solution. Two-nm-thick Ni films were then deposited onto the heated substrates (RDE). For self-assembled NW growth, the procedure of such asymmetric 1-D growth involves the rearrangement of metal atoms from the edge of NWs to the end since the metal atoms impinge from all sides of NWs. The RDE method thus provides deposited atoms sufficient kinetic energy for moving on the Si surface and promotes NW growth.

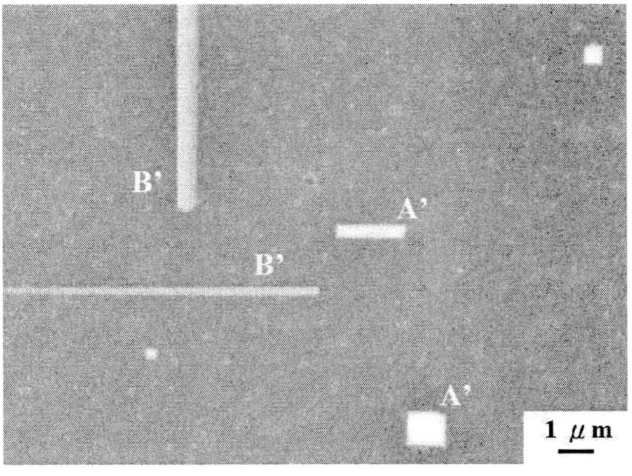

Figure 4.1 Self-assembled epitaxial $NiSi_2$ NWs on (001)Si substrates.

The epitaxial $NiSi_2$ NWs and can be grouped into two types. The two types of NWs are designated as Type-A′ and Type-B′ following the conventional nomenclature that Type-A and Type-B interfaces represent the interfaces between the silicide and the substrate with identical and twinning orientation relationships, respectively. Type-A′ $NiSi_2$ NWs are of the identical crystal orientation to that of the Si substrates. Type-B′ $NiSi_2$ NWs possess one Type-B $NiSi_2$/(111) Si interface with a twinning orientation relationship to the Si substrates. Issues relating to Type-A and Type-B silicide/Si interfaces have been extensively discussed previously.[5,69] The result is similar to that of the Co/Si system.[42] The average aspect ratio of Type-B′ NWs is higher than that of Type-A′ NWs, i.e., about 3.5 for Type-A′ and 25 for Type-B′ NWs. The result indicated that the breaking of the symmetry with the substrate at the twinning interface would promote 1-D growth of silicide NWs.

The formation mechanisms of Type-A′ and Type-B′ $NiSi_2$ NWs are quite different. For Type-A′ islands with isotropic lattice mismatch, the formation of NWs could be attributed to the strain-driven "shape transition" mechanism, as represented by the theoretical models put forward by Tersoff and Tromp, as well as by Jesson *et al.*[70,71] In contrast, the formation of Type-B′ $NiSi_2$ NWs involved the mechanism of endotaxy.[42] The Type-B interface breaks the symmetry of the surface and leads to the asymmetric growth of Type-B′ islands. NWs. The growth of both Type-A′ or Type-B′ $NiSi_2$ NWs involves the rearrangement of Ni atoms from the edge of elongated NWs to the end since the Ni atoms impinge from all sides. Ideally, the deposition rate of Ni atoms during the RDE should be low enough to provide sufficient time for epitaxial islands to release the elastic strain by means of shape transition. However, the deposition rate cannot be arbitrarily decreased in the limited space of the deposition chamber. Thus a NME method which can efficiently diminish the flux of metal atoms to the Si surface was been utilized to further promote NW growth. Previous work revealed that for the Ni/Si_3N_4/Si system, nitride served as a pure physical barrier to diminish the Ni flux.[72] In addition to the flux adjustment, one main mechanism of OME or NME is to allow the metal-rich phase to be skipped and lead to the formation of epitaxial $NiSi_2$ phase directly.[73,74] This mechanism is closely related to the NW growth since only the epitaxial phase could undergo the shape transition. Figure 4.2 shows a SEM image of epitaxial $NiSi_2$ NWs on nitride-capped (001)Si substrates by RDE at

730 °C. The aspect ratios can be improved to 30 and 200 for Type-A′ or Type-B′ $NiSi_2$ NWs, respectively.

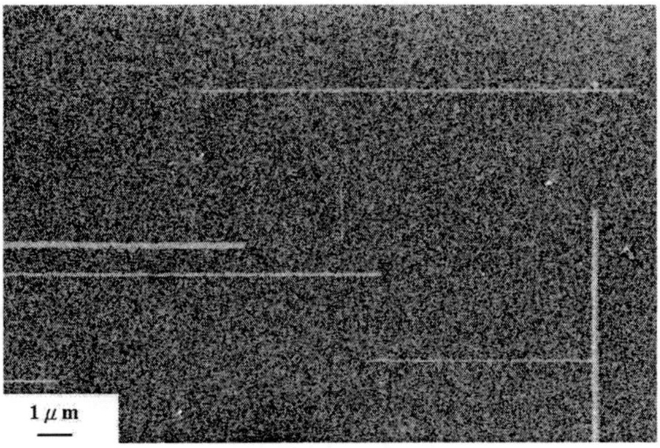

Figure 4.2 Formation of high–aspect ratio epitaxial $NiSi_2$ NWs on (001)Si substrates by combining RDE and NME methods.

The Type-A′ $NiSi_2$ NW can be identified from the $NiSi_2$ (002) dark-field image, as shown in Fig. 4.3. Figure 4.4 shows a Transmission Electron Microscope (TEM) image of Type-B′ $NiSi_2$ NWs. The inset is an atomic-resolution cross-section image of a NW, viewed along the long direction, showing the twinning orientation relationship with respect to silicon. In particular, after twinning operation, $(\bar{1}11)$ $NiSi_2$ is parallel to (511)Si. Both types of $NiSi_2$ NWs have long directions parallel to [110] or $[1\bar{1}0]$ direction.

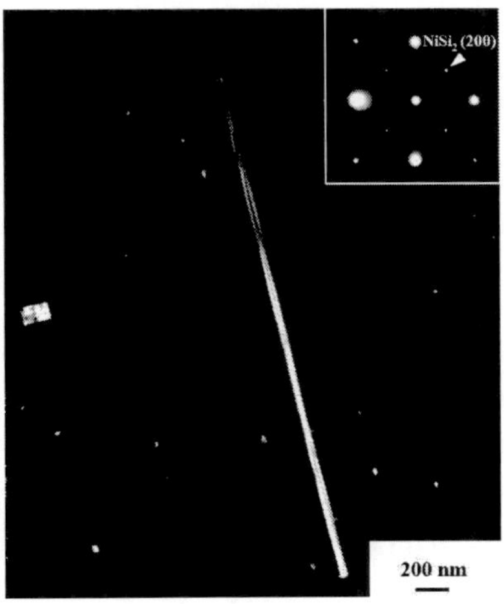

Figure 4.3 $NiSi_2$ [002] dark-field image of a Type-A′ NW. Inset is the corresponding diffraction patterns.

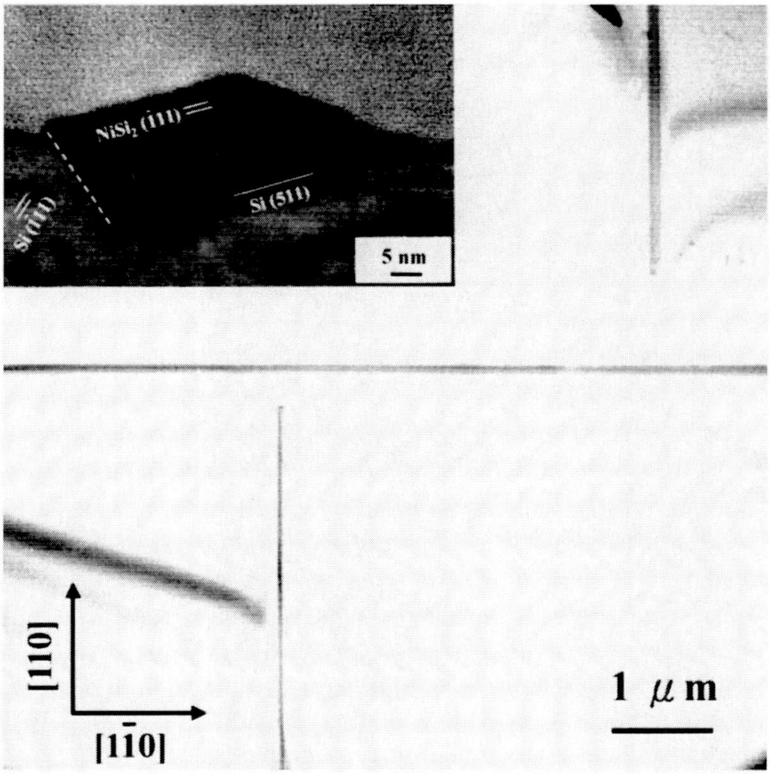

Figure 4.4 Bright-field TEM image of Type-B′ $NiSi_2$ NWs. The inset is an atomic-resolution image showing the cross-section of a NW along the elongated direction. The dotted line denotes the Type-B $NiSi_2$/(111)Si twin-related interface.

4.2.2 Epitaxial α-$FeSi_2$ Nanowires with Length Tunability

Formation of self-assembled α-$FeSi_2$ NWs on (001)Si substrates can be achieved by the RDE and NME methods, as mentioned above. Figures 4.5a and b show TEM images of the resulting structures on blank (001)Si substrate and nitride-capped(001) Si substrate by RDE at 650 °C followed by post-annealing at the same temperature for 10 min, respectively.[37] On blank Si substrates, most silicide islands were formed in the shape of equiaxed islands with a wide range of the size distribution. Some rod-like islands with relatively small size can also be found. For the nitride-capped (001)Si substrate, planview TEM images show that most of the silicide islands formed in the NW shape with a considerable length/width aspect ratio and atomically smooth boundaries. The length, width, and aspect ratio of these NWs in average are 160 nm, 5 nm, and 32, respectively. Figure 4.6 shows a TEM image of the resulting structures on nitride-capped Si substrates by RDE at 650 °C followed by prolonged annealing at the same temperature for 1.5 h. The length and the aspect ratio of the NWs were found to increase significantly to 2 μm and 200, respectively, with a width of 5–10 nm. Although there is no reported data on the diffusion of Fe through thin Si_3N_4 layer, it is thought that the diffusion rate is very low as evidenced by the presence of a high density of Fe particles on top of the thin layer. The amount of Fe atoms arriving at the Si_3N_4/Si interface is therefore only a fraction of the total deposited Fe atoms for as-deposited samples. During the post-annealing, the Fe atoms on the nitride surface or inside nitride layers would arrive at the Si_3N_4/Si interface and attach to the silicide islands continuously with an ultra-low flux. This is in contrast to the rapid mass transport of Ni atoms passing through the Si_3N_4 layers in the Ni/Si_3N_4/Si system, as mentioned above, but could allow the tunability on length of NWs.

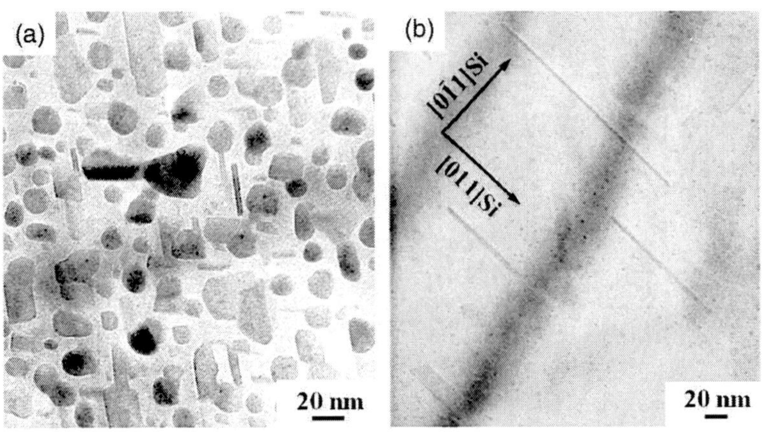

Figure 4.5 TEM images of α-$FeSi_2$ on (a) blank (100)Si substrate and (b) nitride-capped (100)Si substrate by RDE at 650 °C followed by post-annealing at the same temperature for 10 min.

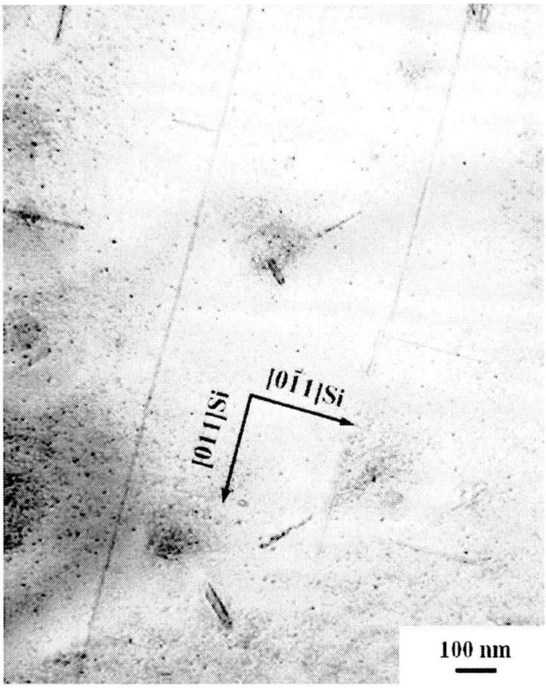

Figure 4.6 TEM images of α-$FeSi_2$ on (a) nitride-capped (100)Si substrate by RDE at 650 °C followed by prolonged post-annealing for 1.5 h.

Figure 4.7a and the inset show TEM image of a NW and corresponding selected-area diffraction pattern on nitride-capped Si substrate. The NW was grown in [011]Si direction. The indexed pattern is shown in Fig. 4.7b. The NWs were identified to be α-$FeSi_2$ grown epitaxially on Si substrate with the epitaxial relationships of [010] α-$FeSi_2$//[010]Si and (100)α-$FeSi_2$//(100)Si. α-$FeSi_2$ is of tetragonal structure with $a = b = 0.2698$ nm and $c = 0.5141$ nm.[75] The multiple lattice mismatch between α-$FeSi_2$ and Si is very small since twice of a and b has a value of 0.5396 nm, which is close to the lattice constant of Si (0.5403 nm). For the epitaxial relationships described above, the lattice mismatches along [001]α-$FeSi_2$/[001]Si and [010] α-$FeSi_2$/[010]Si are –5.33% and –0.64%, respectively. A straightforward calculation indicates that the lattice mismatches parallel and perpendicular to the growth direction are both −2.97%. Figure 4.7c shows an atomic-resolution cross-section image of a NW viewed along the [011] Si direction. The image further reveals that the NW grew into the Si substrate with two coherent {112}α-$FeSi_2$/{111}Si interfaces, which are likely energetically favorable.

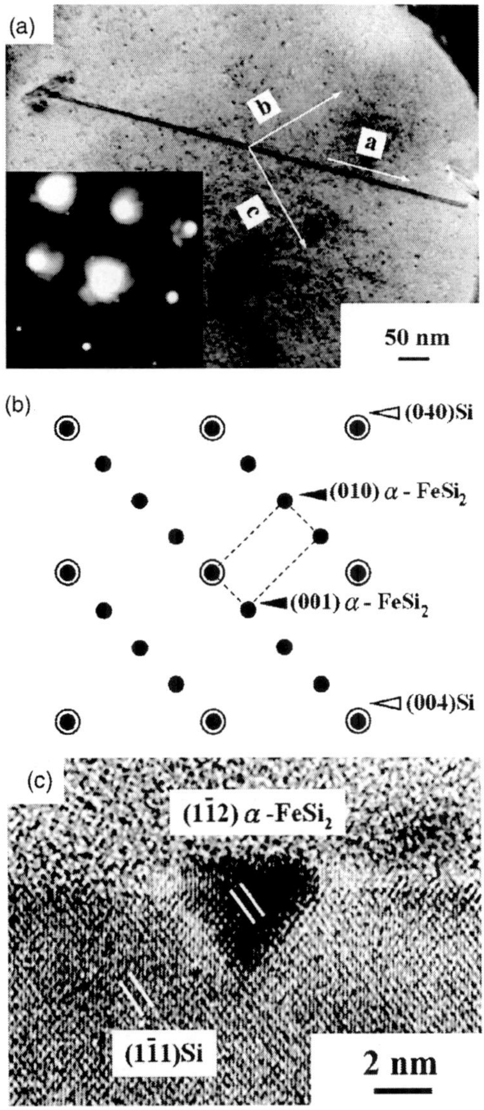

Figure 4.7 (a) Bright-field image of a single α-$FeSi_2$ NW with selected-area diffraction pattern as inset; (b) the indexed pattern of the inset of (a) and (c) cross-section image of a NW. Labels a, b, and c in (a) represent [011]Si, [001]Si, and [010]Si directions, respectively.

Self-assembled epitaxial silicide NWs can be fabricated on Si substrates by the RDE method, which provides deposited metal atoms sufficient kinetic energy for moving on the Si surface during NW growth. In addition, the use of NME method can further promote NW growth by diminishing the flux of metal atoms. Formation of epitaxial $NiSi_2$ and α-$FeSi_2$ NWs with tunable aspect ratios can be achieved by such concepts. Such tunable nature is essential for the application on future interconnects.

4.2.3 Growth of High-Density Titanium Silicide Nanowires in a Single Direction on a Silicon Surface

NW-based materials have been synthesized for a wide range of materials such as metals, semiconductors, and oxides.[38,76–78] Metallic silicide NWs have been studied as electrical interconnect and contact. The silicide NWs exhibited novel physical and electrical properties as promising elements for future nanoscale devices. Understanding the formation mechanisms is a crucial step to the successful implementation of NWs for practical employment. *In situ* UHV-TEM is an extremely valuable method for clarifying the growth mechanism of silicide NWs since it can monitor in nanoscale the dynamic process as it occurs.

The interaction mechanisms of titanium silicide NWs on Si (111) substrate were investigated by *in situ* UHV-TEM.[43] The $TiSi_2$ NWs were oriented along three equivalent Si$< 1\bar{1}0 >$ directions and possessed small width and high aspect ratio. During continuous Ti deposition, the length of NWs increased slowly, as visualized by the increase in the number of Moiré fringes, while the width did not change appreciably at the same time. When two NWs grew in the same direction and eventually collided with each other, they would merge to form a single NW with the combined length of the two NWs, as shown in Figs. 4.8a–d. On the other hand, for two NWs in different directions colliding with each other at the end, one of them collapsed to the other along the side of the NW and was absorbed by the other one, as shown in Figs. 4.8e–h. The merging of the two NWs in different directions occurred as a result of minimizing the total surface energy.

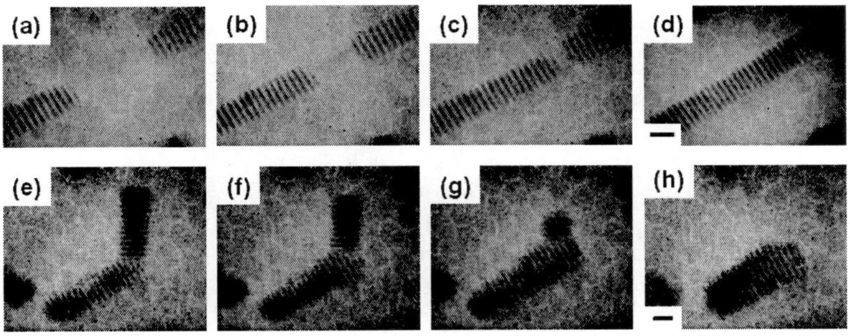

Figure 4.8 *In situ* TEM images of the interactions of titanium silicide NWs at 700 °C. The images of NWs in the same direction were obtained at elapsed times of (a) 0 min, (b) 9 min, (c) 18 min, and (d) 19 min. The images of NWs in different directions were obtained at elapsed times of (e) 0 sec, (f) 25 sec, (g) 49 sec, and (h) 80 sec. The scale bar correspond to 6 nm in (d) and 4 nm in (h).

Figure 4.9 shows that as one $TiSi_2$ NW continued to grow and encounter the midsection of an inclined NW, it tended to recede rapidly and the other one became thicker at the midsection, as marked in Fig. 4.9b by the arrow. It indicates that the silicide atoms migrated from NWs "B" to "A". According to the Gibbs–Thomson equation,[79] the concentration is strongly related to the curvature along the NW ends. As a result, the concentrations of constituent atoms (Ti and Si) at the curved end of NW "B" are higher than those at the flat side of NW "A", resulting in the diffusion of Ti and Si atoms form NW "B" to "A". Owing to the supply of silicide for the NW "B" taking a relatively long time to arrive from the far end to the contact area, receding of the NW was observed.

Therefore, in order to grow a high density of long and high–aspect ratio NWs, it is highly desirable to grow NWs along a single direction. This important insight has been fruitfully exploited to focus on the growth of titanium silicide NWs on Si(110), since crystallographically the NWs are favored to grow only in the $[1\bar{1}0]$ direction, as shown in Fig. 4.10. The results demonstrate the powerfulness of the *in situ* UHV-TEM observation. Most importantly, the achievement in controlled growth represents a significant advance in realizing the vast potential for applications of silicide NWs in nanodevices.

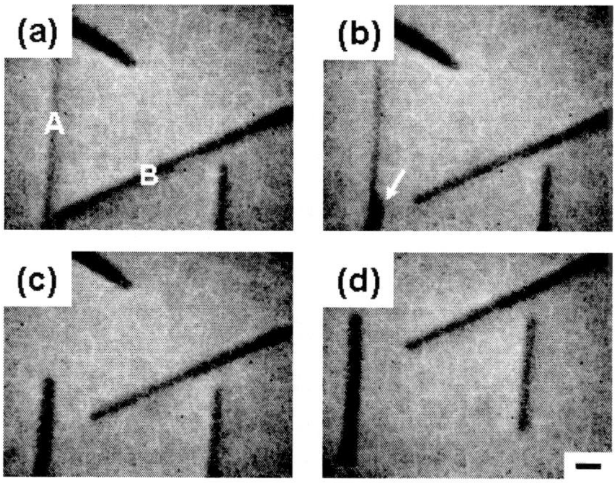

Figure 4.9 *In situ* TEM images showing the interaction of one titanium silicide NW with another at its midsection at 700 °C. The images were obtained at elapsed times of (a) 0 sec, (b) 16 sec, (c) 56 sec, and (d) 81 sec. The scale bar corresponds to 30 nm.

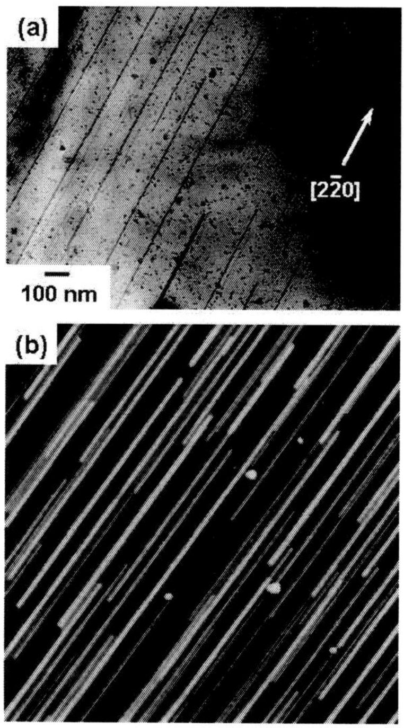

Figure 4.10 Images of titanium silicide NWs on Si(110). (a) TEM and (b) AFM (1.5 μm × 1.5 μm) images of titanium silicide NWs grown on Si(110) along one [1–10] direction.

4.3 Growth of Free-Standing Silicide Nanowires and Their Properties

The fabrication is mainly using metal halides as precursors to grow free-standing silicide NWs with high densities and high aspect ratios in large areas on different substrates.

4.3.1 Growth of Single-Crystal Nickel Silicide Nanowires with Excellent Electrical Transport and Field-Emission Properties

As the microelectronic devices entering into nano era, the requirements of the metallization in complementary metal oxide semiconductor (CMOS) device integration are important to develop new approach for future applications.[35,80–83] Among metal silicides, the low-resistivity NiSi is currently the most promising material because of its electrical properties and appropriate work function. On the other hand, Ni-rich silicides, such as Ni_2Si and $Ni_{31}Si_{12}$, have gained much attention for use in fully silicided (FUSI) gates because of their higher work functions (~4.8 eV) which can effectively improve the p-type MOS (PMOS) device performance.[84]

Several growth methods have been adopted to form vertical Ni silicide NWs. Freestanding NiSi NWs were first synthesized from chemical vapor deposition (CVD) process by Decker *et al.*[85] Wu *et al.*[34] coated evaporated Ni film onto free standing silicon NWs to form NiSi with an annealing process. In addition, NiSi,[85–87] Ni_2Si,[40,88] Ni_3Si_2,[89] and Ni_3Si[51] NWs were synthesized by pyrolysis of silane gas,[85,87–89] sputtering silicon onto Ni surfaces[9] by metal-induced growth (MIG) or using iodine as transport reagent in chemical vapor transport (CVT) methods.[40,51] Recently, a new method by nanocontact reaction between Ni and Si NWs to form NiSi NWs[54] has been reported. In addition, Ni-silicide NWs were obtained by means of solid-state reactions of the Si NWs with subsequently deposited Ni films.[90] As Ni atoms are known to be the dominant diffusing species in the Ni–Si reactions, it was thought that enhanced Si condensation and diminished supply of Ni might facilitate the growth of Si-rich silicide. In this section, we report the growth of free-standing $NiSi_2$ and $Ni_{31}Si_{12}$ NWs on Ni foil.

4.3.1.1 Well-aligned epitaxial $Ni_{31}Si_{12}$ nanowire arrays

Pure Ni foils (99.9% purity), cleaned in acetone, de-ionized water, hexane, and dilute HNO_3, consecutively, were used as the substrates. After the cleaning process, the Ni foils were placed in an alumina boat and inserted into a quartz tube furnace pumped down to about 1.0×10^{-5} Torr in pressure. The flows of 100 sccm (sccm denotes cubic centimeter per minute at STP) Ar and 25 sccm H_2 were kept constant and introduced as the carrier gas with the pressure in the quartz tube fixed at 1 Torr. The upstream and downstream ends of the furnace were heated from room temperature to 1100 °C and 750 °C, respectively. Silicon powders (99.9%) in an alumina boat were positioned at the 1100 °C zone upstream in the quartz tube as the source. Ni foils were located downstream at the end of 750 °C zone. The reaction time was set to be 3 h unless otherwise mentioned. The furnace was then allowed to cool down to room temperature.

X-ray diffraction (XRD) spectra (Fig. 4.11a) of the samples grown at the 750 °C Wen-WeiWu zone indicate that all the diffraction peaks can be ascribed to hexagonal $Ni_{31}Si_{12}$ (space group *P321*(150), the lattice constants a = 0.6671 nm and c = 1.2288 nm, JCPDS Card No. 71-0638) NWs and Ni substrate. No diffraction peaks from any other Ni silicide phase and impurities were detected. The diffraction peaks are strong and sharp indicating the high degree of crystallinity of as-grown samples. Figure 4.11b shows a SEM image of the $Ni_{31}Si_{12}$ NWs which were uniformly grown at the 750 °C Wen-WeiWu zone on grains of $Ni_{31}Si_{12}$ thin film formed on the Ni foil substrate. The length and diameter of $Ni_{31}Si_{12}$ NWs are 3–5 μm and 50–80 nm, respectively. As the growth time was prolonged to 6 h, the length and

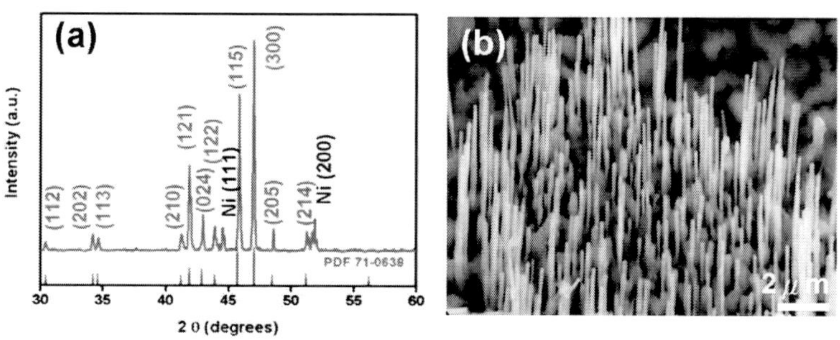

Figure 4.11 (a) XRD spectrum of the $NiSi_2$ NWs and (b) SEM image of well-aligned $Ni_{31}Si_{12}$ NW arrays.

diameter of $Ni_{31}Si_{12}$ NWs were increased to 6–7 μm and 120–130 nm, respectively. If the growth temperature was raised to 850 °C, vertically aligned Ni_2Si NWs were grown instead.[45]

Figure 4.12a shows TEM image of a single 80 nm $Ni_{31}S_{12}$ NW. The growth direction of $Ni_{31}Si_{12}$ NW is [100]. High-resolution TEM (HRTEM) image in Fig. 4.12b reveals that the NW is single-crystalline and defect-free. From the analysis of the EDS spectrum (not shown), the ratio of Ni and Si is close to 2.8:1.

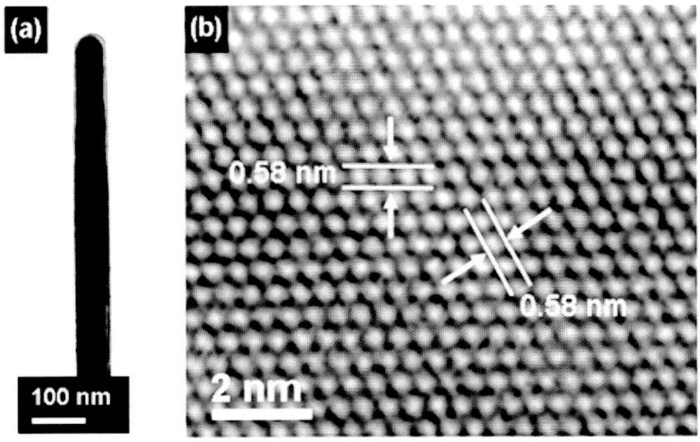

Figure 4.12 (a) TEM image of a $Ni_{31}Si_{12}$ NW and (b) HRTEM image of the NW after filtering noise.

The cross-section TEM image of Fig. 4.13a of a sample prepared with the aid of focused ion beam (FIB) apparatus and corresponding selected-area electron diffraction (SAED) patterns confirms that the NW and underlying film are of the same phase and growth direction of $Ni_{31}Si_{12}$. Fig. 4.13b shows the cross-section TEM image of the $Ni_{31}Si_{12}$/Ni interface and the corresponding SAED patterns of $Ni_{31}Si_{12}$ thin film and Ni substrate. The crystallographic orientation relationships between $Ni_{31}Si_{12}$ and Ni are

$$[010]Ni_{31}Si_{12}//[112]Ni, \text{ and } (006)\ Ni_{31}Si_{12}//(11\bar{1})\ Ni$$

In Fig. 4.13a, the orientation of $Ni_{31}Si_{12}$ thin film is (100). At the $Ni_{31}Si_{12}$/Ni interface, Ni(11-1) is parallel to $Ni_{31}Si_{12}$(006) with the interplanar spacings of 0.20344 and 0.2048 nm, respectively, and a small lattice mismatch of 0.66%. Furthermore, in the directions of $Ni_{31}Si_{12}$ [010] and Ni [112] perpendicular to the zone axis, the interplanar spacings are 0.57772 and 0.5756 nm, respectively. The lattice match is as small as 0.47%. Therefore, the $Ni_{31}Si_{12}$ NWs and the underlying silicide thin film are well matched to the grains of Ni substrate.

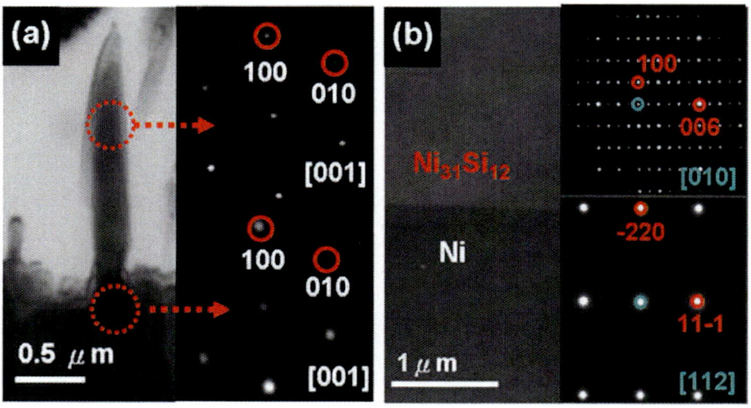

Figure 4.13 Cross-section images of (a) $Ni_{31}Si_{12}$ NW and underlying film with the corresponding SAED patterns and (b) $Ni_{31}Si_{12}$ film and underlying Ni substrate with the corresponding SAED patterns.

Electrical transport properties of the $Ni_{31}Si_{12}$ NWs have been measured by both two- and four-terminal *I–V* measurements. Pt was deposited as metal contacts with a FIB system[91] on the 500 nm SiO_2 film on Si substrate. From the linear *I–V* relationship shown in Fig. 4.14a, the contacts were found to be ohmic. The resistivities of the vertical $Ni_{31}Si_{12}$ NWs were measured to be 83 and 51 μΩ-cm by two- and four-terminal electrical measurements, respectively. The maximum current density (J_{max}) of $Ni_{31}Si_{12}$ NW, obtained by two-terminal measurement as shown in Fig. 4.14b, is about 9.8×10^7 A/cm^2.

For electron field emission (FE), NWs are ideal objects because of their special one-dimensional geometry. Figure 4.15 shows the current density (J) as a function of the applied field (E) in a current density–electric field (J–E) plot and a $\ln(J/E^2) - 1/E$ plot of the well-aligned $Ni_{31}Si_{12}$ NWs with a 200 μm separation between the anode and emitting surface. The turn-on field is defined as the applied field attained to a current density of 1 μA/cm and was found to be about 1.0 V/μm. In accordance with the Fowler–Nordheim (F–N) relationship, FE current from a metal or semiconductor is attributed to the tunneling of electrons from the material into vacuum through a potential barrier under the influence of an applied field. Here is a simplified equation to express the F–N relationship,

$$J = (A\beta^2E^2/\Phi)\exp(-B\Phi^{3/2}/\beta E)$$

where J is the current density, E is the applied electric field, and Φ is the work function. A and B are constants, corresponding to 1.56×10^{-10} [A(eV)V^{-2}] and 6.83×10^3 [eV$^{-3/2}$ (μm^{-1})] = 6.83×10^9 [V(eV)$^{-3/2}$m^{-1}], respectively. The F–N plot is shown in the inset in Fig. 4.5. The field-enhancement factor β was calculated to be about 3190 for $Ni_{31}Si_{12}$ NWs from the slope of the $\ln(J/E^2) - 1/E$ plot with a work function value of about 4.80 eV for $Ni_{31}Si_{12}$ NWs.[92] The high FE performance is ascribed to the aligned geometry, especially the long $Ni_{31}Si_{12}$ NWs. In addition, epitaxy between Ni silicide NWs, Ni silicide films and the metal substrates (Ni foil) also assists the electron transport on the side of emitting surface to enhance the FE performance.

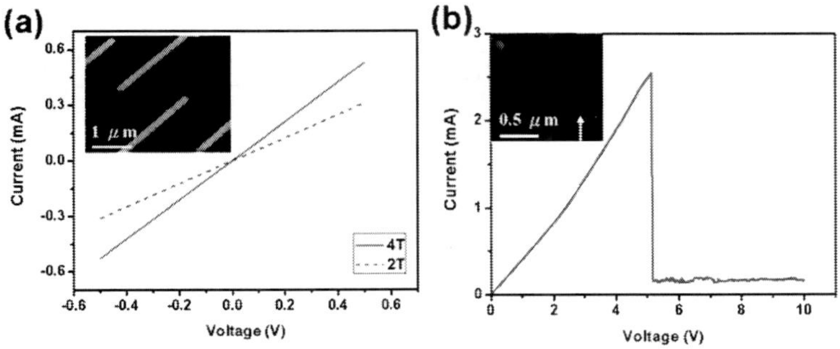

Figure 4.14 (a) I–V curve of $Ni_{31}Si_{12}$ NWs from four-terminal measurement and (b) two-terminal I–V measurement to obtain J_{max} at high applied voltages. Insets show the SEM images of the samples for I–V measurements.

The Si vapor from the source zone would react with the substrate of Ni foil to form single-crystal Ni silicide ($Ni_{31}Si_{12}$) films (Fig. 4.13a) first at the end zone in the furnace. The $Ni_{31}Si_{12}$ films would in turn serve as the catalytic surfaces to assist Si vapor to condense and grow $Ni_{31}Si_{12}$ NWs epitaxially on thin films.[38,93] The phase formation of Ni silicides was found to depend on the ratio of Ni and Si.[94,95] Ni atoms are known to be the dominant diffusing species in the Ni–Si reactions.[54] Consequently, Ni atoms from the Ni substrate can be considered as the excess component. As a result, Ni-rich silicide phase (Ni_2Si, $Ni_{31}Si_{12}$ and Ni_3Si) are prone to form. On the other hand, the supply of Si vapor would diminish with the distance from the Si source. It correlates well with the finding that Ni_2Si and $Ni_{31}Si_{12}$ phases were formed at the zones close and farther away from the source zone downstream, respectively.

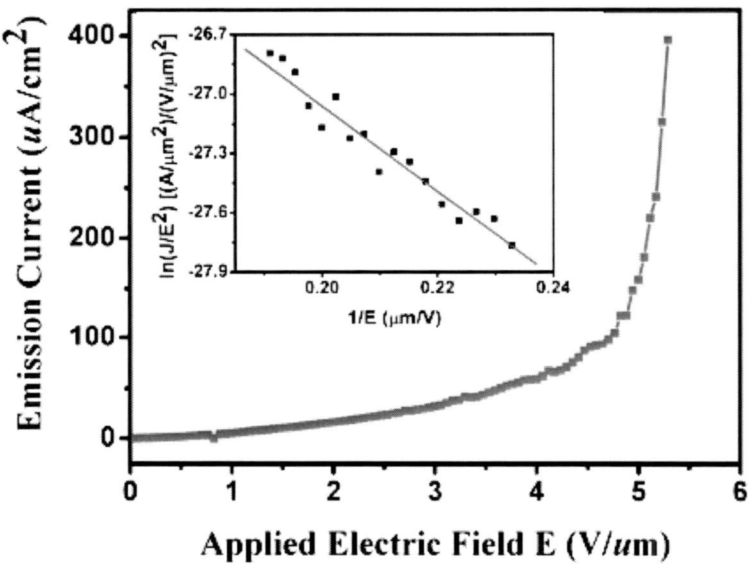

Figure 4.15 FE curve of $Ni_{31}Si_{12}$ NWs as a function of applied electric field. Insets show the corresponding $\ln(J/E^2) - 1/E$ plot.

In order to clarify whether larger Ni grains can lead to the growth of well-aligned Ni silicide NWs over larger areas, the Ni foils were annealed at 1100 °C for 6 h in Ar and H_2 (4:1) ambient to enlarge the size of Ni grains to about ~30 μm. As the temperature of the upstream and downstream of the furnace were increased to 1100, and 750 °C, vertically well aligned $Ni_{31}Si_{12}$ were found to grow over ~30 μm in diameter areas at 750 °C zones. It is conceivable that if a single-crystal Ni foil is used as the substrate, extensive growth over desired large area can be achieved.

In summary, vertically well-aligned single-crystal $Ni_{31}Si_{12}$ NW arrays have been successfully synthesized with a simple vapor transport and condensation method. Highly oriented and large area arrays of NWs were epitaxially grown on $Ni_{31}Si_{12}$ films, which were in turn preferentially formed on Ni foil substrates at the 750 °C zone. The length and width of Ni silicide NWs were found to increase with the growth time. The extent of the growth of NWs is correlated to the size of underlying Ni grains. The NWs can carry very high currents and possess excellent FE properties. The results show that the well-aligned single-crystal $Ni_{31}Si_{12}$ NWs are potentially useful as high-performance field emitters.

4.3.1.2 Growth of free-standing single-crystal $NiSi_2$ nanowires

Pure Ni foils (99.9% purity), cleaned in acetone, D.I. water, hexane, and dilute HNO_3, consecutively, were used as the substrates. After the cleaning process, the Ni foils were placed in an alumina boat and inserted into a three-zone quartz tube furnace with the pressure pumped down to about 1.0×10^{-5} Torr. Through mass flow controllers, the flow of 50 sccm Ar was kept constant and introduced as the carrier gas. A controlling valve was used to keep the pressure of the quartz tube fixed at 1 Torr. The three zones of the furnace were heated from room temperature to 1100 °C, 800 °C and 800 °C, respectively. Silicon powders (99.9%) in an alumina boat were positioned at the 1100 °C zone upstream in the quartz tube as the source. Ni foils were located downstream at the middle and end of zones at 800 °C. In addition, $NiCl_2$ powders were placed in front of the Ni substrates upstream. The reaction time was set to be 3 h unless otherwise mentioned. The furnace was then allowed to cool down to room temperature.

XRD spectrum (Fig. 4.16a) of the samples grown at the 800 °C zone indicate that all the diffraction peaks can be ascribed to cubic $NiSi_2$ (space group *Fm3m*(225), the lattice constants a = 0.5416 nm, JCPDS Card No. 43-0989) NWs and Ni substrate. No diffraction peaks from any other Ni silicide phase and impurities were detected. The diffraction peaks are strong and sharp indicating the high degree of crystallinity of as-grown samples. Figure 4.16b shows SEM image of the $NiSi_2$ NWs which are uniformly grown. The lengths and diameters of $NiSi_2$ NWs are several tens of micrometers and 40–80 nm, respectively.

Figure 4.17a shows low-magnification TEM image of a single 40 nm $NiSi_2$ NW. The growth direction of $NiSi_2$ NW is $[1\bar{1}0]$. HRTEM image of Fig. 4.17b also reveals that the NW is single-crystalline and defect-free. Inset is the corresponding fast Fourier transform (FFT) pattern confirming the phase formed and the growth direction. Figure 4.17c shows an EDS spectrum indicating that the ratio of Ni and Si is close to 1:2. The Cu and C peaks come from the carbon film covering Cu grid.

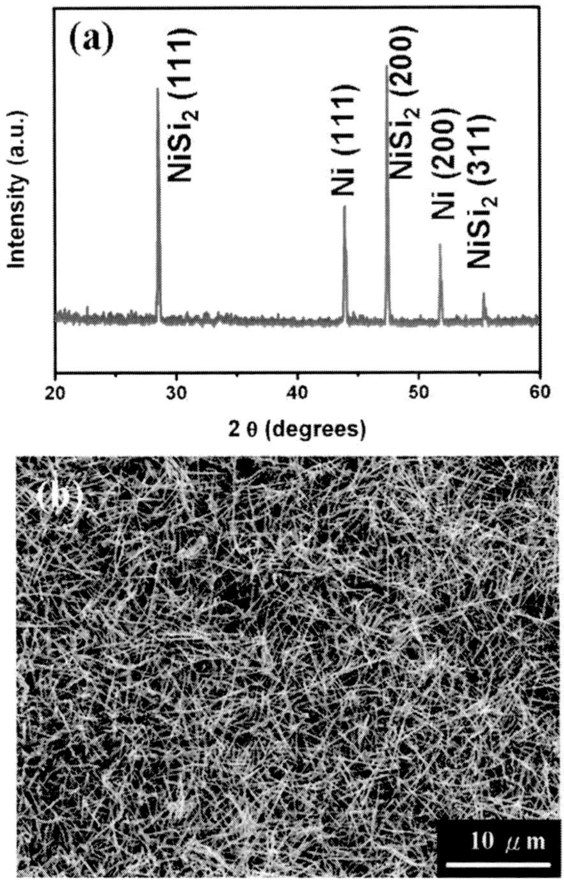

Figure 4.16 (a) XRD spectrum of the $NiSi_2$ NWs and (b) SEM image of the $NiSi_2$ NWs.

Electrical transport properties of the same $NiSi_2$ NWs have been measured by both two- and four-terminal I–V measurements. The Pt electrodes were fabricated in a focused-ion beam system (FIB, FEI Nova 600) using 30 kV Ga^+ ion beam onto NW on 500 nm SiO_2 film on Si substrate. The current involved was very small (0.99 pA) to minimize the damage to the NWs and implantation of Ga^+ ions.

The Pt electrode may contain some impurities. In addition, we have measured the electric resistances of the Pt electrode and found the resistances are very small (~100–200 Ω) by two-terminal measurement. For comparison, the resistances of the $NiSi_2$ NW obtained by two- and four-terminal measurement are about 400 and 1400 Ω, respectively. Four-terminal measurements were conducted to eliminate the impedance contribution of the contact and contact resistances.

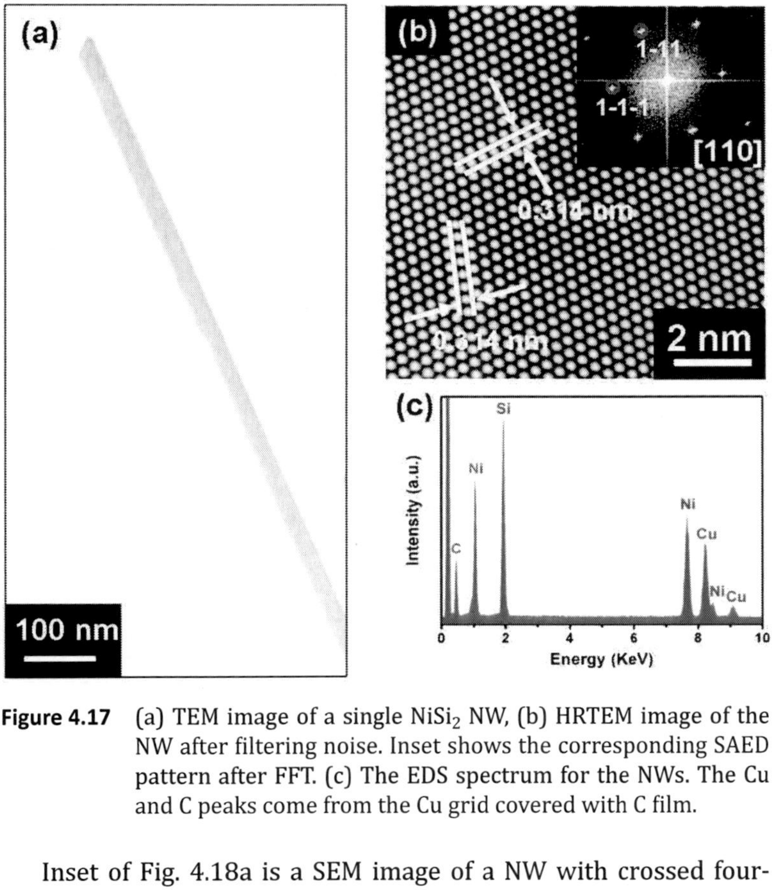

Figure 4.17 (a) TEM image of a single $NiSi_2$ NW, (b) HRTEM image of the NW after filtering noise. Inset shows the corresponding SAED pattern after FFT. (c) The EDS spectrum for the NWs. The Cu and C peaks come from the Cu grid covered with C film.

Inset of Fig. 4.18a is a SEM image of a NW with crossed four-terminal contacts. The contacts were found to be ohmic. The resistivities of the free-standing $NiSi_2$ NWs were measured to be 62 and 30 μΩ-cm by two- and four-terminal electrical measurements, respectively. The low value is lower than that of thin $NiSi_2$ film (~35 μΩ-cm)[1,96] as the NWs are single-crystalline and defect-free.

The maximum current density (J_{max}), obtained by two-terminal measurement and shown in Fig. 4.18b, is as high as 3.0×10^8 A/cm^2. The failure occurred in the middle of the NW indicating the breakdown occurred by melting via self-heating, reflecting the high-quality of single-crystalline $NiSi_2$ NWs.

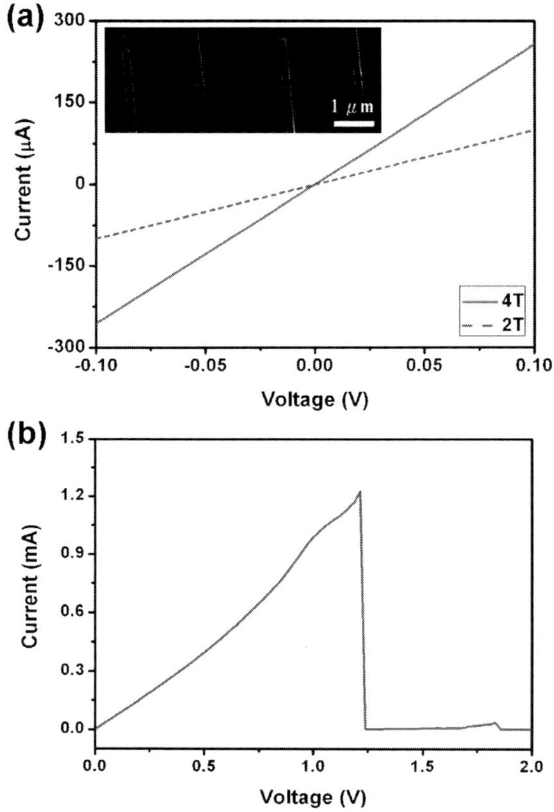

Figure 4.18 (a) *I–V* curve of $NiSi_2$ NWs from four-terminal measurement. Inset shows the SEM image of the sample. (b) J_{max} obtained at high applied voltage in two-terminal measurement.

For electron FE, NWs are ideal objects because of their special one-dimensional geometry. Figure 4.19 shows the current density (J) as a function of the applied field (E) in a current density–electric field (J–E) plot and a $\ln(J/E^2) - 1/E$ plot of the free-standing $NiSi_2$ NWs with a 200 μm separation between the anode and emitting surface. The turn-on field was defined as the applied field (E) attained to a current density of 1 μA/cm^2 and was found to be about 1.5 V/μm for $NiSi_2$. The F–N plot is shown in the inset of Fig. 4.19. The field-enhancement factor β can be calculated to be about 2970 for $NiSi_2$ NWs from the slope of the $\ln(J/E^2) - 1/E$ plot with a work function value of about 4.7 eV for $NiSi_2$ NWs.[97]

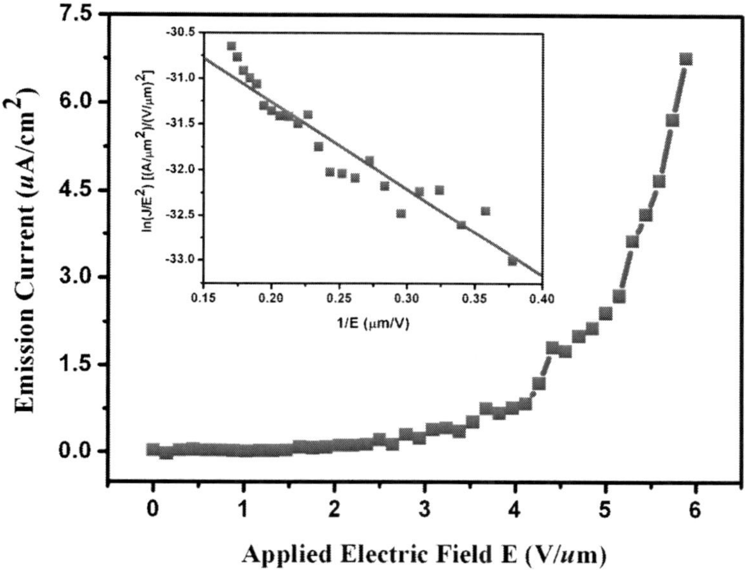

Figure 4.19 FE curve of $NiSi_2$ NWs as a function of applied electric field. Inset shows the corresponding $\ln(J/E^2) - 1/E$ plot.

In addition, since the NWs are a bit disorderedly attached on the Ni foil (as shown in Fig. 4.16b, and their lengths are not uniform, the measured emission current may only come from the protruded NWs. Furthermore, screening effect may also strongly affect the emission process.[98–101] As a result, the F–N plot is obviously nonlinear especially at high electric fields.

The phase formation of Ni silicides was found to depend on the ratio of Ni and Si. Ni atoms are known to be the dominant diffusing species in the Ni–Si reactions. The $NiSi_2$ phase nucleates first because it is the thermodynamically most stable phase in the thin film reactions of Ni and Si if the Si compositions are excessive.[82,84] While using Si powders as the sole source to grow Ni silicide NWs on Ni substrate, the NWs are exclusively of Ni-rich silicide phases.[45] By adding $NiCl_2$ powders in front of Ni substrate, the Si vapor from the source zone would react with the $NiCl_2$ vapor and substrate of Ni foil to form single-crystal $NiSi_2$ NWs at the middle zone in the furnace. The probable reaction pathways are[40,57,102]

$$2NiCl_2(g) + Ni(s) + 7Si(g) \leftrightarrow 3NiSi_2(s) + SiCl_4(g) \quad (1)$$

$$NiCl_2(g) + 2SiCl_4(g) \leftrightarrow NiSi_2(s) + 5Cl_2(g) \quad (2)$$

$$Ni(s) + Cl_2(g) \leftrightarrow NiCl_2(g) \quad (3)$$

It is worthwhile to mention that the $NiCl_2$ vapor plays two main roles. First of all, the vapor of $NiCl_2$ can react with the Si vapor to form the vapor of $SiCl_4$,[57,58,102,103] which is a highly reactive compound and can enhance the Si condensation. Second, the surplus chlorine gas can react with Ni substrate to form the vapor of $NiCl_2$, which can also react, with the vapor of Si to form $SiCl_4$. Consequently, both the enhanced condensation of Si and diminished supply of Ni contributed to the growth of Si-rich NWs. To further clarify the growth mechanism, different amounts of $NiCl_2$ powders (0.1–0.3 mg) were placed in front of the Ni substrates upstream. If the amounts of the $NiCl_2$ powders were more than 0.3 mg, the excess chlorine gas would lead to predominant etching so that no NWs were grown on Ni foils. On the other hand, if the quantity of $NiCl_2$ powders was less than 0.1 mg, its effect to assist the growth of Si-riched silicide NWs was much diminished. The optimal weight range was found to be between 0.15 and 0.2 mg.

In summary, free-standing single-crystal $NiSi_2$ NWs have been successfully synthesized on Ni foils with a chemical vapor reaction and condensation method. The growth of Si-rich silicide NWs was achieved with the addition of $NiCl_2$ powders in front of the Ni substrate upstream. The presence of $NiCl_2$ gas enhanced the condensation of Si and suppressed the supply of Ni atoms. The $NiSi_2$ NWs are several tens of micrometers and 40–80 nm in lengths and diameters, respectively. The resistivity of the $NiSi_2$ NWs was measured to be about 30 $\mu\Omega$-cm by four-terminal electrical measurement. The maximum failure current density of $NiSi_2$ NWs was measured to be as high as 3.0×10^8 A/cm^2. The turn-on field was measured to be about 1.5 V/μm and the field-enhancement factor β was calculated to be about 2970. The excellent electron transport and FE properties point to potential applications of $NiSi_2$ NWs in nanoelectronics.

4.3.2 Cobalt Silicide Nanostructures: Synthesis, Electron Transport, and Field-Emission Properties

One-dimensional nanostructures are expected to play an important role as interconnects and functional units in fabricating electronic, optoelectronic, and electromechanical devices.[83,104–111] Metal silicides constitute an important group of compounds that are used in silicon-based devices.[34,80,82] Cobalt silicide is known to be a high melting point material with very low resistivity, which offers the possibility of electrical interconnections in nanoscale.[112] Cobalt silicides have been shown to be promising cold cathode materials for emission due to the high thermal and chemical stability.[113] A number of metal silicide NWs, including $TaSi_2$, $TiSi_2$, Ni silicides NWs, were also reported to possess excellent FE properties.[38,45,46,87,114]

Various effective strategies have been developed for synthesis of silicide NWs.[43,44,112,115] $TiSi_2$ NWs were grown in a single direction on silicon surface by RDE.[43] CoSi and FeSi NWs were synthesized via CVD of single-source precursor and serve as building block of magnetic semiconducting $Fe_{1-x}Co_xSi$ NWs suitable for silicon-based spintronics.[41,44,116] $TaSi_2$ NWs have been synthesized on a Si substrate by annealing $NiSi_2$ films in an ambient containing Ta vapor.[38] Single-crystalline metal silicide NWs synthesized by metal halide precursors were reported recently.[40,57,58,117–119]

In this section, we describe the growth of Co_2Si and CoSi nanostructures by a spontaneous CVT growth method.[47] Both reaction temperature and vapor flow rate were found to critically influence the growth of nanostructures. In addition, a variety of structures and morphologies of cobalt silicide nanostructures, such as single-stem CoSi and Co_2Si NWs, 3-D CoSi NW networks, and aloe-like Co_2Si NWs, were grown under suitable conditions. The FE properties of various structures and morphology of CoSi NWs were compared.

Single-crystal Si (001) wafers were used as substrates. Prior to the loading into the two-zone furnace at the atmosphere pressure with an argon (Ar) flow, the wafer was cleaned with 1%-buffered HF. The samples were placed at the downstream and the cobalt chloride powder at the upstream. The powders were heated at 600 °C and the heating zone was set at temperatures of 800–1000 °C downstream. The two heating centers were at a distance of 20 cm apart. The precursor vapors were carried by the Ar flow and reacted with silicon substrates placed downstream for 2 h. The furnace was then allowed to cool down to room temperature.

In a previous work, the pathways for $CoCl_2$ vapor to react with Si substrate to form CoSi NWs were proposed as follows:[57]

$$2CoCl_2(g) + 3Si(s) \leftrightarrow 2CoSi(s) + SiCl_4(g) \quad (1)$$

$$CoCl_2(g) + SiCl_4(g) \leftrightarrow CoSi(s) + 3Cl_2(g)$$

The reactions for the formation of Co_2Si NWs are likely to follow similar pathways.

$$2CoCl_2(g) + 2Si(s) \leftrightarrow Co_2Si(s) + SiCl_4(g) \quad (2)$$

$$2CoCl_2(g) + SiCl_4(g) \leftrightarrow Co_2Si(s) + 4Cl_2(g)$$

The growth processes of cobalt silicide (CoSi, Co_2Si) NWs have two steps. First, the silicide particles formed on the substrate. Secondly, self-catalyzed NWs formed from the silicide particles.[102]

The optimum synthesis condition for single-stem CoSi NWs was realized with the heating zone set at 900 °C and Ar flow rate maintained at 300 sccm. A string of Si substrates was placed horizontally in the furnace. The temperatures of three regions in the heating zone were measured with a thermal couple attached to the samples. Apparently, the nanostructures of cobalt silicides (CoSi, Co_2Si) were influenced by a slight variation in vapor flux with the location in the furnace. Region A is the closest to the source boat, and the vapor flux is expected to be the highest. The region at the heating center with a decreased vapor flux and the highest temperature is referred to as region B.

For samples in the region A near the source boat encountered the highest vapor pressure at a temperature range of 750–830 °C. A high density of single-stem CoSi NWs was formed at 750 °C, as shown in Fig. 4.20a The diameters and lengths of the synthesized NWs are 40–80 nm and tens of micrometer, respectively. The TEM image of a single-stem NW grown at 750 °C in region A is illustrated in Fig. 4.20b The diffraction spots in the SAED pattern (not shown) can be ascribed to the cubic B20 type CoSi. The CoSi NW networks were synthesized at the particular zone with a higher temperature (780 °C) than that single-stem NW growth. The high yield of 3-D NW networks can be observed from the low-magnification SEM image and the morphology of 3-D NW networks can be clearly

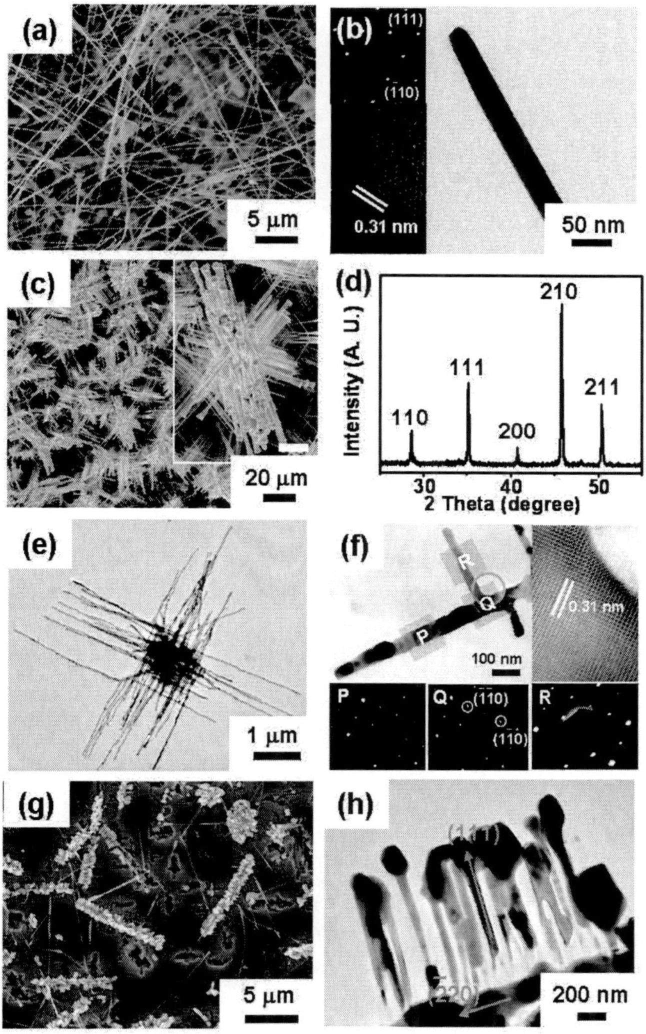

Figure 4.20 (a) Representative SEM image of CoSi NWs grown at 750 °C in region A. (b) TEM image of a CoSi NW grown at 750 °C in region A. (c) Low-magnification SEM images of the CoSi NW networks grown at 780 °C in region A. Inset: a high-magnification SEM image. The scale bar is 5 m. (d) XRD pattern from a sample of CoSi NW networks grown at 780 °C in region A. The TEM images (e) showing a complete 3-D NW network and (f) a broken NW network segment grown at 780 °C in region A. The insets are the SAEDs from the regions P, Q, and R. (g) SEM image of Co_2Si NWs formed in region B. (h) TEM images of aloe-like nanostructures.

demonstrated in a high-magnification image, as shown in Fig. 4.20c The size of each network is about 10–30 μm. The constituents of 3-D network structure are NWs with diameters around 100 nm in width and they intercross with each other to form the 3-D networks. The peaks of the XRD from the 3-D NW networks grown at 750 °C in region A shown in Fig. 4.20d can all be ascribed to the CoSi phase. The TEM images of a complete 3-D NW network and a broken NW network segment are shown in Figs. 4.20e and f, respectively. The angle of the junction is nearly equal to 90°. The SAED patterns were obtained from regions P, Q, and R, indicating that the entire network is single-crystalline CoSi. The NWs P, and R are grown along $[\bar{1}\,\bar{1}0]$ and $[1\bar{1}0]$ directions, respectively, and rotate by 90° with each other. HRTEM image of the intersections of the network further revealed that no dislocations were present to influence the branching growth, as shown in the inset of Fig. 4.20f.

The substrate in the region B was positioned at the comparatively high-temperature zone (880–900 °C). Two kinds of nanostructures were produced in the same area: single-stem NWs and aloe-like NWs. The diameter and length of the single-stem NWs are 50–100 nm and several micrometers, respectively, as illustrated in Fig. 4.20g SAED pattern (not shown) reveals that the NW is single-crystalline and could be ascribed to the orthorhombic Co_2Si structure. The size of each aloe-like NWs and the diameter of its branch are about 10 μm and 80–120 nm, respectively. The TEM image of aloe-like NW is shown in Fig. 4.20h The SAED patterns (not shown) obtained for the trunk, the head, and the branch of the aloe-like NWs can also be ascribed to orthorhombic Co_2Si structure. The branches and trunk of the aloe-like NWs were found to grow along 111 and $[\bar{2}20]$ directions, respectively.

FE measurements were carried out in a vacuum chamber pumped to ~1.0 × 10^{-5} Torr in pressure at room temperature and with an anode to sample distance of 180 μm. The turn-on field and threshold field are defined to be the fields required to produce a current density of 10 μA/cm^2 and 100 μA/cm^2, respectively. The three types of CoSi nanostructure are the high-density single-stem NWs, the 3-D NW networks, and the straggled single-stem NWs. The FE data were taken from 1 cm^2 of size samples and represented the average effects from different nanostructures. For 3-D NW networks, the distances between the NW emission tips to the anode were less than 180 μm. For the same samples, the measurements were made repeatedly.

Stable and reproducible J–E characteristics were only obtained for single-stem CoSi NWs. The emission current density, J, as a function of applied field, E is depicted in Fig. 4.21. The turn-on fields of the high-density single-stem NWs and the 3-D NW networks are 1.42 V/μm and 2.20 V/μm. The threshold field of the high-density single-stem NWs is 2.05 eV. The values are much lower than the reported values of $TaSi_2$, $TiSi_2$, and Ni silicides NWs.[38,45,46,87,114] The turn-on field of the straggled single-stem NWs was not stable and the best profile is shown in Fig. 4.21. Compared to the 3-D network structure, the turn-on field is lower and the emission current is higher for the single-stem NWs. The F–N plot of $\ln(J/E^2) - 1/E$ of the single-stem CoSi NWs is shown in the inset of Fig. 4.21. By determining the slope of the F–N plot, the field enhancement factor β of single-stem NWs was calculated to be 10^3, which is higher than other silicide NW emitters mentioned above.[38,45,46,87,114] The FE of single-stem NWs is more efficient, homogenous, and stable, as illustrated in Fig. 4.21.

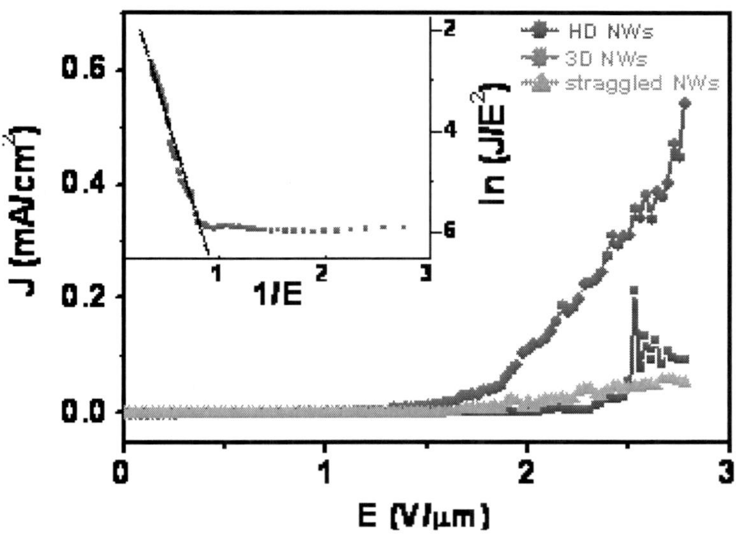

Figure 4.21 FE curves of the three type nanostructures: high-density single-stem NWs (HD NWs), the 3-D NWs networks, and the straggled single-stem NWs. Inset: the F–N plot of high-density single-stem NWs.

In addition to the FE properties, the electrical properties of the CoSi single-stem NWs were also measured. The electrical measurements were carried out for single NWs with various diameters. Pt electrodes were deposited at both ends of NWs with a focus ion beam system. The resistivity reduces with decreasing radius of NWs as illustrated in Fig. 4.22. The resistivities and diameters of the NWs are 287, 181, 100, 27 μΩ-cm and 43, 34, 23, 15 nm, respectively. The increase in conductivity with the decreasing diameter of NWs is rather intriguing. It is attributed to the enhanced surface conductivity since the surface-to-volume ratio increases with the decreasing diameter. A previous work showed that the resistivity of Pt_2Si NWs is about half the reported resistivity of thick Pt_2Si films. Additional conduction paths through surface or interface states on NWs were suggested to lead to the decrease in resistivity.[120]

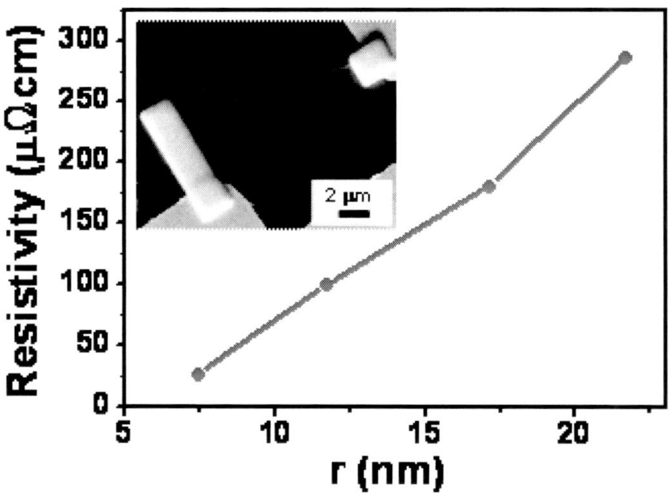

Figure 4.22 The curve depicting resistivity versus diameter of NWs. Insets: the SEM images of the measurement structure.

A variety of cobalt silicide (CoSi, Co_2Si) nanostructures were grown by a CVT method. Both reaction temperature and vapor flow rate were found to critically influence the growth of nanostructures. Notably, novel three-dimensional (3-D) CoSi NW networks and aloe-like Co_2Si NWs were grown under specific conditions in addition to NWs. FE and electrical property measurements of CoSi NWs show excellent performance with low turn-on field and good conductance. The CoSi NWs are potentially useful in nanoelectronics devices.

4.3.3 Synthesis and Properties of the Low-Resistivity $TiSi_2$ Nanowires Grown with Metal Fluoride Precursor

Among metallic silicides, C54-$TiSi_2$ is well known for its excellent electrical properties and thermal stability. Xiang *et al.* synthesized the $TiSi_2$ NWs on silicon by physical vapor deposition (PVD) using titanium powder as the source.[114] On the other hand, an amorphous oxide layer was found to form outside the $TiSi_2$ NWs and affected the electrical and FE properties.[121] Zhou *et al.* synthesized the $TiSi_2$ nanonets by the CVD with premixed SiH_4 and $TiCl_4$ precursor in the hydrogen gas to reduce the influence of the oxide layer and relatively low resistivity of nanonets was found.[122,123] Metal halide precursors [57,58,102,124] had the advantages of lower melting point and higher activity than pure metal powders. They facilitated the growth of metal silicide NWs by a simple vapor transport and condensation method. Seo *et al.* fabricated the CoSi and $CrSi_2$ NWs with thermal decomposition of $CoCl_2$ and $CrCl_2$ precursors on silicon substrate, respectively.[57,58] Quyang *et al.* synthesized the FeSi NWs with the $FeCl_3$ precursor.[102] In the section, the single-crystalline C54-$TiSi_2$ NWs were synthesized by a simple vapor transport and condensation method in air with titanium tetrafluoride (TiF_4) precursor. Remarkable FE and electrical transport properties were obtained for the C54-$TiSi_2$ NWs.[48]

The $TiSi_2$ NWs were grown on the silicon substrate in a two-zone quartz tube furnace. TiF_4 powder (0.3 g) was used as the source and the Si (001) served as the substrate, placed at the upstream zone 800 °C and downstream zone of 900 °C, respectively. Before reaction, the TiF4 power was hated at 100 °C for 30 min to remove the residual hydrate in the powder. The 100 sccm Ar as carrier gas transported the precursor vapor to react with silicon substrate at the atmosphere. The reaction time was set to be 3 h. Since TiF_4 is a good etcher of silicon, the precursor can clean the substrate and inhibit the formation of oxide during the reactions. In order to clarify the growth mechanism, the reaction time was varied ranging from 0.5 to 3 h.

Figures 4.23a and b show high-density NWs grown on the silicon substrate at 900 °C for 3 h. The diameters and lengths of the NW are 20–50 nm and 1–2 μm, respectively. It can be seen from Fig. 4.23c that the NWs grown above a seed layer of about 1 μm in thickness on the silicon substrate. All peaks in XRD spectrum shown in Fig. 4.23d can be ascribed to the orthorhombic structure C54-$TiSi_2$ (JCPDS card, No-71-0187, the lattice constant a = 0.8267 nm, b = 0.48 nm, c = 0.855 nm) and no silicon signal was detected.

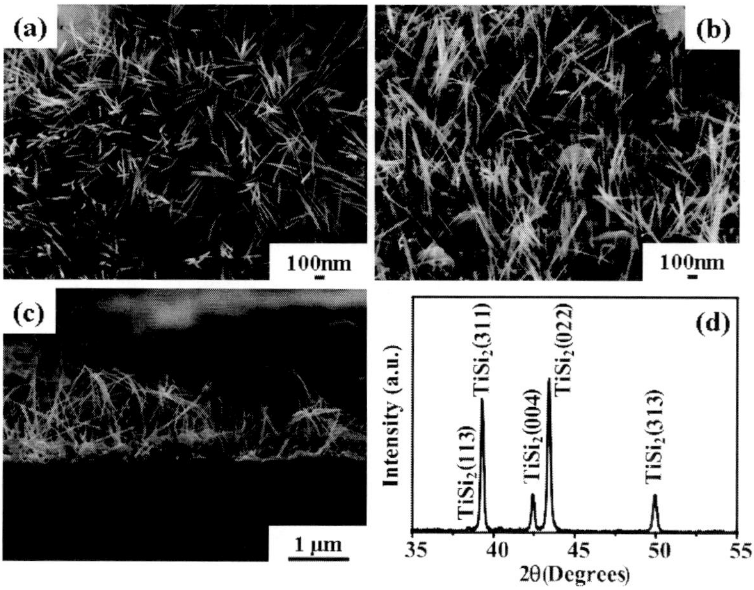

Figure 4.23 SEM images showing the (a) top and (b) tilted view of the $TiSi_2$ NWs grown on Si(001) at 900 °C for 3 h. (c) Cross-section SEM image of as-grown NWs. (d) XRD spectrum of the $TiSi_2$ NWs.

Figure 4.24a shows the TEM image of a NW and inset is the corresponding SAED pattern. The pattern can be ascribed to the single-crystalline orthorhombic C54-$TiSi_2$ in 101 zone axis and the growth direction is along $[10\bar{1}]$. The results obtained from the analysis of the HRTEM image shown in Fig. 4.24b are consistent with that of SAED. In addition, from the TEM/EDS (Fig. 4.24c) spectrum of the NW, both Ti and Si signals were detected and the composition ratio was determined to be close to 1:2.

In order to clarify the growth mechanism of NWs, the growth was conducted for different reaction time. Figures 4.25a–c show SEM images of the structures grown for 0.5, 1.5 and 3 h, respectively. As shown in Fig. 4.25a, a $TiSi_2$ seed layer was grown on silicon substrate after 0.5 h. Subsequently, protruded structure emerged after 1.5 h growth, as shown in Fig. 4.25b. A high-density $TiSi_2$ NWs, shown in Fig. 4.25c, was formed on the seed layer after 3 h growth. The XRD spectra of the products formed with 0.5–3 h growth time (Fig. 4.25d) indicated that only C54-$TiSi_2$ phase was formed. As the silicon signal was not detected, it was inferred that the silicon substrate was covered a C54-$TiSi_2$ seed layer, consistent with that observed from the cross-section image shown in Fig. 4.23c.

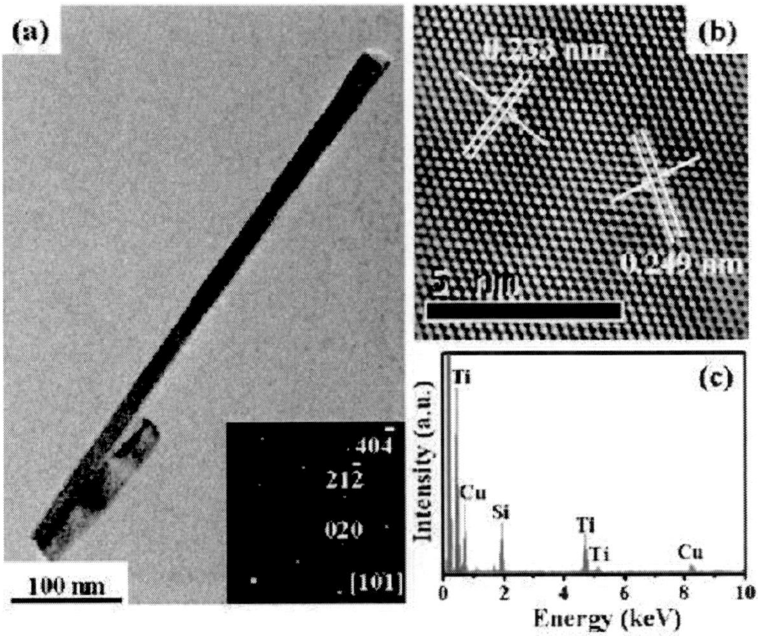

Figure 4.24 (a) TEM image and the corresponding electron diffraction pattern of a $TiSi_2$ NW. The growth direction is along $[10\bar{1}]$. (b) HRTEM image of the single-crystal $TiSi_2$ NW and (c) TEM/EDX spectrum of a $TiSi_2$ NW.

The reactions of TiF_4 precursor vapor with the silicon substrate at the high-temperature zone to form the $TiSi_2$ NWs may follow two reaction pathways.

$$TiF_{4(g)} + 3Si_{(s)} \longleftrightarrow TiSi_{2(s)} + SiF_{4(g)}$$

$$TiF_{4(g)} + 2SiF_{4(g)} \longleftrightarrow TiSi_{2(s)} + 6F_{2(g)}$$

No oxide layer was detected outside the $TiSi_2$ NWs confirming the advantage of metal fluoride precursors to inhibit the formation of oxide.

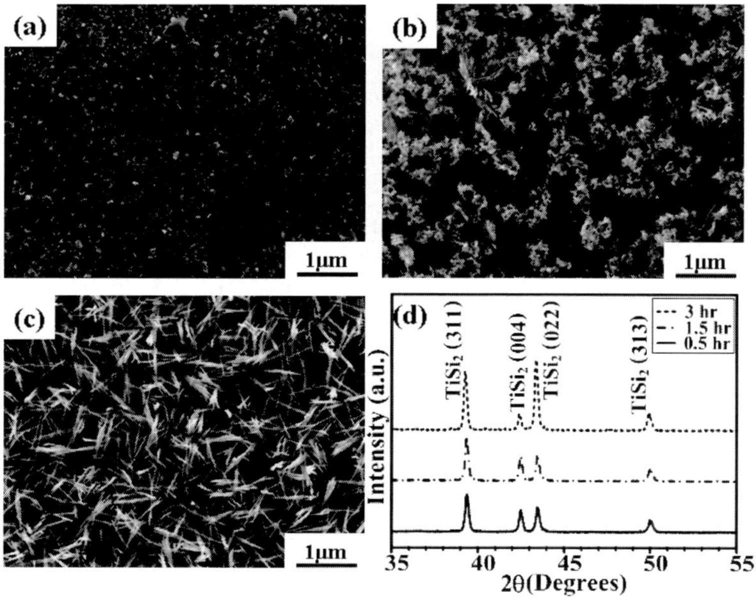

Figure 4.25 SEM image of the $TiSi_2$ NWs grown on the silicon at 900 °C for (a) 0.5, (b) 1.5, and (c) 3 h. (d) XRD spectra of the NWs grown for different periods of time.

Figure 4.26 shows the current density (J) as a function of the applied electric field (E). The turn-on field of the $TiSi_2$ NWs is 8 V/μm when the current density attained 1 μA/cm^2. The lower value compared to that previously reported is correlated to the absence of oxide layer outside the NWs.[114] For FE, the current versus electric field relationship can usually expressed by the simplified F–N equation. Utilizing the F–N equation and substituting the work function of the $TiSi_2$ NWs (Φ = 4.6 eV), the field amplification factor was calculated to be about 800. The inset of Fig. 4.26 for $\ln(J/E^2)$ versus $1/E$ exhibits a straight line, indicating that the FE is consistent with the electron tunneling behavior.

Figure 4.27 shows the plot of emission current density versus time of the $TiSi_2$ NWs under an electric field of 9 V/μm indicating that the emission current is very steady over the test period of 15 h. The emission current exhibits higher stability than that reported previously.[114] Furthermore, the current density was not fluctuated due to the gas absorption in the beginning. SEM images of $TiSi_2$

NWs before and after applying the electrical field of 9 V/μm for 15 h showed no obvious change, confirming the robustness of the NWs.

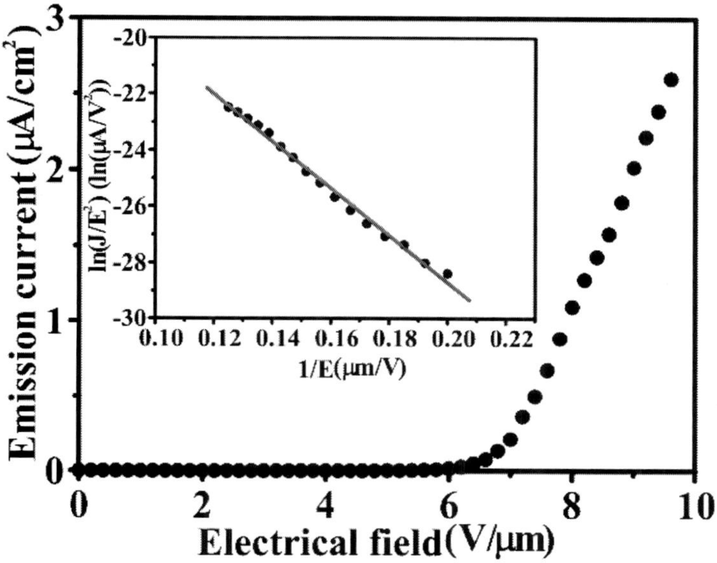

Figure 4.26 FE curve of the $TiSi_2$ NWs grown on the Si at 900 °C for 3 h. Inset shows the corresponding $\ln(J/E^2) - 1/E$ plot.

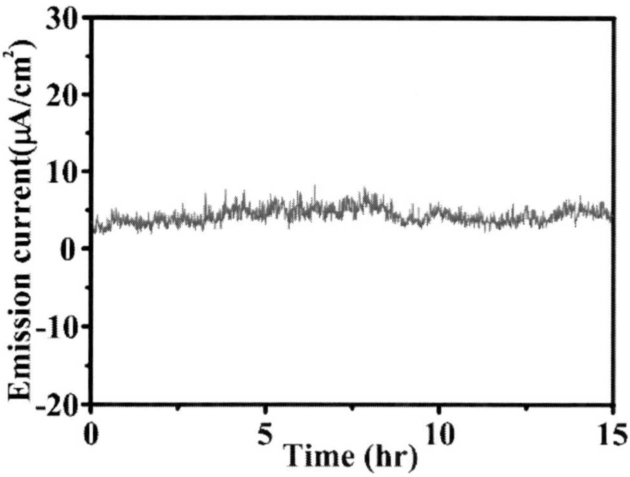

Figure 4.27 Plot of the emission current density versus stressing time under an electric field of 9 V/μm for up to 15 h.

In this study, the two-probe measurement was performed. The measurement configuration is shown in the inset of Fig. 4.28. The length and diameter of the shown NW are 530 and 50 nm, respectively. The *I–V* curve seen in Fig. 4.28 indicates the presence of an ohmic contact and a resistance of about 80 Ω at 300 K. The resistivity of the C54-$TiSi_2$ NW was calculated to be 30 μΩ-cm and is lower than that previously reported.[121] However, it is still higher than the bulk resistivity (15–20 μΩ-cm) and is attributed to the contact resistance. To eliminate the effect of contact resistance, four-probe measurement would be desirable. However, owing to relative shortness of the grown NWs, the conduction of four-probe measurement was not feasible. The maximum transport current (J_{max}) of the C54-$TiSi_2$ NW was measured for the 50 nm-diameter NW to be 5.71 mA before failure. The J_{max} can be calculated to be as high as 3×10^8 A/cm^2. The high value indicates the high quality of single-crystalline C54-$TiSi_2$ NWs. The SEM image shown in the inset of Fig. 4.29 reveals that the breakdown occurred at the location close to the middle of the NW, suggesting that the failure is due to the self-heating.[34]

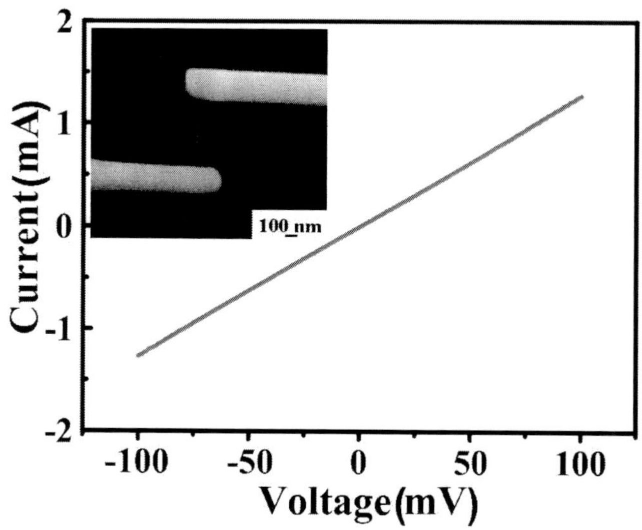

Figure 4.28 Electron transport properties of the $TiSi_2$ NWs at 300 K obtained with two-probe measurement. Inset is the SEM image of the NW placed on the silicon oxide and connected by the deposited platinum.

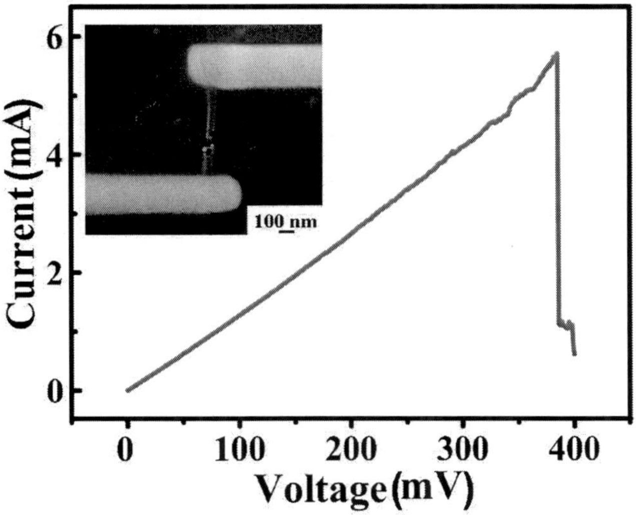

Figure 4.29 Plot of transport current versus applied voltage of the $TiSi_2$ NWs at large applied voltages. Inset is the SEM image of the NW after breakdown.

In conclusion, single-crystalline C54-$TiSi_2$ NWs have been successfully synthesized on silicon substrate with TiF_4 precursor by a vapor transport and condensation method in air. For the FE properties, $TiSi_2$ NWs exhibited the turn-on voltage of 8 V/μm and the field amplification factor of about 800. The FE current under the stress of high electric field was found to be very stable. Two-probe measurement yielded a low resistivity of value of 30 μΩ-cm for the $TiSi_2$ NWs. The failure current density was calculated to be as high as 3×10^8 A/cm^2. The facile growth combined with the excellent electrical and FE properties of the $TiSi_2$ NWs is conductive to their potential applications in the interconnecting electronic architectures and silicon-based FE devices.

4.3.4 Ti_5Si_4 Nanobats with Excellent Field-Emission Properties

Among various NWs, silicide NWs had excellent properties of low resistivity,[41,85,102,125–127] high stability, desirable magnetism,[128–129] and high compatibility with the processing of silicon-based microelectronics.[59,83,130] As the dimensions of the microelectronic devices continue to shrink, nanostructured silicides will be required in the future. For the sub-100 nm devices, $TiSi_2$, $CoSi_2$, and NiSi have been the competing silicides as contacts and interconnects.[80] Consequently, the research on the silicide NWs is of significant and fundamental importance. Understanding the formation mechanism of silicide NWs is also a crucial step toward practical application.[43,54,55] Among the titanium silicide NWs, C54-$TiSi_2$ has been one of the promising FE materials due to its thermal and chemical stability.[114] From the phase diagram, there are five titanium silicide phases. However, scarce related research has been conducted on the Ti-rich phase including Ti_3Si, Ti_5Si_4, and Ti_5Si_3 and the basic properties of these phases were largely unknown.[131–134] In this section, the titanium foil was served as the substrate to synthesize the Ti-rich silicide phase bat-shaped nanorods (nanobats). Electrical resistivity, work function, and FE properties of the nanobats were measured. The growth of the titanium silicide nanostructure on the metal substrate is advantageous in the FE applications compared with the silicon substrate since the voltage drops between the nanostructures and the substrate are nil.

The Ti_5Si_4 nanobats were grown on the titanium through a vapor transport and condensation method in the two zones alumina tube. Titanium foil (0.25 mm thickness and 99.6 % purity) was cut to 1 cm × 1 cm pieces as the substrates. The samples pieces were dipped in HNO_3 for 30 sec to remove the native oxide layer on titanium and followed by cleaning sequentially in ethanol and DI water. 2 g silicon powder (99.8 % purity) was used as the source. The Si powder and substrate were kept at 1100 °C and 900 °C, respectively. The carrier gas was a mixture of 95 sccm Ar mixed and 5 sccm H_2 with a total pressure of 1 Torr. Hydrogen was used to reduce the residual oxygen in the chamber to inhibit the formation of oxide. The growth time ranged from 1 to 3 h.

A high density of nanobats was grown on the Ti substrate as shown in Fig. 4.30a and b. The diameters of the head and tail are 25 nm and 15 nm, respectively, and the average length is about 350 nm. The XRD spectrum peaks can all be ascribed to the tetragonal structure Ti_5Si_4 and Ti substrate, as shown in Fig. 4.30c.[52]

Figure 4.31a shows the TEM image of a nanobat. The head was not planar and appeared to be curved. Inset is a SAED pattern. The pattern was analyzed to correspond to [110] zone axis of the tetragonal Ti_5Si_4 phase with lattice constants a = 0.671 nm and c = 1.217 nm (JCPDS card, No-89-2434) with the growth direction along [001]. Analysis of the lattice image, shown in Fig. 4.31b is consistent with that of SAED analysis. In addition, from the EDS spectrum shown in Fig. 4.31c, Both Ti and Si were detected to be present in the nanobat. In contrast to the Cu nanobats with five-fold symmetry and multiple twins, the Ti_5Si_4 nanobats are single-crystalline.[135]

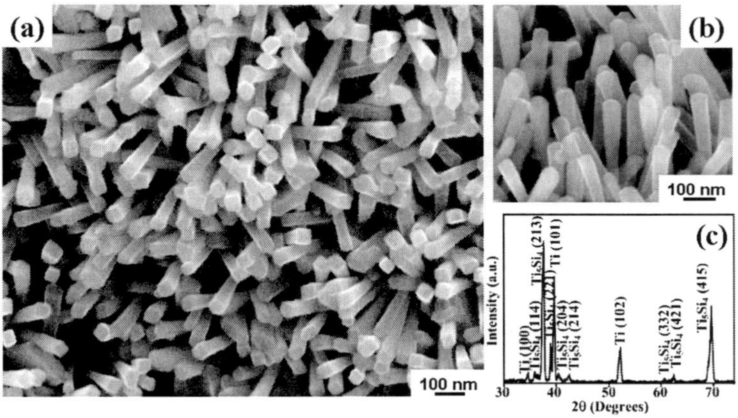

Figure 4.30 (a) Top-view and (b) side-view Field Emission Scanning Electron Microscope (FESEM) images of the Ti_5Si_4 nanobats grown on the titanium foil at 900 °C for 3 h. (c) XRD spectrum of the Ti_5Si_4 nanobats.

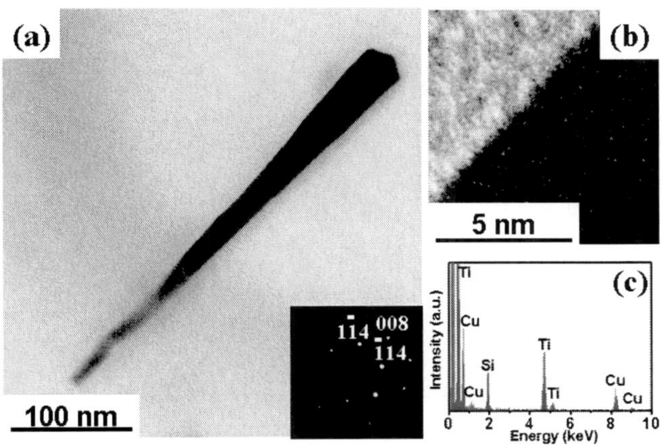

Figure 4.31 (a) TEM image and diffraction pattern of a Ti_5Si_4 nanobat. The growth direction is along [001]. (b) HRTEM image of a single-crystal Ti_5Si_4 nanobat and (c) TEM/EDX spectrum.

In order to clarify the growth mechanism, the growth time was varied from 1 to 3 h. SEM image in Fig. 4.32a shows the formation of protruded bumps after 1 h. Bat-shaped nanorods emerged after 2 h, as shown in Fig. 4.32b. After 3 h reaction, the nanorods grew longer and the bat-shaped morphology maintained as shown in Fig. 4.32c. A schematic plot of growth process is shown in Fig. 4.32d. The variation in the diameters of the nanobats is attributed to the fluctuation of the size protruded bumps, as seen on the substrate after 1 h reaction at 900 °C (Fig. 4.32a).

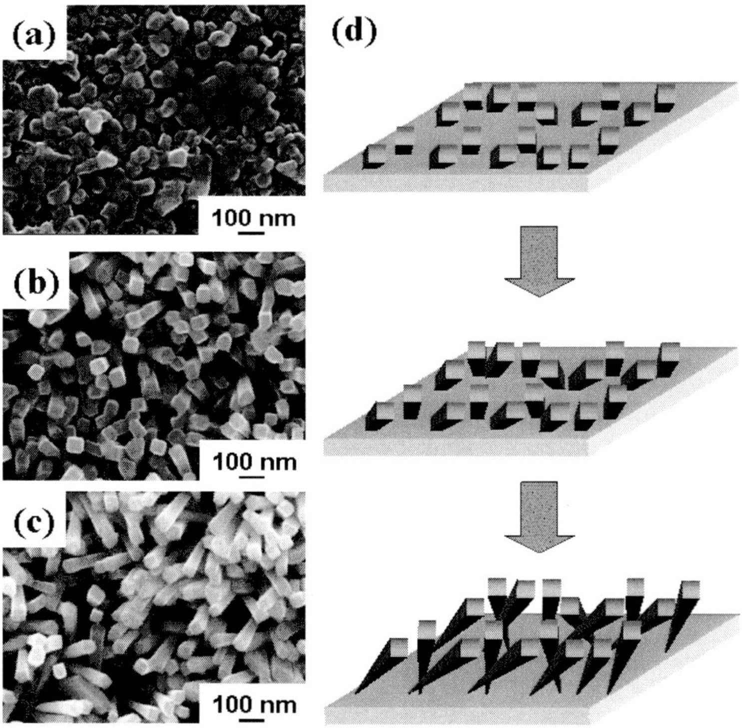

Figure 4.32 FESEM image of the Ti_5Si_4 nanobats grown on the titanium foil at 900 °C for (a) 1, (b) 2, and (c) 3 h. (d) The schematic plots of the formation of the nanobats.

Initially, the silicon vapor reacted with the titanium substrate and formed the Si–Ti alloy islands. Continuous feeding of silicon to the quasi-liquid Si–Ti islands led to the 1-D growth of the Ti_5Si_4 phase. This growth process can be defined as self-catalytic growth. On the other hand, as the 1-D growth continued, the supply of silicon vapor was diminished with the consumption of Si powder. As a result, the dimension of the roots of nanorods became smaller which led to bat-shaped morphology. The mechanism is similar to that proposed for TiSi nanopins in a previous work.[136]

The FE properties of the Ti_5Si_4 nanobats were measured on the parallel plate with a spacing of 180 m and the contact area was 0.25 cm^2 at a pressure of 1×10^{-5} Torr. The emission current was recorded with the applying voltage at a step of 5 V. Figure 4.33 shows the current density as a function of the applied voltage. The turn-on voltage of the Ti_5Si_4 nanobats was assessed to be 1.47 V/μm with the current density reaching 1 A/cm^2. The $\ln(J/E^2) - 1/E$ plot showed the straight line, indicating that it was consistent with the F–N behavior. The calculation was performed using the simplified F–N equation:

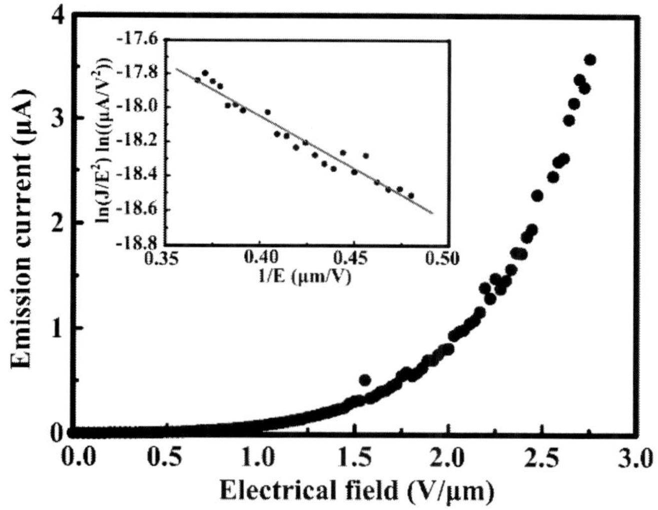

Figure 4.33 *I–V* curves of FE of the Ti_5Si_4 nanobats grown on the titanium foil. Inset is the F–N plot of the emission of the nanobats.

Utilizing the F–N equation and substituting the work function of the Ti_5Si_4 nanobats (Φ = 5.6 eV), The field amplification, calculate to be 1350, was higher than or comparable to those of other 1-D nanomaterials.[137,138] The remarkable FE properties of Ti_5Si_4 nanobats are attributed to the growth on the titanium foil in addition to their intrinsic properties. Ti is a good conductor, compared to silicon, the electron transport on the side of emitting surface was assisted by the nil low voltage across the nanobat/Ti interfaces to enhance the FE performance.[45,46] The low turn-on voltage and high field-enhancement factor are advantageous for application as the FE sources of flat plate display and vacuum nanoelectronics devices.

For electron transport property measurement, Pt was selected as the contact metal with a work function of 5.6 eV, which is the same as that of the Ti_5Si_4 (5.6 eV), and served as good ohmic contacts. The nanobats are relatively short so that the four-probe measurement was not amenable. The diameters of the nanobats were selected to be larger than the average value for the ease of positioning in the electrical measurements. As shown in the inset of Fig. 4.34, the head, tail and length of the measured nanobat are 90, 70, and 300 nm, respectively with an average diameter of 80 nm. The *I–V* plot seen in Fig. 4.34 indicates the presence of an ohmic contact and a resistance of about 276 Ω was obtained for the nanobat at 300 K. The resistivity of the Ti_5Si_4 was therefore calculated to be 348 μΩ-cm. Other measurements yielded essentially the same value of the resistivity.

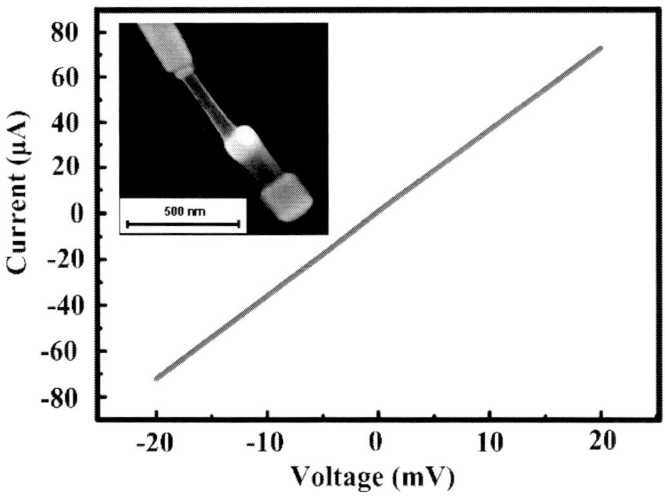

Figure 4.34 Electrical transport property of the Ti_5Si_4 nanobats at 300 K with two-probe measurement. Inset is the SEM image of a nanobat on the silicon oxide connected by the deposited platinum.

In summary, Ti_5Si_4 nanobats have been successfully synthesized on titanium foil by a vapor and condensation method without catalysts. The growth mechanism of nanobats was clarified to be of self-catalytic growth. The bat-shaped morphology is attributed to the diminishing supply of consumed Si powder source. The electrical resistivity (348 μΩ-cm) and work function (5.6 eV) of the Ti_5Si_4 phase were measured for the first time. For FE properties, Ti_5Si_4 nanobats possess a low turn-on voltage of 1.47 V/μm and a field amplification factor as high as 1350. The growth of the titanium silicide nanostructures on the metallic substrate is advantageous for FE applications since the voltage drop across the nanobat/substrate interface is nil. The excellent FE properties of Ti_5Si_4 nanobats may lead to applications as emitting sources in flat-panel displays and vacuum nanoelectronics devices.

4.4 Formation of Silicide/Si/Silicide Nano-Heterostructures from Si Nanowires

The fabrication of silicide NWs here is to use the contact between metals and Si NWs, and annealing. Since metals are the dominant diffusion species in the reactions, metal atoms will diffuse into Si. Silicide will start nucleation and growth at where supersaturation is reached, transforming Si NWs to silicide NWs. If nucleation and growth occur at several places, Si NWs would become silicide/Si/silicide heterostructure NWs. From Si NWs, Wu *et al.* successfully fabricated NiSi NWs and NiSi/Si/NiSi heterostructure NWs with remarkable electrical transport properties.[34] In this section, we will also discuss point-contact reactions between metal and Si NWs, where silicide NWs and silicide/Si/silicide heterostructure NWs can be made.

4.4.1 Controlled Growth of Atomic-Scale Si Layer with Huge Strain in the Nano-Heterostructure NiSi/Si/NiSi through Point-Contact Reaction between Nanowires of Si and Ni and Reactive Epitaxial Growth

As the end of the roadmap approaches for VLSI of transistors on Si wafers, nanoscale transistor devices based on Si NWs are expected to have significant potential and wide interest. To realize this potential, nanoscale device elements such as ohmic contacts and gates on

Si NWs must be developed,[80] the formation of which requires a systematic study of chemical reactions in the nanoscale. The point-contact reactions between Si NWs and Ni NWs were investigated.[53,54] Ni and Si were selected for the study because the monosilicide NiSi is one of the three silicides that have the lowest resistivity for applications in shallow junction devices.[139] In addition, single-crystal NiSi NWs can be scaled to ultrasmall dimension without degradation of electrical conductivity and exhibit remarkably high maximum transport current.[34] Si NWs were prepared on a Si wafer by the VLS method with nano Au dots as nucleation sites for single-crystal Si NWs with a 111 growth direction.[140,141] Polycrystalline Ni NWs were synthesized via the anodic aluminum oxidation (AAO) method and stored in isoproponal.[142] The Ni NWs and Si NWs have diameters ranged from 10 nm to 40 nm and lengths of a few microns. The Si and Ni NWs were stored in solutions. To prepare point-contact samples, droplets of both solutions are put on Si grids having a square opening covered with a window of glassy Si_3N_4 films, as shown in Fig. 4.35a. The thickness of the glassy film is about 20 nm so that it is transparent to the electron beam in the microscope and does not interfere with the images of Si and Ni NWs. The samples were dried under light bulbs.

Figures 35b–f show a time-lapsed series of *in situ* TEM images capturing the growth of a bamboo-type NiSi grain within a straight Si NW at 700 °C. Based on the HRTEM images and selected-area diffraction patterns, we determined the structure of the material to be single-crystal NiSi. Figure 4.35g is a schematic diagram depicting the growth of NiSi, in which the Ni atoms dissolve and diffuse interstitially in Si[143] and stop at the ends of the Si NW, thereby nucleating growth of NiSi to form a NiSi/Si/NiSi heterostructure. In point-contact reaction, growth of single-crystal NiSi started from both ends of the Si NW with an activation energy of 1.25 eV/atom rather than from the point-contact with an activation energy of 1.7 eV/atom. This is attributed to the ease of nucleation and fast diffusion of Ni atoms in Si towards the ends.[54]

Figure 4.36 shows the linear growth behavior of the NiSi NW in the Si NW of 20 nm in diameter over the temperatures ranged from 500 °C to 650 °C. The activation energy of the epitaxial growth was determined to be 1.25 eV/atom, compared to the activation energy of interstitial diffusion of Ni in Si of about 0.47 eV/atom,[144] indicating that the growth may be interfacial-reaction-controlled.

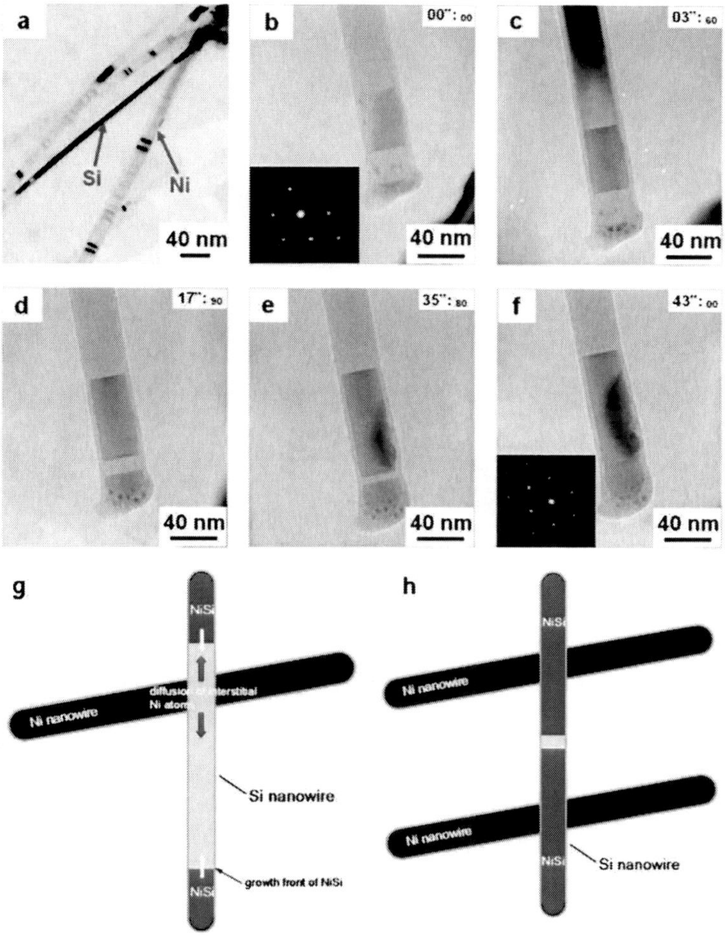

Figure 4.35 Overview of NiSi formation within Si NWs by point-contact reaction. (a) A TEM image of Si and Ni NWs dispersed on a Si_3N_4 membrane. (b to f) Sequence of *in situ* TEM images depicting the growth of a bamboo-type grain of NiSi within a Si NW at 700 °C. Inset in (b): Selected-area diffraction pattern showing the [1–1 0] zone axis of Si. Inset in (f): Selected-area diffraction pattern showing the [1–1 2] zone axis of NiSi. The time of the image capture is given in the rectangular box at the upper-right corner. The first two numbers are in unit of seconds and the following two smaller numbers are in unit of 1/100 second. (g) A schematic illustration of NiSi growth within a Si NW. (h) A schematic illustration of growth of a NiSi/Si/NiSi heterostructure where the length of the Si section is controlled at the atomic scale.

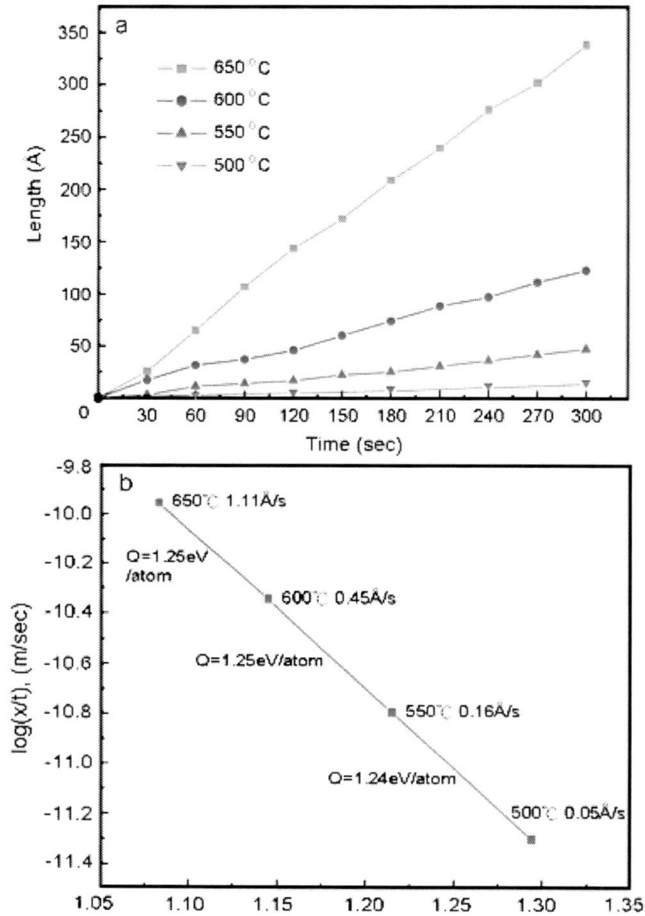

Figure 4.36 Kinetic analysis of the NiSi epitaxial growth within a Si NW of 20 nm in diameter. (a) Plot of the NiSi NW length versus reaction time at various temperatures, illustrating a linear growth rate. The lines are drawn as guides. (b) Arrhenius plot of the NiSi epitaxial growth, from which the activation energy was determined.

To identify the phase and structure of the NiSi NW, its HRTEM image is provided in Fig. 4.37a. The inset is the corresponding diffraction pattern with a zone axis of [113], which verifies that the wire is a single-crystal of NiSi. Compositional analysis has also confirmed that it is NiSi. Figure 4.37b is a HRTEM image showing an epitaxial interface that is atomically sharp between Si and NiSi within a Si NW. Figures 4.37a and b demonstrate that point-contact reaction enables the formation of nano-NiSi in perfect single-crystal and atomically flat NiSi/Si interface without misfit dislocations, which is important for electrical contact properties.[53]

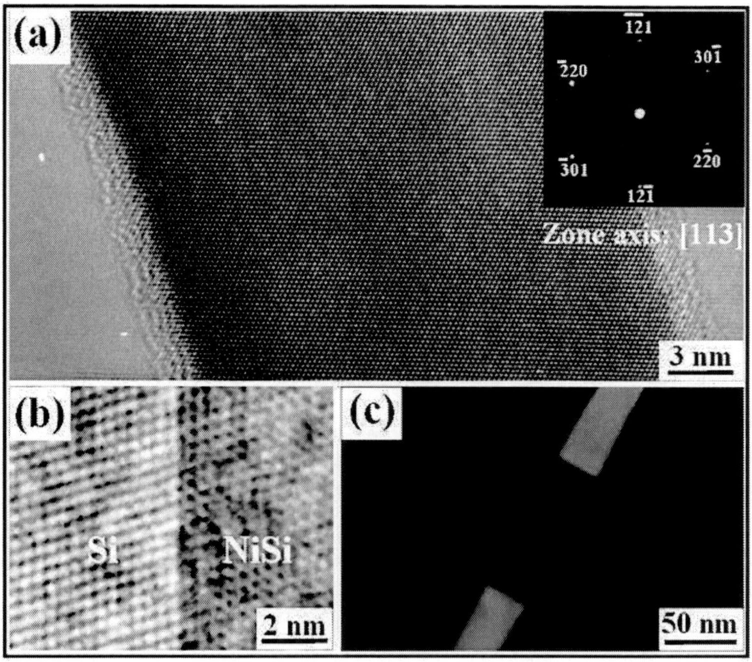

Figure 4.37 (a) HRTEM image of a NiSi NW. The inset is the corresponding diffraction pattern with a [113] zone axis. (b) HRTEM image of an epitaxial interface between NiSi and Si within a Si NW. (c) High-angle annular dark-field image of the NiSi/Si/NiSi heterostructure. The bright image is NiSi and the dark image is Si.

The atomistic mechanism of the epitaxial growth involving a moving interface is of interest. Reactive epitaxial growth is unlike molecular beam epitaxy by deposition, in which atomic layers are built on the substrate surface and film growth requires no breaking of the substrate bonds. In the present case, the covalent Si–Si bonds must be broken and transformed into metallic Ni–Si bonds in order for NiSi to grow and have a moving interface. Although the thermal energy at 700 °C is sufficient to break the covalent Si–Si bonds, the interstitial Ni atoms are crucial in the bond-breaking process.[145]

Figures 4.35g and h depict the growth of NiSi from both ends of a Si NW. It leads to the formation of a NiSi/Si/NiSi nano-heterostructure as shown in Figs. 4.38a and b. The darker regions at the two ends are NiSi and the lighter region in the middle is Si. If annealing is stopped before the entire NW transforms into NiSi, as in Fig. 4.38c, a nano-heterostructure is formed as shown in Figs. 4.38a and b. The nanoscale epitaxial growth rate of single-crystal NiSi has been measured using high-resolution lattice imaging videos. Based on the growth rates, the remaining length of the Si region between the two NiSi regions can be controlled as illustrated in Fig. 4.35h. Two Ni NWs can be patterned or deposited with a given spacing over a Si wire, and then the reaction can be utilized to bring the two NiSi grains as close as possible. At 500 °C, the reaction rate can be controlled down to atomic scale. In Figs. 4.38d–g, a set of lattice images of the nano-heterostructure of NiSi/Si/NiSi with 10, 8.1, 5 and 2 nm lengths of Si is shown. By measuring the length of the Si region and counting the number of (111) lattice planes within the region, the strain can be determined and it is found that the Si is highly compressed. Figure 4.38h shows the relationship between the compressive strain and the length of the Si region in the nano-heterostructure NiSi/Si/NiSi at room temperature. In the process of NiSi formation, the diffusion of Ni atoms into Si lattice could have led to volume expansion. However, due to the confinement of the oxide on the Si NW, NiSi coming from both sides compressed the middle Si region, resulting in compressive stress on Si in the length direction. As a result, the strain increased when the length of the Si region decreased. The strain can be controlled because the length of the Si region can be controlled. Since one-dimensional nano-heterostructures may have potential applications in nano-electronic devices,[35] the strain will affect carrier mobility in the Si region.

In summary, nano-heterostructures of NiSi/Si/NiSi, in which the length of the Si region can be controlled down to 2 nm, have been produced using *in situ* point-contact reaction between Si and Ni NWs in an UHV-TEM. The Si region was found to be highly strained, more than 12%. It is proposed that the reaction is assisted by interstitial diffusion of Ni atoms within the Si NW and is limited by the rate of dissolution of Ni into Si at the point-contact interface. The rate of incorporation of Ni atoms to support the growth of NiSi has been measured to be 7×10^{-4} sec per Ni atom. Based on the rate, the consumption of Si can be controlled and, in turn, the dimensions of the nano-heterostructure down to less than 2 nm, thereby far exceeding the limit of conventional patterning process. The controlled huge strain in the controlled atomic scale Si region, potential gate of Si-NW-based transistors, is expected to significantly impact the performance of electronic devices.

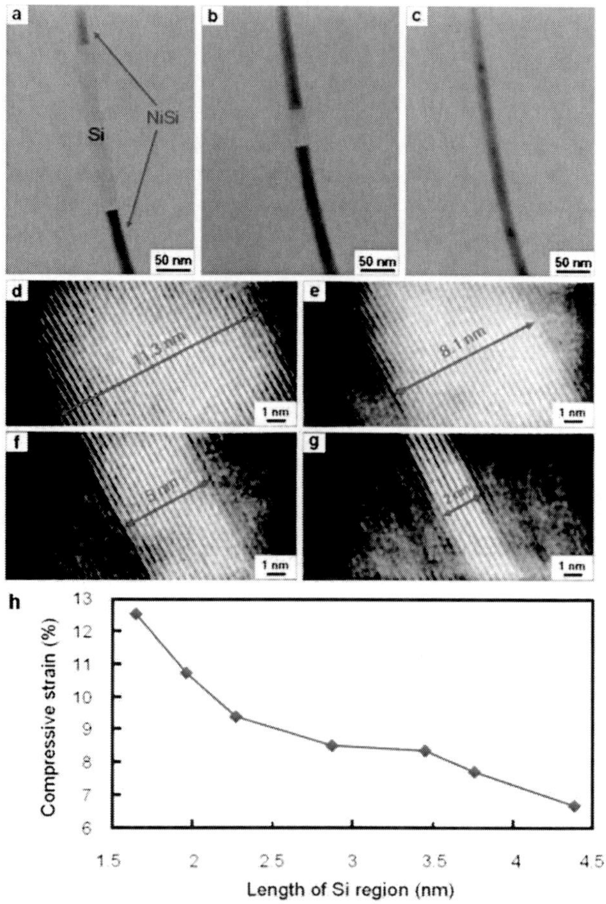

Figure 4.38 *In situ* TEM images showing the formation of NiSi/Si/NiSi heterostructures within a Si NW, and compressive strain in the Si region. (a and b) *In situ* TEM images of the NiSi/Si/NiSi heterostructure. The bright area is Si and the dark area is NiSi. (d–g) HRTEM images of NiSi/Si/NiSi heterostructures. The bright and dark portions of the lattice images correspond to Si and NiSi, respectively. (h) Plot of the compressive strain versus the length of the Si region in the nano-heterostructure NiSi/Si/NiSi.

4.4.2 Repeating Events of Nucleation in Epitaxial Growth of Nano $CoSi_2$ and NiSi in Nanowires of Si

Nanostructures in Si NWs have been studied for basic components in electronic and optoelectronics devices, especially biosensors.[83,146,147] Well-defined nanoscale building blocks such as ohmic contacts and gates on Si NWs must be developed in order to be assembled into functional electronic structures.[34,59,80] It requires a systematic study of chemical reactions in the nanoscale to form these circuit components. When a Si NW touches a metal NW, a point-contact is formed and the chemical reaction between them starts from the point of contact. The point-contact reactions between Si NWs and Ni and Co NWs or nanodots were investigated by *in situ* HRTEM. The metallic NiSi and $CoSi_2$ may serve as the source-drain contacts to the Si in the heterostructure; it is the first step to produce a nanoscale field-effect transistor.

The formation of CoSi and $CoSi_2$ in Si NWs was observed at 700 and 800 °C, respectively, by point-contact reactions between nanodots of Co and NWs of Si by *in situ* HRTEM. The $CoSi_2$ has undergone an axial epitaxial growth in the Si NW and a stepwise growth mode was found, as shown in Figs. 4.39a–d. We observed that the stepwise growth occurs repeatedly in the form of an atomic step sweeping across the $CoSi_2$/Si interface, as shown in Fig. 4.39e. It appears that the growth of a new step or a new silicide layer requires an independent event of nucleation. We are able to resolve the nucleation stage and the growth stage of each layer of the epitaxial growth in video images. In the nucleation stage, the incubation period is measured to be much longer than the period needed to grow the layer across the silicide/Si interface.[55] So the epitaxial growth consists of a repeating nucleation and a rapid stepwise growth across the epitaxial interface. This is a general behavior of epitaxial growth in NWs. The nucleation of silicides in Si NWs is supply limited and source-limited reaction by point-contact reaction rather than diffusion-controlled or interfacial controlled reaction. Besides, the axial heterostructure of $CoSi_2$/Si/$CoSi_2$ with sharp epitaxial interfaces has been obtained, as shown in Fig. 4.40,[55] which are promising as high-performance transistors based on intrinsic Si NWs.

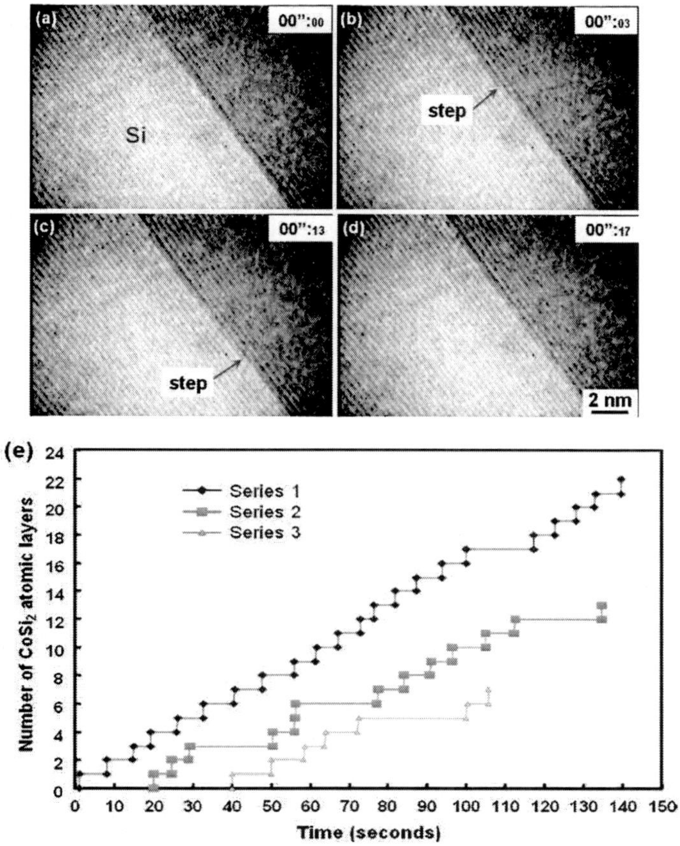

Figure 4.39 *In situ* HRTEM image sequences of growing $CoSi_2$/Si epitaxial interfaces within Si NWs. (a–d) The growth of one atomic layer has occurred, and the step is indicated by the arrow. The left-up and right-down regions are close to center and edge of NW, respectively. (e) Three sets of stair-type growth curve of $CoSi_2$.

The HRTEM videos show that the overall axial growth rate of the Ni and Co silicide layers is linear. However, the linear curve can be decomposed into many stair-steps with the step height equal to an atomic layer thickness and the step width equal to the incubation time of nucleating a new step. After a step is nucleated, it propagates very rapidly across the Si/silicide interface. The overall reaction rate is limited by the incubation time of nucleation, so it is a nucleation-limited reaction, neither diffusion-limited nor interfacial-reaction-limited.

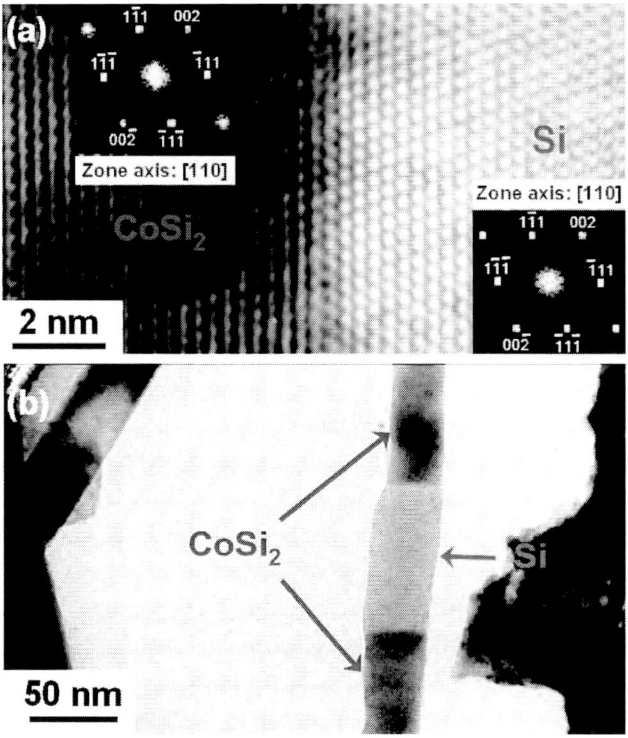

Figure 4.40 (a) HRTEM image of $CoSi_2$–Si interface. (b) A TEM image of nanoheterostructure of $CoSi_2$/Si/$CoSi_2$ form within a Si NW. The light area is Si and the dark area is $CoSi_2$. The insets in panel (a) are the FFT pattern confirming the [110] Si zone axis and the [110] $CoSi_2$ zone axis, respectively. {002} spots in the Si pattern are from double diffraction.

Repeating events of homogeneous nucleation in epitaxial growth of $CoSi_2$ and NiSi silicides in Si NWs by using *in situ* HRTEM was observed as shown in Fig. 4.41.[56] The growth of every single atomic layer requires nucleation. Heterogeneous nucleation is prevented because of non-microreversibility between the oxide/Si and oxide/silicide interfaces so homogeneous nucleation occurs. The incubation time of homogeneous nucleation is determined to 3 sec and 6 sec for NiSi and $CoSi_2$, respectively. Moreover, the calculated and the measured nucleation rates are in good agreement. Using Zeldovich factor to estimate the number of molecules in the critical nucleus; it is about 10 and reasonable. A very high supersaturation is found for the homogeneous nucleation.

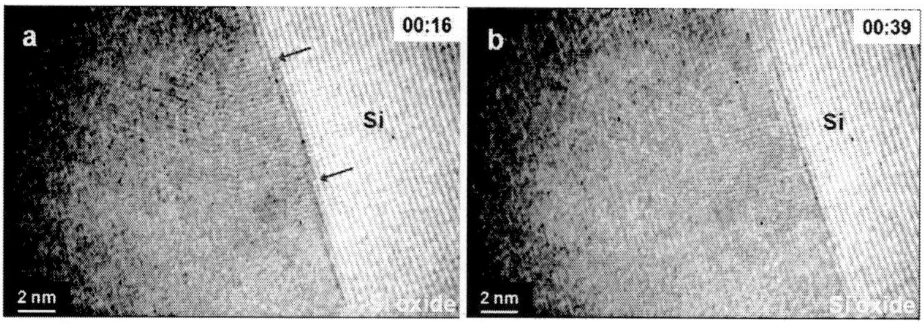

Figure 4.41 HRTEM images of motion of steps on epitaxial NiSi/Si interface at 450 °C. (a) One NiSi atomic layer grows from the middle region of the interface. (b) After two steps, it reaches to the oxide ends, the layer growth ends, and the interface becomes flat without steps.

4.4.3 Reactions between Si Nanowires and Pt Pads

4.4.3.1 Formation of PtSi nanowire and PtSi/Si/PtSi nanoheterostructures

Substantial efforts have been made for the development of nanoscale transistors based on silicon NWs[148,149] due to their potential to replace conventional planar metal-oxide-semiconductor field-effect transistors (MOSFET) in integrated circuits[150,151] or to open new opportunities in flexible macroelectronics[152–154] and highly sensitive biosensors.[155,156] Making reliable electrical contact to individual NW devices is one of the key factors that determine the device performance and reliability. Ni silicide has been reported[34,53,54] and is being used as a contact material for silicon NW transistors.[34,35,54] Being chemically stable in ambient or oxidizing environment, platinum is an interesting interconnecting material for nanoscale electronics where surface to volume ratio is large and the surface chemical stability is important. Additionally, the formation of platinum silicide (PtSi) is of interest because it has a very low barrier height of ~0.2 eV on the valence band of silicon and is an attractive choice for ohmic contacts to p-channel Si NW transistors.[157] Here, we report the formation of single-crystal PtSi NWs, PtSi/Si/PtSi NW heterostructures, and nanodevices from such heterostructures.

Silicon NWs were prepared on a silicon wafer by the vapor–liquid–solid method using Au clusters as the catalyst.[52,158,159] The silicon NWs are typically single-crystals with diameters ranging from 30 to 60 nm, lengths greater than 10 μm, and growth axes along <111> directions. The Si NW device with platinum contacts was fabricated on Si/Si_3N_4 substrate using e-beam lithography and e-beam evaporation (Fig. 4.42a). Prior to platinum deposition, the sample was etched in buffered hydrofluoric acid for 5 s to remove native oxide in the contact region. To allow platinum to diffuse into the silicon NW and form partially (Fig. 4.42b) or fully silicidized (Fig. 4.42c) PtSi NW, the device was annealed in forming gas at 520 °C with a rapid thermal processor for various periods of time. Before annealing, a SEM image of the device shows a uniform contrast along the axis of the NW (Fig. 4.42d). After the device was annealed in forming gas for 30 s, clear contrast developed along the wire axis with two brighter sections emerging from both ends of the NW near the platinum pads (Fig. 4.42e). A darker section remains between the two brighter sections, corresponding to the unreacted silicon NW. These results suggest that a PtSi/Si/PtSi NW heterostructure is formed through reaction between the silicon NW and two platinum

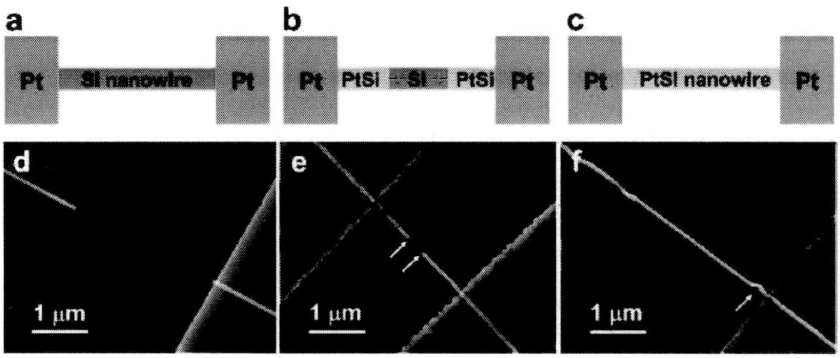

Figure 4.42 Formation of PtSi NW and PtSi/Si/PtSi nanoheterostructures. (a–c) Schematic illustrations depicting growth of a PtSi/Si/PtSi nanoheterostructure and PtSi NWs through controlled reaction between silicon NWs and platinum contact pads. (d) SEM image of a silicon NW device with two platinum contact pads before reaction. (e) SEM image of a PtSi/Si/PtSi nanoheterostructure obtained through partial silicidation of the silicon NW after annealing at 520 °C for 30 s. Two arrows highlight two PtSi/Si interfaces. (f) SEM image of a PtSi NW after complete silicidation by annealing at 520 °C for 60 s. The arrow highlights a curved section near the contact pads due to strain resulting from volume expansion.

contact pads. This is attributed to the fact that many platinum atoms are able to dissolve into silicon through the contacts between silicon NW and platinum pads so that supersaturation can be reached; thereby, nucleation and growth of platinum silicide occur at both platinum pads. Upon further annealing, the two brighter sections (PtSi) converged, and the darker section (Si) disappeared. At this point, all silicon is consumed and a silicon NW is fully transformed into a PtSi NW (Fig. 4.42f). Occasionally, a curved section is seen near the contact pads (Fig. 4.42f), which can be attributed to strain resulting from the volume expansion during the silicidation process.[59]

To prepare TEM samples, silicon NWs devices were prepared on silicon grids with a square opening covered with a window of a glassy Si_3N_4 film. A HRTEM (HRTEM) image (Fig. 4.43a) shows that there are clean interfaces between PtSi/Si with an approximately 2 nm silicon oxide shell surrounding both the silicon and platinum silicide regions, suggesting the growth of platinum silicide NW is confined in the preformed native silicon oxide shell.

4.4.3.2 Epitaxial relationship of PtSi formation within a silicon nanowire

Based on the TEM studies, the silicide material is identified to be single-crystal PtSi phase with an orthorhombic structure with lattice constants a = 0.5939 nm, b = 0.5596 nm, and c = 0.3604 nm. A HRTEM image and its FFT (Fig. 4.43b and insets) show the Si/PtSi epitaxial interface is parallel to the (111) plane of Si as well as the (101) plane of PtSi. Thus, the growth direction of PtSi is normal to the (101) plane. The crystallographic orientation relationships between Si and PtSi are Si [1–10]//PtSi [010] and Si (111)//PtSi (101).

Across the epitaxial interface, the largest lattice mismatch is between the Si (002) plane (d = 0.271 nm) and the PtSi (200) plane (d = 0.297 nm) with a relative mismatch of about 8.8%. Notably, despite this exceedingly large latticemismatch, no apparent dislocations or other defects are seen across the interface. These results demonstrate that a sharp and atomically smooth interface is obtained in our Si/PtSi NW heterostructure. Through controlled reactions, the length of the silicon region in NW PtSi/Si/PtSi heterostructures can be precisely controlled down to the sub-10 nm regime (Fig. 4.43c).

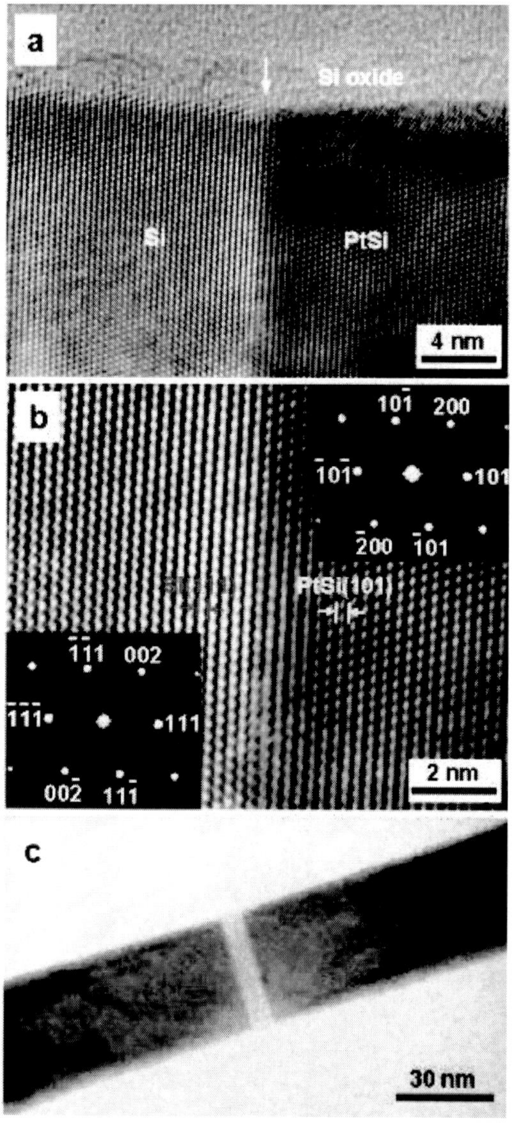

Figure 4.43 Epitaxial relationship of PtSi formation within a silicon NW. (a) A HRTEM image showing clean and sharp interfaces between PtSi and Si. The arrow highlights the Si/PtSi interface. (b) A higher magnification image of a PtSi/Si interface illustrating the epitaxial relationship. The insets are FFT patterns confirming the Si [1–10] zone axis and PtSi [010] zone axis, respectively. (c) A TEM image of a PtSi/Si/PtSi heterostructure with only an 8 nm silicon region.

4.4.3.3 PtSi/i–Si/PtSi nanowire heterostructures as high-performance p-channel enhancement mode transistors

PtSi can be used as nearly perfect ohmic contact for p-channel Si NW transistors (Fig. 4.44a). Additionally, the formation of atomically clean Si/PtSi interfaces prevents Fermi level pinning and helps to maintain a low Schottky barrier determined by work function rather than by interface states.[160] This has enabled us to fabricate high-performance NW transistors from intrinsic silicon naowires using PtSi as the source and drain contacts. To make the device (Fig. 4.44b), a 7 nm thick layer of HfO_2 was deposited on the surface of the partially silicidized PtSi/Si/PtSi device using atomic layer deposition process to form gate dielectrics, and a top Cr/Au electrode was defined with e-beam lithography and deposited using e-beam evaporation.

Electrical transport measurement on the device shows nearly perfect transistor characteristics. Figure 4.44c shows drain current (I_D) versus drain-source bias voltage (V_D) relations at various gate voltages (V_G) in steps of 0.25 V. The device shows typical p-channel enhancement mode (normally off) transistor behavior.[160] I_D increases linearly with V_D at low V_D and saturates at higher V_D. It is important to note that a nearly perfect linear relationship is observed in the I_D–V_D plot in the low bias region, which suggests that a satisfactory Ohmic contact is achieved for hole transport in the intrinsic silicon NW device due to a very low Schottky barrier. The plot of I_D versus V_G (Fig. 4.44d) at constant V_D = –0.2, –0.4, –0.6, –0.8, and –1.0 shows little current can pass through the device when the gate voltage is below a threshold voltage of approximately –2.0 V, and I_D increases nearly linearly when the gate voltage increases in the negative direction beyond the threshold voltage.

The nanoscale field-effect transistors based on PtSi/Si/PtSi NW heterostructures with atomically sharp interfaces, where the source and drain contacts are defined by the metallic PtSi NW regions, and the gate length is defined by the Si NW region, have been found to have a normalized transconductance of 0.3 mS/μm, field-effect hole mobility of 168 cm^2/V-s, and on/off ratio >10^7, demonstrating the best performing device from intrinsic silicon NWs.

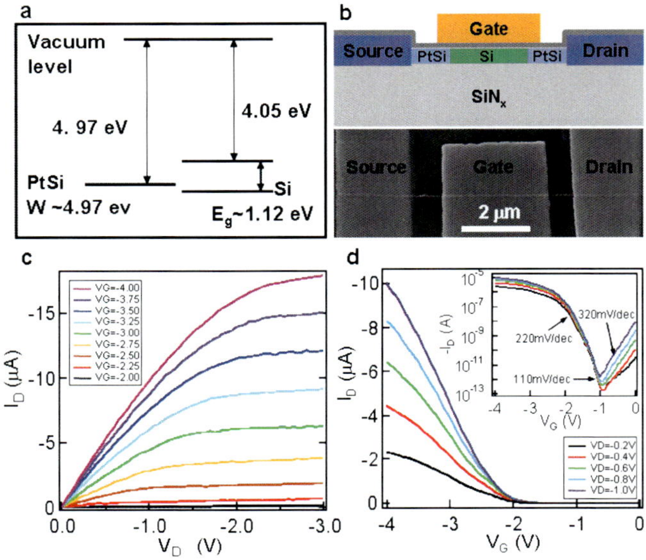

Figure 4.44 PtSi/i-Si/PtSi NW heterostructures as high-performance p-channel enhancement mode transistors. (a) Relative energy band alignment between PtSi and Si. (b) (top) Schematic of PtSi/i-Si/PtSi NW heterostructure device with HfO_2 as gate dielectrics. (bottom) SEM images of a device (scale bar, 2 μm). (c) Drain current (I_D) versus drain-source voltage (V_D) at increasing negative gate voltages (V_{GS}) in steps of 0.25 V starting from the bottom at V_G = −2.0 V. (d) I_D versus V_G at V_D = −0.2, −0.4, −0.6, −0.8, −1.0 V. The inset shows $-I_D$ versus V_G in the exponential scale, highlighting the on/off ratio >10^7, and ambipolar transport with a subthreshold swing of 110–220 mV/decade for hole transport and 320 mV/decade for electron transport.

4.5 Conclusions

Metal silicides were introduced into microelectronics devices in late 1970s as gates, contacts and interconnects. In the 1980s, almost all transition metal silicides were found to grow epitaxially on silicon to some extent. In the 1990s, amorphous interlayers of a few nm in thickness were observed to occur in all refractory metal/Si and a number of rare-earth metal and platinum group metal and crystalline silicon systems.

As the integrated circuits industry moves into the nano era, scaling down the metal silicide contacts and gates have become an important issue. A large number of epitaxial and free-standing silicide NWs were grown. In this chapter, the growth and properties of silicide NWs are reviewed. Applications of silicide NWs are covered wherever appropriate. It is foreseen that silicide NWs will not only be the key building blocks for the future nanoelectronics devices, possible applications in spintronics, thermoelectrics, and FE devices may also be realized.

Acknowledgments

The work was supported by the National Science Council through grant no. 97-2120-M-007-003 and 96-2628-E-007-018-MY3. The authors would like to acknowledge the contributions from Profs. K.N. Tu, Y. Huang, K. C. Lu, and Drs. S. Y. Chen, C. Y. Lee, and C. M. Chang.

References

1. L. J. Chen, *Silicide Technology for Integrated Circuits*, The Institution of Electrical Engineers, London, UK, (2004).
2. K. N. Tu, J. W. Mayer, *Thin Films—Interdiffusions and Reactions*, J. M. Poate, K. N. Tu, J. W. Mayer (Eds.), Wiley, New York, (1978).
3. F. Nava, K. N. Tu, O. Bisi, *Mater. Sci. Rep.* 9 (1993) 141.
4. S. P. Murarka, *Silicides for VLSI Applications*, Academic Press, New York, 1983.
5. L. J. Chen, K. N. Tu, *Mater. Sci. Rep.* 6 (1991) 53.
6. K. N. Tu, R. D. Thompson, B. Y. Tsaur, *Appl. Phys. Lett.* 38 (1981) 626.
7. G. Ottaviani, K. N. Tu, J. W. Mayer, *Phys. Rev. Lett.* 44 (1980) 284.
8. G. Ottaviani, K. N. Tu, J. W. Mayer, *Phys. Rev. B* 24 (1981) 3354.
9. H. C. Cheng, T. R.Yew, L. J. Chen, *J. Appl. Phys.* 57 (1985) 5246.
10. M. S. Fung, H. C. Cheng, L. J. Chen, *Appl. Phys. Lett.* 47 (1985) 1312.
11. S. W. Lu, C. W. Nieh, L. J. Chen, *Appl. Phys. Lett.* 49 (1986) 1770.
12. F. Nava, K. N. Tu, E. Mazzega, M. Michelini, G. Queirolo, *J. Appl. Phys.* 61 (1987) 1085.
13. U. Koster, P. S. Ho, J. E. Lewis, *J. Appl. Phys.* 53 (1982) 7436.

14. S. Saitoh, H. Ishiwara, T. Asano, S. Furukawa, Japan. *J. Appl. Phys.* 20 (1981) 1649.
15. W. J. Chen, L. J. Chen, *J. Appl. Phys.* 71 (1992) 653.
16. T. L. Lee, L. J. Chen, *J. Appl. Phys.* 73 (1993) 5280.
17. C. S. Liu, L. J. Chen, *J. Appl. Phys.* 74 (1993) 3611.
18. J. B. Lai, C. S. Liu, L. J. Chen, J. Y. Cheng, *J. Appl. Phys.* 78 (1995) 6539.
19. S. M. Chang, H. Y. Huang, H. Y. Yang, L. J. Chen, *Appl. Phys. Lett.* 73 (1999) 224.
20. L. Eckertova, *Physics of Thin Films*, Plenum Press, New York, (1986).
21. R. T. Tung, J. M. Poate, J. C. Bean, J. M. Gibson, D. C. Jacobson, *Thin Solid Films* 93 (1982) 77.
22. S. L. Cheng, H. Y. Huang, Y. C. Peng, L. J. Chen, B. Y. Tsui, C. J. Tsai, S. S. Guo, *Appl. Phys. Lett.* 74 (1999) 1406.
23. L. J. Chen, *Mater. Sci. Rep.* 29 (2000) 115.
24. W. W. Wu, T. F. Chiang, S. L. Cheng, S. W. Lee, L. J. Chen Y. H. Peng, H. H. Cheng, *Appl. Phys. Lett.* 81 (2002) 820.
25. W. W. Wu, S. L. Cheng, S. W. Lee, L. J. Chen, J. *Vac. Sci. Technol. B* 21 (2003) 2147.
26. L. J. Chen, J. W. Mayer, K. N. Tu, T. T. Sheng, *Thin Solid Films* 93 (1982) 91.
27. K. F. Kelton, A. L. Greer, C. V. J. Thompson, *Chem. Phys.* 79 (1983) 6261.
28. D. W. Pashey, *Epitaxial Growth*, J. W. Matthews (Eds.), Academic Press, New York, (1975).
29. J. W. Mayer, J. M. Poate, K. N. Tu, *Science* 190 (1975) 228.
30. W. K. Chu, S. S. Lau, J. W. Mayer, H. Muller, K. N. Tu, *Thin Solid Films* 25 (1975) 393.
31. K. N. Tu, W .K. Chu, J. W. Mayer, *Thin Solid Films* 25 (1975) 403.
32. C. M. Lieber, *MRS* Bull. 28 (2003) 486.
33. F. Patolsky, B. P. Timko, G. Zheng, C. M. Lieber *MRS Bull.* 32 (2007) 142.
34. Y. Wu, J. Xiang, C. Yang, W. Lu, C. M. Lieber, *Nature* 430 (2004) 61.
35. W. M. Weber, L. Geelhaar, A. P. Graham, E. Unger, G.S. Duesberg, M. Liebau, W. Pamler, C. Cheze, H. Riechert, P. Lugli, F. Kreupl, *Nano Lett.* 6 (2006) 2660.
36. S. Y. Chen, L. J. Chen, *Appl. Phys. Lett.* 87 (2005) 253111.
37. S. Y. Chen, H. C. Chen, L. J. Chen, *Appl. Phys. Lett.* 88 (2006)193114.

38. Y. L. Chueh, M.T. Ko, L. J. Chou, L. J. Chen, C. S. Wu, C. D. Chen, *Nano Lett.* 6 (2006) 1637.
39. C. J. Kim, K. Kang, Y. S. Woo, K. G. Ryu, H. Moon, J. M. Kim, D. S. Zang, M. H. Jo, *Adv. Mater.* 19 (2007) 3637.
40. Y. P. Song, A. L. Schmitt, S. Jin, *Nano Lett.* 7 (2007) 965.
41. A. L. Schmitt, M. J. Bierman, D. Schmeisser, F. J. Himpsel, S. Jin, *Nano Lett.* 6 (2006) 1617.
42. Z. He, D. J. Smith, P. A. Bennett, *Phys. Rev. Lett.* 93 (2004) 256102.
43. H. C. Hsu, W. W. Wu, H. F. Hsu, L. J. Chen, *Nano Lett.* 7 (2007) 885.
44. A. L. Schmitt, L. Zhu, D. Schmeiaer, F. J. Himpsel, S. Jin, *J. Phys. Chem. B* 37 (2006) 18142.
45. C. Y. Lee, M. P. Lu, K. F. Liao, W. W. Wu, L. J. Chen, *Appl. Phys. Lett.* 93 (2008) 113109.
46. C. Y. Lee, M. P. Lu, K. F. Liao, W. F. Lee, C. T. Huang, S. Y. Chen, L. J. Chen, *J. Phys. Chem. C* 113 (2009) 2286.
47. C. I. Tsai, P. H. Yeh, C. Y. Wang, H. W. Wu, U. S. Chen, M. Y. Lu, W. W. Wu, L. J. Chen, Z. L. Wang, *Cryst. Growth Des.* 9 (2009) 4514.
48. C. M. Chang, Y. C. Chang, Y. A. Chung, C. Y. Lee, L. J. Chen, *J. Phys. Chem.* C 113 (2009) 17720.
49. M. T. Chang, C. Y. Chen, L. J. Chou, L. J. Chen, *ACS Nano* 3 (2009) 3776.
50. H. K. Lin, Y. F. Tzeng, C. H. Wang, N. H. Tai, I. N. Lin, C. Y. Lee, H. T. Chiu, *Chem. Mater.* 20 (2008) 2429.
51. Y. Song, S. Jin, *Appl. Phys. Lett.* 90 (2007) 173122.
52. C. M. Chang, Y. C. Chang, C. Y. Lee, P. H. Yeh, W. F. Lee, L. J. Chen, *J. Phys. Chem. C* 113 (2009) 9153.
53. K. C. Lu, K. N. Tu, W. W. Wu, L. J. Chen, B. Y. Yoo, N. V. Myung, *Appl. Phys. Lett.* 90 (2007) 253111.
54. K. C. Lu, W. W. Wu, H. W. Wu, C. M. Tanner, J. P. Chang, L. J. Chen, K. N. Tu, *Nano Lett.* 7 (2007) 2389.
55. Y. C. Chou, W. W. Wu, S. L. Cheng, B. Y. Yoo, N. Myung, L. J. Chen, K. N. Tu, *Nano Lett.* 8 (2008) 2194.
56. Y. C. Chou, Y. C, W. W. Wu, L. J. Chen, K. N. Tu, *Nano Lett.* 9 (2009) 2337.
57. K. Seo, K. S. K. Varadwaj, P. Mohanty, S. Lee, Y. Jo, M. H. Jung, J. Kim, B. Kim, *B. Nano Lett.* 7 (2007) 1240.
58. K. Seo, K. S. K. Varadwaj, D. Cha, J. In: J. Kim, J. Park, B. Kim, B. J. *Phys. Chem. C.* 111 (2007) 9072.

59. Y.C. Lin, K. C. Lu, W. W. Wu, J. Bai, L. J. Chen, K. N. Tu, Y. Huang, *Nano Lett.* 8 (2008) 913.
60. A. L. Schmitt, J. M. Higgins, J. R. Szczech, S. Jin, *J. Mater. Chem.* 20 (2010) 223.
61. C. Preinesberger, S. Vandre, T. Kalka, M. Dahne-Prietsch, *J. Phys. D*, 31 (1998) L43.
62. Y. Chen, D. A. A. Ohlberg, G. Medeiros-Ribeiro, Y. A. Chang, *Appl. Phys. Lett.* 76 (2000) 4004.
63. J. Nogami, B. Z. Liu, M. V. Katkov, C. Ohbuchi, N. O. Birge, *Phys. Rev., B* 63 (2001) 233305.
64. Y. Chen, D. A. A. Ohlberg, R. S. Williams, *J. Appl. Phys.* 91 (2002) 3213.
65. M. Stevens, Z. He, D. J. Smith, P. A. Bennett, *J. Appl. Phys.* 93 (2003) 5670.
66. W. C. Yang, H. Ade, R. J. Nemanich, *J. Appl. Phys.* 95 (2004) 1572.
67. S. H. Brongersma, M. R. Castell, D. D. Perovic, M. Zinke-Allmang, *Phys. Rev. Lett.* 80 (1998) 3795.
68. J. D. Carter, G. Cheng, T. Guo, *J. Phys. Chem.* 108 (2004) 6901.
69. K. C. R. Chiu, J. M. Poate, J. E. Rowe, T. T. Sheng, A. G. Gullis, *Appl. Phys. Lett.* 38 (1981) 988.
70. J. Tersoff, R. M. Tromp, *Phys. Rev. Lett.* 70 (1993) 2782.
71. D. E. Jesson, G. Chen, K. M. Chen, S. J. Pennycook, *Phys. Rev. Lett.* 80 (1998) 5156.
72. F. Edelman, E. Y. Gutmanai, A. Katz, R. Brener, *Appl. Phys. Lett.* 53 (1988) 1186.
73. R. K. K. Chong, M. Yeadon, W. K. Choi, E. A. Stach, C. B. Boothroyd, *Appl. Phys. Lett.* 82 (2003) 1833.
74. R. T. Tung, *Appl. Phys. Lett.* 68 (1996) 3461.
75. B. Egert, G. Panzner, *Phys. Rev. B* 29 (1984) 2091.
76. X. Y. Kong, Z. L. Wang, *Nano Lett.* 3 (2003) 1625.
77. J. Q. Hu, Y. Bando, Z. W. Liu, T. Sekiguchi, D. Golberg, J. H. Zhan, J. *Am. Chem. Soc.* 125 (2003) 11306.
78. E. A. Stach, P. J. Pauzauskie, T. Kuykendall, J. Goldberger, R. R. He, P. D. Yang, *Nano Lett.* 3 (2003) 867.
79. K. N. Tu, J. W. Mayer, L. C. Feldman, *Electronic Thin Film Science*, Macmillan Publishing, New York, 1992.
80. L. J. Chen, *JOM* 57, 24 (2005).

81. Y. J. Hu, J. Xiang, G. C. Liang, H. Yan, C. M. Lieber, *Nano Lett.* 8 (2008) 925.

82. S. L. Zhang, M. Ostling, *Crit. Rev. Solid State Mater. Sci.* 28 (2003) 1.

83. L. J. Chen, *J. Mater. Chem.* 17 (2007) 4639.

84. J. A. Kittl, A. Lauwers, A. Veloso, T. Hoffmann, S. Kubicek, M. Niwa, M. J. H. van Dal, M. A. Pawlak, S. Brus, C. Demeurisse, C. Vrancken, P. Absil, S. Biesemans, *IEEE Electron Device Lett.* 27 (2006) 966.

85. C. A. Decker, R. Solanki, J. L. Freeouf, J. R. Carruthers, D. R. Evans, *Appl. Phys. Lett.* 84 (2004) 1389.

86. J. Kim, W. A. Anderson, Y. J. Song, G. B. Kim, *Appl. Phys. Lett.* 86 (2005) 253101.

87. C. J. Kim, K. Kang, Y. S. Woo, K. G. Ryu, H. Moon, J. M. Kim, D. S. Zang, M. H. Jo, *Adv. Mater.* 19 (2007) 3637.

88. X. Q. Yan, H. J. Yuan, J. X. Wang, D. F. Liu, Z. P. Zhou, Y. Gao, L. Song, L. F. Liu, W. Y. Zhou, G. Wang, S. S. Xie, *Appl. Phys. A* 79 (2004) 1853.

89. J. Kim, D. H. Shin, E. S. Lee, C. S. Han, Y. C. Park, *Appl. Phys. Lett.* 90 (2007) 253103.

90. Z. Zhang, P. E. Hellstrom, M. Ostling, S. L. Zhang, J. Lu, *Appl. Phys. Lett.* 88, (2006) 043104.

91. C. L. Hsin, J. H. He, C. Y. Lee, W. W. Wu, P. H. Yeh, L. J. Chen, Z. L. Wang, *Nano Lett.* 7 (2007) 1799.

92. J. A. Kittl, M. A. Pawlak, A. Lauwers, C. Demeurisse, K. Opsomer, K. G. Anil, C. Vrancken, M. J. H. van Dal, A. Veloso, S. Kubicek, P. Absil, K. Maex, S. Biesemans, *IEEE Electron Device Lett.* 27 (2006) 34.

93. Y. L. Chueh, L. J. Chou, S. L. Cheng, L. J. Chen, C. J. Tsai, C. M. Hsu, S. C. Kung, *Appl. Phys. Lett.* 87, (2005) 223113.

94. C. Y. Kang, P. Lysaght, R. Choi, B. H. Lee, S. J. Rhee, C. H. Choi, M. S. Akbar, J. C. Lee, *Appl. Phys. Lett.* 86 (2005) 222906.

95. F. M. d'lleurle, *Mater. Res. Sot. Symp. Proc.* 25 (1984) 3.

96. J. P. Gambino, E. G. Colgan, *Mater. Chem. Phys.* 52 (1998) 99.

97. P. H. Yeh, L. J. Chen, P.T. Liu, D. Y. Wang, T. C. Chang, *Electrochim. Acta* 52 (2007) 2920.

98. J. S. Suh, K.S. Jeong, J. S. Lee, I. T. Han, *Appl. Phys. Lett.* 80 (2002) 2392.

99. S. H. Jo, J. Y. Lao, Z. F. Ren, R. A. Farrer, T. Baldacchini, J. T. Fourkas, *Appl. Phys. Lett.* 83 (2003) 4821.

100. S. H. Jeong, H. Y. Hwang, K. H. Lee, Y. Jeong, *Appl. Phys. Lett.* 78 (2001) 2052.

101. L. Nilsson, O. Groening, C. Emmenegger, O. Kuettel, E. Schaller, L. Schlapbach, H. Kind, J. M. Bonard, K. Kern, *Appl. Phys. Lett.* 76 (2000) 2071.

102. L. Ouyang, E. S. Thrall, M. M. Deshmukh, H. Park, *Adv. Mater.* 18 (2006) 1437.

103. P. Yang, *Dalton Trans.* 33 (2008) 4387.

104. M. Law, J. Goldberger, P. Yang, *Annu. Rev. Mater. Res.* 34 (2004) 83.

105. W. Lu, C. M. Lieber, *Nat. Mater.* 6 (2007) 841.

106. Y. Xia, P. Yang,Y. Sun, Y. Wu, B. Mayers, B. Gates, Y. Yin, F. Kim, H. Yan, *Adv. Mater.* 15 (2003) 353.

107. J. H. He, C. L. Hsin, J. Liu, L. J. Chen, Z. L. Wang, *Adv. Mater.* 19 (2007) 781.

108. Z. L. Wang, *Adv. Mater.* 15 (2003) 432.

109. C. T. Huang, C. L. Hsin, K. W. Huang, C. Y. Lee, P. H. Yeh, U. S. Chen, L. J. Chen, *Appl. Phy. Lett.* 91 (2007) 093133.

110. B. Tian, X. Zheng, T. J. Kempa, Y. Fang, N. Yu, G. Yu, J. Huang, C. M. Lieber, *Nature* 449 (2007) 885.

111. Y. Qin, X. Wang,Z. L. Wang, *Nature* 451 (2008) 809.

112. S. L. Zhang, U. J. Smith, *Vac. Sci. Technol. A* 22 (2004) 1361.

113. J. D. Lee, S. H. Jin, B. C. Shim, B. G. Park, *IEEE Electron Device Lett.* 22 (2001) 173.

114. B. Xiang, Q. X. Wang, Z. Wang, X.Z. Zhang, L. Q. Liu, J. Xu, D. P. Yu, *Appl. Phys. Lett.* 86 (2005) 243103.

115. K. Kang, S. Kim, C. J. Kim, M. H. Jo, *Nano Lett.* 8 (2008) 431.

116. A. L. Schmitt, J. M. Higgins, S. Jin, *Nano. Lett.* 8 (2008) 810.

117. J. R. Szczech, A. L. Schmitt, M. J. Bierman, S. Jin, *Chem. Mater.* 19 (2007) 3238.

118. K. S. K. Varadwaj, K. Seo, J. In, P. Mohanty, J. Park, B. J. Kim, *Am. Chem. Soc.* 129 (2007) 8594.

119. K. Seo, S. Lee, H. Yoon, J. In, K. S. K. Varadwaj, Y. Jo, M. H. Jung, J. Kim, B. Kim, *ACS Nano* 3 (2009) 1145.

120. D. K. Lim, O. Kubo, Y. Shingaya, T. Nakayama, Y. H. Kim, J. Y. Lee, M. Aono, H. Lee, D. Lee, S. Kim, *Appl. Phy. Lett.* 92 (2008) 203114.

121. C. Zou, X. Zhang, G. Jing, J. Zhang, Z. Liao, D. P. Yu, *Appl. Phys. Lett.* 92 (2008) 253102.

122. S. Zhou, X. Liu, Y. Lin, D.Wang, *Angew. Chem. Int. Ed.* 47 (2008) 7681.

123. S. Zhou, X. Liu, Y. Lin, D.Wang, *Chem. Mater.* 21 (2009) 1023.

124. B. Yang, H. Hu, Y. Wu, T. Luo, X. Yang, Y. Qian, *Chem. Lett.* 34 (2005) 326.

125. L. Dong, J. Bush, V. Chirayos, R. Solanki, J. Jiao, Y. Ono, J. F. Conley, B. D. Ulrich, *Nano Lett.* 5 (2005) 2112.

126. H. Okino, I. Matsuda, R. Hobara, Y. Hosomura, S. Hasegawa, P. A. Bennett, *Appl. Phys Lett.* 86 (2005) 233108.

127. Q. Wang, Q. Luo, C. Z. Gu, *Nanotechnology* 18 (2007) 195308.

128. S. Liang, R. Islam, D. J. Smith, P. A. Bennett, J. R. O'Brien, B. Taylor, *Appl. Phys. Lett.* 88 (2006) 113111.

129. C. Magen, C. Ritter, L. Morellon, P. A. Algarable, M. R. Ibarra, A. O. Tsokol, K. A. Gschneidner, V. K. Pecharsky, *Phy. Rev. B.* 74 (2006) 174413.

130. J. Kim, W. A. Anderson, *Nano Lett.* 6 (2006) 1356.

131. L. J. Chen, J. *Mater. Sci. R* 29 (2000) 115.

132. R. J. Kematick, C. E. Myers, *Chem. Mater.* 8 (1996) 287.

133. M. H. Wang, L. J. Chen, *Appl. Phys. Lett.* 71 (1992) 5918.

134. M. H. Wang, L. J. Chen, *Appl. Phys. Lett.* 59 (1991) 2460.

135. J. H. Wang, T. H. Yang, W. W. Wu, L. J. Chen, C. H. Chen, C. J. Chu, *Nanotechnology* 17 (2006) 719.

136. J. Du, P. Du, P. Hao, Y. Huang, Z. Ren, G. Han, W. Weng, G. J. Zhao, *J. Phys. Chem. C.* 111 (2007) 10814.

137. Y. L. Chueh, L. J .Chou, S. L. Cheng, J. H .He, W. W. Wu, L, J, Chen, *Appl. Phys. Lett.* 86 (2005) 133112.

138. Y. W. Ok, T. Y. Seong, C. J. Choi, K. N. Tu, *Appl. Phys. Lett.* 88 (2006) 043106.

139. F. Nava, K. N. Tu, O. Bisi, *Mater. Sci. Rep.* 9 (1993) 141.

140. R. S. Wagner, W. C. Ellis, *Appl. Phys. Lett.* 4 (1964) 89.

141. A. M. Morales, C. M. Lieber, *Science* 279 (1998) 208.

142. S. Shingubara, O. Okino, Y. Sayama, H. Sakaue, T. Takahagi, *Solid State Electron.* 43 (1999) 1143.

143. F. H. M. Spit, D. Gupta, K. N. Tu, *Phys. Rev. B* 39 (1989) 1255.

144. E. R. Weber, *Appl. Phys. A* 30 (1983) 1.

145. K. N. Tu, *Appl. Phys. Letts.* 27 (1975) 221.

146. V. Schmidt, H. Riel, S. Senz, S. Karg, W. Riess, U. GÖsele, *Small* 2 (2006) 85.

147. F. Patolsky, B. P. Timko, G. Zheng, C. M. Lieber, *MRS Bulletin* 32 (2007) 142.

148. Y. Cui, X. Duan, J. Hu, C. M. Lieber, *J. Phys. Chem. B* 104 (2000) 5213.

149. G. Zheng, W. Lu, S. Jin, C. M. Lieber, *Adv. Mater.* 16 (2004) 1890.

150. Y. Huang, X. Duan, Y. Cui, L. Lauhon, K. Kim, C. M. Lieber, *Science* 294 (2001) 1313.

151. Z. Zhong, D. Wang, Y. Cui, M. W. Bockrath, C. M. Lieber, *Science* 302 (2003) 1377.

152. X. Duan, C. Niu, V. Sahi, J. Chen, J. W. Parce, S. Empedocles, J. Goldman, *Nature* 425 (2003) 274.

153. X. Duan, *MRS Bull.* 32 (2007) 134.

154. A. Javey, S. Nam, R. S. Friedman, H. Yan, C. M. Lieber, *Nano Lett.* 7 (2007) 773.

155. F. Patolsky, G. Zheng, O. Hayden, M. Lakadamyali, X. Zhuang, C. M. Lieber, *Proc. Natl. Acad. Sci. U.S.A.* 101 (2004) 14017.

156. F. Patolsky, B. P. Timko, G. Zheng, C. M. Lieber, *MRS Bull.* 32 (2007) 142.

157. E. Bucher, S. Schulz, M. C. Lux-Steiner, P. Munz, U. Gubler, F. Greuter, *Appl. Phys. A: Mater. Sci. Process.* 40 (1986) 71.

158. Y. Cui, L. J. Lauhon, M. S. Gudiksen, J. Wang, C. M. Leiber, *Appl. Phys. Lett.* 78 (2001) 2214.

159. J. B. Hannon, S. Kodambaka, F. M. Ross, R. M. Tromp, *Nature* 440 (2006) 69.

160. S. M. Sze, *Physics of Semiconductor Devices*, John Wiley & Sons: New York, 1981.

Chapter 5

Formation of Epitaxial Silicide in Silicon Nanowires

Yi-Chia Chou,[a] Kuo-Chang Lu,[b] and King-Ning Tu[a]
[a]*Department of Materials Science and Engineering, University of California, Los Angeles, CA 90095, USA*
[b]*Department of Materials Science and Engineering, National Cheng Kung University, Tainan 701, Taiwan*
kntu@ucla.edu

5.1 Introduction

5.1.1 Overview of Contacts in Microelectronics and Nanoelectronics

In microelectronic industry, metal silicides have been widely used to serve as ohmic contacts, Schottky barriers, gate electrodes, and interconnect in very-large-scale-integration (VLSI) technology.[1–11] The science and technology of processing thin film silicides have been continuesly studied to understand their properties and solve technological problems.[12–28] The more we understand the silicide formation, the better we can monitor the semiconductor

Silicon and Silicide Nanowires: Applications, Fabrication, and Properties
Edited by Yu Huang and King-Ning Tu

ISBN 978-981-4303-46-0 (Hardcover), 978-981-4303-47-7 (eBook)
www.panstanford.com

device performance.[29–37] In the history of the evolution of semiconductor technology, dimensional scaling is always a crucial step in every device generation. Moore's Law predicts that the numbers of transistors being placed in the integrated circuits will double every two years, accompanying the shrinkage of transistor sizes. As the trend approaches the end of the semiconductor roadmap for microelectronic VLSI technology based on Si field-effect transistors, device engineers and scientists have been seeking new methods and materials for nanoscale transistors fabrication in recent years. For this reason, the properties and synthesis of metal silicide in nanoscale has become a field of intensive interests.[38–66]

Silicon nanowire is a potential material to follow the device-scaling owing to their morphology, size, and electrical properties, which are potentially suitable for electronics assembling.[67–68] Nanostructures in Si nanowires have been studied for basic components in electronic and optoelectronic devices, especially biosensors, and hence electrical contacts to Si nanowires are issues of both scientific and technological interests.[69–72] The nucleation and growth of epitaxial silicide in Si nanowires is basic to a better understanding of the kinetics of silicide formation and essential to a better control of the building blocks in the future nanoelectronics. In this chapter, the reactions of Ni and Co silicides formation in Si nanowires by point contacts will be proposed. The study of nucleation and growth of silicide formation on Si wafers and in Si nanowires will be discussed; where homogeneous and heterogeneous nucleation in nanoscale silicide formation will be compared in nanoscale silicide formation.

5.1.2 Introduction to Contacts in Nanoscale Electronics

5.1.2.1 Transition-metal silicides

Silicide films are employed in integrated semiconductor devices as part of the contact that joins an interconnecting line to the Si substrate in a contact window and for interconnections.

In the first application, the electric current flows mainly perpendicular to the silicide layer and across the silicide–Si interface. So the electrical characteristics of the contact between silicide and Si are of primary interest, especially parameters such as barrier heights, ideality factor and contact resistivity. These parameters are highly related to the morphology and microstructures of the contact interface. People would like to adjust these parameters and control their reproducibility and stability particular for rectifying contacts because their properties enter directly into the final performance of a device.

In the second application, electric current flows mainly along the plane of the silicide thin film. So the parameters of interests now are carrier concentrations, motilities and electromigration. A concern for the behavior of silicide layers under high-temperature annealing is impurities.

The reasons silicides are of interests as contact materials and as interconnection from VLSI is not only their good electrical conduction but also the ability of many transition-metal films to form a uniform silicide layer on a Si substrate by a solid-phase-reaction. The reaction temperatures are low so that the silicide contacting process can be applied without affecting preceding processing step. Moreover, silicides also offer a choice in electronic barrier heights to Si. Interest in silicides for interconnection lines arose with the needs to reduce the resistivity of interconnections as the line widths were reduced with the scaling down the device size. Most silicides are good electrical conductors and can withstand the high temperature in a conventional self-aligned technology, which refractory metals cannot. Furthermore, silicides have small grain sizes which are desirable for good line definition upon etching. Finally, they are compatible with poly-Si gate technology and thus adapt readily to existing process technology.

5.1.2.2 One-dimensional nanostructures

Nanostructures are defined as structures with at least one dimension between 1 to 100 nm. One-dimensional (1-D) semiconductor nanostructures are of particular interest because of their potential applications in nanoscale electronic and optoelectronic devices, especially bio-sensors. However, quantum effects play an increasingly prominent role while dimension shrinking to nanoscale.[73] Nanowires have demonstrated interesting electrical transport properties which are different from bulk materials and thin films. This is because, in nanowires, electrons could be quantum-confined laterally and thus could occupy discrete energy levels that are different from the energy bands found in bulk materials and thin films.[74] Due to low electron density and low effective mass, the quantized conductivity is more easily observed in semiconductors, for example, Si and GaAs, than in metals.[75] In order to understand more about the new physics demonstrated by nanowires, much effort has been dedicated to fabricating high-quality semiconductor nanowires because of the importance role of semiconductor materials to the electronics industry. In particular, the 1-D nanomaterials with perfect atomic structures and their assembly into functional devices are the major challenges and goals for nanomaterials studies.

Nanowire based sensing devices can be configured from high-performance field-effect transistors (FETs) by linking specific receptor groups to the surface of the nanowire. A macromolecule species such as a protein, can specifically bind to the receptor when the devices exposed to a solution, as shown in Fig.5.1a[41] This will affect device conductance depending on the net charge of the bio-molecule and the semiconductor type (p or n). Different nanowire elements with distinct surface receptors opens up the potential for multiplexed, real-time sensors, such as the simultaneous detection of proteins, DNA, viruses, and small molecules, as shown in Fig. 5.1b.[41]

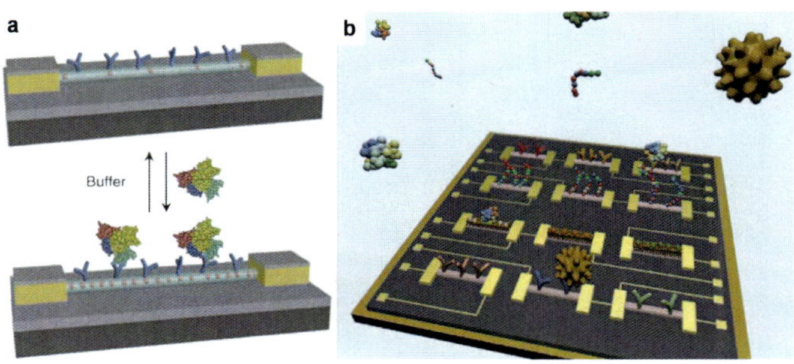

Figure 5.1 (a) Schematic illustration of a nanowire field-effect transistor configured as a sensor with antibody receptors (blue). (b) Schematic diagram of a nanowire device array for multiplexed, real-time sensing of multiple biological species.

The silicide/Si heterostructures can be synthesized by lithography on Si nanowires.[56] The Si region covered by Ni patterns will be transformed to NiSi to have NiSi/Si heterostructure, as shown in Fig. 5.2a. This method can be used to transform single crystalline Si nanowires to produce periodic silicide/Si heterostructures or superlattice. Figure 5.2b is optical dark-field images of a single heterostructure nanowire. Figure 5.2c is a TEM image of a similar NiSi/Si heterostructure.

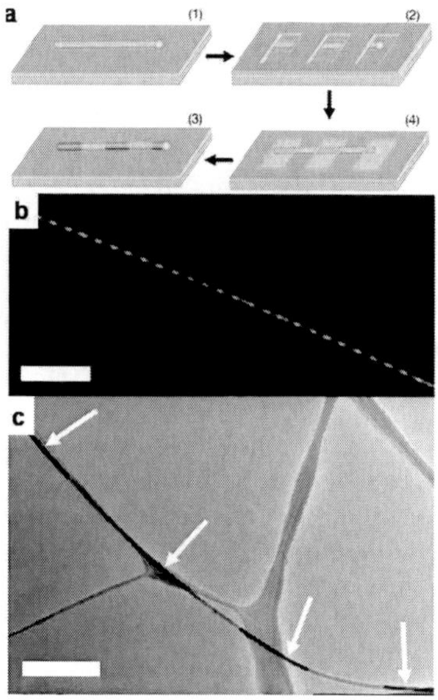

Figure 5.2 Fabrication and structural characterizations of NiSi/Si nanowire heterostructures and superlattices. (a) Fabrication of NiSi/Si nanowire heterostructures and superlattices by lithography. (1) Si nanowire (blue) dispersed on a substrate are (2) coated with photoresist (grey) and lithographically patterned, (3) selectively coated with Ni metal (green) to a total thickness comparable to the Si nanowire diameter, and (4) reacted at 550 °C to form NiSi nanowires. (b) Dark-field optical image of a single NiSi/Si nanowire heterostructure. The bright green segments correspond to silicon and the dark segments to NiSi. Scale bar is 10 mm. (c) TEM image of a NiSi/Si heterostructured nanowire. The bright segments of the nanowire correspond to silicon and the dark segments, which are highlighted with arrows, correspond to NiSi. Scale bar is 1 mm.

5.1.2.3 Si-based nanocircuits in Si nanowires

Well-defined nanoscale building blocks such as ohmic contacts and gates on Si nanowires must be developed in order to be assembled into functional electronic structures.[56,76] It requires a systematic study of chemical reactions in the nanoscale to form these circuit components. Not surprisingly, the formation of silicide nanowire has been investigated recently and widely. NiSi/Si/NiSi nanowire heterostructures have been reported by several groups.[56,66] Silicon nanowire transistors have been fabricated for these heterostructures.[56,66] Chemically doped silicon nanowires were used in order to ensure a reasonable source-drain contact and decent transistor characteristics. On the other hand, PtSi/Si/PtSi transistors based on intrinsic Si nanowires have achieved high performance normally off transistors.[77] The formation of atomically clean Si/silicide interfaces prevents Fermi level pinning and helps to maintain a low Schottky barrier determined by work function rather than by interface states.[78] So it enables the fabrication of high-performance nanowire transistors from intrinsic silicon nanowires using silicides as the source and drain contacts, as shown in Fig. 5.3. In addition, sub-100 nm channel length Ge/Si nanowire transistors contacted with nanoscale $NiSi_xGe_y$ layer showing potential for 2 THz switching speed.[77] It also showed small geometry contact as in the axial heterostructure is advantageous for device performance.[79] The metallic silicides may serve as the source-drain contacts to the Si in the heterostructure; it is the first step to produce a nanoscale field-effect transistor.[38,80]

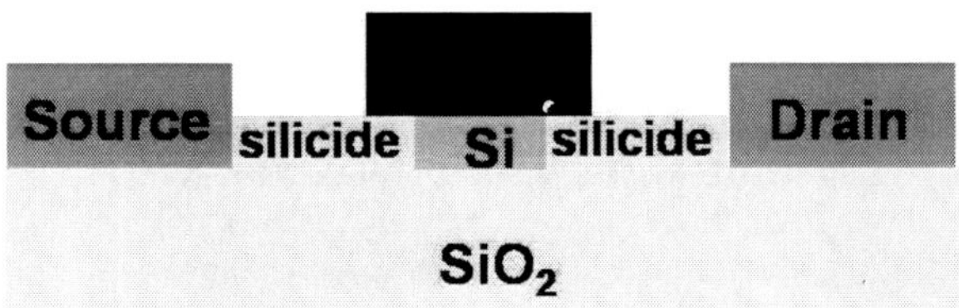

Figure 5.3 Schematic diagram of silicide/i-Si/silicide nanowire heterostructures as high-performance p-channel enhancement-mode transistors with HfO_2 as gate dielectrics.

5.1.3 Introduction to Solid-State Phase Transformations

Most phase transformations begin by nucleation and growth of a new phase in a metastable matrix.[81] Nucleation is a basic subject in all physical and biological phenomena of phase change, and it occurs by a fluctuation of composition and structure in the matrix of the metastable phase. A localized finite fluctuation in the system is called nucleus. Figure 5.4 shows a stable system with the same chemical potentials with the matrix phase β and the new phase α, while the tangent caused by a temperature change shows a fluctuation of composition from X_β and X_α to X_{system} and $X_{nucleus}$.[82] It results in different chemical potentials for both phases and A atoms would diffuse out from the matrix phase β. At the same time, the concentration of A atoms increases and tends to form nuclei as the initial state of phase transformation from β to α. Besides, it has to overcome an energy barrier to become stable because an interface has to be formed between the stable nucleus and the metastable matrix. In the fluctuation process, a spectrum of nuclei or embryos exists in the metastable phase; they grow and shrink simultaneously because of microreversibility. Microreversibility is the principle for elementary reaction in any equilibrium system; the favored reaction path in one direction must have the same frequency as the reverse path in the opposite direction, and the rate constants are

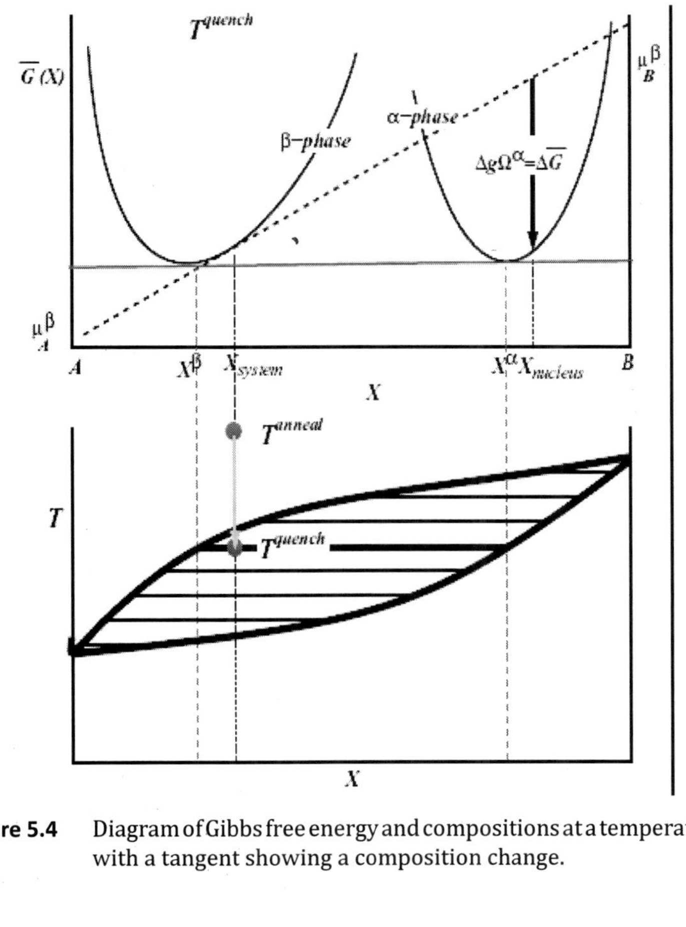

Figure 5.4 Diagram of Gibbs free energy and compositions at a temperature with a tangent showing a composition change.

the equilibrium constant. The critical radius of a nucleus and critical activation energy is highly related to temperature change, in turn the undercooling, and the critical nucleus may shrink back to matrix or keep growing. A stable nucleus is defined as the one that has reached beyond a critical size and will grow continuously. Before critical nuclei form in a sufficient quantity for experimental observation, a waiting period exists and is called the incubation time of nucleation. And it is followed by a steady state nucleation with typically a linear nucleation rate. Incubation time and nucleation rate are important concepts in phase transformations.

To understand the concepts in solid state nucleation, further consideration is necessary. The lattice mismatch between nucleus and matrix can bring an interfacial energy barrier to nucleation. When the nucleating phase has different unit cell volume and shape with the matrix it replaces, elastic energy must be considered as a part of volumetric contribution to nucleation. The interfacial contribution dominates at small nucleus sizes and the elastic contribution dominates at large nucleus sizes. The competition between the two contributions, therefore, would cause a complicated sequence of states for phase change.

Homogeneous nucleation is defined as nuclei forming randomly throughout the entire matrix, and heterogeneous nucleation is defined as nuclei forming at a location of surface or internal imperfection such as defects. While the steady state kinetic theory of fluctuation in size space is rather advanced, it is for homogeneous nucleation only. However, in real events of nucleation, heterogeneous nucleation is encountered inevitably because of the large difference in activation energy between the two.[83–84] Thus there is a gap in our understanding of the basic concepts of nucleation between theory and experiment. To reduce the effect of heterogeneity, nucleation in small droplets was studied, for instance.[85–86] The crystallization of small liquid droplets has been found to require a much larger undercooling than that in a bulk liquid phase. It shows that homogeneous nucleation is indeed much more difficult than heterogeneous nucleation in real thermodynamic systems. We will discuss the formation and kinetics of solid-state phase transformation of transition-metal silicides formation on Si wafer and in Si nanowires. Nucleation events of them will be discussed.

5.2 Introduction to Solid-State Phase Transformation in Thin Film

5.2.1 Thin Film Metal Silicide Formation

Metallurgists have had an interest in transition-metal silicides due to their refractory nature. Much information has been complied which is very useful for the application of these materials in VLSI circuits. Thin film metal silicides, such as C-54 $TiSi_2$, NiSi, and $CoSi_2$ have been used in semiconductor industry as gate, source and drain contacts for a long time. These three silicides have the lowest resistivity among all metallic silicides. The formation of a specific silicide phase on Si is determined by various factors such as annealing temperatures, metal deposition rates and deposition temperatures, metal film thickness, mixture of metal and Si deposition, and so on.[87–93]

5.2.1.1 Phase sequence

A typical initial situation for silicide preparation is depositing a thin metal film on a Si single crystal substrate. The silicide is produced by subjecting the sample to a treatment that will promote the reaction between the film and substrate. The standard treatment is annealing in a steady thermal environment and an inert atmosphere. The phase diagrams for the binary metal-silicide-forming system (Figs. 5.5 and 5.6) show several equilibrium phases; however, only some of these phases are present as detectable growth phases under steady state annealing conditions. Generally, only one phase grows at one time for clean systems. This fact demonstrates that the growth process is strongly controlled by kinetics.

It is possible to predict the first forming phase with a rule proposed by Walser and Bene,[94] which states that the first silicide phase is the highest congruent melting compound next to the lowest melting eutectic in the phase diagram. The Walser–Bene rule is about 80–90% successful in predicting first forming phase. After the metal thin film completely consumed by the first silicide phase formation, a second phase will form at higher temperatures and/or for longer annealing time. The second forming phase can sometimes predict by a rule proposed by Tsaur *et al.*[95] This rule states that the second phase to form is the one that congruently melts and enriched with

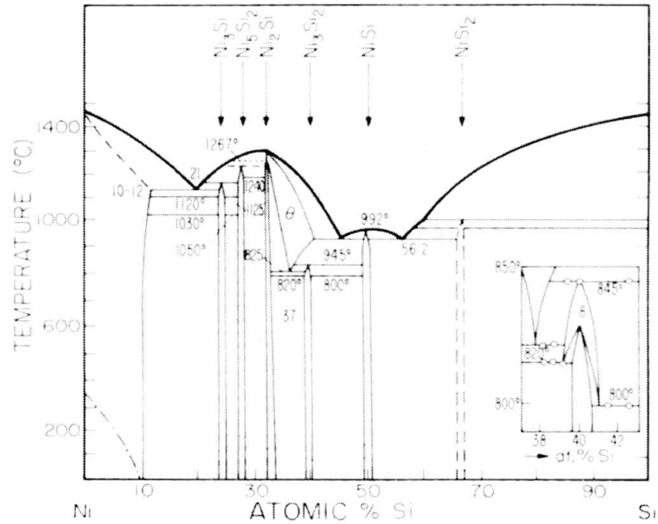

Figure 5.5 Ni–Si system phase diagram.

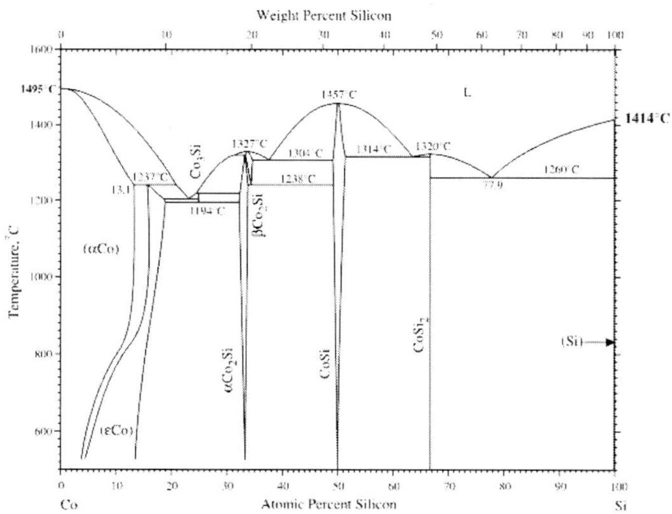

Figure 5.6 Co–Si system phase diagram.

the unreacted element. If the compounds between the first forming phase and other remaining elements are all noncongruently melting compounds, the next forming is the one with smallest temperature difference between peritectic point and the liquid curve above. Subsequent phases appear according to the same rule.

For near-noble metal films, such as Ni and Co, the dominant diffusion species is metal in their reactions with Si.[96] For a sample with thickness of Si is much larger than that of metal, the first silicide phase formed is M_2Si where M is metal, and when all metal film is consumed, MSi forms and then MSi_2 follows. This is because the whole sample is Si-rich when we deposited a metal film on a Si wafer, so the disilicide is the final equilibrium phase on the Si wafer.[97] Hence, according to thermodynamics, the equilibrium silicide phases are respectively $NiSi_2$ and $CoSi_2$ in the binary systems of Ni/Si and Co/Si, however, the first phases formed are Ni_2Si and Co_2Si and second phases formed are NiSi and CoSi. How to predict which is the first phase to form and what is the sequence of phase formation have been the most challenging questions in understanding thin film silicide formation. The phase formation is a kinetic phenomenon; formation energy is not controlling the selection of phases and the selection cannot be predicted by equilibrium phase diagrams.[98] For example, amorphous silicide can become the first phase to form in certain cases. In Ni/Si and Co/Si systems, the silicide phases of $NiSi_2$ and $CoSi_2$ have the fluorite structure with a negligible lattice mismatch to Si at room temperature. So the excellent epitaxy with Si, which is central to the fabrication of semiconductor devices,[99] can be achieved, yet they are not the first phase to form. In silicide formation in Si nanowires, the questions of first phase formation and the sequence of phase formation have no certain answer yet, because there are not enough data for a systematic analysis. However, the phenomenon of single phase formation remains true in nano silicide formation in Si nanowires.

5.2.1.2 Growth kinetics

The solid state reactions for silicide formation under steady state conditions for laterally uniform growth exhibit well-defined kinetics and temperature dependence which takes place over a relative wide range of temperature, and the growth proceeds laterally uniformly with relatively sharp interface. The growth process can be either limited by diffusion, thickness of silicide $\propto \sqrt{\text{anneal time}}$, or limited by interfacial reaction, thickness of silicide $\propto$ anneal time. Examples of diffusion-controlled reaction are the growth of near-noble metal silicides that follows a $\sqrt{t}$ relation, and some early transition-metal silicides are reported to follow linear time dependence. The di-metal and mono-metal silicides of near-noble metals such as Ni and Co all grow by a diffusion-controlled process.

In a point contact reaction, the atomic flux of Ni or Co that supplies the nucleation and growth of silicide is very small. It is unlike that in a plane contact in thin film reactions, where the atomic flux diffusing across the interface between the metal film and the Si substrate is much larger. We expect that the kinetics of these two kinds of reaction should be different. Actually, in the point contact case, it is more similar to a slow deposition of metal atoms on a Si surface to form silicide, where the metal flux can be controlled to be very small. Therefore, here we shall limit the chapter of thin film silicide formation and consider only vapor deposition.

Specifically, in vapor deposition, when a metal atom impinges the Si substrate, there is a probability that the atom diffuses on the substrate and there is also a probability that the atom desorbs from the substrate. In order to characterize the diffusion and desorption, the surface diffusion of the adatom and the residence time of an adatom on the surface have to be considered.

Considering the nucleation of an epitaxial nucleus of the thickness of a single atomic layer, the nucleation of such a circular disc on the flat solid substrate may have misfit strain due to lattice mismatch. However, when we consider the nucleation of a disc of silicide which has the same crystal structure and orientation as the single crystal silicide substrate, we can ignore the strain. More importantly, a positive surface energy comes from the circumference of the disc has to be included in the nucleation event.[100–101] This kind of disc nucleation on a single crystal surface is referred to as homogeneous nucleation. Why homogeneous nucleation can occur in epitaxial growth of silicide in Si nanowires needs to be explained later. This is because in bulk and thin film materials, homogeneous nucleation is rare because it cannot compete with heterogeneous nucleation. Heterogeneous nucleation occurs with a much higher frequency or with a much lower activation energy in most materials because it is enhanced by imperfections, impurities, and external surfaces.[88,102] However, homogeneous nucleation is found in epitaxial silicide growth in Si nanowires.[103]

In the classic nucleation theory, the shape of a standard nucleus by homogeneous nucleation is a sphere in three-dimension. But only a cap-shape of nucleus (a small part of a sphere) is required by heterogeneous nucleation. In nanoscale silicides formation in nanowire of Si, owing to the fact that the growth of the epitaxial silicide is atomic layer by atomic layer, so the nucleation event is two-dimensional and the nucleus has the thickness of a single atomic layer. Thus, the standard shape of the homogeneous nucleus is a circular disc. For heterogeneous nucleation, the nucleus is just a small part of a circular disc. It is worth mentioning that when the disc nucleus has the thickness of an atomic layer, whether its circumference is a surface or a line is more of a theoretical issue to be carefully studied.

5.2.2 Examples of Silicides Formation on Si Wafers

Atoms of Ni are depositing on a Si substrate by molecular beam epitaxy (MBE), chemical vapor deposition (CVD), and so on, interfacial reaction will occur when annealing occurs at a high temperature. The Si substrate serves as a parent phase and the deposition supplies metal atoms to keep the reaction going. The reaction can be interfacial-reaction-controlled or diffusion-controlled depending on growth conditions.[104]

The reaction between metal and Si can occur during metal deposition. The surface of a semiconductor is the starting place of epitaxial silicide nucleation and growth which require the diffusion of metal adatoms to surface steps on the semiconductor surface. The surface steps provide favorable sites for atoms to bond and they serve as a sink for diffusing atoms. The steps are high-energy sites and they are essential in homo-epitaxial and hetero-epitaxial growth.[1] Epitaxial silicide grows in a layer-by-layer fashion to maintain a planar interface.[17,99] If the semiconductor surface is perfectly flat and has no steps, nucleation of a disc is needed.

5.2.2.1 Ni silicides formation

The formation of Ni silicide between a Ni film and Si substrate exhibits a sequential growth of three phases: Ni_2Si, NiSi, and $NiSi_2$.[105] The growth of Ni_2Si is initiated at the interface between Ni and Si at temperatures from 200 °C to 350 °C. The growth kinetics of Ni_2Si follows a parabolic relation between the thickness of Ni_2Si and the annealing time. It was observed that the growth rate of Ni_2Si at 275 °C is two times faster on (100) Si substrate than on (111) Si substrate. In TEM studies, a difference was observed in microstructure of silicide formed on (100) and (111) Si substrate. The activation energy of Ni_2Si formation is 1.5 ± 0.2 eV at the temperature range of 200 to

350 °C; however, the activation energies do not have remarkable difference between silicides grown on (100) and (111) Si.

The transformation of Ni_2Si to NiSi is initiated at the interface between Si and Ni_2Si. The transformation was found to accompany with a stress change from compression in Ni_2Si to tension in NiSi. The formation is very fast at 350 °C, and the phase is stable up to 750 °C. At the temperature higher than 750 °C, epitaxial growth of $NiSi_2$ was found on (111), (110), and (100) Si.

From the thin film experiments, Ni_2Si is the first silicide phase to form.[106] For reaction of mono-layers of metal or metal–silicon mixtures with silicon substrates, the reduced reaction length and stress were found to change the kinetics of the reaction, and energetically, the orientation of the surface can influence the silicide formation.[107–108] For example, the metastable phase θ-Ni_2Si forms at 300 °C with the reaction of certain deposited Ni films on (111) Si but never on (001) Si. Its growth is followed by the subsequent growth of type-A $NiSi_2$ at 450 °C.[109–110]

Epitaxial $NiSi_2$ was found to grow heavily faceted on (001) Si. The interface between $NiSi_2$ and (111) Si is faceted but less than that on (001) Si. The interface was observed to be very rough on a microscopic scale, however, it was quite smooth on atomic scale and at short range.[111] Defect clusters and planar defects were also observed at the interfaces. Besides, at ultra-high-vacuum condition, single-crystal $NiSi_2$ thin films of either type-A or type-B orientation can be grown on (111) Si.[111]

The silicide films were found to grow out from small 3-D silicide islands which are formed at the initial stage of deposition.[1,112–113] A considerable intermixing occurs between the first few mono-layers of Ni and the Si substrate even at room temperature. Under appropriate conditions, $NiSi_2$ can be formed at 450–500 °C using rapid thermal annealing. The Ni–Si reaction precursor for the first few mono-layers of Ni results in a disordered layer containing three-dimensional islands. When the Ni thickness exceeds ~10–15 Å, the islands coalesce and the Ni–Si reaction slows down due to the lack of exposed Si regions.[114] The unreacted Ni is expected .to stay on top of the precursor silicide layer as more Ni is deposited.

5.2.2.2 Co silicides formation

In Co/Si thin film reactions, the first silicide phase is Co_2Si. Under non ultra-high-vacuum (UHV) deposition and annealing, the $CoSi_2$

transition from polycrystalline to epitaxial appeared to be very sluggish and interfacial dislocations at epitaxial silicide/Si interfaces were found with the dislocation spacing of several nm to over 100 nm.[97,115] Besides, single-crystalline $CoSi_2$ can be formed on (111) Si but not on (001) Si by non-UHV deposition of Co thin films and followed by rapid thermal annealing.[116]

Under UHV condition, epitaxial heterostructures, $Si/CoSi_2/Si$, have been fabricated.[117–118] A series of *in situ* low energy electron diffraction (LEED), angle-resolved ultraviolet photoemission spectroscopy (ARUPS), and X-ray photoelectron spectroscopy (XPS) measurements showed that at room temperature the Co/(111)Si interface consists of a very thin silicide layer, an intermediate silicide phase, covered by an unreacted Co film and a few Si atoms dispersed in the metallic matrix and at the top surface. Upon heating to high temperature about 600 °C, sequential silicide formation (Co_2Si, CoSi, $CoSi_2$) occurred and finally the epitaxial $CoSi_2$ was obtained on the (111) Si surface.[119–122] *In situ* TEM study showed that the silicide reaction occurs from isolated nucleation sites and progresses laterally across the surface.[123] Diffusion through the $CoSi_2$ lattice may be difficult at the usual $CoSi_2$ formation temperature (400–600 °C) and this may have an important effect on the morphology of thin $CoSi_2$ layers.[124–125]

5.2.3 Summary

In the phase diagram of metal–Si systems, there are more than three silicides present. Nevertheless, not all equilibrium phases are present as dominant phases during silicide formation in thin film systems. Some of the phases do not grow to macroscopic dimensions greater than the few hundred Å which can be detected by the available techniques. The examples shown in Section 2.2 were chosen as representative of the three broad classes of silicide formation observed in structures composed of metal thin films on single crystal Si: the metal-rich silicides, M_2Si; the monosilicides, MSi; and the disilicides, MSi_2.

Table 5.1 shows the general pattern of silicide formation. The silicides listed in the right-hand columns are chosen as specific examples of the phase formation for the elements shown in the left-hand column. The phase formation proceeds from left to right in each row with more metal-rich silicide phases forming first. The final represents the end phase between Si and the metal shown in the equipment phase diagram. Besides, for refractory metals, many phases exist in their phase diagram, yet only the disilicide is found.

Table 5.1 Silicide formation by contact reactions

Elements	Silicides		
Mg	Mg_2Si (only phase)		
Pt, Pd	Pt_2Si	PtSi (end phase)	
Ni, Co	Ni_2Si	NiSi	$NiSi_2$ (end phase)
Ti, (Zr), Hf		HfSi	$HfSi_2$
Fe, Rh, Mn			
V, Nb, Ta			VSi_2
Cr, Mo, W			

5.2.3.1 Metal-rich silicides

Table 5.2 shows a survey of metal-rich silicides formation. The growth kinetics of Ni_2Si and Co_2Si follows a parabolic law with an activation energy around 1.5 eV. The formation temperature of Ni_2Si and other metal-rich silicides such as Pd_2Si, Pt_2Si, and Mg_2Si starts forming at around 200 °C, with notable exception of Co_2Si which starts forming at 350 °C. The reason of the difference of the forming temperatures may caused by an alternative preparation technique.

5.2.3.2 Monosilicides

The formation of monosilicides is shown in Table 5.3. It also follows a parabolic growth rate but with an activation energy at the range of 1.6 to 2.5 eV. The formation temperatures for monosilicides is higher than that for metal-rich silicides and generally occurs above 350 °C. A notable exception is PdSi which form above 700 °C; this implies that its precursory phase Pd_2Si is very stable. In the cases of Ni, Co, and Pt, it is possible to have metal-rich silicides form as a layer between the metal and the monosilicide. However, in some other cases, the monosilicide is the first growth phase observed in the silicide.

5.2.3.3 Disilicides

Table 5.4 summarizes the formation of the disilicide phase observed in the growth of silicide layers. The formation temperature is around

Table 5.2 Formation of metal-rich silicide M_2Si

Silicode	Formation temperature, °C	Activation energy of growth, eV	Growth rate	Crystal structure	Melting point °C	Density,[a] g cem^{-3}	References
Ni_2Si	200~350	1.5	$t^{1/2}$	orthorhombic $PbCl_2$	1318	7.23	15, 16, 19
Pd_2Si	100~700	1.3–1.5	$t^{1/2}$	hexagonal Fe_2P	1330		4, 13, 14, 17 29, 30, 31
Pt_2Si	200~500	1.1–1.6	$t^{1/2}$	tetragonal $CuAl_2$	1100		32, 33, 34
Mg_2Si	≥200			cubic CaF_2	1102		35
Co_2Si	350~500	1.5	$t^{1/2}$	orthorhombic $PbCl_2$	1332		36,37

Table 5.3 Formation of monosilicide MSi

Silicode	Formation temperature, °C	Activation energy of growth, eV	Growth rate	Crystal structure	Melting point °C	Density,[a] g cem^{-3}	References
PtSi (end phase)	≥300	1.6	$t^{1/2}$	orthorhombic MnP	1229		1, 32, 33 34, 38
PdSi (end phase)	≥700			orthorhombic MnP	1100		31
NiSi	350~750			orthorhombic MnP	992		18, 19, 39
CoSi	425–500	1.9	$t^{1/2}$	cubic B20	1460	6.5	36, 37
FeSi	450–550	1.7	$t^{1/2}$	cubic B20	1410		40
RhSi	377			cubic B20			41, 42, 41
HfSi	550–700	2.5	$t^{1/2}$	orthorhombic FeB			10
TiSi	500			orthorhombic FeB	1760	4.21	23
MnSi	400–500			cubic B20	1275		43

Table 5.4 Formation of disilicide MSi_2

Silicode	Formation temperature, °C	Activation energy of growth, eV	Growth rate	Crystal structure	Melting point °C	Density,[a] g cem^{-3}	References
$TiSi_2$	600			orthorhombic	1540	3.9	23, 44
$ZrSi_2$	700			orthorhombic	1700		1, 43
$HfSi_2$	750			orthorhombic	1950		10
VSi_2	600	2.9, 1.8	t and $t^{1/2}$	hexagonal	1750		20, 21
$NbSi_2$	650			hexagonal	1930		27, 45
$TaSi_2$	650			hexagonal	2200		46, 46*a*
$CrSi_2$	450	1.7	t	hexagonal	1550	–4.91	43, 44
$MoSi_2$	525	3.2	t	tetragonal	2050		44, 47, 47*a*
WSi_2	650	3	t and $t^{1/2}$	tetragonal	2165		48, 47*a*
$NiSi_2$	750			cubic (CaF_2)	993		18
$CoSi_2$	550			cubic (CaF_2)	1326		36, 37
$FeSi_2$	550			tetragonal	1212	4.54	40
$MnSi_2$	800			tetragonal	1150		43

700 °C for most cases. The activation energy of formation is high, ranging from 1.7 to 3.2 eV. The disilicide the Ni and Co are found to grow epitaxially on Si substrates. Both $NiSi_2$ and $CoSi_2$ have very close lattice parameters with Si which results in small lattice mismatch.

In addition, the disilicides of the refractory metals tends to have a linear growth curve in the early stage of reaction while the disilicide is the first forming phase. The presence of a thin oxide layer between the metal and Si can impede silicide formation, as in the case of Cr and Si.

5.3 Nanoscale Silicide Formation by Point Contact Reaction between Ni/Co and Si Nanowires

5.3.1 Introduction

For the development of nanoscale transistors based on Si nanowires,[39–41] the contact formation of silicide nanowires, either by substrate growth[56,61,66,128] or by free-standing growth,[129–133] and nanowire heterostructures has been investigated.[56,66,134,135] Furthermore, the NiSi/Si/NiSi nanowire heterostructures have been demonstrated to exhibit very good transport properties.[56,66] We have selected Ni and Co for the reaction with Si for our study since Ni and Co silicides possess very interesting and important properties. Single crystalline CoSi has ferromagnetic properties which were the focus of intensive research[131,133]. The monosilicide NiSi and disilicide $CoSi_2$ are two of the three silicides which have the lowest resistivity for applications in shallow junction devices[2,136]. The growth kinetics of Ni and Co silicides in nanoscale may be different from that in thin film and bulk samples.[2] However, how to achieve a precisely controlled and reproducible nanostructure remains one of the most challenging problems in nanotechnology today. Additionally, the formation of circuit elements in nanodevices requires a thorough study of chemical reactions in the nanoscale. An essential property of nanoscale chemical reactions between nanowires is that the contact between two nanowires is a point contact, the effect of which on chemical reactions is unknown. The growth of Ni and Co silicides, NiSi and CoSi/ $CoSi_2$ was reported. Both $CoSi_2$ and NiSi undergo epitaxial growth with Si and favorable to form sharp silicide/Si interfaces, while CoSi formed rough interface without epitaxial relation with Si. Experiments direct showed that silicide growth is supplied by metal in Co silicide formation cases by *in situ* TEM. A stepwise growth mode was found in silicides epitaxial growth in Si nanowires. It was observed that the stepwise growth occurs repeatedly in the form of an atomic step sweeping across the silicide/Si interface. It appears that the growth of a new step or a new silicide layer requires an independent event of nucleation. We are able to resolve the nucleation stage and the growth stage of each layer of the epitaxial growth in video images. In the nucleation stage, the incubation period is measured, which is much longer than the period needed to grow the layer across the silicide/Si interface. So the epitaxial growth consists of a repeating nucleation and a rapid stepwise growth across the epitaxial interface. This is a general behavior of epitaxial growth in nanowires. NiSi/Si/NiSi and $CoSi_2$/Si/$CoSi_2$ were synthesized by using *in situ* TEM, the potential gate of a Si nanowire transistor,[66] down to atomic scale. Also, the huge strain in the Si region can be controlled, which could be beneficial in the development of high-mobility sub-10-nm silicon devices.[137–138]

5.3.2 Experimental Methods

Si nanowires were prepared on a Si wafer by the vapor–liquid–solid (VLS) method using nano-Au dots as catalyst for the growth of single crystal Si nanowires with [111] growth direction.[139–140] The Si nanowires were detached from the wafer in ethanol solution by ultrasonic vibration. Polycrystalline Ni and Co nanowires were synthesized by electroplating Ni and Co into a porous anodic aluminum oxide (AAO) template structure and then the AAO template was etched away to obtain Ni and Co nanowires. Finally the nanowires were stored in isopropanol solution.[141] Both the metal nanowires and Si nanowires have diameters ranging from 20 to 70 nm and lengths of a few microns. Point contact samples were prepared by dripping droplets of solutions containing Si and metal nanowires on a Cu grid covered with a 50 nm–thick amorphous SiO_2 film, which is transparent to the electron beam and does not interfere significantly with the images of Si and metal nanowires in high resolution transmission electron microscope (HRTEM). Another method to prepare point contact samples is by depositing ~3.8 nm–thick Ni films on the SiO_2 film on the Cu grid and then anneal the film to form Ni nanoparticles. After that, drip droplets of Si nanowire solution to form point contact samples. The samples were dried at room temperature. *In situ* annealing for point contact reactions and high-resolution lattice videos were performed and taken in a JEOL 2000 V ultrahigh vacuum TEM. The vacuum in the sample stage was about 3×10^{-10} Torr. High-resolution lattice imaging and line scan were performed in a JEOL 3000F high-resolution TEM.

5.3.3 Point Contact Reactions between Nanowires of Si and Co

5.3.3.1 CoSi formation by the supply of Co nanodots into Si nanowires

Point contact samples between Si and Co nanowires were prepared by randomly dropping Si and Co nanowires solution on the ~50 nm–thick SiO_2 coating of Cu grids. Upon heating, the Co nanowires were prone to disintegrate and form Co nanodots on the oxidized surfaces of Si nanowires having about 2 nm–thick surface oxide layer and also on the SiO_2 coating of the Cu grid. Most likely the Co particles were formed by evaporation from the Co nanowires and condensation on the SiO_2 coating of the Cu grids.[103] Figure 5.7a,b are the TEM images sequence of point contacts of a Si nanowire and several Co nanodots annealed at 700 °C. Many small Co particles were observed near the Si nanowires and some of them were on the SiO_2 window, as shown in Fig. 5.7. In the Si nanowire, the dark region on the left hand side is Co silicide. From the selected area diffraction pattern, the silicide formed within Si nanowires was CoSi. We did not find a good epitaxial relation between CoSi and Si. As a result, the interface is rough. Comparing with NiSi formation within Si nanowire, NiSi was found to grow epitaxially on (111) Si with the following epitaxial relation as given before; Si[1-10]//NiSi[1-12] and Si(111)//NiSi(31-1). Therefore, the interface between NiSi and Si is atomically sharp.[139] In addition, no misfit dislocations were found across the NiSi/Si interface.

Figure 5.7a,b shows that a Co particle (indicated by a blue arrow) disappeared within 0.07 s and the CoSi grew promptly within the same time period. If we assume the Co particle is spherical, the volume of the Co particle is about 1580 nm^3. When such a volume of Co reacts with Si to form CoSi, the volume of the formed CoSi can be calculated to be 3500 nm^3. Knowing the densities of Co and CoSi, which are 8.9×10^{-18} and 6.65×10^{-18} g/nm^3, respectively, it was calculated that 2.40×10^{-16} mole of Co and 2.67×10^{-16} mole of CoSi participated in the reaction. It suggests that the nano CoSi phase may be highly deficient in Co, about 10 atomic %. In other words, the nano CoSi may contain a large number of vacancies in the sublattice of Co, which can affect the diffusion of Co in the CoSi. Since the mole values are close, we conclude that the rapid disappearance of the Co particle led to the CoSi formation.[103] From the finding that the instant reaction rate was much higher than the average growth rate, we conclude that the dissolution of a large number of Co atoms into the Si occurs in a short time, so the reaction rate is much faster. The consideration is based on the assumption that the Co particle has had a large contact area to the Si nanowire as the oxide layer might be thinner and the bonding might be weaker at the contact location. Although this is a special case, it provides us, however, the important information that the source of Co needed to form the CoSi is the Co particles on the surface of the Si nanowire. Nevertheless, for most of the Co particles, the point contact having a very small contact area will limit the reaction, so the overall growth of CoSi is much slower, as shown in Fig. 5.7c.

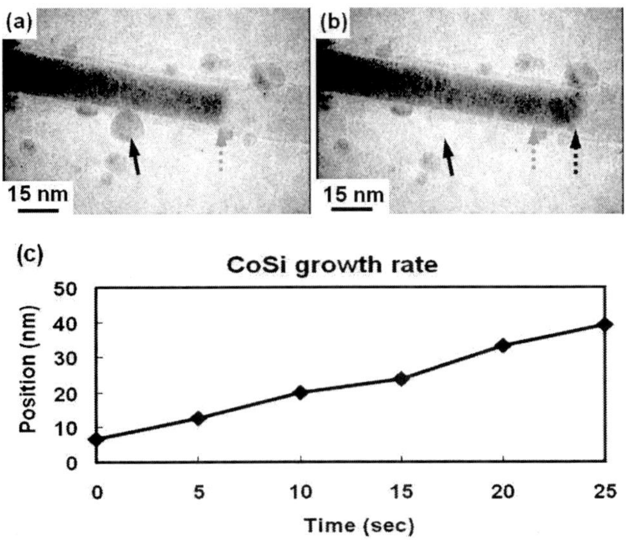

Figure 5.7 *In situ* TEM image sequences showing CoSi growth while Co particle disappears. (a) *In situ* TEM image at 0 s. The Co particle is indicated by the arrow. (b) *In situ* TEM image at 0.07 s. The bright and the dark areas are Si and CoSi, respectively.

5.3.3.2 Epitaxial growth of $CoSi_2$ in Si nanowires

When we annealed the point contact samples at 800 °C, a few small Co particles were formed on the SiO_2 coating surface at the beginning. Then $CoSi_2$ formed within the Si nanowire and more Co particles appeared on the SiO_2 coating surface and some were attached to the surface oxide of the Si nanowires. Figure 5.8a is a TEM image taken *in situ* of Co and Si nanowires in point contact position after the annealing at 800 °C for 13 min. No Co silicide formation was evident at this moment; however, a few small Co particles were formed on the SiO_2 coating surface. After 55 min annealing, the $CoSi_2$ formed within the Si nanowire and more Co particles appeared on the SiO_2 coating surface and some were attached to the surface oxide of the Si nanowires, as shown in Fig. 5.8b. Since the Co source comes from

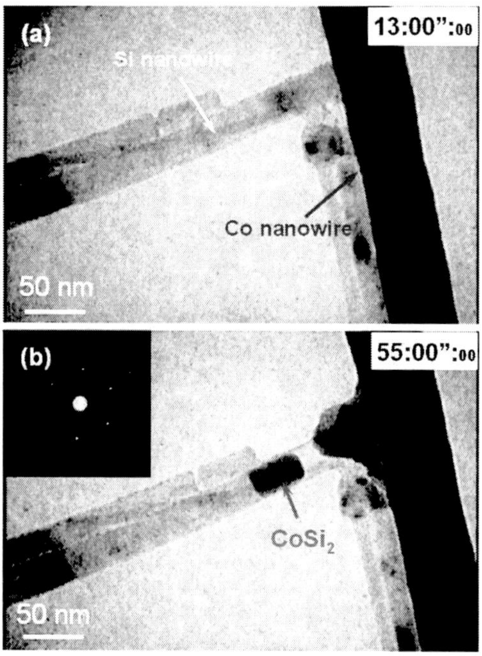

Figure 5.8 Point contact reaction between Co and Si nanowires annealed at 800 °C. (a) After 13 min 800 °C annealing. (b) $CoSi_2$ form after 55 min 800 °C annealing. The time of the image capture is given in the rectangular box at the upper right corners. The first numbers are in units of minutes and the middle numbers are in units of seconds, and the following two small numbers are in units of 1/100 s.

both Co nanoparticles on the SiO_2 coating and the point contact with Co nanowire, $CoSi_2$ tend to nucleate at high Co concentration regions of the Si nanowire as shown in Fig. 5.8b. The selected area diffraction pattern (see inset) shows that the $CoSi_2$ is single crystalline with a CaF_2 structure.

Figure 5.9a is a HRTEM image of a sharp epitaxial $CoSi_2$/Si interface. The darker and brighter regions show the atomic layers in $CoSi_2$

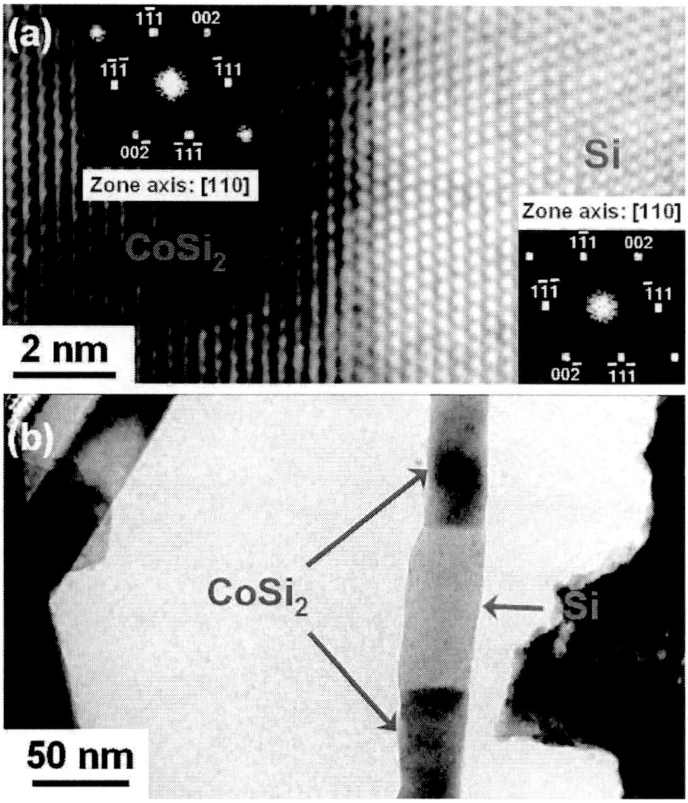

Figure 5.9 TEM images showing a high-resolution $CoSi_2$ and Si interface and $CoSi_2$/Si/$CoSi_2$ heterostructure within a Si nanowire. (a) High-resolution TEM image of $CoSi_2$–Si interface. (b) A TEM image of nanoheterostructure of $CoSi_2$/Si/$CoSi_2$ form within a Si nanowire. The light area is Si and the dark area is $CoSi_2$. The insets in panel (a) are the fast Fourier transform (FFT) pattern confirming the [110] Si zone axis and the [110] $CoSi_2$ zone axis, respectively. {002} spots in the Si pattern are from double diffraction.

and Si, respectively. The crystallographic orientation relationships between the Si and $CoSi_2$ are Si[110]//$CoSi_2$[110] and Si(111)//$CoSi_2$(111) (see the selected area electron diffraction patterns of Si and $CoSi_2$ in the inset),[103] which show the direction[110] on (111) plane of $CoSi_2$ is parallel to the direction of [110] on (111) plane of Si. The Si and $CoSi_2$ are of diamond and CaF_2 crystal structures with lattice parameters of 0.5432 and 0.5365 nm, respectively. Since the structures of $CoSi_2$ and Si are cubic and with small lattice mismatch (1.2%), Si (111) can have a good epitaxial relation with $CoSi_2$ (111); therefore, $CoSi_2$ has a preferential epitaxial orientation with Si to form a sharp interface of low interfacial energy. Since the structures differ for CoSi and $CoSi_2$, which are respectively BCC and CaF_2, the degree for epitaxial growth of CoSi and $CoSi_2$ on Si are different so that there is a large difference for the sharpness of their interfaces. The phase of Co silicide can be determined by selected area diffraction and by the morphology of the interface. It has been reported that $CoSi_2$ is a higher temperature phase and CoSi is a lower temperature phase, in thin film reactions. If we apply the thin film results to our nanowires, it explains our finding of different temperatures for the formation of nano CoSi and $CoSi_2$. Fig. 5.9b is a TEM image of $CoSi_2$/Si/$CoSi_2$ heterostructure with the middle Si region of 90 nm in thickness.

Besides, the Co nanodots appear near the Si nanowires and also on the SiO_2 coating over the Cu grids, and the source of Co comes from Co nanowires evaporation and condensation. In the ultrahigh vacuum TEM chamber and at 700 °C, the partial pressure of Co due to Gibbs–Thomson effect of the Co nanowire will be higher than the equilibrium partial pressure of Co on a flat surface for evaporation and condensation. Yet, the newly formed Co particles have been found to be quite stable since no ripening among them can be detected. It is an interesting subject by itself and we suggest that the stability might be due to the interaction at the triple point between the Co particle and the SiO_2 surface.

The growth rates of $CoSi_2$ within the Si nanowire at 800 °C can be measured from *in situ* HRTEM video, which shows that the $CoSi_2$ grows atomic layer by atomic layer. During the growth of an atomic layer, the growth mode behaves as the moving of steps or kinks across the interface, as shown in Fig. 5.10a–d. From the video, however, we found that there is a long period of stagnation between

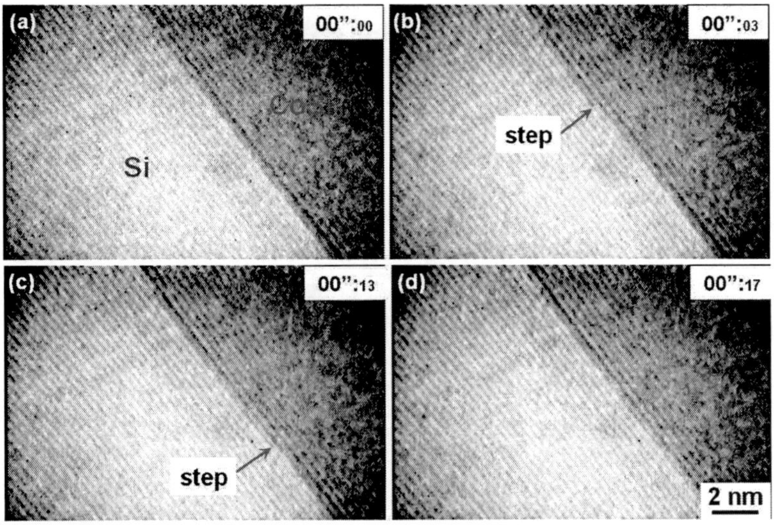

Figure 5.10 *In situ* HRTEM image sequences of growing $CoSi_2$/Si epitaxial interfaces within Si nanowires. (a–d) The growth of 1 atomic layer has occurred, and the step is indicated by the arrow. The left-up and right-down regions are close to center and edge of nanowire, respectively.

the stepwise growths of two successive atomic layers. When we plot the stagnation period as well as the growth period, we obtained the stair-type curve as shown in Fig. 5.11a. It shows that the growth rate of each of the $CoSi_2$ atomic layer is the same, which is ~0.17 s per layer, and it is just the width of the vertical line in the Fig. 5.11a. In between the vertical lines (the horizontal part of the steps in Fig. 5.11a is the stagnation period, which we define as the incubation time of nucleation of a new layer. The average value of the incubation time is about 5 s. In other words, while the motion of a step or a kink across the cross-section of the nanowire is the same, the intervals between the growths of atomic layers are not the same, indicating that the incubation time of the nucleation of a step or a kink on the cross-section can be different, as shown in Fig. 5.11a that the length of the horizontal part are different. For comparison, Fig. 5.11b shows three recorded stair-type of axial epitaxial growth of $CoSi_2$ in Si nanowire.

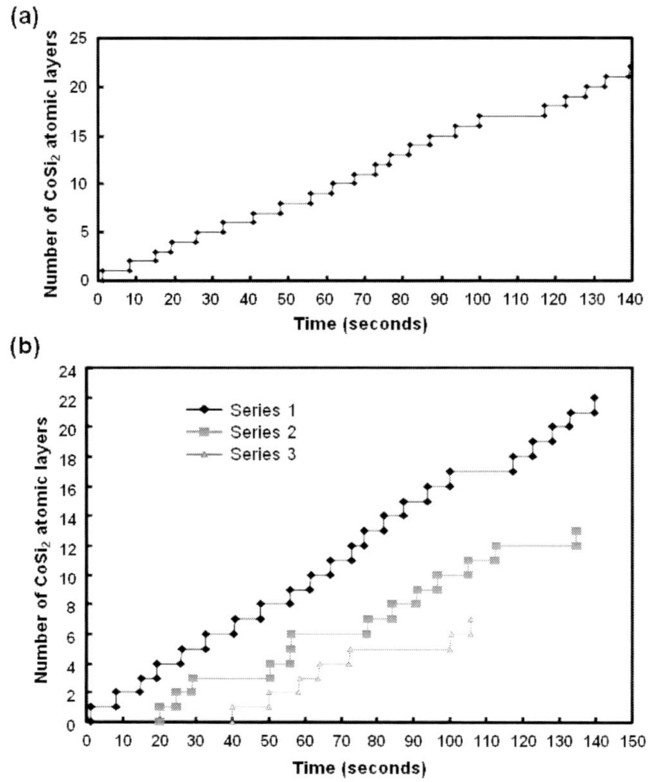

Figure 5.11 Plot of $CoSi_2$ atomic layers as a function of time to show the nucleation time and growth time of each step. (a) One stair-type growth curve of $CoSi_2$. (b) Three sets of stair-type growth curve of $CoSi_2$.

5.3.4 Point Contact Reactions between Nanowires of Si and Ni

5.3.4.1 Formation of NiSi contacts within Si nanowires and NiSi/Si/NiSi nanowire heterostructures as building blocks for field-effect transistors

Silicon nanowires were prepared on a p-type Si wafer by the VLS method using Au nanodots as nucleation sites for single-crystal Si nanowires with a [111] growth direction.[139–140] Ni nanowires were synthesized via the AAO method and stored in isoproponal.[141] To prepare point contact samples, we drip droplets of both solutions on Si discs having a square opening covered with a window of glassy Si_3N_4 film. The samples were dried under light bulbs. Figure 5.12a shows a typical TEM image of randomly oriented Si and Ni nanowires on the window prior to annealing. Both the Si and Ni nanowires are single crystal with a thin surface oxide (~1–5 nm thick). Darker contrast appears on the Si nanowire, resulting from the fact that the zone axis is accurately parallel to Si [110].

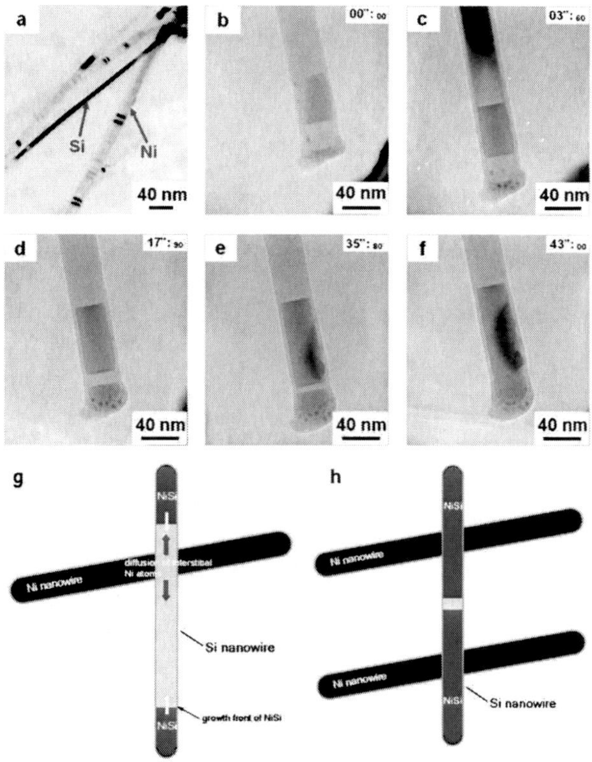

Figure 5.12 Overview of NiSi formation within Si nanowires by point contact reaction. (a) A TEM image of Si and Ni nanowires dispersed on a Si_3N_4 membrane. (b–f) Sequence of *in situ* TEM images depicting the growth of a bamboo-type grain of NiSi within a Si nanowire at 700 °C. The time of the image capture is given in the rectangular box at the upper-right corner. The first two numbers are in unit of seconds and the following two smaller numbers are in unit of 1/100 second. (g) A schematic illustration of NiSi growth within a Si nanowire. (h) A schematic illustration of growth of a NiSi/Si/NiSi heterostructure where the length of the Si section is controlled at the atomic scale.

Figure 5.12b–f shows a time-lapsed series of *in situ* TEM images capturing the growth of a bamboo-type single-crystal NiSi grain within a straight Si nanowire at 700 °C. Figure 5.12g is a schematic diagram depicting the growth of NiSi, in which the Ni atoms dissolve and diffuse interstitially in Si[140] and stop at the two ends of the Si nanowire, thereby nucleating the growth of NiSi to form a NiSi/Si/NiSi heterostructure. Some Ni atoms may be able to diffuse through a grain of silicide; hence, we observed additional and smaller silicide growth at the tip which is attached to the long silicide at the end, as shown in Fig. 5.12d–f. Figure 5.12h depicts two Ni nanowires making contacts to a Si nanowire, in which the NiSi grains have passed the point contacts requiring the Ni atoms to diffuse through the NiSi.

5.3.4.2 Epitaxial relationship between NiSi and Si and atomically sharp interfaces

Figure 5.13a–d show a set of high-resolution TEM images of the NiSi/Si interface taken as the interface advances into the Si. The interface is parallel to the (111) plane of Si as well as the (311) plane of NiSi. Thus, the growth direction of NiSi is normal to the (311) plane. The crystallographic orientation relationships between Si and NiSi, which has an orthorhombic lattice with lattice constants of a = 0.562 nm, b = 0.518 nm, and c = 0.334 nm, are

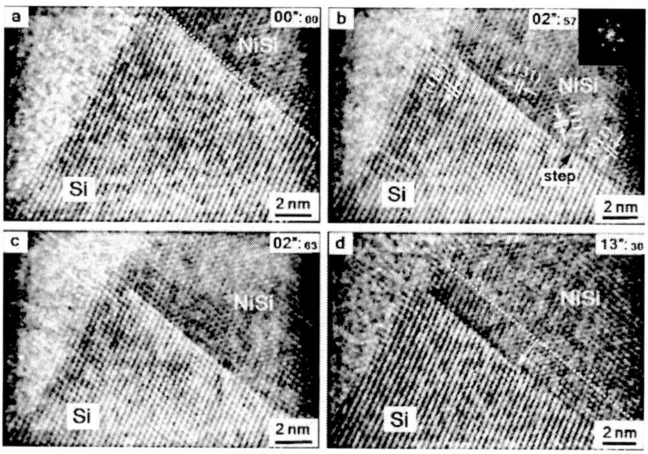

Figure 5.13 (a–d) *In situ* high-resolution TEM image sequence of the growth of the NiSi/Si epitaxial interface. From (a) to (d), growth of five atomic layers has occurred, as indicated by the arrow in (d). The inset in (b) is the corresponding FFT pattern confirming the [1-12] NiSi zone axis.

[1-10] Si//[1-12] NiSi, and (111) Si//(31-1) NiSi

Across the epitaxial interface, the misfit is around 5.6%.[141] However, we were unable to find any misfit dislocations at the NiSi/Si epitaxial interface and this may be due to the difficult nucleation and growth of the dislocations in a nanowire.

5.3.4.3 Kinetic analysis of reactive epitaxial growth of nano-NiSi/Si/NiSi

Figure 5.14 shows the linear growth behavior of the NiSi nanowire in the Si nanowire of 20 nm in diameter over the temperatures ranged from 500 °C to 650 °C. The activation energy of the epitaxial growth was determined to be 1.25 eV/atom, compared to the activation energy of interstitial diffusion of Ni in Si of about 0.47 eV/atom,[142] indicating that the linear growth may be interfacial-reaction-controlled.

Knowing the diameter of the wire and the growth rate, we can estimate the total number of Ni atoms in a given volume of the wire based on the unit cell volume of NiSi and the number of Ni atoms per unit cell. Thus, we have determined that for growth at 700 °C the time needed to incorporate one Ni atom on the growth front of NiSi is about 7×10^{-4} s.[141]

We can estimate the time for Ni atoms to diffuse in a Si nanowire from $x^2 = Dt$, where x is the length of the Si nanowire between the point contact and the Si/silicide interface, D is interstitial diffusivity of Ni atoms in Si nanowires, and t is the time. If we take D to be 10^{-6} cm^2/s, and x to be 1 μm, we have $t = 10^{-2}$ s.[140] Since the diffusion time is much longer than what was estimated for the growth of one atom on the epitaxial interface, the growth cannot be interfacial-reaction-controlled at the epitaxial interface. Instead, we propose that the reaction may be limited by the rate of dissolution of Ni into Si at the point contact interface. Atomic flux by definition is the number of atoms per unit area per unit time, so that the number of Ni atoms diffusing into Si is equal to flux times area times time.[98] If the contact area is a point, the number of Ni atoms diffusing into Si will be extremely small, and thereby the reaction is limited by how fast the Ni atoms can diffuse into the Si nanowire or by how large is the Ni flux. This is a unique feature of point contact reactions.[141] Surface diffusion of Ni on the SiO_2 surface of the Si nanowire was considered to be a kinetic path of the supply of Ni, however, it is slower than the interstitial diffusion of Ni within Si,[140] which dominates the reactions.

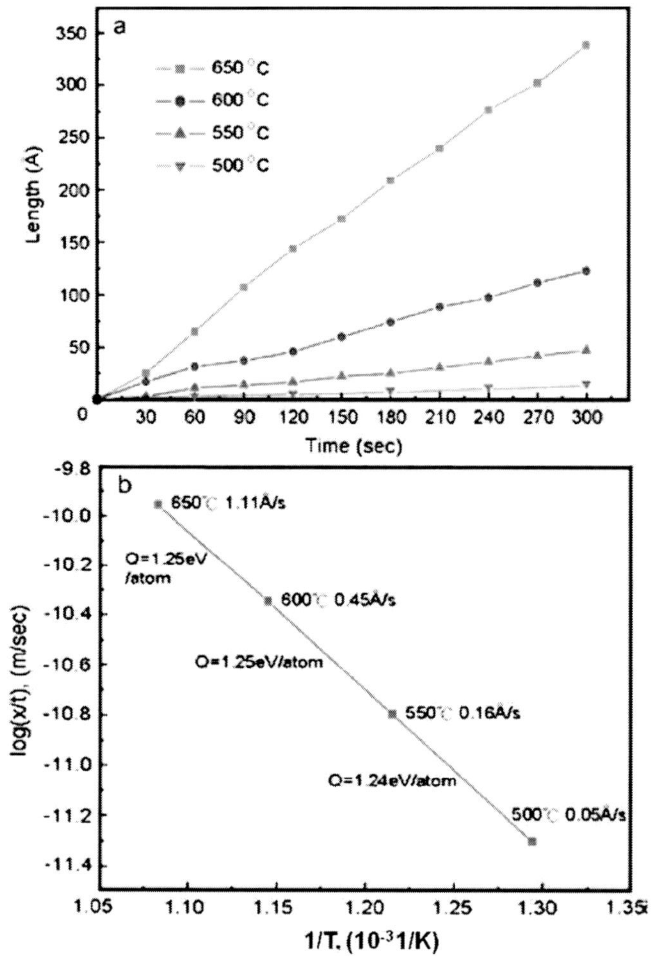

Figure 5.14 Kinetic analysis of the NiSi epitaxial growth within a Si nanowire of 20 nm diameter. (a) Plot of the NiSi nanowire length versus reaction time at various temperatures, illustrating a linear growth rate. The lines are drawn as guides. (b) Arrhenius plot of the NiSi epitaxial growth, from which the activation energy was determined.

Silicide formation occurs at both ends of the Si nanowire, and if annealing is stopped before the entire Si nanowire transforms into NiSi, as shown in Fig. 5.15c, a nanoheterostructure is formed as shown in Fig. 5.15a,b. When we consider the growth of the NiSi in Fig. 5.12h, its growth has passed the point contact and the Ni atoms must diffuse through the NiSi. The growth is found to be slower with an activation energy of 1.7 eV/atom, which is close to the value for NiSi growth in thin film silicide reactions.[32,98] Therefore, in point contact reaction, nano-NiSi growth starts from both ends rather than

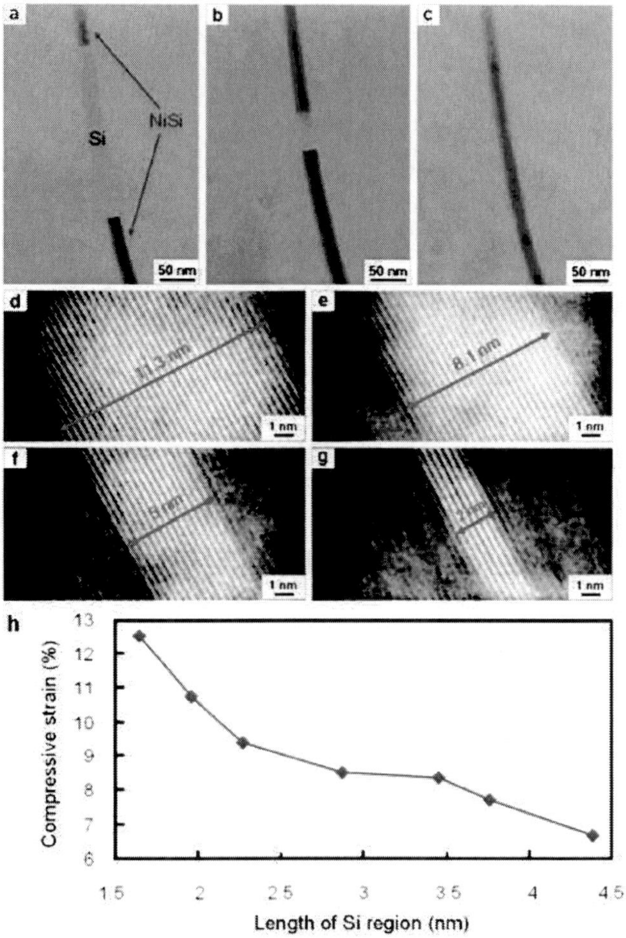

Figure 5.15 *In situ* TEM images showing the formation of our NiSi/Si/NiSi heterostructures within a Si nanowire, and compressive strain in the Si region. (a, b) *In situ* TEM images of the NiSi/Si/NiSi heterostructure. The bright area is Si and the dark area is NiSi. (d–g) High-resolution TEM images of NiSi/Si/NiSi heterostructures. The bright and dark portions of the lattice images correspond to Si and NiSi, respectively. (h) Plot of the compressive strain versus the length of the Si region in the nano-heterostructure NiSi/Si/NiSi.

from the point contact for two reasons. First, the supersaturation of Ni atoms at the point contact is impeded by the limited contact area and the oxide. Second, due to interstitial diffusion within a Si nanowire, it is easier for Ni atoms to diffuse through the entire Si nanowire, leading to grow from the ends, than through NiSi with activation energy of 1.7 eV/atom, contributing to nucleation and growth of NiSi from the location of the point contact.

The atomistic mechanism of the epitaxial growth involving a moving interface is of interest.[141] In the present case, the covalent Si-Si bonds must be broken and transformed into metallic Ni–Si bonds in order for NiSi to grow and have a moving interface. Although the thermal energy at 700 °C is sufficient to break the covalent Si-Si bonds, the interstitial Ni atoms are crucial in the bond-breaking process.[87] It is known that in the interfacial reactions between metal thin films and single crystal Si wafers, near-noble metals such as Ni and Pd can react with Si to form silicide at temperatures as low as 100 °C.[88,111] Near-noble metals diffuse through the silicide (not through Si as we reported here), dissolve interstitially in Si, and assist the breaking of the Si-Si covalent bonds. Furthermore, native oxide on the Si surface is not an effective diffusion barrier to Ni and Pd, and therefore the metal atoms can diffuse through the native oxide and react with the Si below the oxide. Hence, in planar thin film reactions, nucleation and growth of silicide occurs right below the plane of contact. Yet in point contact reactions, owing to the lack of supersaturation below the point of contact because of the very limited number of dissolved Ni atoms and because they can diffuse away quickly, silicide in Si nanowire does not form at the location of the point contact.

5.3.4.4 Fabrication of 2 nm to 200 nm highly strained Si in dimension controlled NiSi/Si/NiSi heterostructures

Based on knowing the growth rates, we can control the remaining length of the Si region between the two NiSi regions, as illustrated in Fig. 5.12h. We can pattern or deposit two Ni nanowires with a given spacing over a Si wire, and then utilize the reaction to bring the two NiSi grains as close as possible. At 500 °C, the reaction rate can be controlled down to atomic scale as shown in Fig. 5.14. In Fig. 5.15d–g, a set of lattice images of the nano-heterostructure of NiSi/Si/NiSi with 11.3, 8.1, 5, and 2 nm lengths of Si is shown.[141]

By measuring the length of the Si region and counting the number of (111) lattice planes within the region, we can determine the strain, and we find that the Si is highly compressed. Figure 5.15h shows the relationship between the compressive strain and the length of the Si region in the nano-heterostructure NiSi/Si/NiSi at room temperature. From the difference in the interplanar spacing of Si and NiSi across the epitaxial interface, Si would be stretched radially at the interface. Thus, on the basis of Poisson's ratio, Si will be compressed axially. While the misfit is 5.6%, but assume a constant unit cell volume, when the *x*- and *y*-axis misfits are 5.6%, the strain in the *z*-axis will be about twice. On the other hand, for fixed strain energy, the average energy distributed per atomic layer in the Si would be larger in a smaller Si region; as a consequence, the strain increased when the length of the Si region decreased. Additionally, due to the confinement of the oxide on the Si nanowire, volume expansion that could have resulted from the diffusion of Ni atoms into Si lattice can only express axially, leading to over 10% of strain at the smallest Si region. The strain can be controlled because we can control the length of the Si region.[141] Since the one-dimensional nano-heterostructures may have potential applications in nanoelectronic devices,[66] the strain could affect carrier mobility in the Si region.

5.3.5 Comparison of Co and Ni Silicides in Si Nanowires

A comparison of growth kinetics of Co silicides and Ni silicides in thin film reactions is shown in Table 5.5.[42] The activation energy of Co silicide formation in thin film reactions is much higher than that of the Ni silicide. If we apply the experience in thin film case to our nanowire case, the formation of Co nano silicide needs higher activation energy, which is perhaps the reason of the slower growth rate of $CoSi_2$ than that of NiSi. The linear growth rates of axial epitaxial NiSi in nanowires of Si in the temperature range of 500–700 °C have been reported.[141] On the basis of the observation shown in Fig. 5.11, we expect that the linear growth of NiSi should show the stair-type behavior, so the nucleation event and the growth event can be studied separately.

Table 5.5 Comparison of Co–Si and Ni–Si Systems

System	Structure	Activation energy (eV/atom)
Co–Si	CoSi BCC	~2.03
	$CoSi_2$ CaF_2	~2.81
Ni–Si	NiSi Orthorhombic	~1.25
	$NiSi_2$ CaF_2	~1.5

In samples annealed at 700 °C and 800 °C, we observed that the Co particles appeared beside the Si nanowires and on the SiO_2 window. These Co particles seemed to be stable in annealing. Here, we eliminate the possibility of surface diffusion in forming Co particles and Co silicides since we found no change in the distribution of the size of Co particles after 50 min at 800 °C. In other words, there is no ripening among the Co particles, as shown in Fig. 5.16a,b.

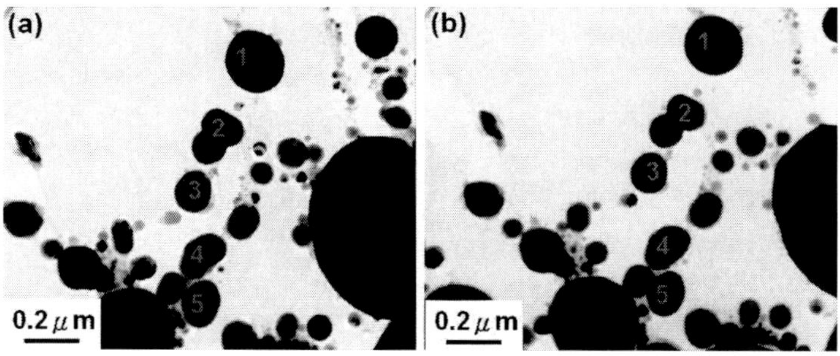

Figure 5.16 TEM images showing that no ripening occurred at 800 °C after 50 min annealing.

5.3.6 Summary

Si nanowires transformed into CoSi and $CoSi_2$ by point contact reaction between Co nanodots and Si nanowires at annealing temperatures of 700 °C and 800 °C, respectively. The $CoSi_2$ has axial epitaxial relation with Si, and HRTEM images show that the interface between $CoSi_2$ and Si is very sharp. A layer by layer stepwise growth was found in which we can resolve the nucleation stage and the growth stage. In addition, lateral growth in the form of an atomic step sweeping across the $CoSi_2$/Si interface was observed. The repeating nucleation stages of $CoSi_2$ in the epitaxial reaction have been observed and the incubation time recorded. In point contact reactions, we propose that it can be limited by the supply of Co atoms or the dissolution of Co into the Si via the point contact area. Single crystal $CoSi_2$/Si/$CoSi_2$ heterostructure with sharp interfaces can be made by this method. The heterostructures are promising as high-performance transistors based on intrinsic Si nanowires.

Si nanowires were found to transform to NiSi at 450–700 °C and to $NiSi_2$ at 800 °C. Both NiSi and $NiSi_2$ have epitaxial growth relation with Si and form very sharp silicide/Si interface. Heterostrucures of NiSi/Si/NiSi and $NiSi_2$/Si/$NiSi_2$ have been obtained with flat interfaces by *in situ* TEM and annealing. Moreover, a regular array of Ni patterning sample was prepared to synthesize multiple heterostructure of silicide/Si/silicide in a Si nanowire.

5.4 Homogeneous Nucleation of Nanoscale Silicide Formation

5.4.1 Introduction

The studies on nanoscale silicide formations in Si nanowires in the recent ten years show several surprising findings which are quite different from thin film silicide reactions. In this section, we will focus on the nucleation studies of epitaxial silicide formation in nanoscale and make a comparison with that of thin film silicide formation.

In the chapter, the finding of repeating events of homogeneous nucleation in epitaxial growth of $CoSi_2$ and NiSi silicides in the axial direction of nanowire of Si will be discussed. Using high-resolution TEM, we observed that the growth of every single atomic layer of the silicides requires nucleation. It is homogeneous nucleation because heterogeneous nucleation has been suppressed due to the surface oxide of the Si nanowire. Owing to the repeating events, we can measure the distribution of incubation time of nucleation. Knowing the incubation time, we can calculate the steady state rate of homogeneous nucleation, that is, the number of stable critical nucleus per unit area per unit time, when the activation energy of nucleation is measured. Since it is homogeneous nucleation, we are able to apply Zeldovich factor to obtain the number of molecules in forming the critical nucleus in the steady state process. In this report, we have correlated the theory and experiment of homogeneous nucleation.

5.4.2 Results and Discussions

5.4.2.1 Stepwise growth and repeating events of nucleation

The epitaxial growth mode of NiSi, $NiSi_2$ and $CoSi_2$ in (111) Si nanowires was found to be the same.[103,143] The growth in the axial direction occur atomic layer by atomic layer with the moving of steps or kinks across the epitaxial interface as shown in Fig. 5.17a–b and d–e. The growth rate of the silicide, at 450 °C to 750 °C for NiSi and at 800 °C for $CoSi_2$, was measured from *in situ* HRTEM video. Figure 5.17a,b shows two consecutive HRTEM images of the motion of one NiSi atomic layer across the NiSi/Si interface, and the stepwise growth direction is the step motion in the radial direction of the wire. Similarly, Fig. 5.17d,e shows two images of step motion of $CoSi_2$. During the growth of an atomic layer, the growth mode is the moving of steps or kinks across the epitaxial interface. However, there is a long period of stagnation before the next stepwise growths of silicide can take place. When we plot the stagnation period as well as the growth period from HRTEM videos, we obtained the stair-type curves as shown in Fig. 5.17c,f. It shows that the growth rate or the time of growth of each silicide atomic layer is the same, which is about 0.06 s per layer for NiSi and about 0.17 s per layer for $CoSi_2$, and we note that it is just the width of the vertical line in the stair-type curves. In between the vertical lines (the horizontal part of the stair steps) is the stagnation period, which we define as the incubation time of nucleation of a new layer. Since nucleation is a fluctuation phenomenon, the incubation periods might be different, caused by fluctuation inside the Si nanowires so that the incubation time of nucleating each atomic layer can vary. Accordingly, the plots of the distribution of incubation time are shown in the inset in Fig. 5.17c,f. The average value of the incubation time of NiSi is about 3 s and that of $CoSi_2$ is about 6 s.

The average growth rate of $CoSi_2$ along the axial direction is 0.0365 nm/s. However, the radial growth rate, or the step velocity, can be calculated to be about 135 nm/s. It is remarkable that the radial growth rate is about 3700 times faster than that of the axial growth rate. This is because the measured average axial growth rate has included the incubation time of nucleation of every step.

Without including the incubation time of nucleation, on the basis of the measured radial growth rate of $CoSi_2$ of 135 nm/s, the axial growth rate of each $CoSi_2$ atomic layer would have been about 1.82 nm/s. This axial growth rate should have been obtained by using the following equation of stepwise growth in MBE,

$$V = Nvh$$

where V is axial growth rate, v is radial growth rate, N is number of steps per unit length, and h is the height of the step. However, this equation fails in describing the epitaxial growth in nanowires when there is a long period of stagnation between each silicide layer growth.

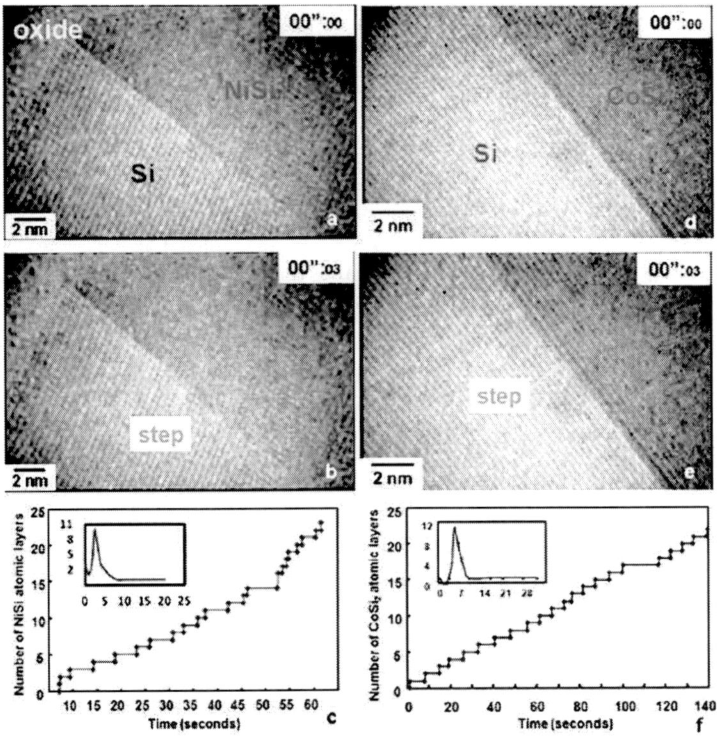

Figure 5.17 High-resolution TEM images of motion of a step on epitaxial silicide/Si interfaces and their growth curves. (a, b) Epitaxial NiSi/Si interfaces. (d, e) Epitaxial $CoSi_2$/Si interfaces. The direction of the atomic layer motion is upward from the center of the nanowire to the edge in NiSi and downward in $CoSi_2$. The first two numbers are in units of seconds and the following two smaller numbers are in units of 1/100 s. (c, f) The stair-type growth curves for NiSi and $CoSi_2$, respectively. The insets are the distribution curves of incubation periods of nucleation.

The HRTEM videos from *in situ* TEM experiments show that the overall growth rate of $CoSi_2$ and NiSi layers is linear,[141] yet we can decompose the linear curve into many stair-steps where the step height equals to an atomic layer thickness and the step width equal to the incubation time of nucleating a new step.[103] Therefore, nucleation stages and growth stages can be separated and the repeating nucleation events can be used to study nanoscale silicide nucleation in Si nanowires experimentally and theoretically.

5.4.2.2 Supply limit reaction

The incubation time of nucleation is required to create a new critical nucleus for the formation of a new silicide layer. The incubation time varies slightly. When a critical nucleus was created, it grew very quickly across the Si/silicide interface and consumed nearly all the supersaturated Ni or Co atoms in the Si nanowire. In repeating nucleation, it has to wait for enough Ni or Co atoms to diffuse into Si nanowires to reach the supersaturation in order to nucleate a new critical nucleus. So the supply limited dissolution of Ni or Co into the Si at the point contact may become rate limiting. The overall reaction rate is limited by the incubation time of nucleation, so it is a nucleation-limited reaction, neither diffusion-limited nor interfacial-reaction-limited.

In bulk and thin-film interfacial reactions, there are diffusion-limited and interfacial-reaction-limited reactions. In point contact reactions, we shall consider supply limited reaction since it can be the slowest kinetic process in the reaction.[143] Let J be the dissolution flux from a nanowire or a nanoparticle of Ni or Co into a Si nanowire, and the unit of J is the number of atoms/cm^2-s. Let δA be the area of a point contact. Thus in a period of t, the number of metal atoms dissolved into the Si nanowire is $J(\delta A)t$. The growth of silicide will consume the dissolved metal atoms. If we assume the axial or cross-sectional silicide/Si interface area is A, and v is the linear growth rate, we have by mass conservation that

$$J(\delta A)t = C(\upsilon t)A \tag{5.1}$$

where C is the concentration of metal in the silicide, and we have

$$J\left(\frac{\delta A}{A}\right) = C\upsilon \tag{5.2}$$

We note that Eq. 5.2 is similar to the conventional flux equation of $J = C\langle v\rangle$ where $\langle v\rangle$ is drift velocity, except that the flux in Eq. 5.2 is limited by δA when it is a point in point contact reaction. A constant reaction rate is expected and the rate constant will be limited by δA. In the experiments, the contact region of metal nanodots or nanowires and Si nanowire is a very small area and it only allows a very limited amount of flux of metal atoms to diffuse into the Si nanowire per unit area per unit time, so the nanosilicide phase formation depends on the supply of metal atoms.

5.4.2.3 Homogeneous nucleation: Experimental observations

Where is the nucleation site of the new disc? The fast radial growth of NiSi and $CoSi_2$ silicides seems to follow its nucleation from the center of the epitaxial silicide/Si interface rather than from the edge of the nanowire. Due to surface oxide of the Si nanowire, it is likely that the energy of the interface of silicide/SiO_2 is higher than that of Si/SiO_2, so the frequency of heterogeneous nucleation at the edge is low, and the nucleation of a disc at the center of the epitaxial interface becomes possible.

By the observations from *in situ* HRTEM recordings,[103,143] NiSi and $CoSi_2$ atomic layers grow toward both ends of the oxide wall of Si nanowires and some are observed as growing from the middle of the interface. In other words, the step moves toward the edge of the nanowire, rather than away from the edge. The fast radial growth of NiSi and $CoSi_2$ atomic layers starts from the center rather than from the edge of the silicide/Si interface. We assume that the energy of the oxide/silicide interface is higher than that of the oxide/Si interface because of the existence of the native surface oxide of the Si nanowires. This is a reasonable assumption since we found that when a step approaches the edge of the Si nanowire, it will slow down before it transforms the edge from the oxide/Si interface to the oxide/silicide interface because the latter is a high-energy barrier. This is shown in Fig. 5.18a,b, where we can see a curvature of the untransformed Si near the oxide/Si edge because several of the silicide layers have greatly slowed down their growth within atomic distance in approaching the edge. The insets in Fig. 5.18a,b are sketches of the observed curvature of the silicide layers. The schematic diagrams of the triple points are shown in Fig. 5.18c,d. Furthermore, we kept the electron beam at the edge region and waited for heterogeneous nucleation to take place, but we were unable to observe one. Figure 5.19 shows that one NiSi atomic layer disc nucleates and grows from the middle region of the interface and spreads toward both ends of the oxide wall of the Si nanowire. This behavior is repeated for every atomic layer growth, which is a direct evidence of homogeneous nucleation of silicide in the epitaxial interface in Si nanowires.

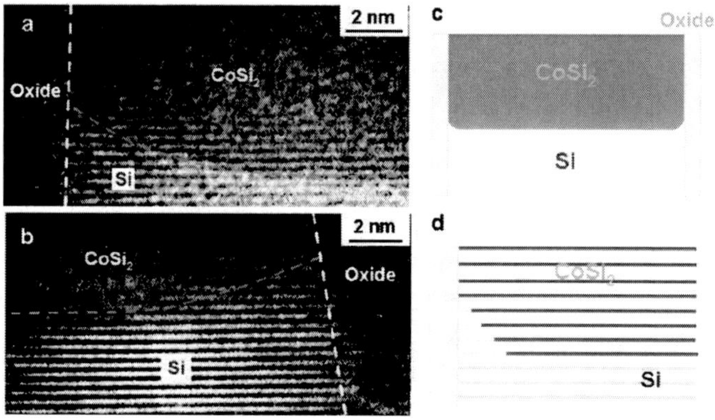

Figure 5.18 High-resolution TEM images and schematic diagrams of the triple points of oxide, Si, and $CoSi_2$. (a, b) The delayed $CoSi_2$ layer growth near the oxide edge at right side and left side, respectively, has caused a curved interface. (c) A cross-sectional schematic diagram of the curved interface at the triple points of oxide, Si, and $CoSi_2$. (d) An enlarged schematic diagram of the curved interface at the triple point.

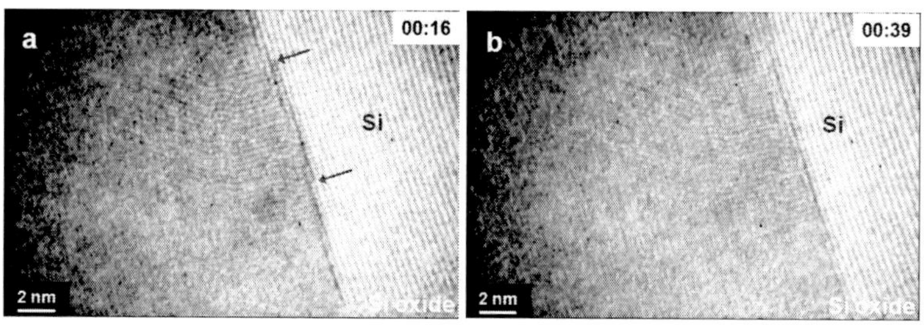

Figure 5.19 High-resolution TEM images of motion of steps on epitaxial NiSi/Si interface at 450 °C. (a) One NiSi atomic layer grows from the middle region of the interface and two steps are formed. (b) The motion of the two steps is toward both ends of oxide. After two steps reaches to the oxide ends, the layer growth ends and the interface becomes flat without steps.

The heterogeneous nucleation of a step at the edge is depicted in Fig. 5.20a. To form the step, the low energy oxide/Si interface will be replaced by the high-energy oxide/silicide interface, which is energetically unfavorable. There is no microreversibility, so the heterogeneous nucleation is suppressed. In Fig. 5.20b, a schematic diagram of a heterogeneous nucleus is assumed with a wetting angle larger than 90°. At the triple point, we consider

$$\gamma_{\text{silicide/oxide}} \geq \gamma_{\text{Si/oxide}} + \gamma_{\text{Si/silicide}} \cos(180 - \theta),$$

where γ represents the surface energy per unit area of the interfaces. We note that the epitaxial interface between Si and silicide is a low energy interface. When the inequality is satisfied and θ = 180°, heterogeneous nucleation will not occur and homogeneous nucleation of a circular disk in the center of the nanowire becomes possible, and the cross-section is depicted in Fig. 5.20c.

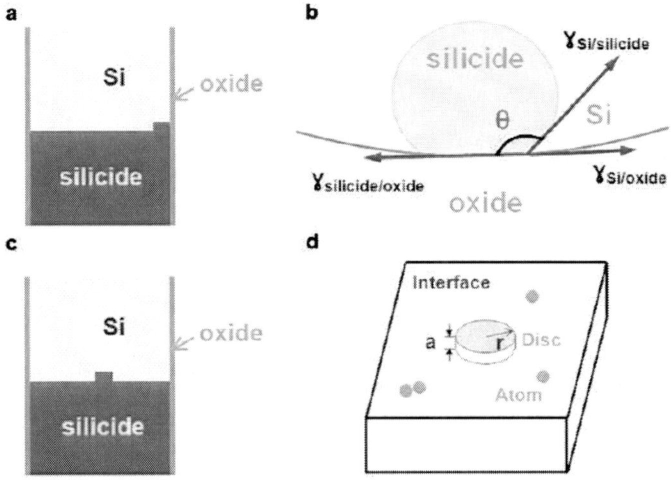

Figure 5.20 Schematic diagrams of the cross-section of the epitaxial interface between silicide and Si. (a) It depicts a heterogeneous nucleation of a step at the right-hand-side corner. (b) Triple point configuration of a heterogeneous nucleus. (c) Cross-section of homogeneous nucleation of a disk in the center of the silicide/Si interface. (d) Schematic diagram of the nucleation of a circular disk on the interface.

In homogeneous nucleation of a circular disk as depicted in Fig. 5.20d, the net change in energy is $\Delta G = 2\pi r a \gamma - \pi r^2 a \Delta G_s$. The critical nucleus has a size $r_{crit} = \gamma/\Delta G_s$ and the activation energy in nucleating the critical disk is $\Delta G^* = \pi r_{crit} a \gamma$, where ΔG_s is the gain in free energy of formation of the silicide per unit volume, γ is the interfacial energy per unit area of the circumference of the disk, and a is atomic height.

5.4.2.4 Homogeneous nucleation: Correlation between experiments and theory

Knowing the activation energy of formation of the critical disc, the probability of nucleation of the metastable critical nucleus can be calculated, that is, the number of critical nuclei per unit area per unit time. Experimentally what we have measured in Fig. 5.17c,f is one stable critical nucleus on the cross-section of the Si/silicide nanowire in the period of one incubation time. Thus we can make a comparison between theoretical analysis and experimental measurement of the rate of homogeneous nucleation. We have measured the nucleation rate of

$$I_{\text{stable-crit}} = \frac{1}{\pi R^2 \tau_i} \tag{5.3}$$

where R is the radius of the Si nanowire, and τ_i is the incubation time. Taking the diameter of Si nanowire to be 30 nm and the incubation time to be 3 s, we have $I_{\text{stable-crit}} = 4.7 \times 10^{10}$ stable nuclei/cm^2-s for the case of NiSi.

In the steady state reaction, it indicates that during one period of incubation, it must have dissolved from the point contact nearly the same amount of Ni atoms to supply the growth of one atomic layer of the silicide. For simplicity, we assume there are 10^{15} atoms per atomic layer per cm^2, and the flux of Ni needed to grow an atomic layer is

$$J_{\text{Ni}} = \frac{10^{15}}{2x3} = 1.67x10^{14}\text{atoms/cm}^2\text{-s}$$

where in the denominator the factor of 2 is because of concentration of Ni in NiSi is half and the factor 3 is from the incubation time. While we can regard this to be a flux of Ni atoms being deposited onto the silicide/Si interface, we note that not all these Ni atoms will involve directly in the nucleation process. This is because, similar to thin film deposition on a substrate, we should consider the adatoms on the interface, and we assume that only the adatoms are taking part in the nucleation process. However, the adatoms have a residence time, τ_{des}, on the interface because of desorption.

$$\tau_{\text{des}} = \frac{1}{\nu_s}\exp\frac{\Delta G_{\text{des}}}{kT} \tag{5.4}$$

where ν_s is the vibrational frequency of an adatom, ΔG_{des} is the activation energy of desorption of an adatom, and kT is thermal energy. Thus $J_{Ni}\tau_{des}$ is the effective number of adatoms per unit area involved in the nucleation process. Then, the equilibrium concentration of critical nucleus can be given as

$$C_{crit} = J_{Ni}\tau_{des}\exp\left(-\frac{\Delta G^*}{kT}\right) \tag{5.5}$$

where C_{crit} has the unit of number of nuclei per unit area. On the basis of assumption of thermally activated process of fluctuation of subcritical nucleus, the steady state homogeneous nucleation rate has been given as

$$I^{s}_{n^*} = \beta^*_n C_{crit} Z = \beta^*_n C_o e^{-\frac{\Delta G^*_n}{KT}}\left[-\frac{1}{2\pi KT}\left(\frac{\partial^2 \Delta G_n}{\partial n^2}\right)_{n^*}\right]^{1/2} \tag{5.6}$$

where β_{n^*} is the frequency of atomic jump toward an critical nucleus that converts it into a stable nucleus, and $C_{crit} = C_0\exp(-\Delta G_{n^*}/kT)$ is the equilibrium concentration of critical size nucleus, which we note is the same as Eq. 5.5.

The Zeldovich factor "Z" has been included in the nucleation rate equation as a kinetic factor that stands for the percentage of critical size nuclei that become stable. This is because the nucleus which has overcome the nucleation barrier may not definitely become a stable nucleus until at the least one more atom has joined it, otherwise most of them may shrink back to subcritical size. The Zeldovich factor is less than 1 in all real cases.

By assuming a circular disc shape of nucleus of atomic height, the Zeldovich factor can be rewritten as

$$Z = \left[\frac{1}{4\pi KT}\cdot\frac{\Delta G^*}{(n^*)^2}\right]^{1/2} \tag{5.7}$$

where ΔG^* and n^* are respectively the activation energy in forming the critical nucleus and the number of molecules in it. Knowing the activation energy of NiSi to be 1.25 eV/atom,[139] we can calculate n^* at a given value of Z.

Table 5.6 shows the n^* of NiSi at 700 °C and $CoSi_2$ at 800 °C with different Zeldovich factors. For $CoSi_2$, we took the activation energy from thin film study at 800 °C. Typical experimental value of Z factor is about 0.05. Table 5.6 lists the value of n^* to be about 10 for both silicides. The n^* value of $CoSi_2$ is higher than that of NiSi, and it may be one of the reasons that the temperature of reaction of $CoSi_2$ is higher than that of NiSi.

Table 5.6 The required molecules to form a stable silicide nucleus

	$CoSi_2$ at 800 °C	NiSi at 700 °C
Z factor	1 0.1 0.05	1 0.1 0.05
n^*	1.6 16 31	1.1 11 22

Note: n^* is the number of molecules required to form a stable nucleus

Since we know the experimentally measured steady state nucleation rate as given by Eq. 5.3, we can check it by using Eqs. 5.5 and 5.6. We have

$$I_{\text{stable-crit}} = \beta_n^* J_{\text{Ni}} \tau_{\text{des}} \exp(-\frac{\Delta G^*}{kT})Z = \beta_n^* J_{\text{Ni}} \frac{1}{\nu_s} \exp(-\frac{\Delta G^* - \Delta G_{\text{des}}}{kT})Z$$

$$= \nu_0 J_{\text{Ni}} \frac{1}{\nu_s} \exp(-\frac{\Delta G^* - \Delta G_{\text{des}} + \Delta G_\beta}{kT})Z$$

where we assume that $\beta_{n^*} = \nu_0 \exp(-\Delta G_\beta/kT)$, and ν_0 is Debye frequency of vibration and ΔG_β is the activation energy of adding an atom to the critical nucleus. It is worthwhile mentioning that the basic nature of the parameter of β_{n^*} is microreversibility. In order to maintain the equilibrium distribution of subcritical size embryos in nucleation, the frequency of adding and subtracting atoms among the embryos is high. Hence we can assume $\Delta G^* >> \Delta G_\beta$, so we can ignore ΔG_β.

To evaluate the products on the right-hand side of the last equation, we cancel ν_0 against ν_s owing to the fact that both are Debye frequency of atomic vibration. For ΔG_{des}, it is known from epitaxial growth of Si on Si, where ΔG_{des} = 1.1 eV/atom. For the desorption of Ni, the activation energy should be lower and we assume ΔG_{des}=0.7 eV/atom. Then we take the measured ΔG^*=1.25 eV/atom and Z = 0.1. Since J_{Ni} = 1.67 × 10^{14} atoms/cm^2-s, the products on the right-hand side at T = 700 °C is 3 × 10^{10} nuclei/cm^2-s, which is in good agreement with the measured nucleation rate of 4.7 × 10^{10} nuclei/cm^2-s.

We caution that there is some uncertainty about ΔG_{des}. While we have no measured data, we note that even if we give it a high uncertainty by taking $\Delta G_{des} = 0.7 \pm 0.2$ eV/atom, it will only change the outcome by a factor about 10. Since this is the first attempt to correlate theory and experiment on homogeneous nucleation, it is expected to have a large uncertainty.

5.4.2.5 Homogeneous nucleation: Supersaturation

Nucleation requires supersaturation or under-cooling. We can calculate the supersaturation in the nucleation of NiSi.[143] The solubility of Ni in Si at 700 °C is about 10^{15}–10^{16} Ni-atoms/cm^3. Since there are 2.5×10^{22} Si atoms/cm^3, the equilibrium concentration of Ni in Si is about 10^{-7}–10^{-8}. When we have dissolved half of a monolayer of Ni (which has a layer thickness of 0.3 nm) into a Si nanowire of 3 μm long before homogeneous nucleation occurs, the concentration of Ni is 0.5×10^{-4}, so the supersaturation is about 10^3, which seems very large. Since Ni atoms are dissolved into nanowire of Si, the solubility can be increased due to Gibbs–Thomson effect by a factor of $\exp(\gamma\Omega/rkT)$. To calculate this factor, we take $\Omega = a^3$ and $a = 0.3$ nm as atomic diameter, $r = 15$ nm, and $kT = 0.084$ eV at 973 K. When we let $\gamma a^2 = 1$–2 eV, we obtain the factor to be 1.25 to 1.58, respectively, which is small as compared to the estimated supersaturation of 1000.

Below we show that the high supersaturation is reasonable for homogeneous nucleation. Considering the equilibrium solubility of Ni over the critical disc, we have from the Gibbs–Thomson equation that $n_{crit}/n_0 = \exp(\gamma\Omega/r_{crit}kT)$, where n_{crit} and n_0 are the equilibrium solubility of Ni above a flat silicide/Si interface and that above a critical nucleus, respectively.

$$\text{Thus, } r_{crit} = \frac{\gamma\Omega}{kT\ln\left(\frac{n_{crit}}{n_0}\right)}$$

$$\text{and } \Delta G^* = \pi r_{crit} a\gamma = \frac{\pi\gamma^2 a\Omega}{kT\ln\left(\frac{n_{crit}}{n_0}\right)} = \frac{\pi\gamma^2 a^4}{kT\ln\left(\frac{n_{crit}}{n_0}\right)}$$

In the last step of the above equation, we have taken $\Omega = a^3$. Since we know $\Delta G^* = 1.25$ eV/atom, we can calculate the ratio of n_{crit}/n_0, provided that we know γa^2, the interfacial energy per cross-sectional area of an atom of the epitaxial interface of silicide/Si. For reference, we know that on (111) surface of Si, each surface atom has a broken bond, so the surface energy is about 1 eV/atomic area. Here we shall take $\gamma = 0.5$ eV/a^2 (which is about 800 erg/cm^2) and $kT = 0.084$ eV at 973 K, we have $n_{crit}/n_0 = 1800$ from the equation below. For $\gamma = 0.4$ eV/a^2, we have $n_{crit}/n_0 = 120$.

$$\Delta G^* = \frac{\pi(\gamma a^2)^2}{kT\ln(\frac{n_{crit}}{n_0})} = \frac{3.14x0.25}{0.084\ln(\frac{n_{crit}}{n_0})}$$

Since ΔG^* is inversely proportional to $ln(n_{crit}/n_0)$, if we take $n_{crit}/n_0 = 1000$, we obtain $\Delta G^* = 1.35$ eV/atom; $n_{crit}/n_0 = 100$ implies $\Delta G^* = 2$ eV/atom; and $n_{crit}/n_0 = 10$ implies $\Delta G^* = 4.1$ eV/atom. It shows that for a low supersaturation just over unity, the activation energy will be extremely high. This is the reason why in most real events of nucleation at a low supersaturation, it is heterogeneous rather than homogeneous nucleation.

5.4.3 Summary

The formation of NiSi, CoSi and $CoSi_2$ in Si nanowires by point contact reactions has been investigated *in situ* by ultrahigh vacuum high-resolution transmission electron microscopy. Si nanowires can be transformed into NiSi, CoSi and $CoSi_2$ by point contact reaction between Ni and Co nanowires or nanodots with Si nanowires at a high annealing temperature. The NiSi and $CoSi_2$ has axial epitaxial relation with Si, and HRTEM images show that the interface between the silicides and Si is very sharp. Nevertheless, CoSi doesn't grow epitaxial with Si so the CoSi/ Si interface is rough. We have demonstrated the controlled growth of NiSi/Si/NiSi nanoheterostructures with the middle Si region as small as 2 nm in length through point contact reaction. Controlled strain up to more than 11% in the Si region was observed. Single crystal $CoSi_2$/Si/$CoSi_2$ heterostructure with sharp interfaces can be made by this method too. In addition, it has been shown that the growth of single-crystal NiSi within a Si nanowire starts from both ends of the nanowire. We attributed it to the ease of nucleation and fast interstitial diffusion of Ni atoms to the ends of the nanowire.

We propose that point contact reaction is limited by supply of metal atoms or the rate of dissolution of metal into Si via the point contact area; in other words, we show that the growth mechanism of NiSi in nanoscale is fundamentally different from that in thin film reactions which are either diffusion-controlled or interfacial-controlled reactions.

Direct evidence from *in situ* TEM videos shows that the single-crystal grains of NiSi have a linear growth rate, however, it can be decomposed into a stair-type growth of layer-by-layer, in which we can resolve the nucleation stage and the growth stage. In addition, the lateral growth in the form of an atomic step sweeping across the $CoSi_2$/Si and NiSi/Si interface was observed. The repeating nucleation stages of NiSi and $CoSi_2$ in the epitaxial reaction have been observed. They lead to axial growth of epitaxial silicide with a stair-type of growth mode. The growth of every new atomic layer of silicide requires an independent event of nucleation accompanied by a long incubation time. The nucleation stage and the growth stage of each layer of NiSi and $CoSi_2$ can be separated. The distribution of incubation time of nucleation has been measured and it has enabled us to determine experimentally the steady state nucleation rate per unit area per unit time.

The nucleation is homogeneous rather than heterogeneous. This is because the nanowire of Si has native oxide and the oxide has prevented heterogeneous nucleation to occur because the energy of the SiO_2/Si interface is lower than that of the SiO_2/silicide interface. The number of molecules required to form a stable NiSi or $CoSi_2$ nucleus for homogeneous nucleation has been calculated to be about ten, on the basis of Zeldovich factor approach and the measured activation energy of nucleation. The calculated and the measured homogeneous nucleation rates are in good agreement. The supersaturation needed for the homogeneous nucleation is about 1000.

5.5 Conclusion

Si nanowires transformed into Co silicides (CoSi and $CoSi_2$) and Ni silicides (NiSi and $NiSi_2$) by point contact reaction between Co/Ni nanodots or nanowires and Si nanowires while annealing. Co and Ni were found to diffuse into Si nanowires. The $CoSi_2$, NiSi, and $NiSi_2$ have axial epitaxial relation with Si, and HRTEM images show that the interfaces between them and Si are very sharp. The

HRTEM video shows that the overall growth rate of NiSi and $CoSi_2$ layers is linear, yet we can decompose the linear curve into many stair-steps where the step height equals to an atomic layer thickness. There was a stagnation period for a new layer to form, and the time interval between every step was varying. It takes an incubation time to create a new step. When a step was created, it propagated very quickly across the silicide/Si interface. So the overall reaction rate was limited by the nucleation stage, and it is linear with time. In nucleation, it might need to wait for enough Co/Ni atoms to diffuse into Si nanowires to reach supersaturation in order to nucleate a silicide step. So the supply limited dissolution of metal into the Si at the point contact may become rate limiting.

Single crystal $CoSi_2$/Si/$CoSi_2$, NiSi/Si/NiSi, and $NiSi_2$/Si/$NiSi_2$ heterostructures with sharp interfaces were made by in situ annealing. The heterostructures are promising as high-performance transistors based on intrinsic Si nanowires. The strain in the Si region of the heterostructures which influence carrier mobility is of interests and to be studied.

Homogeneous nucleation is rare except in theory. We observed repeating events of homogeneous nucleation in epitaxial growth of $CoSi_2$ and NiSi silicides in nanowires of silicon by using high-resolution TEM. The growth of every single atomic layer requires nucleation. Heterogeneous nucleation is prevented because of non-microreversibility between the oxide/Si and oxide/silicide interfaces. We determined the incubation time of homogeneous nucleation. The calculated and the measured nucleation rates are in good agreement. We used Zeldovich factor to estimate the number of molecules in the critical nucleus; it is about 10 and is reasonable. A very high supersaturation is found for the homogeneous nucleation.

Furthermore, we observed both heterogeneous and homogeneous nucleation of epitaxial silicide of $NiSi_2$ in Si nanowires grown in [110] direction by *in situ* observation in HRTEM. Owing to the excellent lattice match between Si and $NiSi_2$, a giant epitaxial step of 2 to 8 nm wide forms at the interface between Si and $NiSi_2$ during the growth of the latter. The step formation results in two vertical silicide/Si interfaces in TEM observations. Heterogeneous nucleation of $NiSi_2$ occurs at the intersection where the step meets the vertical interfaces. However, we have also observed the epitaxial growth of $NiSi_2$ having a single interface with the Si, i.e., without a giant step. Homogeneous nucleation of $NiSi_2$ occurs on the single interface. Incubation time of heterogeneous nucleation of $NiSi_2$ has been measured by high-resolution video to be much shorter than that of homogeneous nucleation. The overall growth rate of $NiSi_2$ for the case of heterogeneous nucleation is faster than that for the case of homogeneous nucleation. Kinetic analysis of both types of nucleation is presented for a direct comparison in order to have a better understanding of the nucleation events.

References

1. K.N. Tu, and J. W. Mayer, *Thin Films—Interdiffusions and Reactions*, J.M. Poate, K.N. Tu, and J.W. Mayer (Eds.), Wiley, New York (1978).
2. F. Nava, K.N. Tu, and O. Bisi, *Mater. Sci. Rep.* **9**, 141 (1993).
3. S.P. Murarka, *Silicides for VLSI Applications*, Academic Press, New York (1983).
4. K.N. Tu, R.D. Thompson, and B.Y. Tsaur, *Appl. Phys. Lett.* **38**, 626 (1981).
5. G. Ottaviani, K.N. Tu, and J.W. Mayer, *Phys. Rev. Lett.* **44**, 284 (1980).
6. I. Ohdomari, and K.N. Tu, *J. Appl. Phys.* **51**, 3735 (1980).
7. M. Eizenberg, H. Foell, and K.N. Tu, *J. Appl. Phys.* **52**, 861 (1981).
8. G. Ottaviani, K.N. Tu, and J.W. Mayer, *Phys. Rev. B* **24**, 3354 (1981).
9. I. Ohdomari, M. Hori, T. Maeda, A. Ogura, H. Kawarada, T. Hamamoto, K. Sano, K.N. Tu, M. Wittmer, I. Kimura, and K. Yoneda, *J. Appl. Phys.* **54**, 4679, (1983).
10. H.K. Liou, E.S. Yang, and K.N. Tu, *Appl. Phys. Lett.* **63**, 911 (1993).
11. K.N. Tu, W.H. Hammer, and J.O. Olowolafe, *J. Appl. Phys.* **51**, 1663 (1980).
12. F. Nava, K.N. Tu, E. Mazzega, M. Michelini, and G. Queirolo, *J. Appl. Phys.* **61**, 1085 (1987).
13. U. Koster, P.S. Ho, and J.E. Lewis, *J. Appl. Phys.* **53**, 7436 (1982).
14. S. Saitoh, H. Ishiwara, T. Asano, and S. Furukawa, *Japan. J. Appl. Phys.* **20**, 1649 (1981).
15. L. Eckertova, *Physics of Thin Films*, Plenum Press, New York (1986).
16. R.T. Tung, J.M. Poate, J.C. Bean, J.M. Gibson, and D.C. Jacobson, *Thin Solid Films* **93**, 77 (1982).
17. L.J. Chen, J. W. Mayer, K.N. Tu, and T.T. Sheng, *Thin Solid Films* **93**, 91 (1982).

18. H. Föll, P.S. Ho, and K.N. Tu, *J. Appl. Phys.* **52**, 250 (1981).
19. R.T. Tung, *Phys. Rev. Lett.* **52**, 461 (1984).
20. R.T. Tung, *J. Vacuum Sci. Technol. B* **2**, 465 (1984).
21. K.N. Tu, and R. Rosenberg (Eds.), *Analytical Techniques for thin films. Vol.27 of Treatises on Materials Science and Technology*, Academic Press, Boston, (1988).
22. K.N. Tu, and R. Rosenberg (Eds.), *Analytical Techniques for thin films. Vol. 24 of Treatises on Materials Science and Technology*, Academic Press, New York, (1988).
23. K.N. Tu, *J. Appl. Phys.* **48**, 3370 (1977).
24. G. Ottaviani, K.N. Tu, J.W. Mayer, and B.Y. Tsaur, *Appl. Phys. Lett.* **36**, 331 (1980).
25. R.D. Thompson, B.Y. Tsaur, and K.N. Tu, *Appl. Phys. Lett.* **38**, 535 (1981).
26. R.D. Thompson, K.N. Tu, and G. Ottaviani, *J. Appl. Phys.* **58**, 705 (1985).
27. K.N. Tu, *Thin Solid Films* **140**, 71 (1986).
28. O. Bisi, L.W. Chiao, and K.N. Tu, *Surface Science* **152/153**, 1185 (1985).
29. K.F. Kelton, A.L. Greer, and C.V.J. Thompson, *Chem. Phys.* **79**, 6261 (1983).
30. D.W. Pashey, *Epitaxial Growth*, J.W. Matthews (Eds.), Academic Press, New York, 1975.
31. W.K. Chu, H. Krautle, J.W. Mayer, H. Muller, M-A. Nicolet, and K.N. Tu, *Appl. Phys. Lett.* **25**, 454 (1974).
32. J.W. Mayer, J.M. Poate, and K.N. Tu, *Science* **190**, 228 (1975).
33. W.K. Chu, S.S. Lau, J.W. Mayer, H. Muller, and K.N. Tu, *Thin Solid Films* **25**, 393 (1975).
34. K.N. Tu, W.K. Chu, and J.W. Mayer, *Thin Solid Films* **25**, 403 (1975).
35. E. Bauer, *Z. Krist.* **110**, 395 (1958).
36. G.A. Bassett, *Phil. Mag.* **3**, 1042 (1958).
37. D.W. Pashley, M.J. Stowee, M.H. Jacobs, and T.J. Law, *Phil. Mag.* **10**, 127 (1964).
38. C.M. Lieber, *MRS Bull.* **28**, 486 (2003).
39. Y. Cui, and C.M. Lieber, *Science* **291**, 851 (2001).
40. Y. Xia, P. Yang, Y. Sun, Y. Wu, B. Mayers, B. Gates, Y. Yin, F. Kim, and H. Yan, *Adv. Mater.* **15**, 353 (2003).

41. F. Patolsky, B.P. Timko, G. Zheng, and C.M. Lieber, *MRS Bull.* **32**, 142 (2007).

42. M.C. McAlpine, R.S. Friedman, and C.M. Lieber, *Nano Lett.* **3**, 443 (2003).

43. Z. Zhong, F. Qian, D. Wang, and C.M. Lieber, *Nano Lett.* **3**, 343 (2003).

44. X. Duan, Y. Huang, and C.M. Lieber, *McGraw-Hill Yearbook of Science and Technology*, E. Geller, J. Weil, D. Blumel, and A. Rappaport (Eds.), p. 272–276, McGraw Hill, 2003.

45. Y. Cui, Z. Zhong, D. Wang, W.U. Wang, and C.M. Lieber, *Nano Lett.* **3**, 149 (2003).

46. X. Duan, Y. Huang, R. Agarwal, and C.M. Lieber, *Nature* **421**, 241 (2003).

47. X. Duan, Y. Huang, Y. Cui, and C.M. Lieber, Nanowire nanoelectronics assembled from the bottom-up, in *Molecular Nanoelectronics*, M.A. Reed and T. Lee (Eds.) p. 199–227, American Scientific Publishers, 2003.

48. Z. Zhong, D. Wang, Y. Cui, M.W. Bockrath, and C.M. Lieber, *Science* **302**, 1377 (2003).

49. M.C. McAlpine, R.S. Friedman, S. Jin, K. Lin, W.U. Wang, and C.M. Lieber, *Nano Lett.* **3**, 1531 (2003).

50. Y. Cui, X. Duan, Y. Huang, and C.M. Lieber, *Materials, Properties and Devices*, Z.L. Wang (Ed.), p. 3–68, Kluwer Academic/Plenum Publishers, 2003.

51. G. Zheng, W. Lu, S. Jin, and C.M. Lieber, *Adv. Mater.* **16**, 1890 (2004).

52. N.A. Melosh, A. Boukai, F. Diana, B. Gerardot, A. Badolato, P.M. Petroff, and J.R. Heath, *Science* **300**, 112 (2003).

53. L. Qin, S. Park, L. Huang, and C.A. Mirkin, *Science* **309**, 113 (2005).

54. S. Jin, D. Whang, M.C. McAlpine, R.S. Friedman, Y. Wu, and C.M. Lieber, *Nano Lett.* **4**, 915 (2004).

55. L.J. Lauhon, M.S. Gudiksen, and C.M. Lieber, *Phil. Trans. R. Soc. Lond. A* **362**, 1247 (2004).

56. Y. Wu, J. Xiang, C. Yang, W. Lu, and C.M. Lieber, *Nature* **430**, 61 (2004).

57. Y. Huang, and C.M. Lieber, *Pure Appl. Chem.* **76**, 2051 (2004).

58. Y. Wu, J. Xiang, C. Yang, W. Lu, and C.M. Lieber, *Nature* **430**, 1 (2004).

59. A.M. Mohammad, S. Dey, K.-K. Lew, J.M. Redwing, and S.E. Mohney, *J. Electrochem. Soc.* **150**, G577 (2003).

60. M.Y. Lee, and P.A. Bennett, *Phys. Rev. Lett.* **75**, 4460 (1995).

61. B. Liu, Y. Wang, S. Dilts, T.S. Mayer, and S.E. Mohney, *Nano Lett.* **7**, 818 (2007).
62. A.L. Schmitt, J.M. Higgins, J.R. Szczech, and S. Jin, *J. Mater. Chem.* **20**, 223 (2010).
63. H.V. Kanel, *Mater. Sci. Rep.* **8**, 193 (1992).
64. E.G. Colgan, C. Cabral, Jr., and D.E. Kotecki, *J. Appl. Phys.* **77**, 614 (1995).
65. C.P. Li, N. Wang, S.P. Wong, C.S. Lee, and S.T. Lee, *Adv. Mater.* **14**, 218 (2002).
66. W.M. Weber, L. Geelhaar, A.P. Graham, E. Unger, G.S. Duesberg, M. Liebau, W. Pamler, C. Cheze, H. Riechert, P. Lugli, and F. Kreupl, *Nano Lett.* **6**, 2660 (2006).
67. V. Schmidt, H. Riel, S. Senz, S. Karg, W. Riess, and U. Gosele, *Small* **2**, 85 (2006).
68. T. Shan, and S.J. Fonash, *ACS Nano* **2**, 429 (2008).
69. R.S. Friedman, M.C. McAlpine, D.S. Ricketts, D. Ham, and C.M. Lieber, *Nature* **434**, 1085 (2005).
70. J.W. Peng, S.J. Lee, G.C.A. Liang, N. Singh, S.Y. Zhu, G.Q. Lo, and D.L. Kwong, *Appl. Phys. Lett.* **93**, 073503 (2008).
71. W.M. Weber, L. Geelhaar, A.P. Graham, E. Unger, G.S. Duesberg, M. Liebau, W. Pamler, C. Che`ze, H. Riechert, P. Lugli, and F. Kreup, *Nano Lett.* **6**, 2660 (2006).
72. F. Patolsky, and C.M. Lieber, *Mater. Today* **8**, 20 (2005).
73. H.Y. Peng, X.T. Zhou, N. Wang, Y.F. Zheng, L.S. Liao, W.S. Shi, C.S. Lee, and S.T. Lee, *Chem. Phys. Lett.* **327**, 263 (2000).
74. N. Wang, Y. Cai, and R.Q. Zhang, *Mater. Sci. Eng. R* **60**, 1 (2008).
75. A.T. Tilke, F.C. Simmel, H. Lorenz, R.H. Blick, and J.P. Kotthaus, *Phys. Rev. B* **68**, 075311 (2003).
76. Y. Huang, X.F. Duan, Y. Cui, L.J. Lauhon, K.H. Kim, and C.M. Lieber, *Science* **294**, 1313 (2001).
77. Y.C. Lin, K.C. Lu, W.W. Wu, B.Y. Yoo, N.V. Myung, L.J. Chen, K.N. Tu, and Y. Huang, *Nano Lett.* **8**, 913 (2008).
78. Sze, S.M. *Physics of Semiconductor Devices*, John Wiley & Sons, New York, 1981.
79. Y. Hu, J. Xiang, G. Liang, H. Yan, and C.M. Lieber, *Nano Lett.* **8**, 925 (2008).
80. E. Lind, A.I. Persson, L. Samuelson, and L.E. Wernersson, *Nano Lett.* **6**, 1842 (2006).

81. D. Turnbull, *Solid State Phys.* **3**, 225 (1965).
82. R.W. Balluffi, S.M. Allen, and W.C. Carter, *Kinetic Processes in Materials*, Wiley-Interscience, NJ, 2005. In MIT course website http://pruffle.mit.edu/~ccarter/3.21/Lecture_24/.
83. F.G. Shi, and J.H. Seinfeld, *Mater. Chem. Phys.* **37**, 1 (1994).
84. B.J. Kim, J. Tersoff, S. Kodambaka, M.C. Reuter, E.A. Stach, and F.M. Ross, *Science* **322**, 1070 (2008).
85. D. Turnbull, and R.E. Cech, *J. Appl. Phys.* **21**, 804 (1950).
86. J.H. Perepezko, J.L. Sebright, P.G. Hockel, and G. Wilde, *Mater. Sci. Eng. A* **326**, 144 (2002).
87. K.N. Tu, *Appl. Phys. Letts.* **27**, 221 (1975).
88. L.J. Chen, K.N. Tu, *Mater. Sci. Rep.* **6**, 53 (1991).
89. U. Koster, K.N. Tu, P.S. Ho, *Appl. Phys. Lett.* **31**, 634 (1976).
90. D. Turnbull, *Solid State Phys.* **3**, 225 (1956).
91. U. Koster, D.R. Campbell, K.N. Tu, *Thin Solid Films* **3**, 129 (1978).
92. M. Eizenberg, G. Ottaviani, K.N. Tu, *Appl. Phys. Lett.* **37**, 87 (1980).
93. M. Wittmer, K.N. Tu, *Phys. Rev. B* **29**, 2010 (1984).
94. R. Walser and R. Bene, *Appl. Phys. Lett.* 28, 624 (1976).
95. B.Y. Tsaur, S.S. Lau, M-A Nicolet, and J.W. Mayer, *Appl. Phys. Lett.* **38**, 922 (1981).
96. R.D Thompson, and K.N. Tu, *Thin Solid Films* **53**, 4372 (1982).
97. K.N. Tu, A.M. Gusak, and I. Sobchenko, *Phys. Rev. B* **67**, 245408 (2003).
98. K.N. Tu, Chapter 7: Metal-Silicon Reaction, *Advances in Electronic Materials*, Metal Park, OH, 1986.
99. L.J. Chen, J.W. Mayer, and K.N. Tu, *Thin Solid Films* **93**, 135 (1982).
100. K.C. Russell, *Adv. Colloid Interface Sci.* **13**, 205 (1980).
101. K.F. Kelton, *J. Non-Cryst. Solids* **274**, 147 (2000).
102. U. Gosele, and K.N. Tu, *J. Appl. Phys.* **53**, 3252 (1982).
103. Y.C. Chou, W.W. Wu, L.J. Chen, and K.N. Tu, *Nano Lett.* **9**, 2337 (2009).
104. L.A. Clevenger, C.V. Thompson, R.C. Cammarata, and K.N. Tu, *Appl. Phys. Lett.* **52**, 795 (1988).
105. K.N. Tu, E.I. Alessandrini, W.K. Chu, H. Krautle, and J.W. Mayer, *Jap. J. Appl. Phys.* **2**, 669 (1974).
106. R.C. Cammarata, C.V. Thompson, and K.N. Tu, *Appl. Phys. Lett.* **51**, 1106 (1987).

107. F.M. d'lleurle, *Mater. Res. Sot. Symp. Proc.* **25**, 3 (1984).
108. R.T. Tung, *J. Vacuum Sci. Technol. A* **5**, 1840 (1987).
109. H. Foell, P.S. Ho, and K.N. Tu, *Phil. Mag. A* **45**, 32 (1982).
110. O. Bisi, L.W. Chiao, and K.N. Tu, *Phys. Rev. B* **30**, 4664 (1984).
111. R.T. Tung, J.M. Gibson, and J.M. Poate, *Phys. Rev. Lett.* **50**, 429 (1983).
112. J.M. Gibson, J.L. Bastone, R.T. Tung, and F.C. Unterwald, *Phys. Rev. Lett.* **60**, 1158 (1988).
113. J.M. Gibson, and J.L. Bastone, *Surface Sci.* **208**, 317 (1989).
114. E.J. van Loenen, J.W.M. Frenken, and J.F. van der Veen, *Appl. Phys. Lett.* **45**, 41 (1984).
115. S.S Lau, J.W. Mayer, and K.N. Tu, *J. of Appl. Phys.* **49**, 4005 (1978).
116. H.C. Cheng, I.C. Wu, and L.J. Chen. *Appl. Phys. Lett.* **50**, 174 (1987).
117. S. Saitoh, H. Ishiwara, T. Asano, and S. Furukawa, *Appl. Phys. Lett.* **37**, 203 (1980).
118. F. Hellman, and R.T. Tung, *Phys. Rev. B* **37**, 10786 (1988).
119. E. Chainet, M. de Crescenzi, J. Derrien, T.T.A. Nguyen, and R.C. Cinti, *Surface Sci.* **168**, 801 (1986).
120. K.N. Tu, G. Ottaviani, R.D. Thompson, and J.W. Mayer, *J. Appl. Phys.* **53**, 4406 (1982).
121. K.M. Liang, L.J. Chen, A. Markov, K.N. Tu, G.U. Singco, L.T. Shi, and C. Farrell, *Mater. Chem. Phys.* **38**, 250 (1994).
122. F.G. Si, and K.N. Tu, *Phys. Rev. Lett.* **74**, 4476 (1995).
123. J.M. Gibson, J.L. Bastone, and R.T. Tung, *Appl. Phys. Lett.* **51**, 45 (1987).
124. R.T. Tung, and J.L. Bastone, *Appl. Phys. Lett.* **52**, 648 (1987).
125. D.A. Smith, K.N. Tu, and B.Z. Weiss, *Ultramicroscopy* **23**, 405 (1987).
126. H.C. Hsu, W.W. Wu, H.F. Hsu, and L.J. Chen, *Nano Lett.* **7** (2007) 885.
127. Y.L. Chueh, M.T. Ko, L.J. Chou, L.J. Chen, C.S. Wu, and C.D. Chen, *Nano Lett.* **6** (2006) 1637.
128. A.L. Schmitt, M.J. Bierman, D. Schmeisser, F.J. Himpsel, and S. Jin, *Nano Lett.* **6** (2006) 1617.
129. A.L. Schmitt, L. Zhu, D. Schmeisser, F.J. Himpsel, and S. Jin, *J. Phys. Chem. B* **110** (2006) 18142.
130. Y. Song, A.L. Schmitt, and S. Jin, *Nano Lett.* **7** (2007) 965.
131. K. Seo, K.S.K. Varadwaj, P. Mohanty, S. Lee, Y. Jo, M.H. Jung, J. Kim, and B. Kim, *Nano Lett.* **7** (2007) 1240.

132. B.T. Björk, B.J. Ohlsson, T. Sass, A.I. Persson, C. Thelander, M.H. Magnusson, K. Deppert, L.R. Wallenberg, and L. Samuelson, *Appl. Phys. Lett.* **80** (2002) 1058.

133. J. Xiang, A. Vidan, M. Tinkham, R.M. Westervelt, and C.M. Lieber, *Nature Nanotech.* **1** (2006) 208.

134. L.J. Chen, *JOM* **57**, 24 (2005).

135. M. Leong, B. Doris, J. Kedzierski, K. Rim, and M. Yang, *Science* **306** (2004) 2057.

136. R. He, and P. Yang, *Nature Nanotech.* **1** (2006) 42.

137. R.S. Wagner, and W.C. Ellis, *Appl. Phys. Lett.* **4** (1964) 89.

138. J.B. Hannon, S. Kodambaka, F.M. Ross, R.M. Tromp, *Nature* **440** (2006) 69.

139. S. Shingubara, O. Okino, Y. Sayama, H. Sakaue, and T. Takahagi, *Solid-State Electronics* **43** (1999) 1143.

140. F.H.M. Spit, D. Gupta, and K.N. Tu, *Phys. Rev. B.* **39** (1989) 1255.

141. K.C. Lu, W.W. Wu, H.W. Wu, C.M. Tanner, J.P. Chang, L.J. Chen, and K.N. Tu, *Nano Lett.* **7** (2007) 2389.

142. E.R. Weber, *Appl. Phys. A.* **30** (1983) 1.

143. Y.C. Chou, W.W. Wu, S.L. Cheng, B.Y. Yoo, N. Myung, L.J. Chen, and K.N. Tu, *Nano Lett.* **8** (2008) 2194.

Chapter 6

Interaction between Inverse Kirkendall Effect and Kirkendall Effect in Nanoshells and Nanowires

A. M. Gusak and T. V. Zaporozhets
Department of Physics, Cherkasy National University,
Cherkasy 18017, Ukraine
gusak@cdu.edu.ua

6.1 Introduction

Ernst Kirkendall published his last paper[1] (of three papers in total) in 1947, and after that left science for the remaining 58 years of his life.[2] Nevertheless, his results became the classics of diffusion studies, they are still widely used and discussed. There are three Kirkendall-named concepts familiar to anybody involved in materials science—Kirkendall shift, Kirkendall voiding and inverse Kirkendall effect—all of them related to difference of diffusivities between different components in the alloy. New wave of interest to these phenomena was generated recently (in 2004) by a new method of producing hollow nanoparticles and nanotubes.[3–10] Briefly, this method is based on Kirkendall voiding in diffusion-controlled reactions at

Silicon and Silicide Nanowires: Applications, Fabrication, and Properties
Edited by Yu Huang and King-Ning Tu

ISBN 978-981-4303-46-0 (Hardcover), 978-981-4303-47-7 (eBook)
www.panstanford.com

nanolevel, at suppressed Kirkendall shift. Stability of such hollow nanostructures is determined by interplay of capillary effect and inverse Kirkendall effect and controlled by diffusion as well.[11-14]

Void formation upon reactive diffusion in the samples with closed geometry (cylindrical or spherical) has been known for rather long time. It is a common feature of interdiffusion in binary systems and is most frequently related to the difference of partial diffusivities of the components. Such difference leads to difference of intrinsic atomic fluxes (in the lattice reference frame) creating the vacancy flux directed toward a faster component. Since the vacancy flux has some spatial distribution (with maximum in the vicinity of initial contact zone), it has nonzero divergence, leading to local accumulation of vacancies at the fast component side and depletion with vacancies at the slow component side. If vacancy sinks at the fast component side are not effective enough then extra vacancies should gather into voids (Frenkel effect or Kirkendall voiding). Such voiding should be especially pronounced in spherical and cylindrical samples when the faster component is situated inside (Fig. 6.1a). Such behavior should be typical at any interdiffusion—with formation of solid solution or with intermediate compound (IMC) formation.

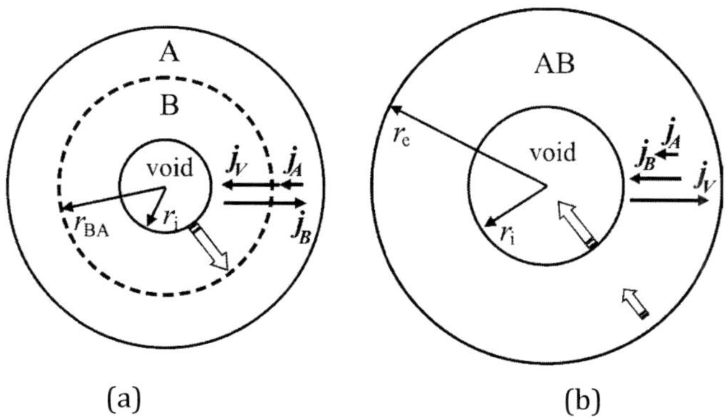

Figure 6.1 Schematic diagram of the (a) formation and (b) collapse of hollow nanoparticle. At the first stage (formation) the vacancy flux is directed inside, opposite to the flux of faster component B. At the second stage (collapse), when the chemical potential gradient becomes smaller, the Gibbs–Thomson effect prevails: vacancy flux changes direction and flows from inner surface (with higher vacancy concentration) to external surface (with lower one). Vacancy flux generates the opposite fluxes of A and B, with B-flux being larger, leading to redistribution–segregation of faster component B at the inner surface.

This possibility was realized by Aldinger,[15] who obtained hollow shells of a BeNi alloy after annealing of Be microparticles coated by Ni. Moreover, Geguzin *et al.* reported the formation of hollow intermetallic wire by the reaction Cd + Ni → $Cd_{23}Ni_5$ in the Cd wire covered by Ni.[16,17] Authors of Refs. 16 and 17 prescribed this phenomenon to the large step of molar volume in the intermetallic formation (about 8%). Masumura *et al.* numerically estimated the accumulation of vacancies in the central part of spherical particle.[18]

In 2004, void formation in spherical samples was rediscovered at nanolevel.[3–5,8–9] The hollow nanoshells of cobalt, iron, copper, and nickel oxides and sulfides have been obtained by means of reaction of metallic nanopowders with oxygen or sulfur. Contrary to the results in Refs. 16 and 17, these results have been explained by the Kirkendall voiding (also called Frenkel effect): out-diffusion of metal through the forming spherical layer of compound is faster than in-diffusion of oxygen or sulfur through the same phase; this inequality of fluxes generates the inward flux of vacancies, meeting inside and forming the void in the central part of the system (Fig. 6.1a). Using Kirkendall name to denote the interdiffusion-induced voiding sometimes leads to confusion, since Kirkendall voiding and Kirkendall shift are in fact two different effects, and moreover, two competing effects, competing consequences of the same reason (difference of partial diffusivities). To emphasize the competing character of Frenkel voiding and Kirkendall shift, Geguzin[16] introduced special notations for different vacancy sinks/sources—K-sinks and F-sinks. K-sinks are, for example, the kinks at dislocations. Vacancies can be annihilated or generated at these K-sinks, changing locally the number of lattice sites and generating a dislocation climb and corresponding lattice shift (Kirkendall effect). F-sinks are voids. Vacancies can join these F-sinks without changing the lattice sites number. Functioning of F-sinks leads to voiding instead of Kirkendall shift. If one applies a comparatively low pressure of about 5–10 MPa, which does not change the jump frequency considerably but is sufficient to suppress the void nucleation and/or growth, then one will suppress Frenkel effect and will observe "pure" Kirkendall effect—Kirkendall shift of markers indicating the shift of lattice. Indeed, such experiments were made by Geguzin group and demonstrated increase of Kirkendall shift about twice with almost full suppression of voiding.[16]

In hollow nanoshell formation we have opposite phenomenon—almost pure Frenkel effect (Kirkendall voiding) with fully or partially suppressed Kirkendall shift. Most probably, this happens due to peculiarities of interdiffusion in nanosystems: (1) there is little place for dislocations in nanoparticle so that K-sinks are just lacking; (2) common Kirkendall shift in radial direction leads to tangential deformation and corresponding stresses in spherical or cylindrical samples, which turn to be large in case of nano-systems, and suppress the shift. Therefore, the incoming vacancy flux (arising due to outcoming fast component flux) leads to supersaturation with vacancies in the "core." If core is small enough and temperature is high, this can lead directly to formation of central single void. Otherwise, formation of multiple tiny voids near interface core-compound may lead to the bridge-like structure (remaining core-linked with growing compound shell by kind of bridges providing the outcoming bulk and surface fluxes of core element).[4,7,14]

No doubt, application of any structure depends on its stability. Immediately after publishing of paper in Science,[3] Tu and Gösele came to an almost obvious conclusion that any hollow structure should be unstable since shrinkage of hollow shell is energetically favorable (total surface and surface energy decreases).[11] Mechanism of shrinkage is related to capillary, or Gibbs–Thomson effect (effect of Laplace pressure on the boundary vacancy concentrations): vacancy concentration on the inner boundary of nanoshell should be higher than that on the external boundary. This difference of boundary concentrations leads to outward vacancy flux meaning shrinking.

We, together with Tu and Gösele, predicted one more important factor of shrinkage kinetics, in which, we argued that, contrary to shrinking of single-component shell, shrinking of binary compound shell is effected also by inverse Kirkendall effect.[12] Indeed, outward vacancy flux generates opposite fluxes of two components, which are different due to different mobilities (Fig. 6.1b). In its turn, it leads to segregation of faster component near the inner boundary. Corresponding concentration gradient makes probability of vacancy jump higher for inward direction (exchange with segregated fast diffusant atoms inside). Mathematically it means the first term in the two-term expression for vacancy flux,

$$\Omega J_{\mathrm{V}} = (D_{\mathrm{B}} - D_{\mathrm{A}})\frac{\partial c_{\mathrm{B}}}{\partial r} - D_{\mathrm{V}}\frac{\partial c_{\mathrm{V}}}{\partial r} \tag{6.1}$$

(c_{B} and c_{V} are the local fractions of sites occupied by atoms B and of vacant sites, respectively; D_{B}, D_{A}, and D_{V} are the partial diffusivities of main components and vacancies, related by Frenkel relation,

$$D_V = \frac{c_A D_A + c_B D_B}{c_V \varphi} \tag{6.2}$$

where φ is the thermodynamic factor, common for both components of binary alloy:

$$\varphi = \frac{c_A \partial \mu_A}{kT \partial c_A} = \frac{c_B \partial \mu_B}{kT \partial c_B} \qquad (\mu_A \text{ and } \mu_B \text{ are the chemical potentials})$$

The first term of Eq. 6.3 is opposite in sign to the vacancy gradient term and, hence, substantially suppresses the vacancy flux. Therefore, the shrinkage rate of compound shell appears to be controlled by the slow species. Shrinkage of compound shells was experimentally proved by Nakamura *et al.*[9,10] Actually, this situation is more complicated than predicted in Ref. 12. Namely, the hollow nanoparticles of copper oxide and nickel oxide in air indeed shrink to form the oxide compact particles without voids. Yet, if the mentioned oxide nanoparticles are kept in vacuum, the shrinking proceeds simultaneously with reduction of metal: oxygen from oxide evaporates to vacuum, and the resulting particles after shrinking are just copper or nickel balls. Shrinkage of Fe_3O_4 hollow nanoballs and nanotubes proceeds simultaneously with conversion to gamma-Fe_2O_3. Moreover, shrinkage of central void may be accompanied by formation of tiny secondary voids in between central void and external boundary.[10]

We also analyzed the formation of hollow compound nanoshell.[13] We demonstrated that the Gibbs–Thomson effect, leading to shrinkage of "ready" compound shells, should be important at formation stage as well—it sometimes may even suppress the nanoshell formation.

In both studies,[12,13] compound shells were treated with very narrow homogeneity range. For this case the steady state approximation for both vacancies and main components worked well. We also analyzed formation and collapse of hollow shell in the system with full solubility or, at least, with broad homogeneity range.[14] In this case, formation and collapse appeared to be the two stages of one whole process—at the first stage (when the chemical driving forces are still big), Kirkendall voiding wins over curvature effect, then, after arriving at sufficiently small concentration gradient, the curvature-driven shrinking should take the revenge. We also used phenomenological and Monte Carlo (MC) modeling, giving possibility to observe both stages (formation and collapse) in one run.

We will try to make a review of the mentioned results and to add comparison of spherical case with cylindrical case. Logics of our presentation differ from actual historical sequence of developments in this field. As far as we know from personal communication of late Ulrich Gösele, the first experiments with formation of hollow nanostructures had been made by chemists who were rather far from Kirkendall effect. After introducing the Kirkendall voiding idea, the mechanism of hollow particle production became transparent. Yet, before Tu and Gösele introduced the next idea (Gibbs–Thomson effect for vacancies, 2005) nobody was thinking about the problems of stability and life time of the novel objects. Now it looks strange because the theory of caking based on the same effect had been known since the end of the fiftees (as a direct application of the Lifshits-Slyozow-Wagner (LSW) ripening theory[19]). Only after this we introduced the idea of inverse Kirkendall effect,[12] making slow species responsible for shrinking kinetics. One more aspect—competition between Kirkendall voiding and Kirkendall shift in nanoparticle formation is analyzed very briefly so far and has no experimental basis (so far)—we think it can become a hot topic for intermediate sizes of hollow particles (larger than nano but less than micron). We build our presentation from more theoretical point of view treating hollow structure, first of all, as something already produced by experimenters. Therefore, we start with stability analysis, and formation problems go later.

At first, we will formulate the main regularities of three "Kirkendall-related" effects and curvature-related effect (Section 6.2). In Section 6.3, we discuss the problem of stability of the hollow particles and nanotubes for different types of particles (pure elements, chemical compounds, solid solutions). In Section 6.4, we come to the problem of hollow structure formation keeping in mind that the factors influencing the shrinking, can well be important at the formation stage as well. It means, in principle, that at some conditions, formation may be impossible, and we formulate the criteria of hollow nanoparticle formation. In Section 6.5, we describe (by numeric phenomenologic calculations as well as by atomistic MC simulation) both formation and shrinking as two stages of full "life cycle" of core–shell structure characterized by cross-over from formation to shrinking. Practical use of hollow structures is realistic only if the shrinking stage is much longer than the formation stage (e.g., due to lower temperature).

6.2 Basic Notions

In this section we remind shortly the basic notions necessary for analysis and understanding of hollow shell formation and collapse competition of Kirkendall shift with Frenkel–Kirkendall voiding, with account of capillary effects.

6.2.1 Kirkendall Shift and Frenkel–Kirkendall Voiding in Bulk Samples

Interdiffusion in binary crystalline systems consisting of two chemical elements, proceeds, most often, by the vacancy mechanism. Local concentration of vacancies, in general, is not constant due to divergence of vacancy fluxes. Thus, at least from kinetic point of view, one should consider binary system as a ternary one. With some simplifying assumptions (neglecting correlation effects, like Manning's corrections, neglecting atomic volumes difference, etc.), one can write down the volume fluxes of this pseudo-ternary system in the lattice reference frame as follows:

$$\Omega j_{\mathrm{B}} = -D_{\mathrm{B}}^{*}\varphi\frac{\partial c_{\mathrm{B}}}{\partial x} + \frac{c_{\mathrm{B}}D_{\mathrm{B}}^{*}}{c_{\mathrm{V}}}\frac{\partial c_{\mathrm{V}}}{\partial x} \tag{6.3}$$

$$\Omega j_{\mathrm{A}} = -D_{\mathrm{A}}^{*}\varphi\frac{\partial c_{\mathrm{A}}}{\partial x} + \frac{c_{\mathrm{A}}D_{\mathrm{A}}^{*}}{c_{\mathrm{V}}}\frac{\partial c_{\mathrm{V}}}{\partial x} \tag{6.4}$$

$$\Omega j_{\mathrm{V}} = (D_{\mathrm{B}}^{*} - D_{\mathrm{A}}^{*})\varphi\frac{\partial c_{\mathrm{B}}}{\partial x} - D_{\mathrm{V}}\frac{\partial c_{\mathrm{V}}}{\partial x} \tag{6.5}$$

where vacancy self-diffusivity D_{V} and thermodynamic factor φ were defined in Introduction. Darken in the Forties was the first who reminded diffusion community that Galileo principle of velocity adding should be taken into account in diffusion as well. Velocity of lattice in the laboratory reference frame is equal to the volume flux density of vacancies:

$$U = \Omega j_{\mathrm{V}} \tag{6.6}$$

Then the fluxes in laboratory reference frame are

$$\Omega J_{\mathrm{B}} = \Omega j_{\mathrm{B}} + c_{\mathrm{B}}U\,,\ \Omega J_{\mathrm{A}} = \Omega j_{\mathrm{A}} + c_{\mathrm{A}}U \tag{6.7}$$

$$\Omega J_{\mathrm{V}} = \Omega j_{\mathrm{V}} + c_{\mathrm{V}}U = (1 + c_{\mathrm{V}})\Omega j_{\mathrm{V}} \cong \Omega j_{\mathrm{V}} \tag{6.8}$$

Thus, vacancy flux is practically the same in all reference frames (due to small vacancy concentration and respectively small convective term), but fluxes of main components can be very different in the lattice and in laboratory reference frames.

The main difference between the third component (vacancies) and the main two components is that the vacancies can be generated and annihilated without any nuclear or chemical reactions. At fixed composition annihilation and generation should come to dynamic equilibrium at some equilibrium vacancy concentration. Since in interdiffusion process the composition changes all the time at any place of diffusion zone, the vacancy concentration is, strictly speaking, never equilibrium. Moreover, difference of partial diffusivities (look at the first term in the right-hand side of Eq. 6.5 always generates a vacancy flux in the diffusion zone. Existence of sinks and sources of vacancies can be taken into account in two ways. First way: we can treat the places where vacancies can be created or annihilated (sinks/sources) as the internal boundaries for which some boundary conditions should be formulated. Alternatively, we can treat sinks/sources as uniformly distributed in the diffusion zone and characterized by some relaxation time τ_V. It means, that a number of vacancies being produced or annihilated (depending on sign) per unit time per unit volume is equal to:

$$\frac{dn_V}{dt} = -\frac{1}{\Omega}\frac{c_V - c_V^{eq}}{\tau_V} \tag{6.9}$$

So, the second Fick's law for vacancy diffusion in this relaxation time approximation has the following form:

$$\frac{\partial c_V}{\partial t} = -div(\Omega J_V) - \frac{c_V - c_V^{eq}}{\tau_V} \tag{6.10}$$

where the vacancy flux density is determined by Eqs. 6.1 and 6.8. Equivalent parameter is $L_V = \sqrt{D_V \tau_V}$ which can be interpreted as the mean free path of vacancy migration). If the characteristic width of diffusion zone is significantly larger than L_V then the second way looks preferable. Creation or annihilation of vacancies may proceed in two ways—with or without change of the lattice sites number:

A. If vacancy comes to the kink of the edge dislocation (to the edge of extra-plane) and annihilates there, dislocation climbs and the number of lattice sites decreases by one (extra-plane looses one site). The total volume in this case also decreases approximately by one atomic volume. If, in another place, vacancy is generated at the kink of edge dislocation, number of sites increases by one (extra-plane gains one site). The total volume in this case increases approximately by one atomic volume. Such kind of process (dismantling of extra-planes in one place and construction of extra-planes in other place) should lead to the shift of lattice as a whole between these places which can be observed as the shift of "frozen-in" markers (Kirkendall shift).

B. If vacancy attaches to the already existing void (pore) inside the crystal, the total volume remains practically the same since the empty sites are just redistributed. The growth of voids due to supersaturation with vacancies caused by the difference of atomic mobilities, is called Kirkendall voiding or Frenkel effect. If vacancies are generated at the dislocations at the side of slower diffusant and disappear at the side of faster diffusant at the growing voids, it will lead to eventual volume increase (swelling) but not to Kirkendall shift. Voiding was extensively studied in microelectronics but, instead of interdiffusion, in electro- and thermomigration conditions, when the main reason of vacancy flux is an electron wind (transfer of momentum to migrating atoms from the drifting electrons, mainly in Al, Cu, and Sn)[20–23] or the temperature gradient (mainly in solders).[24] Kirkendall (Frenkel) voiding under interdiffusion is observed almost always but, strangely, is clearly insufficiently investigated theoretically. We regard it to be a challenge for modeling community to develop a predictive model of interdiffusion with account of competition between Kikendall shift and Kirkendall–Frenkel voiding.

It is important to understand that the diffusion itself and the diffusion-driven Kirkendall shift has different space and time scales. Diffusion in its strict sense—atomic migration—proceeds at atomic scale. Kirkendall shift is caused by generation of vacancy in one place and annihilation of vacancy in another place, distance between these places is not less than the vacancy mean free migration path, which can well be of the order of micron. In other words, sinks and sources of vacancies which are the movement generators, are spatially separated, so that Kirkendall shift is, in principle, a non-local effect. It may be important for the systems with characteristic sizes less than about micron. For interdiffusion in the bulk diffusion couples at high temperatures and absence of Frenkel voiding the story is now more or less complete:[25]

(1) after some initializing period of about vacancy relaxation time τ_V, during the annealing time interval $\tau_V << t << \tau_V/c_V^{eq}$, Kirkendall shift is almost zero, redistribution of the components is governed by the Nernst–Planck interdiffusivity,

$$D_{NP} = \frac{D_A D_B}{c_A D_A + c_B D_B} \tag{6.11}$$

controlled by the slow diffusant (it tends to zero if one of partial diffusivities tends to zero);

(2) at the transient stage, $t \sim \tau_V / c_V^{eq}$, diffusion is not parabolic and non-local, with non-locality distance L_V. It means that the vacancy flux, Kirkendall shift velocity, and the main components fluxes in laboratory reference frame in some place are determined by some integral of concentration gradient over some region about L_V size:

$$\Omega J_V = \frac{1}{2L_V}\int_0^{\infty} \exp\left(-\frac{|x'-x|}{L_V}\right)(D_B - D_A)\frac{\partial c_B}{\partial x'}dx' \tag{6.12}$$

$$\Omega J_B = -\Omega J_A$$

$$= -\tilde{D}_{Darken}\left((1-w)\frac{\partial c_B}{\partial x} + \frac{w}{2L_V}\int_0^{\infty} \exp\left(-\frac{|x'-x|}{L_V}\right)(D_B - D_A)\frac{\partial c_B}{\partial x'}dx'\right) \tag{6.13}$$

where

$$D_{Darken} = c_B D_A + c_A D_B \tag{6.14}$$

$$w = 1 - D_{NP}/\tilde{D}_{Darken} = 1 - \frac{D_A D_B}{(c_A D_A + c_B D_B)(c_B D_A + c_A D_B)}$$

$$= \frac{c_A c_B (D_A - D_B)^2}{(c_A D_A + c_B D_B)(c_B D_A + c_A D_B)} \tag{6.15}$$

(3) at the latest stage, $t >> \tau_V/c_V^{eq}$, diffusion standard parabolic law, characterized by interdiffusivity and shift velocity predicted by Darken theory:

$$U = \Omega J_V = (D_B - D_A)\frac{\partial c_B}{\partial x} \tag{6.16}$$

$$\Omega J_B = -\Omega J_A = -\tilde{D}_{\text{Darken}} \frac{\partial c_B}{\partial x} = -(c_B D_A + c_A D_B) \frac{\partial c_B}{\partial x} \tag{6.17}$$

Yet, even at this last stage of interdiffusion the redistribution of components at microscale, inside regions free from sinks/sources, is governed by second Fick's law with Nernst–Planck interdiffusivity from Eq. 6.11, as have been shown in Refs. 26 and 27.

Mathematically, Eqs. 6.12 and 6.13 are transformed into Eqs. 6.16 and 6.17 if $L_V \to 0$ transforming peak-like function $\frac{1}{L_V}\exp\left(-\frac{|x'-x|}{L_V}\right)$ into Dirac delta-function $\delta(x - x)$.

Three years after his paper, Stephenson suggested three analogous stages using concept of viscosity instead of nonequilibrium vacancies.[25,28]

The last fundamental experimental finding concerning Kirkendall shift was made by the group of Prof. Frans van Loo at the very beginning of new millennium,[29–31] in which bifurcations of Kirkendall planes were discovered experimentally and instabilities of Kirkendall planes were predicted theoretically and then found experimentally.

Quantitative investigation of Frenkel effect is yet to be done.

6.2.2 Inverse Kirkendall Effect

Inverse Kirkendall effect was discovered by the community investigating alloys under irradiation (see e.g., Refs. 32 and 33). Such alloys often demonstrated segregation near the external surfaces or near pore surfaces. Actually, inverse Kirkendall effect involves two interrelated "sub-effects".

First, generated by some external reasons directed vacancy flux, due to inequality of partial diffusivities, causes counter-fluxes of both components but with different velocities, leading to segregation of faster component at the side from which vacancies flow. Indeed, Eqs. 6.3–6.5 can be easily rearranged as:

$$\Omega j_B = -D_{NP} \frac{\partial c_B}{\partial x} - \frac{c_B D_B^*}{c_V D_V} \Omega j_V \tag{6.18}$$

$$\Omega j_A = -D_{NP} \frac{\partial c_A}{\partial x} - \frac{c_A D_A^*}{c_V D_V} \Omega j_V \tag{6.19}$$

Last terms in the right-hand sides of Eqs. 6.18 and 6.19 are responsible for the mentioned counter-fluxes.

Second, the above mentioned segregation means formation of concentration gradient which, according to le Chatelier's principle buffer principle and according to Eq. 6.1, makes the vacancy flux less.

To illustrate the first "sub-effect", we can (combining Eqs. 6.3, 6.6, and 6.7) represent the flux of, say, component B in laboratory reference frame in the following form:

$$\Omega J_{\mathrm{B}} = -\frac{D_{\mathrm{A}} D_{\mathrm{B}}}{c_{\mathrm{B}} D_{\mathrm{B}} + c_{\mathrm{A}} D_{\mathrm{A}}} \frac{\partial c_B}{\partial x} - c_{\mathrm{B}} c_{\mathrm{A}} \frac{D_{\mathrm{B}} - D_{\mathrm{A}}}{c_{\mathrm{B}} D_{\mathrm{B}} + c_{\mathrm{A}} D_{\mathrm{A}}} \Omega j_{\mathrm{V}} \tag{6.20}$$

If the vacancy flux is kept constant then the flux of B tends to zero when the following gradient of composition is achieved:

$$\frac{\partial c_{\mathrm{B}}}{\partial x} = -c_{\mathrm{B}} c_{\mathrm{A}} \frac{D_{\mathrm{B}} - D_{\mathrm{A}}}{D_{\mathrm{A}} D_{\mathrm{B}}} \Omega j_{\mathrm{V}} \tag{6.21}$$

This gradient is a measure of steady-state segregation existing as long as vacancy flux is being kept. Actually, it is almost impossible to keep the vacancy flux arbitrarily fixed (e.g., by electromigration) in binary or multicomponent alloy. Reason is very simple: as mentioned above, segregation caused by the vacancy flux, changes the vacancy flux itself. Mathematically it is related to the first term of Eq. 6.5. We will see some examples of second sub-effect in the shrinking kinetics of binary nanoshells (below).

6.2.3 Gibbs–Thomson Effect for Vacancies (Elementary)

As was first realized by Tu and Gösele,[11] effect of curvature on the equilibrium vacancy concentration near the curved surfaces, plays a key role in stability of hollow structures. Let us make a very simple derivation of this effect. Equilibrium vacancy concentration is determined by the vacancy formation energy E_{V}—energy needed to take the atom from inside and place it at the surface of a crystal, leaving vacancy inside:

$$c_{\mathrm{V}} = \exp\left(-\frac{E_{\mathrm{V}} + p\Omega}{kT}\right) \tag{6.22}$$

If crystalline body has spherical external surface with radius r, then one can imagine that the atom taken outside, is "buttered" (dabbed) over all surface (this continuous picture is okay if we create a lot of vacancies). If atomic volume Ω is distributed over spherical surface it increases its radius: $\Omega = d\left(\frac{4}{3}\pi r^3\right) = 4\pi r^2 dr$, so that $dR = \frac{\Omega}{4\pi r^2}$ (in case of cylinder of radius r and height H one has $\Omega = d(\pi r^2 H)$, so that $dR = \frac{\Omega}{2\pi r H}$). Thus, the area of spherical surface increases by $d(4\pi r^2) = 8\pi r dr = \frac{2\Omega}{r}$ (in case of cylinder $d(2\pi r H) = 2\pi H dr = \frac{\Omega}{r}$), and the surface energy increases by $E_{\mathrm{GT}}(r) = \frac{2\gamma\Omega}{r}$ for sphere and by $E_{\mathrm{GT}}(r) = \frac{\gamma\Omega}{r}$ for cylinder (γ is a surface tension), which can be interpreted also as a product of Laplace tension and atomic volume. It means that the vacancy formation energy indeed depends on the curvature:

$$E_{\mathrm{V}} = E_{\mathrm{V}}^{\mathrm{planar}} + \frac{2\gamma\Omega}{r} \text{ (sphere) or } E_{\mathrm{V}} = E_{\mathrm{V}}^{\mathrm{planar}} + \frac{\gamma\Omega}{r} \text{ (cylinder)} \quad (6.23)$$

If vacancy is created near the inner spherical surface (surface of a spherical void), then the atom "distributed over" this inner surface, decreases its radius. Eq. 6.23 remains valid but with negative curvature radius. Combining Eqs. 6.22 and 6.23, one obtains

$$\begin{cases} c_{\mathrm{V}}(r_{\mathrm{e}}) = c_{\mathrm{V}_0} e^{-\frac{L_{\mathrm{GT}}}{r_{\mathrm{e}}}} \approx c_{\mathrm{V}_0}\left(1 - \frac{L_{\mathrm{GT}}}{r_{\mathrm{e}}}\right) \\ c_{\mathrm{V}}(r_{\mathrm{i}}) = c_{\mathrm{V}_0} e^{\frac{L_{\mathrm{GT}}}{r_{\mathrm{i}}}} \approx c_{\mathrm{V}_0}\left(1 + \frac{L_{\mathrm{GT}}}{r_{\mathrm{i}}}\right) \end{cases} \quad (6.24)$$

where L_{GT}—characteristic Gibbs–Thomson length (about 1–10 nm), which is determined by $L_{\mathrm{GT}} \equiv \frac{2\gamma\Omega}{kT}$ for spherical and $L_{\mathrm{GT}} \equiv \frac{\gamma\Omega}{kT}$ for cylindrical particle.

Linear approximation is valid only for the case $r >> L_{\mathrm{GT}}$. Thus, for nanometric particles one should be very careful to use linear approximation.

It is easy to check that for cylindrical surfaces we have the same but without factor 2:

$$E_{\mathrm{V}}^{\mathrm{cyl}} = E_{\mathrm{V}}^{\mathrm{planar}} \pm \frac{\gamma\Omega}{r},\ c_{\mathrm{V}}^{\mathrm{cyl}}(R) = c_{\mathrm{V}}^{\mathrm{planar}} \exp\left(\mp\frac{\gamma\Omega}{kTr}\right) \quad (6.25)$$

6.2.4 Gibbs–Thomson Effect for Basic Components

Till 2009[13] nobody took into account one more curvature effect in nanoshell formation—Gibbs-Thomson effect for composition itself. Namely, curvature changes the thermodynamic driving force of compound formation. We will denote the core material as "B" (metallic core in typical experiments) and external material as "A" (e.g., sulphur in sulfide formation), Fig. 6.2. It is well known for the case of planar layer growth, that for the growth of a single intermediate phase, the difference of the right and the left derivatives would be proportional to the driving force of reaction,

$$\left.\frac{\partial g}{\partial c_{\mathrm{B}}}\right|_{r_{\mathrm{e}}}-\left.\frac{\partial g}{\partial c_{\mathrm{B}}}\right|_{r_{\mathrm{i}}}=\left.\frac{\partial g}{\partial c_{\mathrm{B}}}\right|_{\infty}^{\mathrm{compound/A}}-\left.\frac{\partial g}{\partial c_{\mathrm{B}}}\right|_{\infty}^{\mathrm{compound/B}}=-\frac{\Delta g}{\bar{c}_{\mathrm{B}}\bar{c}_{\mathrm{A}}} \qquad (6.26)$$

Yet, in our case we should take into account the influence of curvature not only on the vacancy concentration but also on the phase equilibrium at the curved boundaries. At the external boundary with radius r_{e} the compound is under additional Laplace compression with additional energy per atom $E_{\mathrm{GT}}(r_{\mathrm{e}})$, so that, taking into account the common tangent rule, we have

$$\left.\frac{\partial g}{\partial c_{\mathrm{B}}}\right|_{r_{\mathrm{e}}}=\left.\frac{\partial g}{\partial c_{\mathrm{B}}}\right|_{\infty}^{\mathrm{compound/A}}+\frac{E_{\mathrm{GT}}(r_{\mathrm{e}})}{c_{\mathrm{B}}(r_{\mathrm{e}})-0}\cong\left.\frac{\partial g}{\partial c_{\mathrm{B}}}\right|_{\infty}^{\mathrm{compound/A}}+\frac{E_{\mathrm{GT}}(r_{\mathrm{e}})}{\bar{c}_{\mathrm{B}}} \qquad (6.27)$$

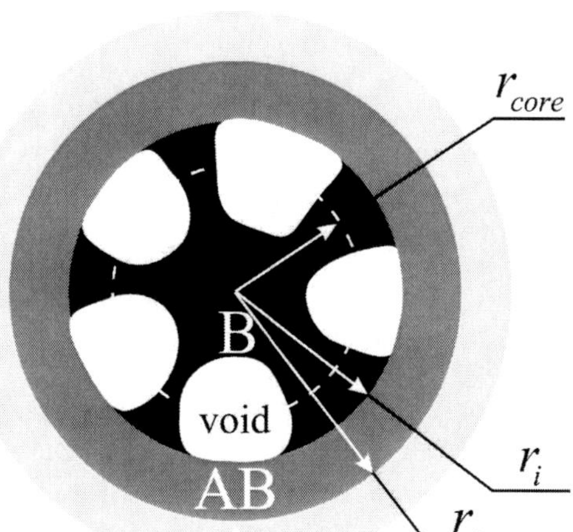

Figure 6.2 Bridge model of hollow shell formation.

At the internal boundary with radius r_i the compound shell is under Laplace tension with negative additional energy per atom $-E_{GT}(r_i)$, so that

$$\left.\frac{\partial g}{\partial c_B}\right|_{r_i} = \left.\frac{\partial g}{\partial c_B}\right|_{\infty}^{\text{compound/B}} + \frac{-E_{GT}(r_i)}{c_B(r_i)-1} \cong \left.\frac{\partial g}{\partial c_B}\right|_{\infty}^{\text{compound/B}} - \frac{E_{GT}(r_i)}{\overline{c}_A} \quad (6.28)$$

Then, instead of Eq. 6.26, we obtain

$$\left.\frac{\partial g}{\partial c_B}\right|_{r_e} - \left.\frac{\partial g}{\partial c_B}\right|_{r_i} = -\frac{\Delta g}{\overline{c}_B\overline{c}_A} + \frac{E_{GT}(r_e)}{\overline{c}_B} - \frac{E_{GT}(r_i)}{\overline{c}_A} \quad (6.29)$$

By the way, last two terms in Eq. 6.29 have opposite signs. It means that the curvature may decrease and may increase an effective driving force of intermediate phase formation in core–shell structure, depending on stoichiometry and on the ratio of external and internal radii. In particular, if stoichiometry is 1:3 then the interchange of core–shell components may lead to switch from suppression to acceleration of phase growth or vice versa. At initial stage of phase formation, when external and inner radii are approximately the same, we can introduce an effective driving force $\Delta g^{ef} = \Delta g - E_{GT}(r_0)\frac{c_A - c_B}{c_A c_B}$ in which the second term can change sign depending on stoichiometry.

6.3 Instability of Hollow Nanostructures

Hollow nanoshells can be produced by various methods. Independently on production method, all such hollow structures should have a common feature—instability. From thermodynamic point of view, nanoshells should shrink because it leads to decrease of surface, and, hence, of total energy. Kinetic mechanism of shrinking should be shown additionally. For the case of nanoshells of a single element the physical picture of shrinking seems very clear. It is related to Gibbs–Thomson effect for vacancies (Section 6.2.3). It is analogous to the LSW theory of caking,[19] when voids are dissolved by vacancy fluxes to external surface. Yet, in the nanoshell case it has a much more symmetrical picture: vacancies are generated at the inner surface (void dissolution), migrate to the external surface, causing the in-flux of atoms and movement of the inner surface, and annihilate at the external surface, causing shrinking of the latter (Fig. 6.3a). In case of a binary alloy or IMC nanoshell, the process is more complicated due to the "inverse Kirkendall effect" (Section 6.2.2), which was first introduced in the problem of irradiation of an alloy with energetic particles. As it is known, the radiation leads to generation of a number of Frenkel pairs "vacancy-interstitial". Interstitials quickly disappear at sinks since their mobility is much higher than that of the vacancies. Vacancies move to sinks much slowly, generating the directed vacancy fluxes. Since in an alloy or intermetallic, one of the components may diffuse much faster than the other ($D_B >> D_A$), the vacancy flux should lead to a redistribution of the components—segregation of B-atoms at the inner boundary, and of A-atoms near the external surface (Fig. 6.3b).

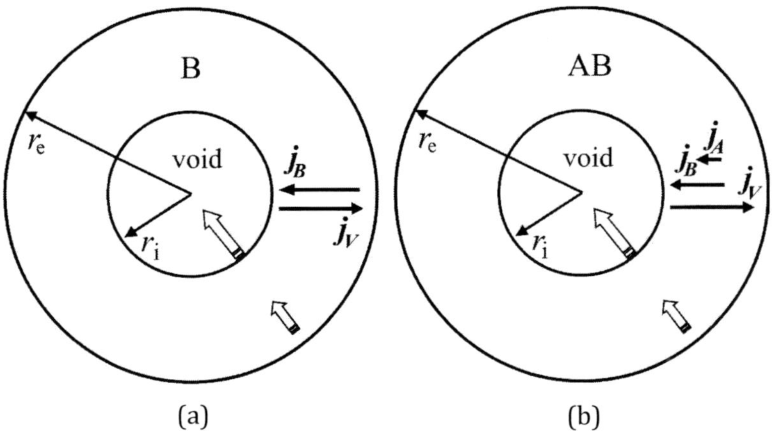

Figure 6.3 Schematic diagram of (a) pure element shell shrinking and of (b) a binary alloy shell shrinking ($D_B > D_A$), accompanied by B-segregation at inner surface.

The arising concentration gradient $\partial c_B/\partial r$ will influence and reduce the vacancy flux and hence will generate a tendency to suppress the shrinking process. We will deduce below that the slower diffusing specie will control the shrinking rate. Moreover, segregation increases with decreasing inner radius. It means that the concentration of B at the inner radius may exceed the homogeneity limit of the IMC, and pure B component can precipitate at the void/IMC interface. After this the shrinking should proceed further, ending with tiny pure B nanoparticle inside an inhomogeneous (with concentration gradient inside) IMC nanoshell. This kinetic process will be discussed in detail elsewhere. To develop the kinetic equations, we shall use the formalism of the classic Darken analysis of interdiffusion and take into account the non-equilibrium vacancies in the interdiffusion. This formalism follows what was developed first by Nazarov and Gurov[34] and later modified by Gurov and Gusak.[25–27]

6.3.1 Shrinking of Pure Hollow Shells

It is energetically favorable for a hollow spherical or cylindrical nanoshell to shrink into a compact nanosphere or nanowire as the interface area is reduced at constant total mass. Herein and in the following we will write equations for spherical (balls) and cylindrical particles (wires) adding, respectively, indexes "s" and "c" to their numbers.

Namely, decrease of energy due to shrinking in these two cases can be formulated as:

$$\gamma(4\pi r_{\mathrm{i0}}^2 + 4\pi r_{\mathrm{e0}}^2) > \gamma 4\pi r_{\mathrm{f}}^2 \tag{6.30s}$$

$$\gamma(2\pi r_{\mathrm{i0}} + 2\pi r_{\mathrm{e0}}) > \gamma 2\pi r_{\mathrm{f}} \tag{6.30c}$$

(γ represents the surface tension; r_{i0} and r_{e0} are the initial inner and external radii of the nanoshell; r_{f} is the final radius of collapsed particle), with conservation of volume implying $(4\pi/3)(r_{\mathrm{e0}}^3 - r_{\mathrm{i0}}^3) = (4\pi/3)r_{\mathrm{f}}^3$ for spherical particles and $2\pi(r_{\mathrm{e0}}^2 - r_{\mathrm{i0}}^2) = 2\pi r_{\mathrm{f}}^2$ for wires.

The proposed atomistic mechanism of shrinking is the vacancy flux from the inner surface (with radius r_{e}) to the external surface (with radius r_{i}). The driving force of the vacancy flux is the difference of vacancy/atoms chemical potentials and corresponding difference of equilibrium vacancy concentrations at the curved interfaces: $c_{\mathrm{V}}(r_{\mathrm{i}}) > c_{\mathrm{V}}(r_{\mathrm{e}})$. This Gibbs–Thomson effect can be expressed, **in linear approximation so far**, as

$$c_{\mathrm{V}}(r_{\mathrm{i}}) = c_{\mathrm{V}}^{\mathrm{eq}}\left(1 + \frac{L_{\mathrm{GT}}}{r_{\mathrm{i}}}\right),\ c_{\mathrm{V}}(r_{\mathrm{e}}) = c_{\mathrm{V}}^{\mathrm{eq}}\left(1 - \frac{L_{\mathrm{GT}}}{r_{\mathrm{e}}}\right) \tag{6.31}$$

with $L_{\mathrm{GT}} = 2\gamma\Omega/kT$ for spherical particles and $L_{\mathrm{GT}} = \gamma\Omega/kT$ for wires, Ω is the atomic volume, and c is the concentration of the fraction of vacant lattice sites; $c_{\mathrm{V}}^{\mathrm{eq}}$ refers to the equilibrium concentration of vacancies with respect to a planar surface. The linear approximation holds if $r_{\rm i} >> L_{\rm GT}$. If the term $L_{\rm GT}/r$ becomes comparable with unity, and we'll need to reformulate Eq. 6.31 using the exponential size dependencies.

6.3.1.1 Model

The vacancy out-flux is generated by the difference of equilibrium vacancy concentrations at two curved boundaries (Fig. 6.3a). The equation governing the vacancy concentration $c_{\rm V}$ (fraction of vacant sites) inside the shell is:

$$\frac{\partial c_{\rm V}}{\partial t} = -\frac{1}{r^2}\frac{\partial}{\partial r}(r^2 \Omega j_{\rm V}) + 0, \quad r_{\rm i}(t) < r < r_{\rm e}(t) \tag{6.32s}$$

$$\frac{\partial c_{\rm V}}{\partial t} = -\frac{1}{r}\frac{\partial}{\partial r}(r \Omega j_{\rm V}) + 0, \quad r_{\rm i}(t) < r < r_{\rm e}(t) \tag{6.32c}$$

$j_{\rm V}$ is a vacancy flux density.

The term "0" in Eq. 6.32 is just to indicate that vacancies in our model are generated and annihilated only at the inner and external surfaces of shell. We do not consider any vacancy sinks/sources (dislocation kinks) inside the nanoshell.

The vacancy flux is governed by Fick's first law:

$$\Omega j_{\rm V} = -D_{\rm V} \,\partial c_{\rm V} / \partial r \tag{6.33}$$

Boundary conditions for vacancies were formulated by Eq. 6.31.

Inner shell/vacuum (or inert medium) and external surfaces move due to vacancy fluxes:

$$\frac{dr_{\rm i}}{dt} = -\Omega j_{\rm V}(r_{\rm i}) = D_{\rm V}\frac{\partial c_{\rm V}}{\partial r}(r = r_{\rm i}) \tag{6.34}$$

$$\frac{dr_{\rm e}}{dt} = -\Omega j_{\rm V}(r_{\rm e}) = D_{\rm V}\frac{\partial c_{\rm V}}{\partial r}(r = r_{\rm e}) \tag{6.35}$$

In steady-state approximation

$$\frac{\partial c_{\rm V}}{\partial t} \approx 0 = -\frac{1}{r^2}\frac{\partial}{\partial r}(r^2 \Omega j_{\rm V}) \tag{6.36s}$$

$$\frac{\partial c_{\rm V}}{\partial t} \approx 0 = -\frac{1}{r}\frac{\partial}{\partial r}(r \Omega j_{\rm V}) \tag{6.36c}$$

which leads to

$$\Omega j_{\rm V}(r_{\rm i}) = -D_{\rm V}\frac{\partial c_{\rm V}}{\partial r} = \frac{K_{\rm V}}{r^2} \tag{6.37s}$$

$$\Omega j_{\mathrm{V}}(r_{\mathrm{i}}) = -D_{\mathrm{V}}\frac{\partial c_{\mathrm{V}}}{\partial r} = \frac{K_{\mathrm{V}}}{r} \tag{6.37c}$$

so that

$$c_{\mathrm{V}}(r) = \frac{K_{\mathrm{V}}}{D_{\mathrm{V}}r} + M_{\mathrm{V}} \tag{6.38s}$$

$$c_{\mathrm{V}}(r) = -\frac{K_{\mathrm{V}}}{D_{\mathrm{V}}}\ln r + M_{\mathrm{V}} \tag{6.38c}$$

where $K_{\mathrm{V}}(t)$ = const, $M_{\mathrm{V}}(t)$ = const with respect to r.

Substituting solution of Eq. 6.40 into the Gibbs–Thomson boundary conditions of Eq. 6.31, we obtain:

$$K_{\mathrm{V}} = c_{\mathrm{V}}^{\mathrm{eq}} D_{\mathrm{V}} L_{\mathrm{GT}} \frac{r_{\mathrm{e}} + r_{\mathrm{i}}}{r_{\mathrm{e}} - r_{\mathrm{i}}} = D^{*} L_{\mathrm{GT}} \frac{r_{\mathrm{e}} + r_{\mathrm{i}}}{r_{\mathrm{e}} - r_{\mathrm{i}}} \tag{6.39s}$$

$$K_{\mathrm{V}} = c_{\mathrm{V}}^{\mathrm{eq}} D_{\mathrm{V}} L_{\mathrm{GT}} \frac{1/r_{\mathrm{e}} + 1/r_{\mathrm{i}}}{\ln r_{\mathrm{e}}/r_{\mathrm{i}}} = D^{*} L_{\mathrm{GT}} \frac{1/r_{\mathrm{e}} + 1/r_{\mathrm{i}}}{\ln r_{\mathrm{e}}/r_{\mathrm{i}}} \tag{6.39c}$$

with D^{*} being the self-diffusivity of the pure element B.

Substituting the expression of Eq. 6.39 into Eqs. 6.34 and 6.35 for the boundary movement, we obtain the simple equations for the shrinking kinetics:

$$\frac{dr_{\mathrm{i}}}{dt} = -\frac{D^{*} L_{\mathrm{GT}}}{r_{\mathrm{i}}^{2}} \frac{r_{\mathrm{e}} + r_{\mathrm{i}}}{r_{\mathrm{e}} - r_{\mathrm{i}}} \tag{6.40s}$$

$$\frac{dr_{\mathrm{i}}}{dt} = -\frac{D^{*} L_{\mathrm{GT}}}{r_{\mathrm{i}}^{2}} \frac{r_{\mathrm{i}}/r_{\mathrm{e}} + 1}{\ln r_{\mathrm{e}}/r_{\mathrm{i}}} \tag{6.40c}$$

$$\frac{dr_{\mathrm{e}}}{dt} = -\frac{D^{*} L_{\mathrm{GT}}}{r_{\mathrm{e}}^{2}} \frac{r_{\mathrm{e}} + r_{\mathrm{i}}}{r_{\mathrm{e}} - r_{\mathrm{i}}} \tag{6.41s}$$

$$\frac{dr_{\mathrm{e}}}{dt} = -\frac{D^{*} L_{\mathrm{GT}}}{r_{\mathrm{e}}^{2}} \frac{r_{\mathrm{e}}/r_{\mathrm{i}} + 1}{\ln r_{\mathrm{e}}/r_{\mathrm{i}}} \tag{6.41c}$$

Eqs. 6.40 and 6.41 are not independent and contain already the conservation of matter (and volume):

$$r_{\mathrm{e}}^{3} = r_{\mathrm{i}}^{3} + r_{\mathrm{f}}^{3} \tag{6.42s}$$

$$r_{\mathrm{e}}^{2} = r_{\mathrm{i}}^{2} + r_{\mathrm{f}}^{2} \tag{6.42c}$$

In principle, set of Eqs. 6.40–6.42 has analytical solution found in Ref. 35. Yet, explicit form of this solution is very bulky and inconvenient for practical use. To obtain a simple analytical estimate, we will use further assumptions. We will distinguish two cases: (A) $r_{i0} << r_f$ (thick shell, small void); (B) $r_{i0} >> r_f$ (initially thin shell, large void).

Case A. In this case inequality $r_i << r_f$ will hold for any intermediate moment. Therefore,

$$r_e = (r_i^3 + r_f^3)^{1/3} = r_f(1 + r_i^3/r_f^3)^{1/3} \approx r_f + r_i^3/3r_f^2 \qquad (6.43s)$$

$$r_e = (r_i^2 + r_f^2)^{1/2} = r_f(1 + r_i^2/r_f^2)^{1/2} \approx r_f + r_i^2/2r_f \qquad (6.43c)$$

Then

$$\frac{dr_i}{dt} \approx -\frac{D^* L_{GT}}{r_i^2} \qquad (6.44s)$$

$$\frac{dr_i}{dt} \approx -\frac{D^* L_{GT}}{r_i^2} \frac{1}{\ln r_f / r_i} \qquad (6.44c)$$

Solution of differential equation for spherical form is obvious:

$$r_i^3 = r_{i0}^3 - 3D^* L_{GT} t \qquad (6.45s)$$

For cylindrical particle shrinking rate differs by the last logarithmic factor making it evidently slower, which can be seen from the following transformed kinetic equation $\frac{dr_i^3}{dt} = -3D^* L_{GT} \frac{1}{\ln r_f / r_i} \underset{r_i \to 0}{\to} 0.$

This equation can be integrated:

$$\int_{r_{i0}/r_f}^{r_i/r_f} \ln x dx^3 = \frac{3D^* L_{GT}}{r_f^3} t \qquad (6.45c)$$

The collapse time for spherical particle can be found directly from Eq. 6.45s if one puts $r_i = 0$:

$$t_{\text{collapse}} \approx \frac{r_{i0}^3}{3D^* L_{GT}} = \frac{kT}{6\gamma\Omega} \frac{r_{i0}^3}{D^*} \qquad (6.46s)$$

for thick cylindrical wire we transform Eq. 6.45c into $\ln\frac{r_i}{r_f} d\left(\frac{r_i}{r_f}\right)^3 = -\frac{3D^* L_{GT}}{r_f^3} dt$ and integrate from the initial size to zero inner radius:

$$t_{\text{collapse}} \approx \frac{r_{i0}^3}{3D^* L_{GT}} \left(\ln\frac{r_f}{r_{i0}} + \frac{1}{3}\right) = \frac{kT}{3\gamma\Omega} \frac{r_{i0}^3}{D^*} \left(\ln\frac{r_f}{r_{i0}} + \frac{1}{3}\right) \qquad (6.46c)$$

Case B. In this case shrinking can be divided into three stages: initial ($r_{\rm i} >> r_{\rm f}$), transient ($r_{\rm i} \sim r_{\rm f}$) and finishing ($r_{\rm i} << r_{\rm f}$). Simple analytic expressions can be derived only for initial and finishing stages. Therefore, we split transient stage into two substages distributing them between above-mentioned two marginal stages.

Initial stage:

$$r_{\rm e} = (r_{\rm i}^3 + r_{\rm f}^3)^{1/3} = r_{\rm i}(1 + r_{\rm f}^3/r_{\rm i}^3)^{1/3} \approx r_{\rm i} + r_{\rm f}^3/3r_{\rm i}^2 \qquad (6.47\text{s})$$

$$r_{\rm e} = (r_{\rm i}^2 + r_{\rm f}^2)^{1/2} = r_{\rm i}(1 + r_{\rm f}^2/r_{\rm i}^2)^{1/2} \approx r_{\rm i} + r_{\rm f}^2/2r_{\rm i} \qquad (6.47\text{s})$$

Then, Eq. 6.40 with account of inequality $r_{\rm i} >> r_{\rm f}$ and expansion in series over small parameter $r_{\rm f}/r_{\rm i}$ gives:

$$\frac{dr_{\rm i}}{dt} \approx -\frac{6D^* L_{\rm GT}}{r_{\rm f}^2} r_{\rm i} \qquad (6.48\text{s})$$

$$\frac{dr_{\rm i}}{dt} \approx -\frac{4D^* L_{\rm GT}}{r_{\rm f}^2} \qquad (6.48\text{c})$$

Solution of differential equation is obvious:

$$r_{\rm i} \approx r_{\rm i0} \exp\left[-\frac{6D^* L_{\rm GT}}{r_{\rm f}^3} t\right] \qquad (6.49\text{s})$$

$$r_{\rm i} \approx r_{\rm i0} - \frac{4D^* L_{\rm GT}}{r_{\rm f}^2} t \qquad (6.49\text{c})$$

The time of initial stage can be estimated directly from Eq. 6.49 if one puts $r_{\rm i} = r_{\rm f}$:

$$t_{\rm collapse}^{\rm initial} \approx \frac{r_{\rm f}^3}{6D^* L_{\rm GT}} \ln \frac{r_{\rm i0}}{r_{\rm f}} \qquad (6.50\text{s})$$

$$t_{\rm collapse}^{\rm initial} \approx \frac{r_{\rm f}^2 r_{\rm i0}}{4D^* L_{\rm GT}} \qquad (6.50\text{c})$$

Finishing stage: Conditions for this stage, $r_{\rm i} << r_{\rm f}$, coincides with conditions for thick shell (Case A). Therefore, conditions (Eq. 6.43) hold as well as differential Eq. 6.44, but with another initial conditions ($r_{\rm f}$ instead of $r_{\rm i0}$). Therefore, solutions differ from Eq. 6.45:

$$r_{\rm i}^3 = r_{\rm f}^3 - 3D^* L_{\rm GT} t \qquad (6.51\text{s})$$

$$\int_1^{r_{\rm i}/r_{\rm f}} \ln x\, dx^3 = -\frac{3D^* L_{\rm GT}}{r_{\rm f}^3} t \qquad (6.51\text{c})$$

The time of finishing stage can be estimated directly from Eq. 6.51 if one puts $r_i = 0$:

$$t_{\text{collapse}}^{\text{finishing}} \approx \frac{r_f^3}{3D^* L_{GT}} \qquad (6.52s)$$

$$t_{\text{collapse}}^{\text{finishing}} \approx \frac{r_f^3}{9D^* L_{GT}} \qquad (6.52c)$$

Total collapse time for Case B is estimated as the sum of initial and finishing stages:

$$t_{\text{collapse}} \approx \frac{r_f^3}{3D^* L_{GT}} \left(\frac{1}{2} \ln \frac{r_{i0}}{r_f} + 1 \right) = \frac{kT}{6\gamma\Omega} \frac{r_f^3}{D^*} \left(\frac{1}{2} \ln \frac{r_{i0}}{r_f} + 1 \right) \qquad (6.53s)$$

$$t_{\text{collapse}} \approx \frac{r_f^2 r_{i0}}{4D^* L_{GT}} + \frac{r_f^3}{9D^* L_{GT}} \approx \frac{r_f^2 r_{i0}}{4D^* L_{GT}} = \frac{kT}{4\gamma\Omega} \frac{r_f^2 r_{i0}}{D^*} \qquad (6.53c)$$

In both cases A and B spheres shrink faster than cylinders.

Actually, in many cases described in experimental papers initial inner radius of the shell and final radius of the compact particle have the same order of magnitude. Hence, both cases A and B lead to the same rough estimate of collapse time:

$$t_{\text{collapse}} \approx \frac{kT}{p\gamma\Omega} \frac{r_f^3}{D^*} \qquad (6.54)$$

where $p \sim 10^1$.

To estimate the time of collapse of a nanoshell, we take typical parameters r_f = 3 nm, γ = 0.5 J/m^2, Ω = 4.10^{-29} m^3, $D_A^* = 3.25 \cdot 10^{-8} \exp\left(-\frac{1.24eV}{kT} \right)$ m^2/s, p = 6 and we found that $T = 300\text{ K} \Rightarrow t_{\text{collapse}} \approx 600\text{ years}$, $T = 600\text{ K} \Rightarrow t_{\text{collapse}} \approx 1\text{ s}$.

6.3.2 Shrinking of Chemical Compound Hollow Shells

Now consider shrinking of binary nanoshell (chemical compound), which has narrow homogeneity range. The general driving force of shrinking is the same for a pure component shell—a decrease of total surface energy (in other formulation—Gibbs–Thomson effect). Yet, the kinetics will be different. The main differences are (1) segregation of fast diffusant at inner surface due to the inverse Kirkendall effect, and (2) the resulting decrease of vacancy flux and of corresponding shrinking rate. In general, the problem of interdiffusion within a region with moving boundaries is rather complicated for rigorous solution. So, we try to find a simplified model, which should be effective, but on the other hand, one should not lose the main feature of the full process. The main simplifications are the steady-state approximations for the fluxes of vacancies and atoms. More "honest" but also much more complicated models were presented in Ref. 12. So, we present here most simple model (assuming steady-state both for components and for vacancies) and shortly compare its results with more rigorous models.

The vacancy out-flux should lead to redistribution of components, that is, segregation of the fast diffusant B atoms at the inner boundary, and segregation of the slow diffusant A atoms near the external surface (the segregation of A will be relatively less because of the larger total area of the external surface). At first the vacancy flux, caused by difference of curvatures and corresponding vacancy concentrations, generates the inward migration fluxes j_A, j_B of both A and B species (shrinking of nanoshell). Migration of B proceeds faster, leading to segregation of B near the inner surface (Fig. 6.3b). The arising concentration gradient of B influences all fluxes: it decreases the vacancy flux, suppressing the shrinking process; it also changes the fluxes of components. Here we use a phenomenological model treating a total local vacancy concentration and total local fluxes of vacancies and atoms. We don't write down separately the fluxes and concentrations at each sublattice. Different mobilities at different sublattices are taken into account implicitly, via different intrinsic diffusivities of species A and B.

6.3.2.1 Basic equations

In the equation for vacancies we consider sources/sinks inside a shell to be absent, therefore

$$\frac{\partial c_V}{\partial t} = -\frac{1}{r^2}\frac{\partial}{\partial r}(r^2 \Omega j_V) + 0, \quad r_i(t) < r < r_e(t) \tag{6.55}$$

The equation for species B with concentration c_B (local atomic fraction of B) is

$$\frac{\partial c_B}{\partial t} = -\frac{1}{r^2}\frac{\partial}{\partial r}(r^2 \Omega j_B), \quad r_i(t) < r < r_e(t) \tag{6.56}$$

Here the expressions for both fluxes should include the contributions from the vacancy gradient, in contrast to classic Darken analysis:[25–27,34]

$$\Omega j_{\mathrm{V}} = (D_{\mathrm{B}} - D_{\mathrm{A}})\frac{\partial c_{\mathrm{B}}}{\partial r} - D_{\mathrm{V}}\frac{\partial c_{\mathrm{V}}}{\partial r} \tag{6.57}$$

$$\Omega j_{\mathrm{B}} = -D_{\mathrm{B}}\frac{\partial c_{\mathrm{B}}}{\partial r} + \frac{c_{\mathrm{B}} D_{\mathrm{B}}^{*}}{c_{\mathrm{V}}}\frac{\partial c_{\mathrm{V}}}{\partial r} \tag{6.58}$$

All fluxes are written down in the lattice reference frame D_{B}, D_{B}^{*}, D_{V} are, respectively, the partial diffusivity of B, the tracer diffusivity of B, and the diffusivity of vacancies inside the nanoshell. In Eqs. 6.57 and 6.58, we have neglected Manning's corrections (vacancy wind effect). It could be done but it would make mathematical equations more cumbersome without a substantial change of the main results. The mentioned diffusivities are interrelated in the following way (if one neglects correlation effects):

$$D_{\mathrm{B}} = D_{\mathrm{B}}^{*}\varphi\,,\; D_{\mathrm{A}} = D_{\mathrm{A}}^{*}\varphi\,,\; \varphi = \frac{c_{\mathrm{A}} c_{\mathrm{B}}}{kT}\frac{\partial^2 g}{\partial c_{\mathrm{B}}^2}\,,\; D_{\mathrm{V}} = \frac{c_{\mathrm{A}} D_{\mathrm{A}}^{*} + c_{\mathrm{B}} D_{\mathrm{B}}^{*}}{c_{\mathrm{V}}} \tag{6.59}$$

Here φ is a thermodynamic factor, g is Gibbs free energy per atom.

As in case of pure shell, the inner and external surfaces move due to vacancy fluxes:

$$\frac{dr_{\mathrm{i}}}{dt} = -\Omega j_{\mathrm{V}}(r_{\mathrm{i}}) = -(D_{\mathrm{B}} - D_{\mathrm{A}})\frac{\partial c_{\mathrm{B}}}{\partial r}(r = r_{\mathrm{i}}) + D_{\mathrm{V}}\frac{\partial c_{\mathrm{V}}}{\partial r}(r = r_{\mathrm{i}}) \tag{6.60}$$

$$\frac{dr_{\mathrm{e}}}{dt} = -\Omega j_{\mathrm{V}}(r_{\mathrm{e}}) = -(D_{\mathrm{B}} - D_{\mathrm{A}})\frac{\partial c_{\mathrm{B}}}{\partial r}(r = r_{\mathrm{e}}) + D_{\mathrm{V}}\frac{\partial c_{\mathrm{V}}}{\partial r}(r = r_{\mathrm{e}}) \tag{6.61}$$

Boundary conditions for vacancies were formulated by Eq. 6.31. Conservation of atoms of B implies that the flux of B across the moving boundaries should be absent:

$$\Omega J_{\mathrm{B}}(r_i) - c_{\mathrm{B}}(r_{\mathrm{i}})\frac{dr_{\mathrm{i}}}{dt} = 0 \tag{6.62}$$

$$\Omega J_{\mathrm{B}}(r_{\mathrm{e}}) - c_{\mathrm{B}}(r_{\mathrm{e}})\frac{dr_{\mathrm{e}}}{dt} = 0 \tag{6.63}$$

6.3.2.2 Main assumptions

Let the segregation be small, so that we can use c_1 instead of c_{B} and $1 - c_1$ instead of c_{A} everywhere except derivatives (c_1—stoichiometric concentrations of B in IMC). Let $L_{\mathrm{GT}}/r << 1$, so that we can use $c_{\mathrm{V}}^{\mathrm{eq}}$ instead of c_{V} everywhere except derivatives. This means that all diffusivities are treated constant.

We'll use the steady-state approximation both for components and for vacancies:

$$\frac{\partial}{\partial r}(r^2 j_{\mathrm{V}}) \approx 0, \ \frac{\partial}{\partial r}(r^2 J_{\mathrm{B}}) \approx 0$$

so that

$$\Omega j_{\mathrm{V}} \approx \frac{K_{\mathrm{V}}}{r^2}(>0), \ \Omega j_{\mathrm{B}} \approx \frac{K_{\mathrm{B}}}{r^2}(<0) \tag{6.64}$$

which immediately leads (with almost constant diffusivities) to the same type of dependence for both vacancies and component B:

$$\frac{\partial c_{\mathrm{V}}}{\partial r} \approx -\frac{L_{\mathrm{V}}}{r^2} \tag{6.65a}$$

$$\frac{\partial c_{\mathrm{B}}}{\partial r} \approx -\frac{L_{\mathrm{B}}}{r^2} \tag{6.65b}$$

From Eq. 6.65a and boundary conditions in Eq. 6.31, we obtain:

$$c_{\mathrm{V}} = M_{\mathrm{V}} + \frac{L_{\mathrm{V}}}{r} \tag{6.66}$$

$$L_{\mathrm{V}} = c_{\mathrm{V}}^{\mathrm{eq}} L_{\mathrm{GT}} \frac{1/r_{\mathrm{i}} + 1/r_{\mathrm{e}}}{1/r_{\mathrm{i}} - 1/r_{\mathrm{e}}}, \ M_{\mathrm{V}} = c_{\mathrm{V}}^{\mathrm{eq}}\left(1 - \frac{2L_{\mathrm{GT}}}{r_{\mathrm{e}} - r_{\mathrm{i}}}\right) \tag{6.67}$$

On the other hand, combining Eqs. 6.60–6.65, we obtain the following set of relations between parameters K_{V}, K_{B}, L_{V}, L_{B}:

$$\begin{aligned} K_{\mathrm{V}} &= -(D_{\mathrm{B}} - D_{\mathrm{A}})L_{\mathrm{B}} + D_{\mathrm{V}} L_{\mathrm{V}} \\ K_{\mathrm{B}} &= D_{\mathrm{B}} L_{\mathrm{B}} - \frac{c_1 D_{\mathrm{B}}^*}{c_{\mathrm{V}}} L_{\mathrm{V}} \\ K_{\mathrm{B}} &= -K_{\mathrm{V}} c_1 \end{aligned} \tag{6.68}$$

This set of algebraic equations, together with Eq. 6.67, has the following solution:

$$\begin{aligned} K_{\mathrm{V}} &= L_{\mathrm{V}} \frac{D_{\mathrm{A}}^* D_{\mathrm{B}}^*}{c_{\mathrm{V}}^{\mathrm{eq}}(c_1 D_{\mathrm{A}}^* + (1-c_1)D_{\mathrm{B}}^*)} \\ &= \frac{D_{\mathrm{A}}^* D_{\mathrm{B}}^*}{c_1 D_{\mathrm{A}}^* + (1-c_1)D_{\mathrm{B}}^*} L_{\mathrm{GT}} \frac{1/r_{\mathrm{i}} + 1/r_{\mathrm{e}}}{1/r_{\mathrm{i}} - 1/r_{\mathrm{e}}} \end{aligned} \tag{6.69}$$

$$K_{\mathrm{B}} = -\frac{c_1 D_{\mathrm{A}}^* D_{\mathrm{B}}^*}{c_1 D_{\mathrm{A}}^* + (1-c_1) D_{\mathrm{B}}^*} L_{\mathrm{GT}} \frac{1/r_{\mathrm{i}} + 1/r_{\mathrm{e}}}{1/r_{\mathrm{i}} - 1/r_{\mathrm{e}}} \tag{6.70}$$

$$L_{\mathrm{B}} = \frac{1}{\varphi} \frac{c_1(1-c_1)(D_{\mathrm{B}}^* - D_{\mathrm{A}}^*)}{c_1 D_{\mathrm{A}}^* + (1-c_1) D_{\mathrm{B}}^*} L_{\mathrm{GT}} \frac{1/r_{\mathrm{i}} + 1/r_{\mathrm{e}}}{1/r_{\mathrm{i}} - 1/r_{\mathrm{e}}} \tag{6.71}$$

To obtain parameters of B-component redistribution,

$$c_{\mathrm{B}} - c_1 = M_{\mathrm{B}} + \frac{L_{\mathrm{B}}}{r} \tag{6.72}$$

we should take the conservation of matter into account:

$$\int_{r_{\mathrm{i}}}^{r_{\mathrm{e}}} (c_{\mathrm{B}}(r) - c_1) r^2 dr = 0 \tag{6.73}$$

Substituting Eq. 6.72 into Eq. 6.73, we obtain:

$$M_{\mathrm{B}} = -\frac{3}{2} L_{\mathrm{B}} \frac{r_{\mathrm{e}}^2 - r_{\mathrm{i}}^2}{r_{\mathrm{e}}^3 - r_{\mathrm{i}}^3}$$

$$= \frac{1}{\varphi} \frac{c_1(1-c_1)(D_{\mathrm{B}}^* - D_{\mathrm{A}}^*)}{c_1 D_{\mathrm{A}}^* + (1-c_1) D_{\mathrm{B}}^*} L_{\mathrm{GT}} \frac{(r_{\mathrm{e}} + r_{\mathrm{i}})^2}{(r_{\mathrm{e}} - r_{\mathrm{i}})(r_{\mathrm{e}}^2 + r_{\mathrm{i}}^2 + r_{\mathrm{e}} r_{\mathrm{i}})} \tag{6.74}$$

Thus, the main results of our model are the following:

1. the kinetics of shrinking is governed by the slow component diffusion ($D_{\mathrm{A}}^* << D_{\mathrm{B}}^*$):

$$\frac{dr_{\mathrm{i}}}{dt} = -\frac{D_{\mathrm{A}}^*}{1 - c_1 + c_1 \dfrac{D_{\mathrm{A}}^*}{D_{\mathrm{B}}^*}} \frac{r_{\mathrm{e}} + r_{\mathrm{i}}}{r_{\mathrm{e}} - r_{\mathrm{i}}} \frac{L_{\mathrm{GT}}}{r_{\mathrm{i}}^2} \approx -\frac{D_{\mathrm{A}}^*}{1-c_1} \frac{r_{\mathrm{e}} + r_{\mathrm{i}}}{r_{\mathrm{e}} - r_{\mathrm{i}}} \frac{L_{\mathrm{GT}}}{r_{\mathrm{i}}^2}$$

$$t_{\mathrm{collapse}} \approx \frac{1}{6(1-c_1)} \frac{r_{\mathrm{f}}^3}{L_{\mathrm{GT}} D^*} \tag{6.75}$$

$$\frac{dr_{\mathrm{e}}}{dt} = \frac{r_{\mathrm{i}}^2}{r_{\mathrm{e}}^2} \frac{dr_{\mathrm{i}}}{dt} \tag{6.76}$$

2. vacancy distribution across the shell is:

$$c_{\mathrm{V}} = c_{\mathrm{V}}^{\mathrm{eq}} \left(1 - \frac{2L_{\mathrm{GT}}}{r_{\mathrm{e}} - r_{\mathrm{i}}} + \frac{L_{\mathrm{GT}}}{r} \frac{r_{\mathrm{e}} + r_{\mathrm{i}}}{r_{\mathrm{e}} - r_{\mathrm{i}}} \right) \tag{6.77}$$

3. redistribution of component B across the shell is:

$$c_{\mathrm{B}}(r)-c_1 = \frac{c_1(1-c_1)}{\varphi}\frac{(D_{\mathrm{B}}^*-D_{\mathrm{A}}^*)}{c_1D_{\mathrm{A}}^*+(1-c_1)D_{\mathrm{B}}^*}\frac{r_{\mathrm{e}}+r_{\mathrm{i}}}{r_{\mathrm{e}}-r_{\mathrm{i}}}\left(\frac{L_{\mathrm{GT}}}{r}-\frac{3L_{\mathrm{GT}}}{2}\frac{r_{\mathrm{e}}^2-r_{\mathrm{i}}^2}{r_{\mathrm{e}}^3-r_{\mathrm{i}}^3}\right) \tag{6.78}$$

In full analogy with case of pure component shell, we can obtain a rough estimate for the shrinking time:

$$t_{\mathrm{collapse}} \approx \frac{2}{p(1-c_1)}\frac{r_{\mathrm{f}}^3}{L_{\mathrm{GT}}D^*} \text{ with } D^* \equiv \frac{D_{\mathrm{A}}^*D_{\mathrm{B}}^*}{c_1D_{\mathrm{A}}^*+(1-c_1)D_{\mathrm{B}}^*} \tag{6.79}$$

parameter p was introduced in Section 6.3.1.

Now, using the data of Ref. 12, we compare some results of our rough "double steady-state model" (further Model S) with more rigorous model with still steady-state for vacancies but non-steady-state for main components (further Model R).

Characteristic profiles of $c_{\mathrm{V}}(r)/c_{\mathrm{V}}^{\mathrm{eq}}$ and $c_{\mathrm{B}}(r)$ for Model S and Model R are shown in Fig. 6.4: profiles for vacancies $c_{\mathrm{V}}(r)/c_{\mathrm{V}}^{\mathrm{eq}}$ practically coincide, and differ for B component, but are of the same order of magnitude.

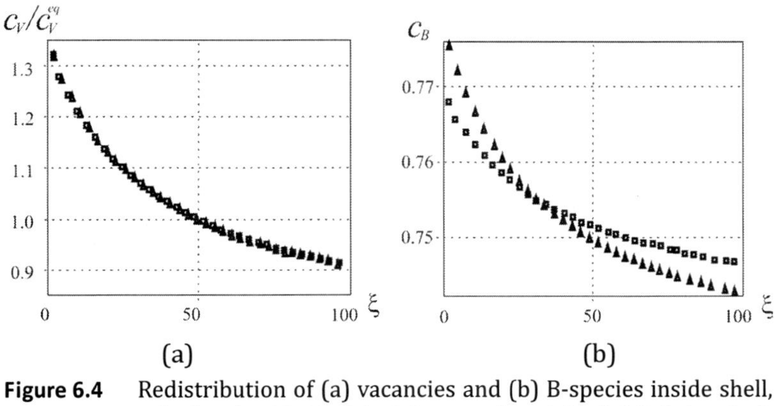

Figure 6.4 Redistribution of (a) vacancies and (b) B-species inside shell, calculated with Models S (squares) and Model R (triangles). Nondimensional space coordinate is $\xi = 100(r-r_{\mathrm{i}})/(r_{\mathrm{e}}-r_{\mathrm{i}})$: (a) profiles of $c_{\mathrm{V}}(r)/c_{\mathrm{V}}^{\mathrm{eq}}$ practically coincide; (b) profiles of $c_{\mathrm{B}}(r)$ inside shell for models S and R differ substantially but not by order of magnitude.

Time dependences on the inner and external radii (typical example shown in Fig. 6.5) can be found up to a certain minimal radius ($3L_{\mathrm{GT}}$ in this case), since at inner radius approaching zero the finite difference methods becomes less accurate. One may stop calculating earlier as the last stage of shrinking (collapse) is very quick which is explained by inverse relation between of the velocity and the inner radius. Thus, the total shrinking time almost equals the time of shrinking to the abovementioned minimal radius.

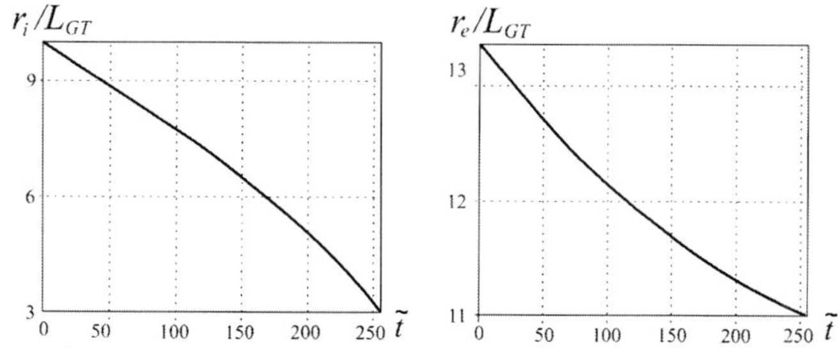

(a) (b)

Figure 6.5 Time dependences of (a) inner and (b) external radii according to Model S ($\tilde{t} = D_B^* t / L_{GT}^2$ is non-dimensional time, L_{GT} is characteristic size introduced in Eq. 6.31.

Dependences of the collapse time on the ratio D_B^* / D_A^* at initial reduced sizes $r_i / L_{GT} = 10, 11, 12, 13, 14, 15$ are shown in Fig. 6.6. Dependencies of the collapse time on the initial inner radius r_i/L_{GT} at $D_B^* / D_A^* = 10,\ 20, 40, 60,\ 80, 100$ are shown in Fig. 6.7. Figures 6.6 and 6.7 demonstrate that the rough approximation of shrinking time in Eq. 6.79 is quite reasonable.

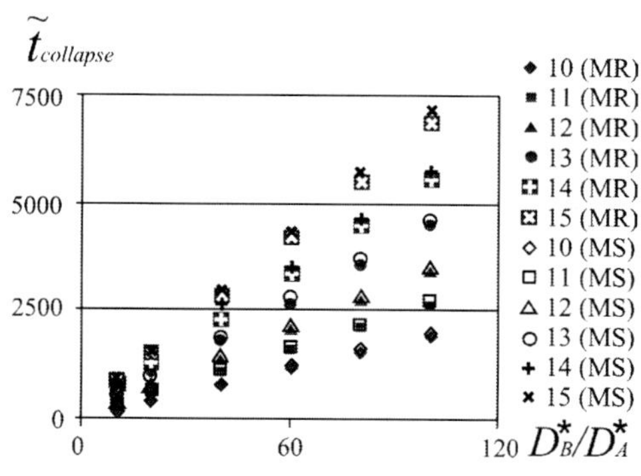

Figure 6.6 Dependence of non-dimensional collapse time $\tilde{t}_{\text{collapse}} = D_B^* t_{\text{collapse}} / L_{GT}^2$ on the ratio of diffusivities D_B^* / D_A^* for Models S and R at different initial nanoshell sizes r_{i0} / L_{GT} (e.g., 11 (MR) calculated in Model R for r_{i0} / L_{GT} = 11).

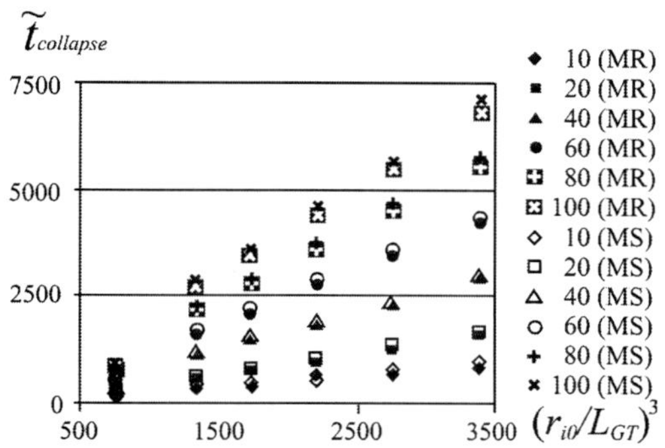

Figure 6.7 Dependence of non-dimensional collapse time $\tilde{t}_{\text{collapse}} = D_{\text{B}}^{*} t_{\text{collapse}} / L_{\text{GT}}^{2}$ on the initial nanoshell size $r_{\text{i0}} / L_{\text{GT}}$ for Models S and R at different diffusivity ratios $D_{\text{B}}^{*} / D_{\text{A}}^{*}$ (e.g., 20 (MS) calculated in Model S for $D_{\text{B}}^{*} / D_{\text{A}}^{*} = 20$).

6.3.3 Instability of Binary (Solid Solution) Hollow Shells

We consider interdiffusion and vacancy fluxes in the spherical hollow nano-particle, consisting of binary solid solution with broad homogeneity range. As was explored in Ref. 12, such shell should shrink due to outward vacancy flux generated by difference of vacancy concentrations at inner and external boundaries. Vacancy flux should lead to redistribution of main components with corresponding feedback of created concentration gradient on the vacancy flux and on corresponding shrinking rate. Contrary to the case of almost stoichiometric compound,[12] in our case of solid solution we cannot take diffusivities and concentrations as almost constant. Thus, one has to solve the coupled non-linear problems of vacancy flux, interdiffusion and boundary motion. The **following approximations** will be used.

A1. Tracer diffusivities of both components are proportional to local vacancy concentration (fraction of empty sites):

$$D_{\text{A}}^{*}(c_{\text{B}}) = c_{\text{V}} K_{\text{A}}(c_{\text{B}}), \; D_{\text{B}}^{*}(c_{\text{B}}) = c_{\text{V}} K_{\text{B}}(c_{\text{B}}) \tag{6.80}$$

here concentration c_{B} is a molar fraction of component B; K_{A}, K_{B} can be called the "partial vacancy diffusivities" (determining the fractions of vacancy exchanges with atoms of each species). At that, the total vacancy diffusivity satisfies the Frenkel relation $D_{\text{V}} = c_{\text{B}} K_{\text{B}} + c_{\text{A}} K_{\text{A}}$, if one neglects the correlation effects.

A2. Composition dependence of both diffusivities is taken as exponential one, which is typical. For simplicity (just to limit the number of model parameters) the logarithms of both diffusivities are treated as linear functions of concentration.

$$K_A(c_B) = K_{A0}\exp(\alpha_A c_B),\ K_B(c_B) = K_{B0}\exp(\alpha_B c_B) \tag{6.81}$$

A3. The diffusion fluxes of main components and of vacancies take into account the cross-terms and are written in the lattice reference frame:

$$\Omega j_A(r) = -D_A^*\varphi\frac{\partial c_A}{\partial r} + \frac{c_A D_A^*}{c_V}\frac{\partial c_V}{\partial r} = +K_A\varphi c_V\frac{\partial c_B}{\partial r} + c_A K_A\frac{\partial c_V}{\partial r} \tag{6.82aA}$$

$$\Omega j_B(r) = -D_B^*\varphi\frac{\partial c_B}{\partial r} + \frac{c_B D_B^*}{c_V}\frac{\partial c_V}{\partial r} = -K_B\varphi c_V\frac{\partial c_B}{\partial r} + c_B K_B\frac{\partial c_V}{\partial r} \tag{6.82B}$$

$$\Omega j_V(r) = (K_B - K_A)\varphi c_V\frac{\partial c_B}{\partial r} - (c_A K_A + c_B K_B)\frac{\partial c_V}{\partial r} \tag{6.82V}$$

φ is a thermodynamic factor, $\varphi = \frac{c_A c_B}{kT}\frac{\partial^2 g}{\partial c_B^2}$, g is a Gibbs free energy per atom of solution. In the model of regular solution (Z—coordination number, E_{AA}, E_{BB}, E_{AB}—pair interaction energies between atoms of corresponding kinds):

$$\varphi = 1 + c_A c_B \underbrace{\frac{2Z}{kT}\left(\frac{\mathrm{E}_{AA} + \mathrm{E}_{BB}}{2} - \mathrm{E}_{AB}\right)}_{-\lambda} \equiv 1 + c_A c_B(-\lambda) \tag{6.83}$$

Here the Manning's corrections due to effect of vacancy wind are neglected. Our preliminary estimates have shown that these corrections, at least for disordered solution, do not change the general picture. In this approximation $c_A K_A + c_B K_B = D_V$ (neglecting the correlation factor). Factor λ is proportional to the mixing enthalpy and is negative if solution formation is energetically favorable.

A4. There is no Kirkendall shift inside the nano-shell. First, as was mentioned in Introduction, there is little place for vacancy sinks/sources in nano-volume. Second, if all atoms of the lattice try to move along radial direction, this shift would immediately generate tangential deformation and corresponding stresses. Boundaries of the shell do move but not due to lattice shift—just some atoms leave one boundary and attach to another boundary.

Therefore we can write down the continuity equations (analogue of second Fick's law) in the lattice reference frame:

$$\frac{\partial c_V}{\partial t} = -\frac{1}{r^2}\frac{\partial}{\partial r}(r^2 \Omega j_V) + 0 \tag{6.84Vs}$$

$$\frac{\partial c_V}{\partial t} = -\frac{1}{r}\frac{\partial}{\partial r}(r \Omega j_V) + 0 \tag{6.84Vc}$$

$$\frac{\partial c_B}{\partial t} = -\frac{1}{r^2}\frac{\partial}{\partial r}(r^2 \Omega j_B) \tag{6.84Bs}$$

$$\frac{\partial c_B}{\partial t} = -\frac{1}{r}\frac{\partial}{\partial r}(r \Omega j_B) \tag{6.84Bc}$$

Zero in Eq. 6.84V was put just to remind about absence or ineffectiveness of vacancy sources and sinks inside the shell.

6.3.3.1 Boundary conditions

Though $c_V(r, t)$ is a nonequilibrium vacancy distribution, its values at the boundaries are equilibrium and fixed by Gibbs–Thomson relation:

$$c_V(r_i) = c_V^{eq} \exp\left(+L_{GT}\frac{1}{r_i}\right),\ c_V(r_e) = c_V^{eq} \exp\left(-L_{GT}\frac{1}{r_e}\right) \tag{6.85}$$

Flux balance equations at the moving inner and external boundaries depend on conditions of experiment. Here we will treat the case when evaporation of atoms from nanoshell is impossible. So, all fluxes outside the shell and inside central void are absent. Vacancy flux from the inner boundary to external boundary generates the movement of both boundaries according to trivial relations:

$$\frac{dr_i}{dt} = -\Omega j_V(r_i),\ \frac{dr_e}{dt} = -\Omega j_V(r_e) \tag{6.86}$$

Boundary conditions (Eqs. 6.85 and 6.86) would be enough for description of shrinking of the single-component shell. In our case of binary solution we should have additional conditions. Boundary concentrations of the main components are not fixed (we don't have analogue of Eq. 6.85 for A or B), but the conservation laws are valid, of course, implying the conditions on fluxes. Sum of three fluxes is zero in the lattice reference frame ($j_V + j_A + j_B = 0$). Thus, two fluxes are independent. It means that one should write down the flux balance equations at both moving boundaries for one of main components, taking into account that fluxes outside the shell equal to zero:

$$(c_{\mathrm{B}}(r_{\mathrm{i}})-0)\frac{dr_{\mathrm{i}}}{dt}=\Omega j_{\mathrm{B}}(r_{\mathrm{i}})-0\,,\ (c_{\mathrm{B}}(r_{\mathrm{e}})-0)\frac{dr_{\mathrm{e}}}{dt}=\Omega j_{\mathrm{B}}(r_{\mathrm{e}})-0 \quad (6.87)$$

Combining Eqs. 6.86 and 6.87, one gets:

$$-c_{\mathrm{B}}(r_{\mathrm{i}})\Omega j_{\mathrm{V}}=\Omega j_{\mathrm{B}}(r_{\mathrm{i}})\,,\ -c_{\mathrm{B}}(r_{\mathrm{e}})\Omega j_{\mathrm{V}}=\Omega j_{\mathrm{B}}(r_{\mathrm{e}}) \quad (6.88)$$

Below we'll use common steady-state approximation for fast diffusing vacancies which reflects the hierarchy of characteristic times—vacancy subsystem is fast enough to adapt to the slow redistribution of main components and to movement of boundaries (except the very last stage of collapse which is very fast):

$$\frac{\partial c_{\mathrm{V}}}{\partial t}\approx 0\Rightarrow r^{2}\Omega j_{\mathrm{V}}(r)=\frac{\Omega J_{\mathrm{V}}^{\mathrm{total}}}{4\pi}=\mathrm{const\ at\ }r \quad (6.89\mathrm{s})$$

$$\frac{\partial c_{\mathrm{V}}}{\partial t}\approx 0\Rightarrow r\Omega j_{\mathrm{V}}(r)=\frac{\Omega J_{\mathrm{V}}^{\mathrm{total}}}{2\pi H}=\mathrm{const\ at\ }r \quad (6.89\mathrm{c})$$

($J_{\mathrm{V}}^{\mathrm{total}}$ is the total flux of vacancies which has to be determined from boundary conditions).

It is convenient now to make the following transformation of variables:

$$t'=t\,,\ \xi=\frac{1/r-1/r_{\mathrm{e}}}{1/r_{\mathrm{i}}-1/r_{\mathrm{e}}}\,,\ 0<\xi<1 \quad (6.90\mathrm{s})$$

$$t'=t\,;\ \xi=\frac{\ln r_{\mathrm{e}}-\ln r}{\ln r_{\mathrm{e}}-\ln r_{\mathrm{i}}}\,,\ 0<\xi<1 \quad (6.90\mathrm{c})$$

$$\frac{\partial}{\partial t}=\frac{\partial}{\partial t'}+\left(\frac{\xi}{r_{\mathrm{i}}^{2}}\frac{dr_{\mathrm{i}}}{dt}+\frac{(1-\xi)}{r_{\mathrm{e}}^{2}}\frac{dr_{\mathrm{e}}}{dt}\right)\frac{r_{\mathrm{e}}r_{\mathrm{i}}}{r_{\mathrm{e}}-r_{\mathrm{i}}}\frac{\partial}{\partial\xi}$$

$$\frac{\partial}{\partial r}=-\frac{[r_{\mathrm{i}}+(r_{\mathrm{e}}-r_{\mathrm{i}})\xi]^{2}}{(r_{\mathrm{e}}-r_{\mathrm{i}})r_{\mathrm{e}}r_{\mathrm{i}}}\frac{\partial}{\partial\xi} \quad (6.91\mathrm{s})$$

$$\frac{\partial}{\partial t}=\frac{\partial}{\partial t'}+\frac{\xi\frac{1}{r_{\mathrm{i}}}\frac{dr_{\mathrm{i}}}{dt}+(1-\xi)\frac{1}{r_{\mathrm{e}}}\frac{dr_{\mathrm{e}}}{dt}}{\ln\frac{r_{\mathrm{e}}}{r_{\mathrm{i}}}}\frac{\partial}{\partial\xi}\,,\ r\frac{\partial}{\partial r}=-\frac{1}{\ln\frac{r_{\mathrm{e}}}{r_{\mathrm{i}}}}\frac{\partial}{\partial\xi} \quad (6.91\mathrm{c})$$

Using Eqs. 6.89, 6.82V, 6.90, and 6.91 one obtains:

$$\frac{\Omega J_V^{\text{total}}}{4\pi}=\frac{r_e r_i}{r_e-r_i}\left(-(K_B-K_A)\varphi c_V\frac{\partial c_B}{\partial \xi}+(c_A K_A+c_B K_B)\frac{\partial c_V}{\partial \xi}\right)\text{, or}$$

$$\frac{\partial c_V}{\partial \xi}=\frac{(K_B-K_A)\varphi}{c_A K_A+c_B K_B}c_V\frac{\partial c_B}{\partial \xi}+\frac{(\Omega J_V^{\text{total}}/4\pi)}{c_A K_A+c_B K_B}\frac{(r_e-r_i)}{r_e r_i} \tag{6.92s}$$

$$\frac{\Omega J_V^{\text{total}}}{2\pi H}=\frac{1}{\ln r_e/r_i}\left(-(K_B-K_A)\varphi\cdot c_V\frac{\partial c_B}{\partial \xi}+(c_A K_A+c_B K_B)\frac{\partial c_V}{\partial \xi}\right)\text{, or}$$

$$\frac{\partial c_V}{\partial \xi}=\frac{(K_B-K_A)\varphi\cdot c_V}{c_A K_A+c_B K_B}\frac{\partial c_B}{\partial \xi}+\frac{\ln r_e/r_i}{c_A K_A+c_B K_B}\frac{\Omega J_V^{\text{total}}}{2\pi H} \tag{6.92c}$$

If $c_B(\xi)$ profile is known, then Eq. 6.91 has a standard form:

$$\frac{dc_V(\xi)}{d\xi}=f(\xi)c_V+\frac{a}{\psi(\xi)} \tag{6.93}$$

with

$$\psi(\xi)=c_A(\xi)K_A(\xi)+c_B(\xi)K_B(\xi),\ f(\xi)=\frac{(K_B-K_A)\varphi}{c_A K_A+c_B K_B}\frac{\partial c_B}{\partial \xi} \tag{6.94}$$

and can be solved with respect to unknown function $c_V(\xi)$:

$$c_V(\xi)=\exp\left(\int_0^{\xi} f(\xi')d\xi'\right)\left(c_V^{\text{eq}}\exp\left(-\frac{L_{GT}}{R_e}\right)\right.$$
$$\left.+\frac{r_e-r_i}{r_e r_i}\frac{\Omega J_V^{\text{total}}}{4\pi}\int_0^{\xi}\frac{d\xi'}{\psi(\xi')}\exp\left(-\int_0^{\xi'} f(\xi'')d\xi''\right)\right) \tag{6.95s}$$

$$c_V(\xi)=\exp\left(\int_0^{\xi} f(\xi')d\xi'\right)\left(c_V^{\text{eq}}\exp\left(-\frac{L_{GT}}{R_e}\right)\right.$$
$$\left.+\ln\frac{r_e}{r_i}\frac{\Omega J_V^{\text{total}}}{2\pi H}\int_0^{\xi}\frac{d\xi'}{\psi(\xi')}\exp\left(-\int_0^{\xi'} f(\xi'')d\xi''\right)\right) \tag{6.95c}$$

Combining Eq. 6.95 with Gibbs–Thomson relations (Eq. 6.85) for boundary vacancy concentrations, we obtain expression for total vacancy flux in terms of unknown B-concentration profile:

$$\frac{\Omega J_{\mathrm{V}}^{\mathrm{total}}}{4\pi}=\frac{r_{\mathrm{e}}r_{\mathrm{i}}}{r_{\mathrm{e}}-r_{\mathrm{i}}}c_{\mathrm{V}}^{\mathrm{eq}}\frac{\exp\left(\frac{L_{\mathrm{GT}}}{r_{\mathrm{i}}}\right)\exp\left(-\int_0^1 f(\xi)d\xi\right)-\exp\left(-\frac{L_{\mathrm{GT}}}{r_{\mathrm{e}}}\right)}{\int_0^1\exp\left(-\int_0^{\xi} f(\xi')d\xi'\right)\frac{d\xi}{\psi(\xi)}} \quad (6.96\mathrm{s})$$

$$\frac{\Omega J_{\mathrm{V}}^{\mathrm{total}}}{2\pi H}=\frac{1}{\ln\frac{r_{\mathrm{e}}}{r_{\mathrm{i}}}}c_{\mathrm{V}}^{\mathrm{eq}}\frac{\exp\left(\frac{L_{\mathrm{GT}}}{r_{\mathrm{i}}}\right)\exp\left(-\int_0^1 f(\xi)d\xi\right)-\exp\left(-\frac{L_{\mathrm{GT}}}{r_{\mathrm{e}}}\right)}{\int_0^1\exp\left(-\int_0^{\xi} f(\xi')d\xi'\right)\frac{d\xi}{\psi(\xi)}} \quad (6.96\mathrm{c})$$

Substituting Eq. 6.91 into Eq. 6.82B, one can express the flux of one of main components in terms of its gradient and of total vacancy flux (which is constant within the layer $\partial\xi$):

$$r^2\Omega j_{\mathrm{B}}(r)=-K_{\mathrm{B}}\varphi c_V r^2\frac{\partial c_{\mathrm{B}}}{\partial r}+c_{\mathrm{B}}K_{\mathrm{B}}r^2\frac{\partial c_{\mathrm{V}}}{\partial r}$$

$$=\frac{r_{\mathrm{e}}r_{\mathrm{i}}}{(r_{\mathrm{e}}-r_{\mathrm{i}})}\frac{K_{\mathrm{A}}K_{\mathrm{B}}\varphi}{c_{\mathrm{A}}K_{\mathrm{A}}+c_{\mathrm{B}}K_{\mathrm{B}}}c_{\mathrm{V}}\frac{\partial c_{\mathrm{B}}}{\partial\xi}-\frac{c_{\mathrm{B}}K_{\mathrm{B}}}{c_{\mathrm{A}}K_{\mathrm{A}}+c_{\mathrm{B}}K_{\mathrm{B}}}\frac{\Omega J_{\mathrm{V}}^{\mathrm{total}}}{4\pi} \quad (6.97\mathrm{s})$$

$$r\Omega j_{\mathrm{B}}(r)=-K_{\mathrm{B}}\varphi c_{\mathrm{V}}r\frac{\partial c_{\mathrm{B}}}{\partial r}+c_{\mathrm{B}}K_{\mathrm{B}}r\frac{\partial c_{\mathrm{V}}}{\partial r}$$

$$=\frac{1}{\ln\frac{r_{\mathrm{e}}}{r_{\mathrm{i}}}}\frac{K_{\mathrm{A}}K_{\mathrm{B}}\varphi c_{\mathrm{V}}}{c_{\mathrm{A}}K_{\mathrm{A}}+c_{\mathrm{B}}K_{\mathrm{B}}}\frac{\partial c_{\mathrm{B}}}{\partial\xi}-\frac{c_{\mathrm{B}}K_{\mathrm{B}}}{c_{\mathrm{A}}K_{\mathrm{A}}+c_{\mathrm{B}}K_{\mathrm{B}}}\frac{\Omega J_{\mathrm{V}}^{\mathrm{total}}}{2\pi H} \quad (6.97\mathrm{c})$$

Thus, steady-state approximation for vacancies, together with Gibbs–Thomson boundary conditions, with account of Eq. 6.96 reduces the system of two differential Eqs. 6.84V and 6.84B to only one but integro-differential equation:

$$\frac{\partial c_{\mathrm{B}}}{\partial t'}=\frac{\xi\frac{d\eta_{\mathrm{i}}}{dt}+(1-\xi)\frac{d\eta_{\mathrm{e}}}{dt}}{\eta_{\mathrm{i}}-\eta_{\mathrm{e}}}\frac{\partial c_{\mathrm{B}}}{\partial\xi}+$$

$$+\frac{(\eta_{\mathrm{e}}+(\eta_{\mathrm{i}}-\eta_{\mathrm{e}})\xi)^4}{(\eta_{\mathrm{i}}-\eta_{\mathrm{e}})^2}\frac{\partial}{\partial\xi}\left(\frac{K_{\mathrm{A}}K_{\mathrm{B}}\varphi(\xi)}{c_{\mathrm{A}}K_{\mathrm{A}}+c_{\mathrm{B}}K_{\mathrm{B}}}c_{\mathrm{V}}(\xi)\frac{\partial c}{\partial}\right.$$

$$\left.-\frac{c_{\mathrm{B}}K_{\mathrm{B}}}{c_{\mathrm{A}}K_{\mathrm{A}}+c_{\mathrm{B}}K_{\mathrm{B}}}(\eta_{\mathrm{i}}-\eta_{\mathrm{e}})\frac{\Omega J_{\mathrm{V}}^{\mathrm{total}}}{4\pi}\right) \quad (6.98\mathrm{s})$$

with $\eta_{\mathrm{i}} = 1/r_{\mathrm{i}}$, $\eta_{\mathrm{e}} = 1/r_{\mathrm{e}}$,

$$\frac{\partial c_{\mathrm{B}}}{\partial t'} = -\frac{\xi \frac{1}{r_{\mathrm{i}}}\frac{dr_{\mathrm{i}}}{dt} + (1-\xi)\frac{1}{r_{\mathrm{e}}}\frac{dr_{\mathrm{e}}}{dt}}{\ln \frac{r_{\mathrm{e}}}{r_{\mathrm{i}}}} \frac{\partial c_{\mathrm{B}}}{\partial \xi} +$$

$$+ \frac{1}{r_{\mathrm{i}}^{2\xi} r_{\mathrm{e}}^{2(1-\xi)}} \frac{1}{\ln \frac{r_{\mathrm{e}}}{r_{\mathrm{i}}}} \frac{\partial}{\partial \xi} \left(\frac{1}{\ln \frac{r_{\mathrm{e}}}{r_{\mathrm{i}}}} \frac{K_{\mathrm{A}} K_{\mathrm{B}} \varphi(\xi)}{c_{\mathrm{A}} K_{\mathrm{A}} + c_{\mathrm{B}} K_{\mathrm{B}}} c_{\mathrm{V}}(\xi) \frac{\partial c_{\mathrm{B}}}{\partial \xi} \right)$$

$$- \frac{c_{\mathrm{B}} K_{\mathrm{B}}}{c_{\mathrm{A}} K_{\mathrm{A}} + c_{\mathrm{B}} K_{\mathrm{B}}} \frac{\Omega J_{\mathrm{V}}^{\mathrm{total}}}{2\pi H} \tag{6.98c}$$

The total vacancy flux in Eq. 6.97 is determined by Eq. 6.96. The conditions for B-profile at internal ($\xi = 1$) and external ($\xi = 0$) boundaries can be found from Eq. 6.89s with account of Eqs. 6.96 and 6.97:

$$\left.\frac{\partial c_{\mathrm{B}}}{\partial \xi}\right|_{\xi=1} = -\frac{r_{\mathrm{e}} - r_{\mathrm{i}}}{r_{\mathrm{e}} r_{\mathrm{i}}} \frac{\Omega J_{\mathrm{V}}^{\mathrm{total}}}{4\pi} \left[\frac{c_{\mathrm{A}} c_{\mathrm{B}} (K_{\mathrm{B}} - K_{\mathrm{A}})}{K_{\mathrm{A}} K_{\mathrm{B}} \varphi} \right]_{\xi=1} \exp\left(-\frac{L_{\mathrm{GT}}}{r_{\mathrm{i}}} \right) \Big/ c_{\mathrm{V}}^{\mathrm{eq}} \tag{6.99s}$$

$$\left.\frac{\partial c_{\mathrm{B}}}{\partial \xi}\right|_{\xi=1} = \ln \frac{r_{\mathrm{e}}}{r_{\mathrm{i}}} \frac{\Omega J_{\mathrm{V}}^{\mathrm{total}}}{2\pi H} \frac{c_{\mathrm{A}} c_{\mathrm{B}} (K_{\mathrm{B}} - K_{\mathrm{A}})}{K_{\mathrm{A}} K_{\mathrm{B}} \varphi} \exp\left(-\frac{L_{\mathrm{GT}}}{r_{\mathrm{i}}} \right) \Big/ c_{\mathrm{V}}^{\mathrm{eq}} \tag{6.99c}$$

$$\left.\frac{\partial c_{\mathrm{B}}}{\partial \xi}\right|_{\xi=0} = -\frac{r_{\mathrm{e}} - r_{\mathrm{i}}}{r_{\mathrm{e}} r_{\mathrm{i}}} \frac{\Omega J_{\mathrm{V}}^{\mathrm{total}}}{4\pi} \left[\frac{c_{\mathrm{A}} c_{\mathrm{B}} (K_{\mathrm{B}} - K_{\mathrm{A}})}{K_{\mathrm{A}} K_{\mathrm{B}} \varphi} \right]_{\xi=0} \exp\left(+\frac{L_{\mathrm{GT}}}{r_{\mathrm{e}}} \right) \Big/ c_{\mathrm{V}}^{\mathrm{eq}} \tag{6.100s}$$

$$\left.\frac{\partial c_{\mathrm{B}}}{\partial \xi}\right|_{\xi=0} = \ln \frac{r_{\mathrm{e}}}{r_{\mathrm{i}}} \frac{\Omega J_{\mathrm{V}}^{\mathrm{total}}}{2\pi H} \frac{c_{\mathrm{A}} c_{\mathrm{B}} (K_{\mathrm{B}} - K_{\mathrm{A}})}{K_{\mathrm{A}} K_{\mathrm{B}} \varphi} \exp\left(\frac{L_{\mathrm{GT}}}{r_{\mathrm{e}}} \right) \Big/ c_{\mathrm{V}}^{\mathrm{eq}} \tag{6.100c}$$

Boundary problem (Eqs. 6.98–6.100) is solved by the explicit finite-difference scheme, with B-profile at previous time step used for calculation of total vacancy flux according to Eq. 6.96 and substituted into Eq. 6.98 for calculating the profile at the next time step. After this the just calculated B-profile is used to calculate the new vacancy profile according to Eq. 6.95. The velocities of boundaries, and new internal and external radii were calculated according to Eq. 6.86, also from the just found value of total vacancy flux. The main results of calculations for spheres and cylinders are following.

One can see that the time law of shrinking at various ratios of partial diffusivities $\kappa = K_{B0}/K_{A0}$ (Fig. 6.8) correlates with time dependence of segregation magnitude (Fig. 6.9) measured as the difference between B-species concentration at external and internal nanoshell boundaries (a— lattice parameter). Also, one can notice a peculiarity in time dependence just after initial stage. Shrinking rate slows down after rather fast initial segregation— $r_i(t)$ dependence has an inflection point. Later we'll see that this peculiarity is not general, and disappears in the cross-over regime when nanoshell collapse follows the nanoshell formation at the same temperature. We believe that the reason is related to different initial conditions. The results shown in the mentioned figures are obtained for rather artificial initial condition, when initial concentration is taken equal everywhere. So, some short period of adjustment follows. If we take a just grown nanoshell, it has inhomogeneous concentration distribution from the very beginning (because competition of Kirkendall, inverse Kirkendall and Gibbs–Thomson effects proceeds at the formation stage as well).

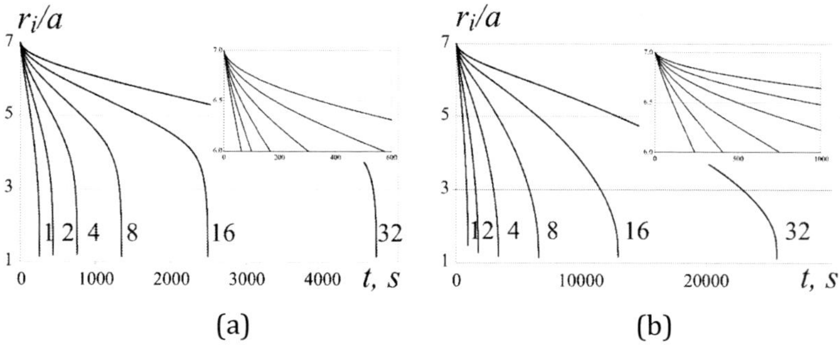

Figure 6.8 Time dependences of void radius for (a) spherical and (b) cylindrical particles during shrinking at various ratio of diffusivities κ = 1, 2, 4, 8, 16, 32 (actually, pre-exponential factors)—phenomenological model. Parameters: $c_V^{eq} = 10^{-4}$, $c_B^{init} = 0.5$, $r_i^{init} = a$, $r_e^{init} = 17a$, $L_{GT} = 10^{-9} m$, $\alpha_A = -6$, $\alpha_B = -2$, $\lambda = 0$.

The time of full collapse depends almost linearly on the ratio κ of pre-exponential factors in Eq. 6.81: $t_{collapse} = m\kappa + b$. The magnitudes of coefficients m, b in this linear approximation change at small mixing enthalpies and reach asymptotic values at large mixing enthalpies (Fig. 6.10).

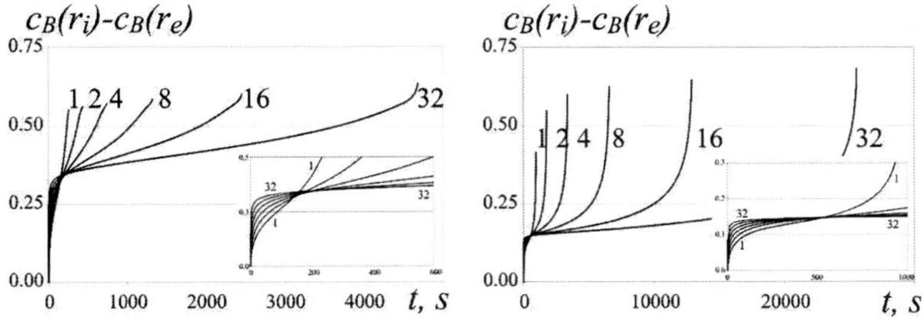

(a) (b)

Figure 6.9 Time dependences of segregation magnitude for (a) spherical and (b) cylindrical particles during shrinking at various ratio of diffusivities κ = 1, 2, 4, 8, 16, 32 (actually, pre-exponential factors) —phenomenological model. Parameters: $c_V^{eq}=10^{-4}$, $c_B^{init}=0.5$, $r_i^{init}=a$, $r_e^{init}=17a$, L_{GT} = 10^{-9} m, $\alpha_A=-6$, $\alpha_B=-2$, $\lambda=0$.

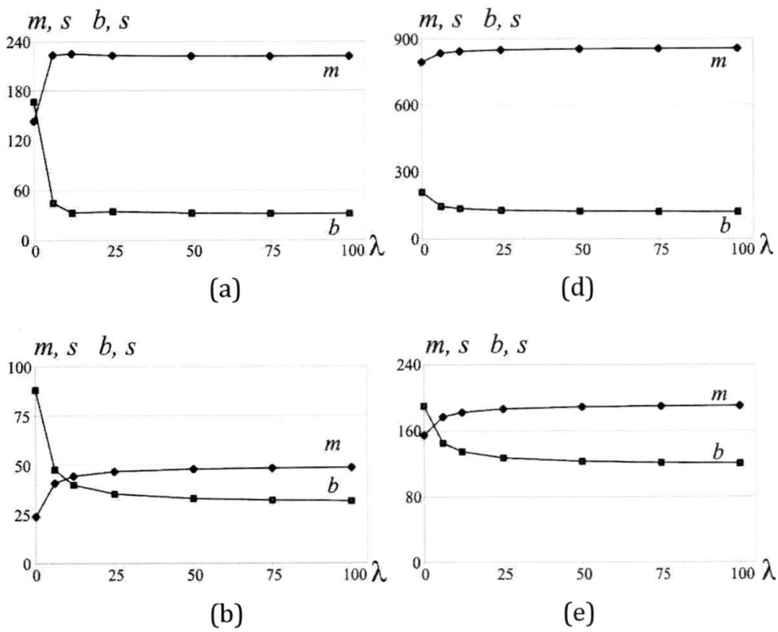

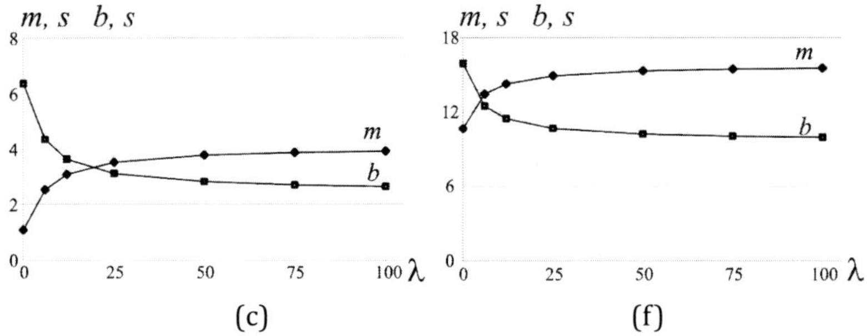

Figure 6.10 Dependences of the parameters m and b of linear approximation $t_{\text{collapse}} = m\kappa + b$ (κ = 1, 2, 4, 8, 16, 32) for collapse time t_{collapse} of spherical (a,b,c) and cylindrical (d,e,f) particles on non-dimensional mixing enthalpy—phenomenological model. Parameters: $c_V^{eq} = 10^{-4}$, $c_B^{init} = 0.5$, $r_i^{init} = a$, $r_e^{init} = 17a$, $L_{GT} = 10^{-9} m$, λ = 0, 6, 12, 25, 50, 100 and (a), (d) $\alpha_A = -6$, $\alpha_B = -2$; (b),(e) $\alpha_A = -3$, $\alpha_B = -2$; (c),(f) $\alpha_A = 2$, $\alpha_B = 3$.

Such linear dependence was predicted by us in the case of compound nanoshells with very narrow homogeneity range:[12]

$$t_{\text{collapse}} \approx \frac{1}{6L_{GT}} r_f^3 \frac{c_A D_B^* + c_B D_A^*}{D_A^* D_B^*} = \frac{1}{6L_{GT}} \frac{r_f^3}{D_B^*} \left((1-c_B)\frac{D_B^*}{D_A^*} + c_B \right) \quad (6.101)$$

(r_f is a final particle radius after collapse).

In our case $D_B^* / D_A^* = \kappa \exp((\alpha_B - \alpha_A) c_B)$. Thus, in the case of narrow compound phase (meaning large mixing enthalpy) one can expect that

$$\frac{m}{b} = \frac{1-c_B}{c_B} \exp((\alpha_B - \alpha_A) c_B) \quad (6.102)$$

One can see that the asymptotic values of this ratio in the present case of solid solution are indeed close to the mentioned analytical prediction (Table 6.1).

6.3.4 Energy Barrier—Does it Really Suppress the Shrinking?

The above stated theory has been mathematically developed[36] and was subjected to some criticism.[35,37] First, the authors[35–37] doubt the possibility to apply the quasi steady-state assumption to the nanoshell shrinkage and, thus, the role of Kirkendall effect in the shrinkage rate. Indeed, if the nanoshell is a solid solution with wide homogeneity range, the quasi steady-state approximation is not appropriate. But for the nanoshell with a narrow homogeneity interval, and, therefore, large thermodynamic factor, the quasi steady-state approximation works well.[38] Second, instead of the results on nanoshell time instability the authors[37] assume that hollow nanoshells can metastable due to specific radial vacancy distribution. Grounding upon the results of molecular-dynamic simulation they conclude that the energy of the nanoshell non-monotonically depends on the location of vacancies in it. It is energetically most unfavorable for the vacancy to be located in the bulk between internal and external surfaces. According to Ref. 37, this creates the energy barrier for vacancy migration from the internal surface to the external one and provides the metastability of the system (stability of the nanoshell cavity).

Table 6.1 Asymptotic (at large λ) ratio of linear fit coefficients for dependence of collapse time on diffusivity ratio

α_A	α_B	m/b, numerical model		m/b, analytical
		spherical	cylindrical	prediction (Eq. 6.102)
−6	−2	7.07	7.11	7.39
−3	−2	1.54	1.58	1.65
2	3	1.49	1.56	1.65

It is known that the energy of any spatially limited system with a vacancy depends on the distance from this vacancy to the surface. For example, the exponential spatial dependence both for the energy of vacancy formation and for vacancy migration near the planar free surface was established.[39] The non-monotonic dependence with the maximum in the vicinity of the second atomic plane was obtained.[40] Obviously, such depths of non-monotony appear to be significant only for nano-objects, since the change of the particle's radius by one atomic plane results in the curvature change of several percent; segregation in several atomic planes causes noticeable gradients of chemical potential.

On the other hand, nanosizes allow neglecting the work of vacancy sources and sinks, and, therefore, the movement of lattice

(Kirkendall shift) is also negligible. First, the linear size of the nanoparticle is less than the characteristic distance between vacancy sinks in real metal (~1 μm). Second, even having assumed that a nanoshell has dislocations or grain boundaries, serving as sources/sinks of vacancies, the disappearance/formation of vacancies on such heterogeneities must cause high stresses in the bulk which, in their turn, will prevent further formation/disappearance of vacancies.

Besides, the nanoshell shrinkage leads to the change of radii and, correspondingly, to the change of surface's curvature determining vacancy concentrations on the internal ($c_V(r_i)$) and external ($c_V(r_e)$) surfaces. That's why concentration profile can be quasi steady-state but not equilibrium. So, even knowing the dependence of the system's energy on the location of the vacancy $E_V^f(r)$, one cannot apply the standard equality $c_V(r)=\exp(-E_V^f(r)/kT)$.

For the sake of simplicity, let's consider a one-component hollow nanoshell where vacancy sources/sinks are provided only by internal and external surfaces. At such assumption inside the nanoshell the law of conservation of vacancies is fulfilled:

$$\frac{\partial c_V}{\partial t}=-\frac{1}{r^2}\frac{\partial(r^2\Omega j_V)}{\partial r}+0,\ r_i<r<r_e \qquad (6.103)$$

Beside the entropy term (vacancy gradient) the vacancy flux density contains the drift term, connected with energy spatial dependence:

$$\Omega j_V=-D_V\frac{\partial c_V}{\partial r}+\frac{c_V D_V}{kT}\left(-\frac{\partial E_V}{\partial r}\right) \qquad (6.104)$$

where $E_V(r)$ is the total energy of the system with a vacancy, located at the distance r from the nanoshell's centre (different from the formation energy E_V^f).

Since vacancies are much more mobile than atoms, after the short transition period the spatial vacancy distribution must attain the quasi steady-state regime: $\partial c_V/\partial t=0$. At that, the total flux through an arbitrary sphere is constant, so let's designate $r^2\Omega j_V=\text{const}=\varpi$.

Hereinafter we will consider the vacancy diffusivity constant everywhere. Then, the change of migration energy is negligible and we can treat the dependence of vacancy concentration profile only on the energy profile $E_V(r)$:

$$\frac{\partial c_V}{\partial r} + c_V \frac{\partial (E_V kT)}{\partial r} = -\frac{\omega}{r^2}, \quad \omega = \frac{\varpi}{D_V} \tag{6.105}$$

with boundary conditions

$$c_V(r_i) = \exp\left(-\frac{E_V(r_i) - E_i}{kT}\right), \quad c_V(r_e) = \exp\left(-\frac{E_V(r_e) - E_e}{kT}\right) \tag{6.106}$$

In Eq. 6.106 two different system's energies for each surface were introduced:

- $E_V(r_i)$, $E_V(r_e)$, when the vacancy is still in the nanoshell at the subsurface atomic layer; such a site in the plane 111 has three vacancies, in the plane 100 it has four (Fig. 6.11a);
- E_i, E_e, when the vacancy disappears at the surface step (Fig. 6.11b).

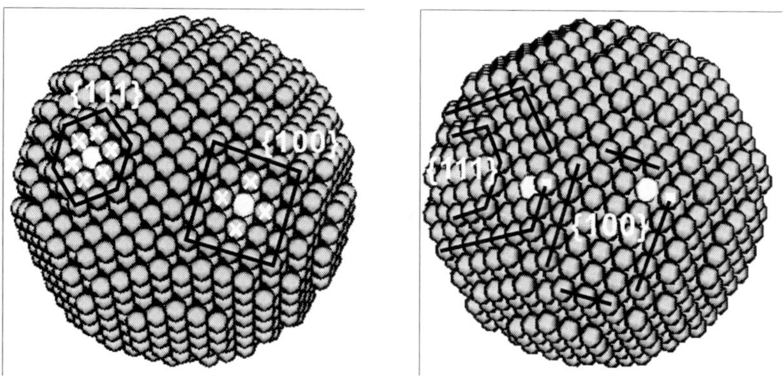

(a) (b)

Figure 6.11 Vacancies (light) at the external surface of the nanoshell inside surface atomic plane; the empty site is considered as a vacancy (has less than a half of empty neighboring sites); (a) if surface vacancy exchanges places with the atom marked by cross, it remains inside the facet with the same number of neighbors and is still considered as a vacancy; (b) if surface vacancy exchanges places with the atom marked by square and finds itself et the edge of surface plane, it is not considered as a vacancy anymore—so, the mentioned exchange describes annihilation of vacancy at the surface step.

Taking into account the steady state of the vacancy flux for the system (Eqs. 6.105 and 6.106)

$$4\pi r^2 \Omega j_V = 4\pi\varpi = 4\pi D_V \frac{\exp\left(\dfrac{E_i - E_e}{kT}\right) - 1}{\displaystyle\int_{r_i}^{r_e} \exp\left(\frac{E_V(r') - E_e}{kT}\right)\frac{dr'}{r'^2}} \tag{6.107}$$

we get radial vacancy concentration profile at $E_i - E_e = 2\gamma\Omega(1/r_i + 1/r_e)$:

$$c_V(r) = \exp\left(-\frac{E_V(r) - E_e}{kT}\right) + \frac{\exp\left(\dfrac{E_i - E_e}{kT}\right) - 1}{\displaystyle\int_{r_i}^{r_e} \exp\left(\frac{E_V(r') - E_e}{kT}\right)\frac{dr'}{r'^2}} \int_{r}^{r_e} \exp\left(\frac{E_V(r') - E_V(r)}{kT}\right)\frac{dr'}{r'^2} \tag{6.108}$$

From Eq. 6.107, it is seen that the vacancy flux always exists at any dependence $E_V(r)$. So, vacancy segregation cannot present the reason for the stability of hollow nanoshell. The delay can be connected only with the time of attainment of the quasi steady-state regime. But it goes without saying it, that vacancy segregation in the nanoshell cannot present anything special, since similar segregation must take place in macroscopic samples (e.g., at sintering of powder mixtures) and, as far as it is known, it in no way prevents, say, void coalescence.

For qualitative estimation of radial vacancy concentration profile we need the corresponding dependence of the system's energy. To find $E_V(r)$ the model in atomic scale was used, which allows us to place the vacancy in the given site and to determine the system's energy directly through atomic interactions. Since the system has free surfaces, it is correct to determine the energy after system's relaxation. For comparison we treated a nanoshell both **without** (all atoms in the sites of ideal lattice) and **with** relaxation (after system has reached equilibrium by the method of molecular statics).

To obtain the dependence of the system's energy on the position of the vacancy r, determined by the radius-vector from the nanoshell's centre, one atom in turn was replaced by a vacancy, and the energy of the system was calculated. Due to the absence of spherical symmetry of the crystal, discreetness of the sites and faceting of the nanoshell's surfaces, the systems with equal distance r from the vacancy to the void's centre can have different energies (Fig. 6.12). And conversely, the systems at different r may have equal energies. In the latter case one more reason arises: the vacancy "feels" the surface up to a certain depth (along with physical reasons the model may reveal artifacts, connected with the cutoff of the many-body Sutton-Chen potential for copper[41]).

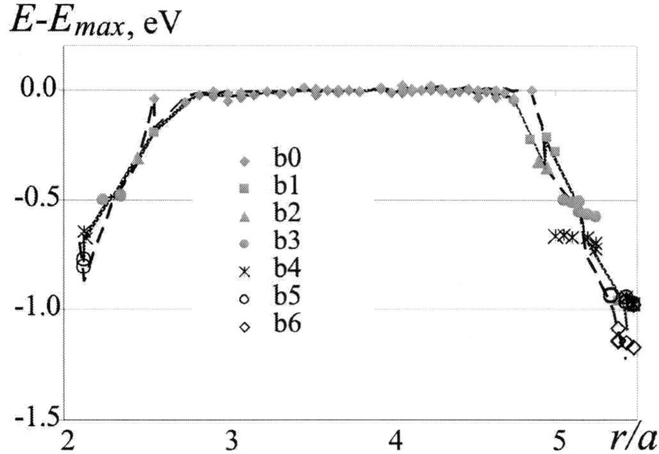

Figure 6.12 Dependence of $E_v(r)$ without relaxation (rigid lattice) and after relaxation (by molecular statics) of the system with $r_i = 2a$ and $r_e = 5.5a$. Dashed line corresponds to rigid lattice case. Solid line is an approximation of data for relaxed system. Data for energies at various vacancy positions are given with indication of vacancy type ("bn" corresponds to vacancy position with n empty neighbor sites).

As seen from Fig. 6.12, energy dependence near the surfaces has localized range of values. We have found that the number of empty sites in the vacancy's nearest neighborhood is common for such regions. In what follows, we mark any site with one empty nearest neighbor site as b1, with two nearest neighbor sites—b2, etc. Atom in the bulk is marked as b0. In particular, the vacancy at the internal surface has no more than six empty neighbors, at the external one—up to nine. Since the energy is the least for the site with the largest number of broken bonds (for complete detachment there is the least number of existing bonds left), then the vacancies with the largest number of empty sites determine minimal energies at internal and external surfaces, E_i and E_e. One may conclude that the system's energy with a subsurface vacancy is strongly influenced by the roughness (imperfection) of the surface which, in its turn, is determined by local curvature, faceting, etc.

For the small system with $r_i = 2a$ and $r_e = 5.5a$ (a—lattice parameter), containing 2639 atoms, an octant with boundary planes (518 atoms in all) was used for determination of dependence. Considering the superposition of the dependences of nanoshell's energies $E_v(r)$ on the position of vacancy for the system before and after relaxation, we notice that relaxation doesn't qualitatively change the radial energy dependence for the system with a vacancy (Fig. 6.12). The slopes become 11% lower and acquire some smoothness. The localization of atoms by the number of broken bonds is preserved. Below we will qualitatively compare the obtained dependences.

Additionally we analyzed five more systems "$r_i/a - r_e/a$" with different internal (r_i) and external (r_e) radii without relaxation (Fig. 6.13). As we should have expected, the influence of the curvature is essential near the surface. In the bulk at the depth of the vacancy location (6–7 Å) the energy of the system reaches the plateau with constant magnitude of E_{bulk} which can be treated as the energy of the system with a vacancy in the bulk. Then, the differences $E_{\text{bulk}} - E_i$ and $E_{\text{bulk}} - E_e$ can be interpreted as the energies of vacancy formation on internal and external surfaces, respectively. The plateau is not observed for the system "2–5.5" (Fig. 6.12), since the thickness of its nanoshell is 12.6 Å, and the subsurface layers are just overlapping. Such depths can be connected with the radius of potential cut-off $a\sqrt{3} = 6.26$ Å, though accounting of farther neighborhood when finding the potential energy doesn't make any essential contribution.

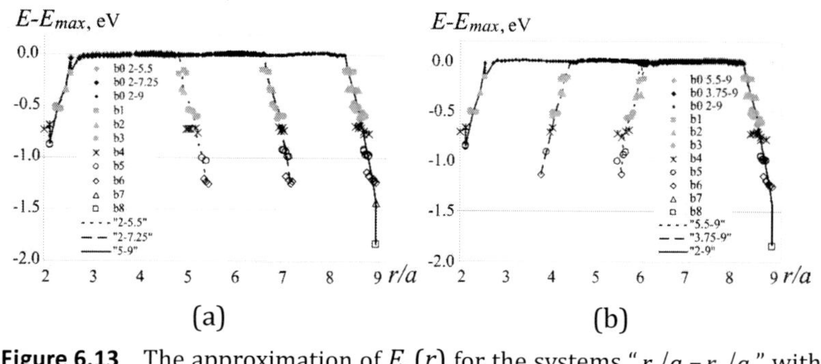

Figure 6.13 The approximation of $E_v(r)$ for the systems "$r_i/a - r_e/a$" with different internal (r_i) and external (r_e) radii without relaxation: (a) for the system "2–5", "2–7.25", "2–9"; (b) for the system "3.75–9", "5.5–9", and "2–9".

Knowing the dependence $E_V(r)$, from the Eq. 6.108 one can find the steady-state radial profile of vacancy concentration (Fig. 6.14).

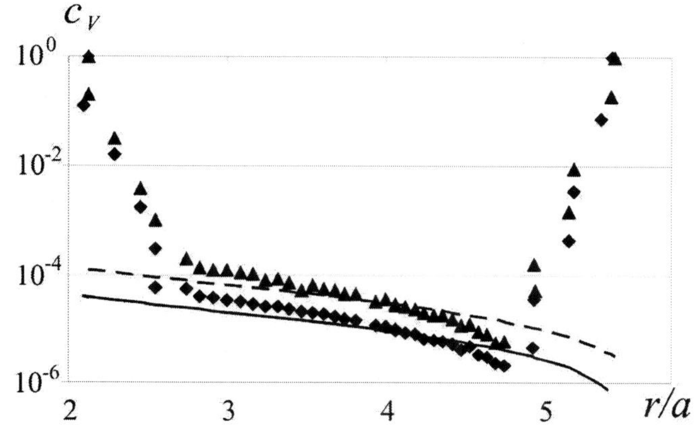

Figure 6.14 Steady state radial dependence of vacancy concentration $c_V(r)$ calculated according to Eq. 6.108 for the system "2–5.5" without relaxation (rhombuses and solid line) and after relaxation (triangles and dashed line). The lines determine the concentration for the profile without segregation $E_V(r) = E_{\text{bulk}}$ (Eq. 6.109).

If we neglect the abovementioned peculiarities of the vacancy subsystem near the curved surface and consider the energy of the system with a vacancy to be equal everywhere, irrespective of the vacancy position (E_{bulk}), we can simplify expression (Eq. 6.108):

$$c_V^0(r) = \exp\left(-\frac{(E_{\text{bulk}} - E_e)}{kT}\right)\left(1 + \frac{1/r - 1/r_e}{1/r_i - 1/r_e}\left(\exp\left(\frac{(E_i - E_e)}{kT}\right) - 1\right)\right) \tag{6.109}$$

The profile calculated according to Eq. 6.109 is depicted as a solid line in Fig. 6.14. Apparently, in this case the concentration dependence is monotonic and there is no surface segregation of vacancies.

To find out how vacancy segregation influences the kinetics of the nanoshell shrinkage it is suffice to compare the vacancy fluxes which correspond to concentration profiles with and without segregation:

$$\frac{\left.\dfrac{dr_i}{dt}\right|^{\text{with segr}}}{\left.\dfrac{dr_i}{dt}\right|^{\text{without segr}}} = \frac{\left(\dfrac{1}{r_i} - \dfrac{1}{r_e}\right)\exp\left(\dfrac{E_{\text{bulk}} - E_e}{kT}\right)}{\displaystyle\int_{r_i}^{r_e} \exp(E_V(r') - E_e)\frac{dr'}{r'^2}} = \eta \tag{6.110}$$

The ratio (Eq. 6.110) for nanoshells of different sizes is indicated in Table 6.1. The shrinkage rate increases 1.5 times on average, and twice as much for the relaxed system. Thus, we may conclude that vacancy segregation near the surface accelerates the shrinkage. In other words, the "barrier" understood by the authors of Ref. 21 always exists for the nanoshells of any thickness (rise of the energy of the system after embedding the vacancy into it). At transition to nano-thicknesses this barrier becomes narrower and smoother which leads to the increase of vacancy flux.

To estimate the correctness of the proposed calculations we used the phenomenological equations for the energy of vacancy formation for a non-planar surface: $E_{\text{bulk}} - E_{\text{i}} = E_{\text{V}}^{\text{f},0} - 2\gamma\Omega/r_{\text{i}}$, where ${}^{\text{f}}E_{\text{V}}^{0}$ is the energy of vacancy formation near the planar surface and γ is the surface tension. So,

$$E_{\text{V}}^{\text{f},0} = \frac{r_{\text{i}}(E_{\text{bulk}} - E_{\text{i}}) + r_{\text{e}}(E_{\text{bulk}} - E_{\text{e}})}{r_{\text{i}} + r_{\text{e}}}, \ \gamma = \frac{E_{\text{i}} - E_{\text{e}}}{2\ a^3/4\ (1/r_{\text{i}} + 1/r_{\text{e}})} \tag{6.111}$$

Substituting the obtained values of E_{i}, E_{e}, r_{i}, r_{e} into (Eq. 6.111), we can compare ${}^{f}E_{\text{V}}^{0}$ and γ, characterizing the proposed model, with experimental data for copper[16,20]: $E_{\text{V}}^{\text{f},0} \approx (1.17 \pm 0.11)\ \text{eV}$, $\gamma \approx 1.7\,\text{J}/\text{m}^2$ (Table 6.2). To our mind, the reason for poor coincidence of the surface tension with the experimental value can be explained by a too rough criterion for choice of surface steps (which determine E_{i}, E_{e}) and neglect of "unrecoverable" sources/sinks of vacancies (those which disappear after creation or annihilation of vacancy).

6.3.2 Conclusions to Section 6.3

Nanoshells, produced in the diffusive reactions of nanoparticles with ambient with simultaneous formation of Kirkendall voids inside, are unstable in principle, but the shrinking time can be very large due to its cubical dependence on the radius of the nanoparticle and exponential dependence on inverse temperature. Mechanism of shrinking is the out-diffusion of vacancies from the void due to curvature effect.

Shrinking of a nanoshell of intermetallic compound with a large difference between diffusivities of components is accompanied by segregation of the fast diffusant at the inner boundary due to the inverse Kirkendall effect. The rate of shrinking is controlled by the slow species, contrary to usual interdiffusion with quasi-equilibrium vacancies.

Table 6.2 Calculated parameters for the system "$-r_e/a$" with internal (r_i) and external (r_e) radii

$r_i/a - r_e/a$	*2–5.5* (MC)	*2–5.5*	*2–7.25*	*2–9*	*3.75–9*	*5.5–9*
Thickness $(r_e - r_i)/a$	3.5	3.5	5.25	7	5.25	3.5
Number of atoms	2, 642	2, 642	6, 319	12, 035	11, 240	9, 396
Formation energy at internal surface $E_i - E_{bulk}$ (eV)	−0.82	−0.82	−0.82	−0.82	−0.98	−1.04
Formation energy at external surface $E_e - E_{bulk}$ (eV)	−1.04	−1.04	−1.07	−1.09	−1.09	−1.09
${}^{f}E_V^0$ (eV) according to Eq. 6.111	1.01	1.13	1.13	1.14	1.18	1.19
γ (J/m^2) according to Eq. 6.111	1.18	1.32	1.44	1.47	0.45	0.73
η according to Eq. 6.110	2.03	1.75	1.33	1.35	1.33	1.51

The steady-state approximation for both vacancies and atoms provides a reasonable and simple scheme for predicting the shrinking kinetics. In order to properly describe the details of segregation, one must take into account the non-steady state redistribution of atoms, but the steady-state for vacancies can be used.

Kinetic MC simulations confirm the main predictions of the phenomenological models.

Both near planar and curved surface an activation barrier for vacancies always exists, it is connected with the rise of energy at embedding the vacancy into it. At transition to nano-thicknesses this barrier becomes narrower and smoother which leads to the increase of vacancy flux. Thus, vacancy segregation near the surface accelerates the nanoshell's shrinkage (instead of suppressing it).

6.4 Formation of Hollow Shells

Formation of hollow nanoshell by solid-state reaction with growth of compound shell differs from formation of hollow nanoshells by interdiffusion in solid solution. Therefore we consider separately these two cases.

6.4.1 Formation of IMC Hollow Shells

For description of hollow shell formation, we will use a so-called "bridge model" (Fig. 6.2). We imagine that during growth of the product phase and of the void the remaining core inside is linked to the growing phase by some "bridges" providing easy diffusion paths (by surface diffusion) from the core to the growing phase. The components can be, in principles, arbitrary. Yet, we will denote the core material as *B*, and external reacting component as *A*.

6.4.1.1 A simple case of the competition between "Kirkendall-driven" and "curvature-driven" effects

We start with a simplified situation and consider the initial stage of compound nanoshell formation, when the shell thickness $r_e - r_i$ is much less than the initial particles radius, r_0. In this case we can use flux equation for planar interfaces. Moreover, let the curvature effect be linear: $L_{GT} << r_0$, so that

$$c_V(r_e) \approx c_{V_0} \cdot \left(1 - \frac{L_{GT}}{r_e}\right), \quad c_V(r_i) \approx c_{V_0} \cdot \left(1 + \frac{L_{GT}}{r_i}\right) \tag{6.112}$$

Then the product of vacancy flux density and atomic volume (which can be called the free volume flux density) by the simple equation taking into account the vacancy concentration gradient caused by Gibbs–Thomson effect:

$$\begin{aligned} \Omega j_V &= (D_B - D_A)\frac{c_B(r_e) - c_A(r_i)}{\Delta r} - D_V \frac{c_V(r_e) - c_V(r_i)}{\Delta r} \\ &\approx -(D_B - D_A)\frac{\Delta c_B}{\Delta r} + D_V c_{V_0} \cdot L_{GT}\left(\frac{1}{r_i} + \frac{1}{r_e}\right)\frac{1}{\Delta r} \end{aligned} \tag{6.113}$$

Here $D_B = D_B^* \varphi$, $D_A = D_A^* \varphi$, where the thermodynamic factor $\varphi = \frac{c_B c_A}{kT}\frac{\partial^2 g}{\partial c_B^2}$, and D_B^*, D_A^* are tracer diffusivities in the compound, respectively. In Eq. 6.113, we have taken $c_{\mathrm{B}}D_{\mathrm{B}}^{*}+c_{\mathrm{A}}D_{\mathrm{A}}^{*}=c_{\mathrm{V}}D_{\mathrm{V}}$ (thus, we neglect a correlation factor here).

Furthermore, $\Delta c_{\mathrm{B}}=c_{\mathrm{B}}(r_{\mathrm{i}})-c_{\mathrm{B}}(r_{\mathrm{e}})$ is the homogeneity range of compound (typically very narrow). Note, that the concentrations c_{B} and c_{A} are mole fractions, so that $c_{\mathrm{B}}+c_{\mathrm{A}}=1$, and $1/\Omega$ is a number of atoms per unit volume.

The first term in the right-hand side of Eq. 6.113 is an inward vacancy flux caused by the difference of diffusivities and leading to void formation. Second term in the same equation is an outward vacancy flux caused by different sign of curvatures at inner and outer boundaries and counteracting the first term. After all metal is consumed, this vacancy concentration difference leads to a collapse of the nanoshell. Evidently, if the curvature effect wins, meaning

$$(D_{\mathrm{B}}-D_{\mathrm{A}})\frac{\Delta c_{\mathrm{B}}}{\Delta r}<c_{\mathrm{V}_0}D_{\mathrm{V}}\cdot L_{\mathrm{GT}}\left(\frac{1}{r_{\mathrm{i}}}+\frac{1}{r_{\mathrm{e}}}\right)\frac{1}{\Delta r}$$

void formation becomes impossible because the vacancy flux should be directed outward. At the initial stage, when $r_{\mathrm{e}}\approx r_{\mathrm{i}}\approx r_0$, this criterion can be rewritten as:

$$r_0<L_{\mathrm{GT}}\cdot\frac{2(c_{\mathrm{B}}D_{\mathrm{B}}^{*}+c_{\mathrm{A}}D_{\mathrm{A}}^{*})}{(D_{\mathrm{B}}-D_{\mathrm{A}})\Delta c_{\mathrm{B}}} \tag{6.114}$$

Criterion in Eq. 6.114 is hard to use since homogeneity range of compound is very narrow and usually unknown. Yet, this difficulty can be easily circumvented. Indeed, we take

$$\begin{aligned}D_{\mathrm{B}}\Delta c_{\mathrm{B}}&=\int_{c_{\mathrm{B}}(r_{\mathrm{e}})}^{c_{\mathrm{B}}(r_{\mathrm{i}})}D_{\mathrm{B}}^{*}\varphi\,dc_{\mathrm{B}}=\int_{c_{\mathrm{B}}(r_{\mathrm{e}})}^{c_{\mathrm{B}}(r_{\mathrm{i}})}D_{\mathrm{B}}^{*}\frac{c_{\mathrm{B}}c_{\mathrm{A}}}{kT}\frac{\partial^2 g}{\partial c_{\mathrm{B}}^2}dc_{\mathrm{B}}\\&\approx\bar{D}_{\mathrm{B}}^{*}\frac{\bar{c}_{\mathrm{B}}\bar{c}_{\mathrm{A}}}{kT}\int_{c_{\mathrm{B}}(r_{\mathrm{e}})}^{c_{\mathrm{B}}(r_{\mathrm{i}})}\frac{\partial^2 g}{\partial c_{\mathrm{B}}^2}dc_{\mathrm{B}}=\bar{D}_{\mathrm{B}}^{*}\frac{\bar{c}_{\mathrm{B}}\bar{c}_{\mathrm{A}}}{kT}\cdot\left(\left.\frac{\partial g}{\partial c_{\mathrm{B}}}\right|_{r_{\mathrm{i}}}-\left.\frac{\partial g}{\partial c_{\mathrm{B}}}\right|_{r_{\mathrm{e}}}\right)=\bar{D}_{\mathrm{B}}^{*}\frac{\Delta g}{kT}\end{aligned} \tag{6.115}$$

Here Δg is a compound formation Gibbs free energy per atom (it has been measured for many compounds). As a result, Eq. 6.115 can be reduced to the following form:

$$r_0<\frac{2\gamma\Omega}{\Delta g}\cdot 2c_{\mathrm{B}}\cdot\frac{(D_{\mathrm{B}}^{*}/D_{\mathrm{A}}^{*})+c_{\mathrm{A}}/c_{\mathrm{B}}}{(D_{\mathrm{B}}^{*}/D_{\mathrm{A}}^{*})-1} \tag{6.116}$$

We will use the non-dimensional parameter $G = \dfrac{2\gamma\Omega}{\Delta g r_0} = \dfrac{E_{\mathrm{GT}}(r_0)}{\Delta g}$ as a measure of the ratio of curvature effect and chemical driving force of reaction. Thus, in our simplified approach the condition of impossibility for hollow shell to form by reaction can be represented as

$$G > \frac{1}{2c_{\mathrm{B}}} \cdot \frac{(D_{\mathrm{B}}^{*}/D_{\mathrm{A}}^{*}) - 1}{(D_{\mathrm{B}}^{*}/D_{\mathrm{A}}^{*}) + c_{\mathrm{A}}/c_{\mathrm{B}}} \tag{6.117}$$

In this simplified analysis we neglected the influence of curvature on the chemical driving force (in other words, on the homogeneity range for main components), nonlinearity of Gibbs–Thomson effect for vacancy concentration at nanoscale, spherical geometry. Moreover, knowledge of the vacancy flux is not sufficient for predicting the reaction rate and the resulting void size. All this will be modified below. Nevertheless, the simplified Eq. 6.117 is rather close to the graphical criterion obtained by a more rigorous analytic and numeric analysis (to be presented in Fig. 6.17).

6.4.1.2 General case

We start from the flux balance for component B at the outer boundary and the flux balance for component A at the inner boundary. Flux balance means that the product of concentration step across the boundary and the boundary velocity is equal to the difference of fluxes on both sides of this moving boundary. For the external boundary the flux balance for B is more convenient since we know for sure that both the concentration of B and the outward flux of B are equal to zero. On the other hand, for the internal boundary the flux balance of A is more convenient since we know almost for sure that the solubility of A in B (e.g., of sulfur in the metal) can be neglected and no component A goes into the void so that the corresponding concentration and flux of A at $r < r_{\mathrm{i}}$ are equal to zero. Thus, in each flux balance equation we will use only one flux (B for external and A for inner boundary).

In the expression for both fluxes (Eq. 6.118 below), we will take into account the input of vacancy gradients due to the Gibbs–Thomson effect—the last term in Eqs. 6.118i and 6.118e. We note that the diffusion of B-flux is directed out of the shell, the diffusion of A-flux is directed into it.

At the external radius r_e we have

$$(0-c_B)\frac{dr_e}{dt}=0-\left(-D_B\frac{\partial c_B}{\partial r}\bigg|_{r_e}+\frac{c_A D_A^*}{c_V}\frac{\partial c_V}{\partial r}\bigg|_{r_e}\right) \quad (6.118i)$$

At the internal radius r_i we have

$$(c_A-0)\frac{dr_i}{dt}=\left(-D_A\frac{\partial c_A}{\partial r}\bigg|_{r_i}+\frac{c_A D_A^*}{c_V}\frac{\partial c_V}{\partial r}\bigg|_{r_i}\right)-0 \quad (6.118e)$$

Concentrations of A and B inside the compound can slightly change in the narrow homogeneity ranges ($\Delta c_B=-\Delta c_A$). We consider that $c_B+c_A\cong 1$, $\partial c_A/\partial r\cong -\partial c_B/\partial r$.

Note that we use the fluxes in the lattice reference frame and intrinsic diffusivities (instead of interdiffusivity) according to our basic approximation of no Kirkendall shift. We neglect the correlation factors of Manning's vacancy wind terms,[42] since they can change results only quantitatively. Both tracer diffusivities are proportional to vacancy concentration. (Here we use one more approximation that we treat only one effective vacancy concentration, without distinguishing between sublattices in the compound). We take

$$D_B^*=c_V K_B,\ D_A^*=c_V K_A, \quad (6.119)$$

where K_B and K_A can be treated as constants (see Eq. 6.81). Thus, we rearrange Eq. 6.118 as

$$\frac{dr_e}{dt}=K_B\left(-\frac{c_V\varphi}{c_B}\frac{\partial c_B}{\partial r}\bigg|_{r_e}+\frac{\partial c_V}{\partial r}\bigg|_{r_e}\right) \quad (6.120e)$$

$$\frac{dr_i}{dt}=K_A\left(\frac{c_V\varphi}{c_A}\frac{\partial c_B}{\partial r}\bigg|_{r_e}+\frac{\partial c_V}{\partial r}\bigg|_{r_e}\right) \quad (6.120i)$$

We note that in the above equations the values of vacancy concentration at the external and internal boundaries can be very different, due to Gibbs–Thomson effect, and the vacancy gradient does not tend to zero. At the same time the concentrations of the main components at external and internal boundaries remain close to constant because of narrow homogeneity range of the compound.

Now we use a steady state approximation both for components and for vacancies inside the compound, which has proved to be valid inside intermediate phases with narrow homogeneity range.[43] From the continuity equation we have

$$\frac{\partial c_{\mathrm{B}}}{\partial t} \approx 0 \Rightarrow div\Omega \vec{j}_{\mathrm{B}} \approx 0,$$

$$\frac{\partial c_{\mathrm{V}}}{\partial t} = div\Omega \vec{j}_{\mathrm{V}} + \sigma_{\mathrm{V}} \approx div\Omega \vec{j}_{\mathrm{V}} \approx 0 \Rightarrow div\Omega \vec{j}_{\mathrm{V}} \approx 0 \tag{6.121}$$

In the last equation we used our assumption about the absence of effective vacancy sinks/sources inside the growing layer of compound.

Writing down the fluxes and divergences in spherical coordinates, we obtain

$$div\vec{j} \approx 0 \Rightarrow \frac{1}{r^2}\frac{\partial}{\partial r}(r^2 j_{\mathrm{r}}) \approx 0 \Rightarrow r^2 j_{\mathrm{r}} \approx \mathrm{const}$$

$$\begin{aligned} r^2 \cdot \Omega j_{\mathrm{B}}(r) &= r^2\left(-c_{\mathrm{V}}(r)\varphi\frac{\partial c_{\mathrm{B}}}{\partial r} + c_{\mathrm{B}}(r)\frac{\partial c_{\mathrm{V}}}{\partial r}\right)K_{\mathrm{B}} \\ &= r_{\mathrm{e}}^2 \cdot \Omega j_{\mathrm{B}}(r_{\mathrm{e}}) = r_{\mathrm{e}}^2 \cdot \frac{dr_{\mathrm{e}}}{dt} \cdot c_{\mathrm{B}} \end{aligned} \tag{6.122}$$

$$\begin{aligned} r^2 \cdot \Omega j_{\mathrm{A}}(r) &= r^2(-c_{\mathrm{V}}(r)\varphi\frac{\partial c_{\mathrm{A}}}{\partial r} + c_{\mathrm{A}}(r)\frac{\partial c_{\mathrm{V}}}{\partial r})K_{\mathrm{A}} \\ &= r_{\mathrm{i}}^2 \cdot \Omega j_{\mathrm{A}}(r_{\mathrm{i}}) = r_{\mathrm{i}}^2 \cdot \frac{dr_{\mathrm{i}}}{dt} \cdot c_{\mathrm{A}} \end{aligned} \tag{6.123}$$

where $\Omega j_{\mathrm{B}} = c_{\mathrm{B}}\frac{dr_{\mathrm{e}}}{dt}$ and $\Omega j_{\mathrm{A}} = c_{\mathrm{A}}\frac{dr_{\mathrm{i}}}{dt}$ are fluxes of the B and A, respectively, and Ω is the volume per atom of the compound.

Eqs. 6.122 and 6.123 can be treated as the set of two **differential** equations for two unknown functions $c_{\mathrm{B}}(r)$ and $c_{\mathrm{V}}(r)$ (treating r_{e}, r_{i} and their time derivatives as known parameters). These equations are non-linear because the vacancy concentration $c_{\mathrm{V}}(r)$ and thermodynamic factor $\varphi(r)$ in the coefficients cannot be treated as constants. On the other hand, Eqs. 6.122 and 6.123 can be treated as the linear set of **algebraic** equations for determining the $\frac{dc_{\mathrm{V}}}{dr}$ and $c_{\mathrm{V}}(r)\varphi(r)\frac{dc_{\mathrm{B}}}{dr}$. By rearrangement, we change Eqs. 6.122 and 6.123 to

$$\begin{cases} \dfrac{dc_{\mathrm{V}}}{dr} = \dfrac{B_2}{r^2} \\ c_{\mathrm{V}}(r)\varphi(r)\dfrac{dc_{\mathrm{B}}}{dr} = -\dfrac{B_1}{r^2} \end{cases} \tag{6.124}$$

with B_1 and B_2 being expressed in terms of the boundary velocities:

$$\begin{cases} B_2 = -\dfrac{c_A}{K_A} r_i^2 \dfrac{dr_i}{dt} - \dfrac{c_B}{K_B} r_e^2 \dfrac{dr_e}{dt} \\ B_1 = c_B c_A \cdot \left(\dfrac{r_e^2}{K_B} \dfrac{dr_e}{dt} - \dfrac{r_i^2}{K_A} \dfrac{dr_i}{dt} \right) \end{cases} \tag{6.125}$$

The solutions of Eq. 6.124 give

$$\begin{cases} c_V(r) = -\dfrac{B_2}{r} + F \\ \varphi \dfrac{dc_B}{dr} = \dfrac{B_1}{r^2 \left(-\dfrac{B_2}{r} + F \right)} \end{cases} \tag{6.126}$$

where F is the integration constant. In the first equation in Eq. 6.126, the boundary conditions at r_e and r_i, taking into account Gibbs–Thomson effect on vacancy concentration, will give

$$\begin{cases} c_V(r_e) = -\dfrac{B_2}{r_e} + F = c_{V_0} e^{-\frac{L_{GT}}{r_e}} \\ c_V(r_i) = -\dfrac{B_2}{r_i} + F = c_{V_0} e^{\frac{L_{GT}}{r_i}} \end{cases} \tag{6.127}$$

Then we integrate the second equation in Eq. 6.126 over radius from the inner boundary to the outer one, and obtain

$$\int_{r_i}^{r_e} \varphi \frac{dc_B}{dr} dr = \int_{r_i}^{r_e} \frac{B_1}{r^2 \left(-\dfrac{B_2}{r} + F \right)} dr \tag{6.128}$$

The integration of the right-hand side of Eq. 6.128 by substituting Eq. 6.127 gives

$$\begin{aligned} \frac{B_1}{B_2} \int_{r_i}^{r_e} \frac{d\left(-\dfrac{B_2}{r} + F \right)}{\left(-\dfrac{B_2}{r} + F \right)} &= \frac{B_1}{B_2} \int_{c_V(r_i)}^{c_V(r_e)} \frac{dc_V}{c_V} = \frac{B_1}{B_2} \cdot \left(\ln \frac{c_V(r_e)}{c_V(r_i)} \right) \\ &= -\frac{B_1}{B_2} L_{GT} \left(\frac{1}{r_e} + \frac{1}{r_i} \right) \end{aligned} \tag{6.129}$$

For the integration of left-hand side of Eq. 6.128 we take into account that the thermodynamic factor $\varphi = \dfrac{c_B c_A}{kT} \dfrac{\partial^2 g}{\partial c_B^2}$, and that the homogeneity interval is very narrow (concentrations are almost constant within the compound layer but the derivatives of g are very far from being constant), we obtain,

$$\int_{r_i}^{r_e} \varphi \frac{dc_B}{dr} dr = \int_{r_i}^{r_e} \frac{c_B c_A}{kT} \frac{\partial^2 g}{\partial c_B^2} \frac{dc_B}{dr} dr$$
$$= \frac{\bar{c}_B \bar{c}_A}{kT} \int_{c_B(r_i)}^{c_B(r_e)} \frac{\partial^2 g}{\partial c_B^2} dc_B = \frac{\bar{c}_B \bar{c}_A}{kT} \cdot \left(\left. \frac{\partial g}{\partial c_B} \right|_{r_e} - \left. \frac{\partial g}{\partial c_B} \right|_{r_i} \right) \quad (6.130)$$

As was shown in Section 6.2.4, the right-hand side of Eq. 6.130 can be expressed in terms of reaction driving force in planar case plus some term determined by curvature:

$$\left. \frac{\partial g}{\partial c_B} \right|_{r_e} - \left. \frac{\partial g}{\partial c_B} \right|_{r_i} = -\frac{\Delta g}{\bar{c}_B \bar{c}_A} + \frac{E_{GT}(r_e)}{\bar{c}_B} - \frac{E_{GT}(r_i)}{\bar{c}_A} \quad (6.131)$$

Substituting Eq. 6.131 into Eq. 6.130, we obtain

$$\int_{r_i}^{r_e} \varphi \frac{dc_B}{dr} dr = \frac{\bar{c}_B \bar{c}_A}{kT} \cdot \left(\left. \frac{\partial g}{\partial c_B} \right|_{r_e} - \left. \frac{\partial g}{\partial c_B} \right|_{r_i} \right)$$
$$= -\frac{\Delta g}{kT} + L_{GT} \left(\frac{\bar{c}_A}{r_e} - \frac{\bar{c}_B}{r_i} \right) \quad (6.132)$$

Substituting Eqs. 6.132 and 6.129 into Eq. 6.128, we obtain

$$\frac{\Delta g}{2\gamma\Omega} - \frac{\bar{c}_A}{r_e} + \frac{\bar{c}_B}{r_i} = \frac{B_1}{B_2} \cdot \left(\frac{1}{r_e} + \frac{1}{r_i} \right) \quad (6.133)$$

After having substituted B_1 and B_2 from Eq. 6.125 into Eq. 6.133 and after having excluded the time, we obtain (after simple but long algebra) the main equation of this section:

$$\frac{dr_e}{dr_i} = -\frac{r_i^2}{r_e^2} \frac{K_B}{K_A} \frac{c_A}{c_B} \frac{\frac{\Delta g}{2\gamma\Omega} - \frac{1}{r_e}}{\frac{\Delta g}{2\gamma\Omega} + \frac{1}{r_i}} \quad (6.134)$$

We will use the following nondimensional parameters: $\xi_i = \frac{r_i}{r_0}$, $\xi_e = \frac{r_e}{r_0}$, $G = \frac{E_{GT}(r_0)}{\Delta g}$, where r_0 is an initial radius of nanoparticle. This parameter G, together with the ratio of mobilities $\frac{K_B}{K_A} = \frac{D_B^*}{D_A^*} = \frac{D_B}{D_A}$ (

$D_{B(A)} = D^*_{B(A)}\varphi$) are the two major parameters of our model. Then Eq. 6.134 can be represented in the following form:

$$\frac{d\xi_e}{d\xi_i} = -\frac{\xi_i^3}{\xi_e^3}\frac{D^*_B}{D^*_A}\frac{1-c_B}{c_B}\frac{\xi_e - G}{\xi_i + G} \tag{6.135}$$

This equation differs from a similar equation by Alivisatos *et al.* by the last term containing $G \neq 0$ (responsible for Laplace pressure and Gibbs–Thomson effect both for vacancies and for main components).[4] When both radii r_e and r_i are large ($r_e, r_i >> 2\gamma\Omega/\Delta g$, $G << 1$), the equations become similar.

Eq. 6.135 can be easily integrated but gives, as a result, the transcendent interrelation between ξ_e and ξ_i. It can be solved numerically by finite difference method. The governing parameters are $D \equiv D^*_B/D^*_A$, c_A and G. The equation has been solved with "initial" condition $\xi_i = 1$, $\xi_e = 1 + \varepsilon$. We made calculations in two cases:

(1) when all pure metal (core) was consumed by reaction:

$$c_A \cdot (\xi_e^3 - \xi_i^3) \geq 1 \tag{6.136}$$

(2) when the inner radius of the compound shell became equal to the radius of the remaining core (void in the hollow shell between the compound shell and the remaining core with connecting bridges shrinks prior to the end of the reaction):

$$\xi_i^3 \leq \frac{r^3_{core}}{r^3_0} = 1 - c_A \cdot (\xi_e^3 - \xi_i^3) \tag{6.137}$$

It is evident from Eq. 6.135 that in the case $G > 1$ the nanoshell formation is impossible (taking into account that ξ_i and ξ_e start from almost 1). This means that in very small particles, $r_0 < 2\gamma\Omega/\Delta g$, the reaction with void formation is impossible. It is interesting that this critical condition coincides with the critical radius for nucleation of a new phase.

Below we present some results for various parameters in Eq. 6.135. A characteristic solution of Eq. 6.135 is represented in Fig. 6.15. Here the reduced inner radius ξ_i was decremented uniformly and arbitrarily, the reduced external radius y was calculated by numeric integration of Eq. 6.135, the reduced radius $\xi_{core} = r_{core}/r_0$ of the remaining core was calculated according to conservation of matter (neglecting difference of atomic volumes in compound and metal): $\xi_{core} = (1 - c_A \cdot (\xi_e^3 - \xi_i^3))^{1/3}$. Calculations were stopped when the core or void disappeared ($\xi_{core} = 0$ or $\xi_i = \xi_{core}$).

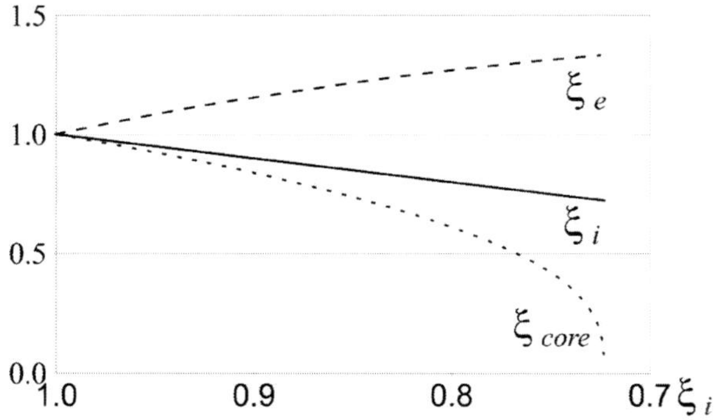

Figure 6.15 Formation kinetics: time dependence of the external ξ_e, internal ξ_i, and core ξ_{core} radii at uniform variation of ξ_i. The condition $\xi_{core} = 0$ means full consumption of initial core. Parameters $D = 10$, $c_B = c_A = 0.5$, $G = 0.7$.

According to analysis of Alivisatos *et al.*, the resulting radius of final void is determined only by ratio of diffusivities and by stoichiometry of the compound.[4] (Ideally, if only metallic atoms migrate, the final void radius should be just equal to the initial radius of metallic particle.) But, according to our model, the resulting ratio of void radius to initial particle size depends on the initial particle size via the G-factor. Indeed, according to our results, the less is the initial radius and the larger is the factor G, the less will be the ratio of final void size to initial particle size $\xi_i^{\text{final}} = r_i^{\text{final}} / r_0$ as shown in Fig. 6.16 Thus, if initial radius is less than some critical one, void cannot be formed at all.

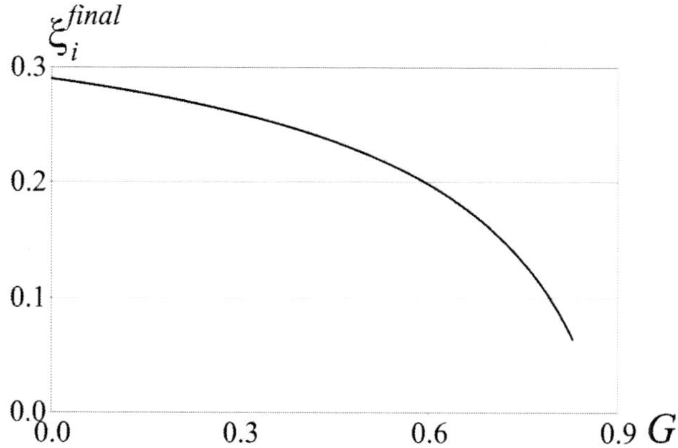

Figure 6.16 Dependence of maximal void radius ξ_i^{final} on parameter G at $D = 10$, $c_B = c_A = 0.5$.

And what happens then? In this case, due to large vacancy gradient, the inner radius converges inside faster than the radius of the remaining metallic core, and abolishes the void. In Fig. 6.17 we divide the region of parameters G and D ($G = 2\gamma\Omega/\Delta g r_0$, $D \equiv D_B^*/D_A^*$) into two subregions (results of our numeric calculations of $\xi_e(\xi_i)$ with various parameters of G and D). For the left and upper subregion the formation of hollow nanoshell is forbidden. This criterion appears to be rather close to approximate analytic criterion in Eq. 6.117, as demonstrated on Fig. 6.17

One can also calculate the absolute rate of void formation in real time[13]:

$$\frac{dr_i}{dt} = -\frac{K_A c_{V0}}{c_A} \frac{\exp\left(\frac{L_{GT}}{r_i}\right) - \exp\left(-\frac{L_{GT}}{r_e}\right)}{r_i^2\left(\frac{1}{r_i} - \frac{1}{r_e}\right)\left(1 - \frac{r_i}{r_e}\frac{r_e/r_0 - G}{r_i/r_0 + G}\right)} \tag{6.138}$$

The formation of hollow nanoshell is determined by the competition of three factors:

(1) Growth of compound phase due to reaction and tendency for vacancies to go inside (due to difference of diffusivities);

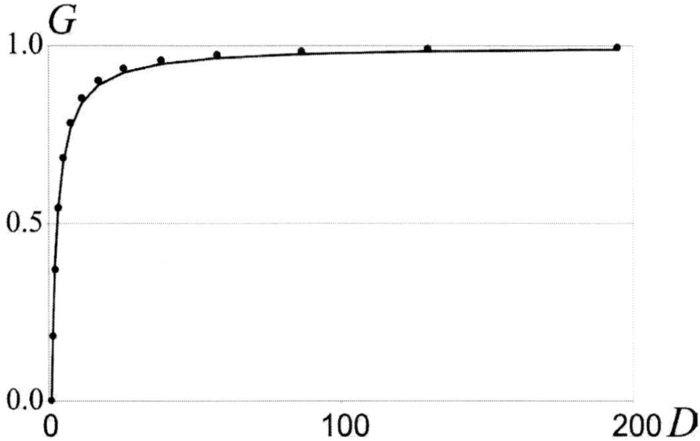

Figure 6.17 The criterion of hollow shell formation points below or/ and to the right of $G(D)$-curve correspond to possible to void formation. Region up or/and left of $G(D)$-curve corresponds to forbidden voiding. Solid line corresponds to approximate analytic criterion (Eq. 6.117), dots correspond to data obtained by numerical solution (Eq. 6.135), $c_B = 0.5$.

(2) Shrinking of shell due to vacancy gradient between inner and outer boundaries (Gibbs–Thomson effect);

(3) Change of driving force by Laplace tension at inner boundary and Laplace compression at external boundary. The second factor should win for very small particles. Critical value for initial radius is very similar to Gibbs nucleation theory, that is $r_0^{\text{cr}} = 2\gamma\Omega/\Delta g$.

6.4.2 Formation of Binary Solution Hollow Shells

To describe the formation of a nanoshell consisting of solid solution with unlimited solubility, we use the model developed in Section 6.3.3. We don't need to change the model equations—we change only initial conditions. Namely, consider a sphere of pure B of radius R_{BA} surrounded by a shell of pure A (Fig. 6.1a). To escape solving the nucleation problem, let's assume that the initial pure sphere A already contains a small void in the centre. Of course, for too big initial core this assumption seems unreasonable since the first voids should nucleate in the vicinity of initial contact between A and B. Yet, for nanoparticles it is natural that initial nanovoids coalesce very fast into single central void. Thus, in our model the initial B-profile is:

$$t' = 0, c_{\text{B}}(\xi) = \begin{cases} 1, & 0 < \xi < \xi_{\text{BA}} \\ 0, & \xi_{\text{BA}} < \xi < 1 \end{cases}, \xi_{\text{BA}} \equiv \frac{1/r_{\text{BA}} - 1/r_{\text{e0}}}{1/r_{\text{i0}} - 1/r_{\text{e0}}} \tag{6.139}$$

Evidently, the curvature effect, determining the kinetics of collapse at the shrinking stage, cannot be neglected at formation stage as well. Formation of a nanoshell is a result of competition between Frenkel effect due to chemical forces (and difference of diffusivities) and Gibbs–Thomson effect. The smaller is the radius, the larger should be the role of curvature-driven effects. Thus, one can expect that for initial void radius smaller than some critical value formation will be impossible. This critical radius can be found from Eq. 6.96 for vacancy flux. Namely, initial pore in the B/A core–shell structure will shrink from the very beginning if the sign of vacancy flux is positive. For this it is necessary that

$$\exp\left(\frac{L_{\text{GT}}}{r_{\text{i}}}\right)\exp\left(-\int_0^1 f(\xi)d\xi\right) > \exp\left(-\frac{L_{\text{GT}}}{r_{\text{e}}}\right) \tag{6.140}$$

which means

$$\int_{c_B(r_e)}^{c_B(r_i)} \frac{(K_B - K_A)(1 - c_A c_B \lambda)}{c_A K_A + c_B K_B} dc_B < L_{GT}\left(\frac{1}{r_i} + \frac{1}{r_e}\right) \tag{6.141}$$

At the very beginning $c_B(r_e) = 0$, $c_B(r_i) = 1$. Thus, formation of a hollow nanoshell from a core–shell structure is forbidden if

$$\frac{1}{r_i} + \frac{1}{r_e} > \frac{1}{L_{GT}} \int_0^1 \frac{(\kappa e^{\alpha x} - 1)(1 - \lambda x(1 - x))}{1 + x(\kappa e^{\alpha x} - 1)} dx \tag{6.142}$$

Here $\kappa = K_{B0}/K_{A0}$, $\alpha = \alpha_B - \alpha_A$, and $x = c_B$

This criterion is not rigorous since it was obtained under two major assumptions: (a) steady-state approximation for vacancies and (b) assumption of single central void preexisting in the center and serving as the sink for all extra vacancies coming into B-rich core. In reality, at the very beginning the largest supersaturation with vacancies should appear just under the initial contact surface B/A, as will be seen in Fig. 6.18 (vacancy profile at initial formation stage). It may lead to several tiny void formation under B/A perimeter, as we'll see later in MC simulation and as was observed in experiments at low temperatures.[4]

In Fig. 6.18, we demonstrate (for the first time, as far as we know) formation and shrinkage in one run. During the formation stage the chemical forces in sharply inhomogeneous system are stronger than curvature driven vacancy gradient. After formation of a partially homogeneous hollow nanoshell, chemical forces become weaker, and curvature driven Gibbs–Thomson effect wins leading to collapse due to outward vacancy flux.

Profiles of B-species in Fig. 6.18 demonstrate that the full homogenization is not reached during nanoshell formation—even at maximal void radius the concentration gradient is not zero, and later it grows back due to inverse Kirkendall effect. Vacancy profile shows formation of supersaturation just under initial B/A interface at the initial stage. Later this nonmonotonic profile transforms into a monotonic one.

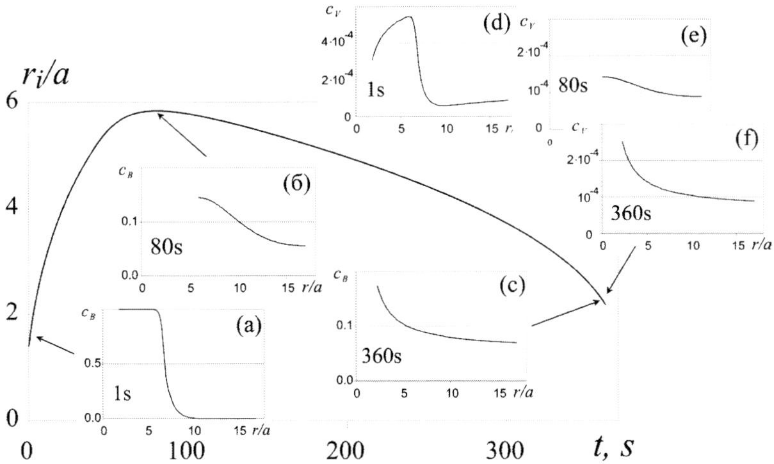

Figure 6.18 Time evolution of initial tiny void radius—at first a fast formation of hollow nanoshell and then its collapse (slow and then faster and faster). Inserts (a), (b), and (c) show the concentration profiles of faster component at formation, crossover, and collapse stages, respectively. Inserts (d), (e), and (f) show the vacancy concentration profiles at the same stages. One can see that the transition from formation to collapse does not mean the full homogenization of the shell. Parameters: $c_V^{eq} = 10^{-4}$, $c_B^{init} = 0.075$, $r_i^{init} = 1.4a$, $r_{BA}^{init} = 7a$, $r_e^{init} = 17a$, $L_{GT} = 0.75 \cdot 10^{-9}$m, $\kappa = 10$, $\lambda = 0$, $\alpha_A = -4.5$, $\alpha_B = -2$, $dt = 10^{-4}$s.

6.4.3 Formation of a Spherical Nano-Shell in Monte Carlo Simulation

Alternatively to the offered phenomenological models, we developed a 3-D MC model of nanoshell evolution for the fcc-structure of binary alloy with vacancies. To be sure that the proposed model is valid and chosen parameters are reasonable, we should check at first, if the hollow nanoshell is indeed formed from the initial core–shell structure.

The difference of diffusivities of A and B species (even in the case of ideal alloy with zero mixing enthalpy) is provided by the probability to choose the atom of the given sort in Metropolis algorithm, determined by the ratio of frequencies

$K = \nu_B/\nu_B$ ($\nu_A \leq \nu_B$). Temperature is, typically for MC simulations, regulated by varying nondimensional ratio $\varphi \equiv \Phi / kT$ of average interaction energy Φ and kT (larger absolute value of φ means lower temperature). In ideal alloy one would have $\varphi_{AA} = \varphi_{AB} = \varphi_{BB} = \varphi$. Instead, we simulate an ordered phase, where interrelation between pair energies corresponds to the average interaction energy

$$c_A^2 \varphi_{AA} + c_B^2 \varphi_{BB} + 2c_A c_B \varphi_{AB} = \varphi \tag{6.143}$$

We consider two cases—low ($\varphi^L = -1.535$)[36] and high ($\varphi^H = -0.98$) temperatures in vacuum.

In this case the difference of mobilities is stipulated not only by different frequencies K but also by thermodynamics and structures of sublattices: in the ordered phase A_1B_3 the component B will have higher mobility even in case of equal frequencies.

To investigate the kinetics of a nanoshell formation a nanoparticle of pure B component (radius $r_{BA} = 15.7a$) was initially enveloped by a pure A component (external radius $r_e = 17a$) (total number of atoms is 82,421) —the chosen ratio of internal r_{BA} and external r_e concentric radii provides necessary volume fractions for the A_1B_3 phase. According to Eq. 6.143 the pair energies for A_1B_3 phase are chosen as $\varphi_{AA} = \varphi_{BB} = 0.7\varphi_{AB}$: at low temperature $\varphi_{AA}^L = \varphi_{BB}^L = -1.32$, $\varphi_{AB}^L = -1.89$ with average energy φ^L; at high temperature $\varphi_{AA}^H = \varphi_{BB}^H = -0.86$, $\varphi_{AB}^H = -1.23$ with average energy φ^H. As a result of simulation the widening of the A/B diffusion zone is observed and small voids start forming at the pure B/alloy interface. In the process of computer experiment the voids expand, leaving between themselves the "bridges" of pure B. At high temperature we observed the subsequent coalescence of voids into a single spherical void in the centre the nanoparticle (Fig. 6.19). These simulation results correlate with the experiments on the formation of cobalt selenides and sulfides.[3,4]

6.4.4 Conclusions to Section 6.4

The formation of hollow nanoshell is determined by the competition of three factors; (1) growth of compound phase due to reaction and tendency for vacancies to go inside (due to difference of diffusivities), (2) shrinking of shell due to vacancy gradient between inner and outer boundaries (Gibbs–Thomson effect), and (3) change of driving force by Laplace tension at inner boundary and Laplace compression at external boundary. The second factor should win for

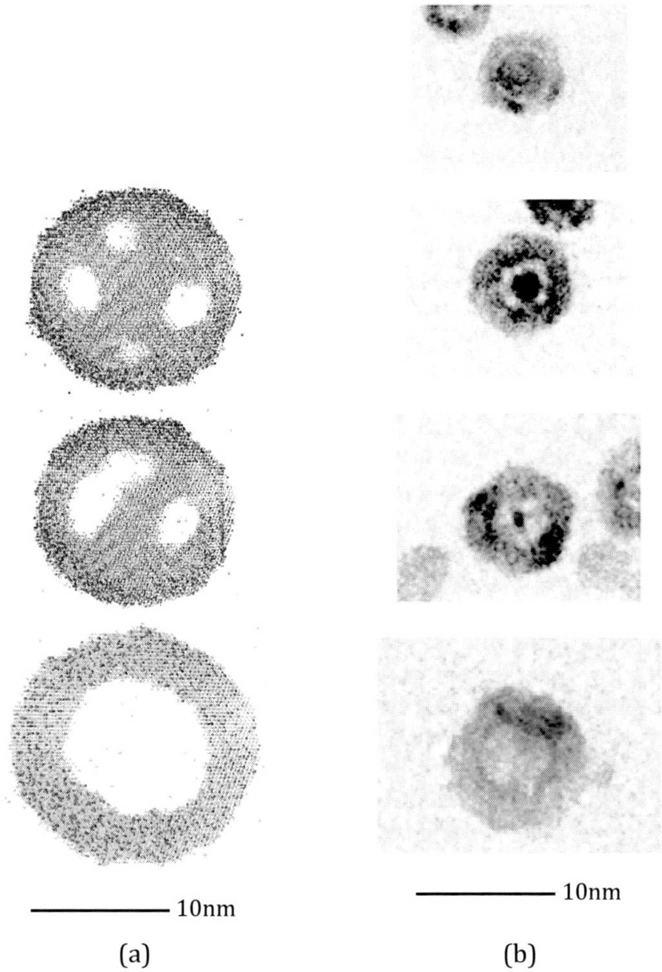

Figure 6.19 Void formation in a nanoparticle: (a) results of MC simulation with $\bar{c}_B = 0.75$, $K = 10$ at high temperature (φ^H) at time points of 167, 1000, 2000, and 10,000 MCS (sample cross-section, light dots—B, dark ones—A), the maximal size of the void attained at 8400 MCS; (b) evolution of CoSe hollow nanocrystals with time by injection of a suspension of selenium in *o*-dichlorobenzene into a cobalt nanocrystal solution at 455 K: 10 s, 1 min, 2 min, and 30 min.[3]

very small particles. Critical value for initial radius is very similar to Gibbs nucleation theory. Nucleation and phase competition, in case of two phase sequential formation of the spherical particles will be considered elsewhere.

6.5 Cross-Over from Formation to Collapse

Each nonequilibrium system has it's "life cycle", like, for example, any star in Universe—from birth to death. Hollow nanoshells also have their life cycle—formation, growth, and coalescence of initial voids, reaching the maximal void size, and then slow shrinking ending with collapse. As in living systems, or as in stars evolution, in hollow shells we can also observe the competition of growth tendency versus shrinking tendency. Growth tendency is due to chemical driving force combined with inequality of diffusivities. Shrinking tendency is due to capillary effect. In case of oxide or sulfide formation the chemical driving forces are, typically, much stronger than capillary forces, until the reagents are consumed. Therefore, in this case the shrinking stage should take much longer time than the formation stage. So, to observe shrinking in reasonable time, one should raise a temperature, in comparison with formation stage. On the contrary, in case of solid solution when the driving force of mixing can be even purely entropic and therefore weak, it can be comparable with capillary effect. This idea was first checked by the authors of this paper by phenomenological and atomistic modeling,[14] and development for nanowires,[44] reported at DIMAT-2008 in Lanzarotte. These results stimulated group of Prof. D. Bake (Debrecen) to make real experiments with formation and collapse of almost ideal metallic systems (first of all, gold-silver). Their experimental results[45–47] proved our theoretic models to be correct.

6.5.1 Phenomenological Model

After detailed analysis of formation and collapse of hollow shells we know that both of these processes are the results of competition of the same factors: (1) chemical driving force—concentration gradient (maximal at formation stage), (2) curvature (Gibbs–Thomson) effect for vacancies providing vacancy concentration difference between inner and external surfaces, (3) Kirkendall vacancy flux due to inequality of partial diffusivities of the main components, (4) inverse Kirkendall effect—segregation of components caused by vacancy flux. Thus, we can treat all "life cycle" of the hollow nanoparticle—from formation to collapse—as one process. At the first stage of this process the factors (1) and (3) are dominating. At second (typically much longer) stage the factors (2) and (4) are dominating.

We investigated the mentioned life cycle of spherical and cylindrical particles in the frame of phenomenological model for binary solid solution with unlimited solubility. Life cycle of spherical particles was first introduced by us,[13] and cylindrical case is presented here and compared with spherical one. Comparison was made at the following parameters: $c_V^{eq} = 10^{-4}$, $\lambda = 0$, $K_{A0} = 10^{-16}\,\mathrm{m^2/s}$, $K_{B0} = 10^{-15}\,\mathrm{m^2/s}$, $\kappa = 10$, $\alpha_A = -4.5$, $\alpha_B = -2$. For the right comparison, one should take into account that the parameter L_{GT}, determined via surface tension γ, for cylinder is twice less than for sphere: $L_{GT}^{sph} = 2L_{GT}^{cyl} = 0.75\ \mathrm{nm}$.

Both formation and collapse were investigated within one run, with fixing of cross-over point as a maximal void radius R_i^{max}. To avoid problems with description of void nucleation, we started with preexisting void of minimal initial radius equal to lattice parameter $R_{i0} = a$. Initial concentration profile was taken as stepwise (spherical or cylindrical core B of radius R_{BA} enveloped by pure A shell). One more parameter—initial external radius of ball or wire, R_{e0}—was changed in logarithmic scale within interval (a, $34a$). Average concentration of the faster component, $\overline{c_B}$, determined by the cubed ratio R_{BA}/R_{e0}, was taken equal to 0.075, 0.1, 0.2, 0.3, 0.4, 0.5, 0.6. Time step in the finite difference scheme depended on the particle size (10^{-5} s < dt < 1 s) and was chosen according to well-known rule—if further decrease of the time step does not change the result.

Time evolution of balls and wires is qualitatively the same (Fig. 6.20a). Similar full life cycle was observed in the experiments for semispherical hollow shells[45] (Fig. 6.20b). The crossover point corresponds to the moment of transition from one stage to another. The stages are respectively characterized by the formation time t_{form} and shrinking time t_{shr} (Fig. 6.20a), whole relation determines the relative stability of the pore.

The radius of the maximal pore R_i^{max} reached at the crossover moment (Fig. 6.20) at the average concentration $\overline{c_B}$ = 0.075 depends almost linearly on the initial particle radius R_{e0} both for spherical and for cylindrical samples (Fig. 6.21a). The deviation from the linear

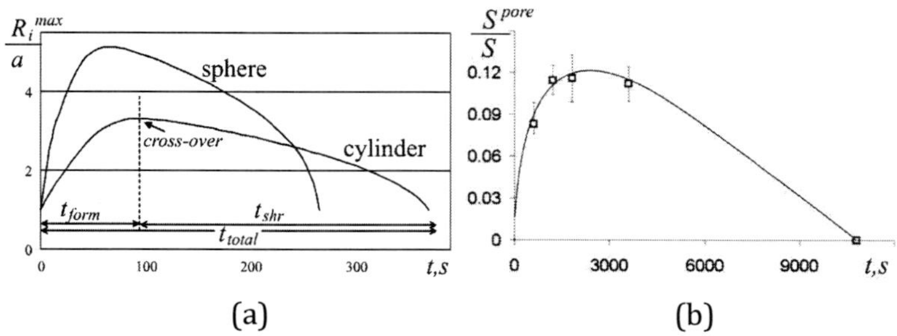

Figure 6.20 Time evolution (from formation to collapse) of voids. (a) Time dependence of reduced void radius calculated by phenomenologic scheme for cylindrical and spherical samples (core B enveloped by shell A)—at first a fast formation stage t_{form}, then a slow shrinking stage t_{shr} (which accelerates just before finish). $R_{\text{e}0} = 15.2\text{a}$, $\overline{c_{\text{B}}} = 0.075$. Total time t_{total} of the two-stage process consists of formation time t_{form} and shrinking time t_{shr}. (b) Experimental time dependence of the ratio of hollow area S^{pore} to full area S for Au–Ag hemispherical hollow shells [45]—ensemble of nanoparticle with sizes scattered between 20 and 50 nm.

law at small initial radii is explained by the existence of the initial centrally symmetric pore along the cylinder axis with the radius a, which is a necessary condition for consistent calculations in the given phenomenological model. It is evident that, for small real core–shell systems with the prevalence of one of the components, the chemical moving forces can appear insufficient for the formation (or even for nucleation)[13] of a pore with an effective radius exceeding a. In other words, the artificial introduction of an initial pore "makes" the system forming a void with a radius not less then a.

At increase of the average concentration of the fast component $\overline{c_{\text{B}}}$ does not result in a change of the linear character of the dependence $R_{\text{i}}^{\text{max}}(R_{\text{e}0})$ (Fig. 6.21b). In the case of large concentrations and small radii, however, the linear law is again violated. It is caused by a reduction of the reserve of chemical moving forces in the same way as at low concentrations. Moreover, the situation worsens due to the non-symmetry of the mobilities of the components (the faster internal component B cannot dissolve in shell A and the radial vacancy flux coming inside the cylinders is very small). If introducing a restriction on cylinders with small radii at the corresponding average concentrations $R_{\mathrm{e0}}^{\min}$ (Table 6.3), then the error of the linear approximation $R_{\mathrm{i}}^{\max} \sim k(c_{\mathrm{B}}) \cdot R_{\mathrm{e0}} + b(c_{\mathrm{B}})$ at a fixed average concentration will not exceed 0.2%. To realize a correct approximation at $\overline{c_{\mathrm{B}}} > 0.8$, one should consider cylinders with $R_{\mathrm{e0}} > 34a$, that is why the Table 6.3 contains data for $\overline{c_{\mathrm{B}}} \leq 0.8$.

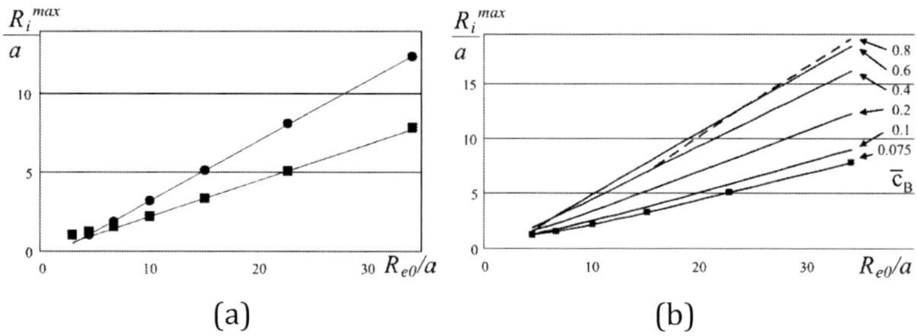

Figure 6.21 Maximal pore radius $R_{\mathrm{i}}^{\max}$ as a function of initial external radius R_{e0} (a) for spheres (circles) and cylinders (squares) at $\overline{c_{\mathrm{B}}} = 0.075$; (b) for cylinders at various average concentrations $\overline{c_{\mathrm{B}}}$.

For the specified diffusion parameters of the model, the slope $k(c_{\mathrm{B}})$ of the approximation straight lines $R_{\mathrm{i}}^{\max}(R_{\mathrm{e0}})$ in the interval $0 < \overline{c_{\mathrm{B}}} \leq 0.8$ is proportional to the square root of the average concentration of the fast component: $k(c_{\mathrm{B}}) \sim 0.75\sqrt{c_{\mathrm{B}}}$ (Table 6.3). The obtained approximate dependence demonstrates that a pore can occupy nearly a half of the initial volume of the fast component with the initial radius R_{BA}: $\pi(R_{\mathrm{i}}^{\max})^2 H \sim \pi 0.56\overline{c_{\mathrm{B}}}(R_{\mathrm{e0}})^2 H \sim 0.56\pi(R_{\mathrm{BA}})^2 H$. This relation yields understated values of the maximal pore radius $R_{\mathrm{i}}^{\max}$ at low concentrations (small amount of the mobile component very rapidly dissolves in the matrix of the slow one). At high concentrations, these values are overestimated (chemical moving forces come to the end much faster than the diffusion homogenization takes place; at some sets of parameters, a thick interlayer of pure component B, which had no time to react, is still observed on the inner surface after the crossover is already reached). The indicated deviations are corrected by taking the linear approximation coefficient b (see Table 6.3).

Table 6.3 Linear approximation coefficients of the dependences $R_i^{max}(R_{e0})$ (Fig. 6.21b) for cylinders at various average concentrations of the fast component $\overline{c_B}$

$\overline{c_B}$	R_{e0}^{min}/a	$k(c_B)$	$b(c_B)$
0.025	15	0.12	0.27
0.075	10	0.22	0.12
0.100	6.75	0.26	0.09
0.200	4.5	0.36	-0.05
0.300	4.5	0.43	-0.14
0.400	3	0.48	-0.26
0.500	3	0.52	-0.45
0.600	4.5	0.56	-0.78
0.700	6.75	0.61	-1.70
0.800	15	0.67	-3.66

To analyze the pore formation efficiency, let us consider the ratio of the volume of the maximal pore at the crossover moment to the volume of the cylinder (sphere) without a void:

$$\Xi_{cyl} = \frac{(R_i^{max})^2 - (R_{i0})^2}{(R_e^{max})^2 - (R_i^{max})^2}\ ;\ \Xi_{sph} = \frac{(R_i^{max})^3 - (R_{i0})^3}{(R_e^{max})^3 - (R_i^{max})^3}$$

One can see from Fig. 6.22a that the pore formation in cylindrical samples is more efficient then in spherical ones. As the initial radius of a particle increases, this efficiency tends to a certain asymptotic value. That is why, at large initial radii, the efficiency is determined by the chemical composition representing a driving force of pore formation (Fig. 6.22b). It is evident that the most efficient pore formation must be observed at equiatomic concentrations.

However, an increase of the cylinder radius results in a shift of the pore formation efficiency toward larger concentrations (0.7, 0.8), Fig. 6.23. To our mind, this peculiarity is related to the radial symmetry of particles. For a particle with fixed outer radius R_{e0}, a rise of the concentration leads to the growth of the radius R_{BA} that determines the area of the reaction surface and, respectively,

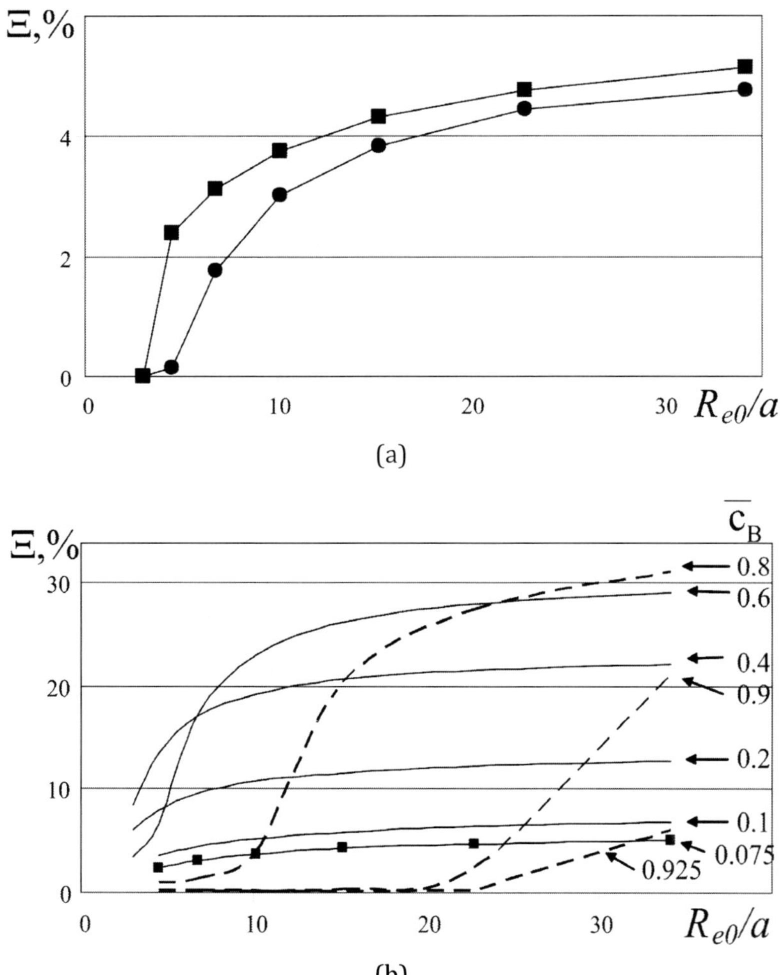

Figure 6.22 Pore formation efficiency Ξ as a function of the initial radius of the particle R_{e0} (a) for spheres (circles) and cylinders (squares) at $\overline{c_B}$ = 0.0.75; (b) for cylinders at various average concentrations $\overline{c_B}$.

provides a higher pore formation efficiency. The effect of the chemical composition on the pore formation efficiency can be traced by considering the dependences of Ξ on the average concentration of the fast component (Fig. 6.23): as the radius of a cylindrical particle increases, the efficiency grows with the gradual shift of the maximum toward higher concentrations. The asymmetry of $\Xi(\overline{c_B})$

is related to the asymmetry of the diffusion core–shell couple and different mobilities of the components.

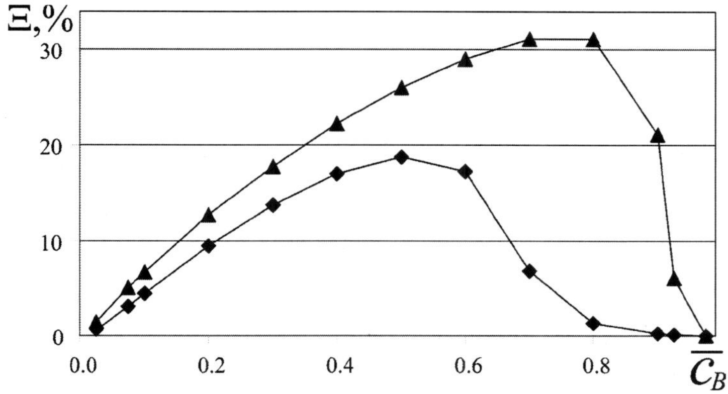

Figure 6.23 Pore formation efficiency Ξ as a function of the average concentration $\overline{c_B}$ at various initial cylinder radii: small ($R_{e0} = 6.75a$, diamonds) and large ($R_{e0} = 34.2a$, triangles).

It is evident that the formation and the collapse of a larger void require more time. The formation time t_{form} and the shrinking time t_{shr} (introduced in Fig. 6.20a) grow according to the power law with increase in the initial cylinder radius R_{e0} (Fig. 6.24). Moreover, a more rapid increase of the shrinking time as compared to the formation one (larger exponent) testifies to the fact that voids in cylinders of larger radii are more stable.

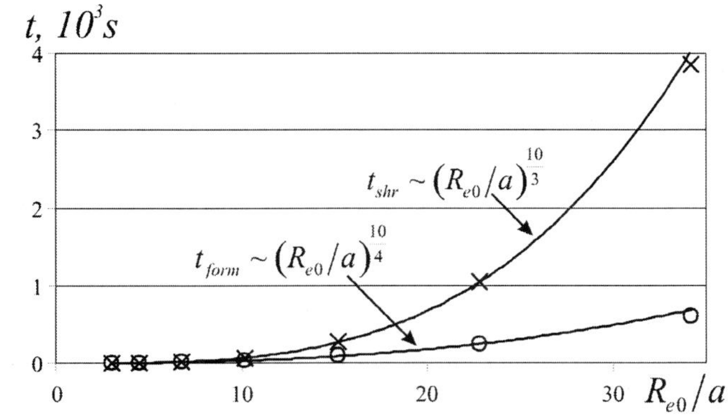

Figure 6.24 Void formation time t_{form} (circle) and shrinking time t_{shr} (crosses) as a function of the initial radius of the particle R_{e0} for cylinders, $\overline{c_B} = 0.075$.

For the quantitative description of the stability of hollow shells (which is important for technological applications of nanotubes), let us analyze the ratio t_{shr}/t_{total} of the absolute time of existence of a void before its collapse t_{shr} to the total lifetime $t_{total} = t_{form} + t_{shr}$. If the stage of void formation is longer than that of its collapse, then $t_{form}/t_{shr} < 0.5$. In the opposite case, $t_{form}/t_{shr} > 0.5$. The closer the proposed ratio to 1, the more stable is the pore. One can see from Fig. 6.25a that, at small initial radii 3a < R_{e0} < 6.75a, a pore collapses easier then it is formed ($t_{form}/t_{shr} < 0.5$). At large radii, the shrinking time is by an order of magnitude higher than the formation time ($t_{shr}/t_{total} \sim 0.9$). In contrast to cylinders, the ratio t_{shr}/t_{total} for spherically symmetric shell has a stronger dependence on the radius of an initial particle, which is explained by the different radius dependences of the volume (square for cylinder and cubic for spheres). As a result, spherical nanoshells are very unstable at low radii (with the shrinking time twice larger than the formation one) reach the ratio $t_{shr}/t_{total} \sim 0.9$ with increase in the radius and even exceed the corresponding characteristic for cylindrically symmetric voids (Fig. 6.25a).

The maximum stability is reached at the average concentration $\overline{c_B} \sim 0.5$ (Fig. 6.25b), in contrast to the maximum efficiency reached at about 0.7 to 0.8 (Fig. 6.22b). Such a non-coincidence at a significant concentration prevalence of the faster component can be due to its excess in the form of a thick interlayer in a vicinity of the inner surface at the crossover moment. In the case of a solid solution, this accelerates the collapse rather than impedes it.

Therefore, results of phenomenological modeling confirm correctness of description of hollow particles formation/shrinking as a competition of Frenkel, Gibbs–Thomson and inverse Kirkendall effects. Formation stage is controlled by two competing processes—Frenkel voiding (caused by the gradient of chemical potential) and Gibbs–Thomson effect (caused by different by value and sign curvatures of inner and external surfaces, leading to gradient of vacancy concentration). Gibbs–Thomson effect appears at once after void nucleation and tends to dissolve the just formed void. Possibility of voiding and its critical size is determined by the particle size and by diffusion parameter.

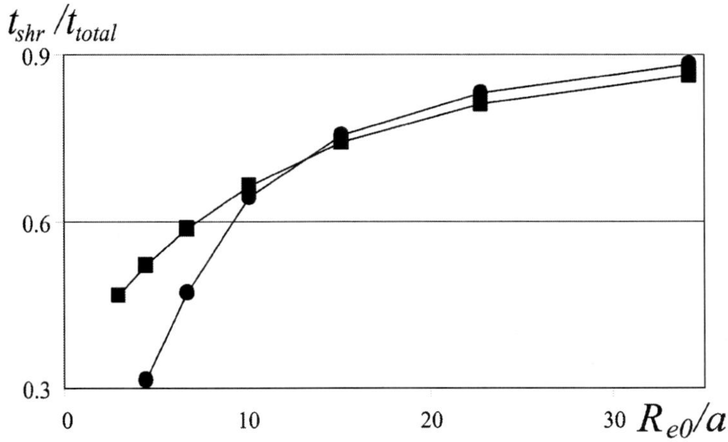

(a)

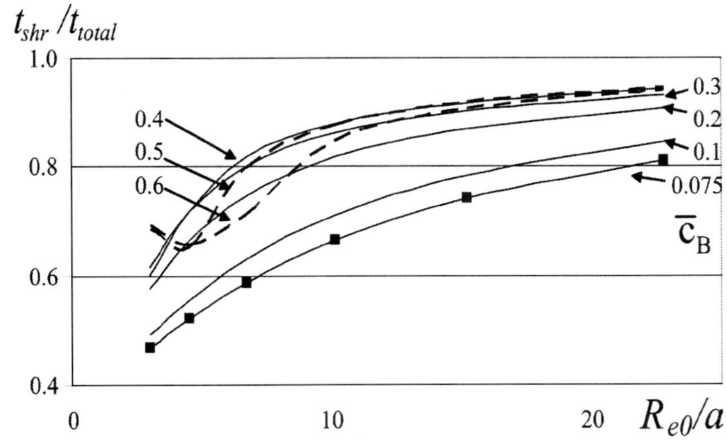

(b)

Figure 6.25 Ratio $t_{\rm shr}/t_{\rm total}$ as a function of the initial radius of the particle $R_{\rm e0}$ (a) for spheres (circles) and cylinders (squares) at $\overline{c_{\rm B}}$ = 0.075; (b) for cylinders at various average concentrations $\overline{c_{\rm B}}$.

At shrinking stage also two effects compete: Gibbs–Thomson effect leads to vacancy flux outwards. This vacancy flux generates unequal counter fluxes of main components leading to segregation of faster component near inner surface (inverse Kirkendall effect) and, respectively, shrinking becomes controlled by slow component.

Phenomenological modeling allows investigation of the full life-cycle of hollow nanostructure including formation and shrinking stages, with cross-over at some maximal void radius. Voiding is more effective in wires than in balls. Ratio of shrinking time to formation time is larger for wires.

6.5.2 Monte Carlo Simulation

In the above mentioned simulations the process practically stopped after formation of hollow nanoshell. It seems quite clear since formation is driven by chemical forces and collapse—by capillary ones. The latter are usually significantly less than chemical forces. Therefore, shrinking of nanovoid is typically a much slower process and can be observed at temperatures higher than for the process of formation. To describe the formation and collapse in one simulation, one needs higher temperatures. We took φ_{AA}^{H}, φ_{BB}^{H}, φ_{AB}^{H}, and initial interrelation of the components' radii $r_{BA} = 7a$, $r_e = 17a$, corresponding to average concentration $\bar{c}_B = 0.075$ (5745 B atoms, 76,676 A atoms) and to $\Phi = 0.9$. The process of void formation proceeded very quickly (during 1238 MCS). At that, the void formed in the centre of the nanoparticle and appeared to be unstable: just after having reached the maximal size (3314 sites) it collapsed (Fig. 6.26). To correlate the simulation results with simulation model we recalculated the number of empty sites into effective void radius R_i.

6.5.2.1 Shrinking and segregation kinetics in Monte Carlo simulation

Now we come back to synergy of curvature driven effects and inverse Kirkendall effect at the shrinking stage. The above described MC-model was used to simulate the collapse of the nanoshell with $r_i = 7a$, $r_e = 17a$ (total number of atoms is 76676 and the number of empty sites in the void is 5745) for ideal alloy at $\bar{c}_B = 0.75$ and K = 10. At high temperature (Φ^H) after 7000 MCS the void has shrunk by 9.58% (Fig. 6.27).

To investigate segregation effects we have used the nanoshells of an ideal alloy with $r_i = 4a$, $r_e = 8a$ at low temperature (Φ^L) with different ratios of frequencies K (10, 50, 100) and concentrations c_B

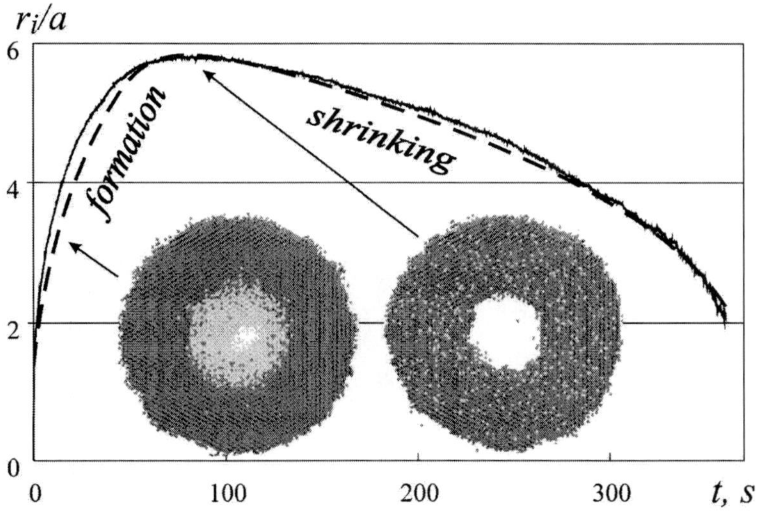

Figure 6.26 Nanoshell formation and collapse in one run—atomistic simulation and phenomenology. MC (solid line) and phenomenological (dashed line) plots are superimposed with respective time rescaling: 1 MCS = 62 ms. MC modeling at high temperature ($\varphi^{\mathrm{H}}_{\mathrm{AA}}, \varphi^{\mathrm{H}}_{\mathrm{BB}}, \varphi^{\mathrm{H}}_{\mathrm{AB}}$), K = 10 and small fraction of a more mobile B component (light dots, $\bar{c}_{\mathrm{B}} = 0.075$): (a) formation stage; (b) cross-over. Phenomenological plot calculated at $c_{\mathrm{V}}^{\mathrm{eq}} = 10^{-4}$, $c_{\mathrm{B}}^{\mathrm{init}} = 0.075$, $r_{\mathrm{i}}^{\mathrm{init}} = 1.4a$, $r_{\mathrm{BA}}^{\mathrm{init}} = 7a$, $R_{\mathrm{e}}^{\mathrm{init}} = 17a$ L = $0.75 \cdot 10^{-9}$ m, K = 10, $\lambda = 0$, $\alpha_{\mathrm{A}} = -4.5$, $\alpha_{\mathrm{B}} = -2$, $dt = 10^{-4}$ s.

(0.25, 0.50, 0.75, 1.00). We may conclude that the larger is K and/or c_{B}, the more intensive are the segregations of a more mobile component near the internal surface (Fig. 6.28) and the slower is the process of the nanoshell collapse (Fig. 6.39).

6.5.3 Conclusions to Section 6.5

Above treated theoretical scheme allows to predict evolution of hollow nanoshells with various average concentrations, at various temperatures. The very first stage of the life cycle—nucleation of the first void in the open nanosystem—metallic core—was recently described by us, jointly with F. Hodaj, in a recent letter.[4]

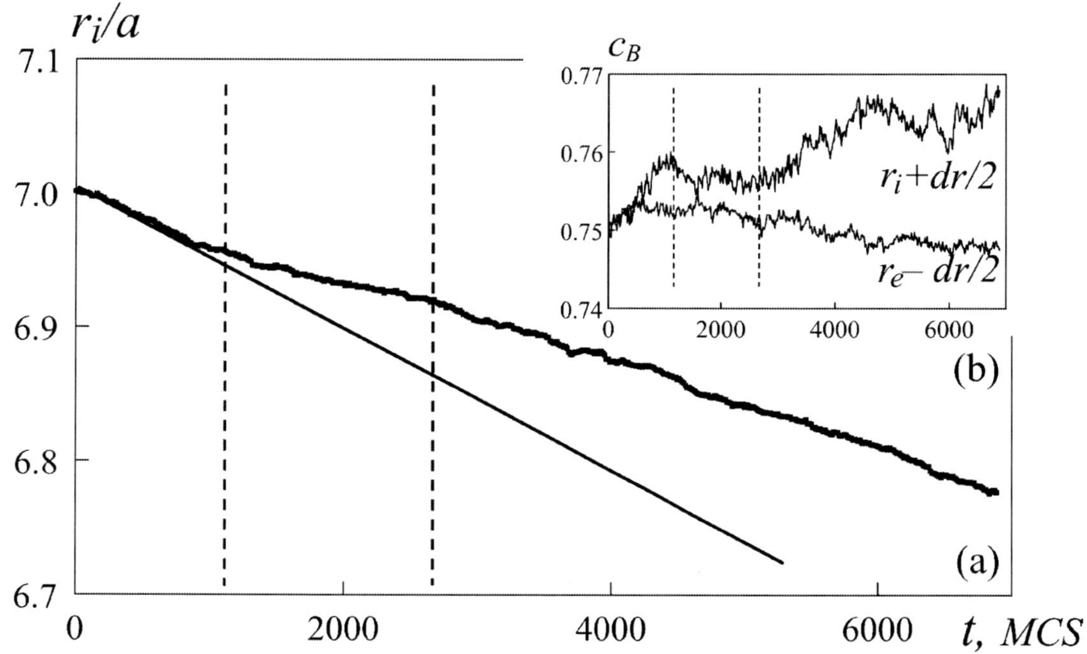

Figure 6.27 MC time dependence of inner radius (a) and of concentration near moving boundaries (b) at initial stage of the shrinkage of the A_1B_3 nanoshell ($r_i = 7a$, $r_e = 17a$, $dr \approx 1.7a$, $K = 10$, Φ^H): segregation of B species near the internal surface leads to slowing down the nanoshell shrinkage.

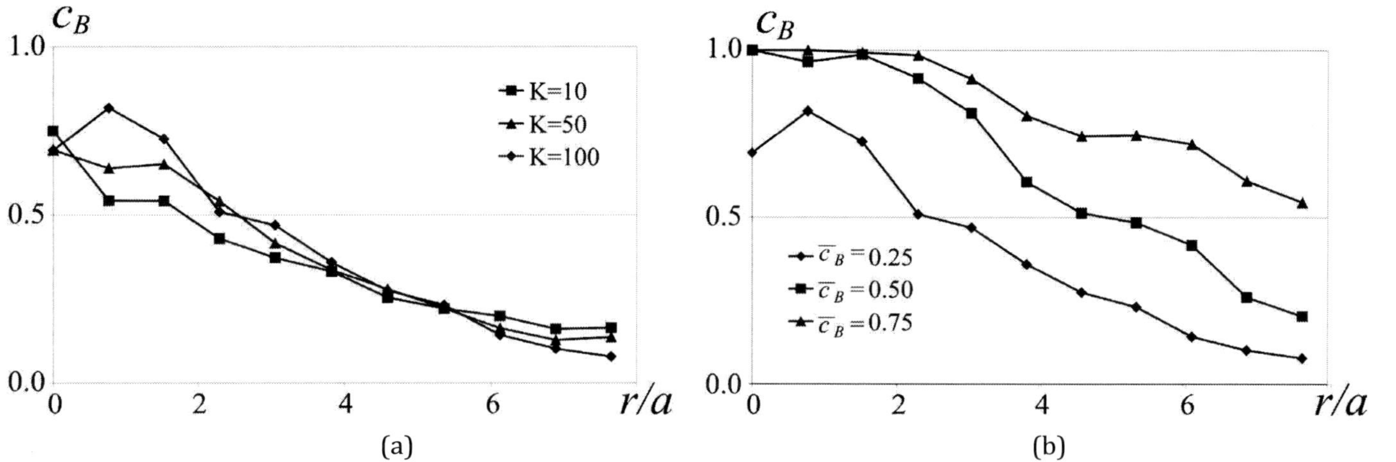

Figure 6.28 Radial concentration profile for a nanoparticle just after collapse: (a) at different ratios of the frequencies of components (average concentration $\overline{c}_B = 0.25$); (b) at different average concentrations (at $K = 100$).

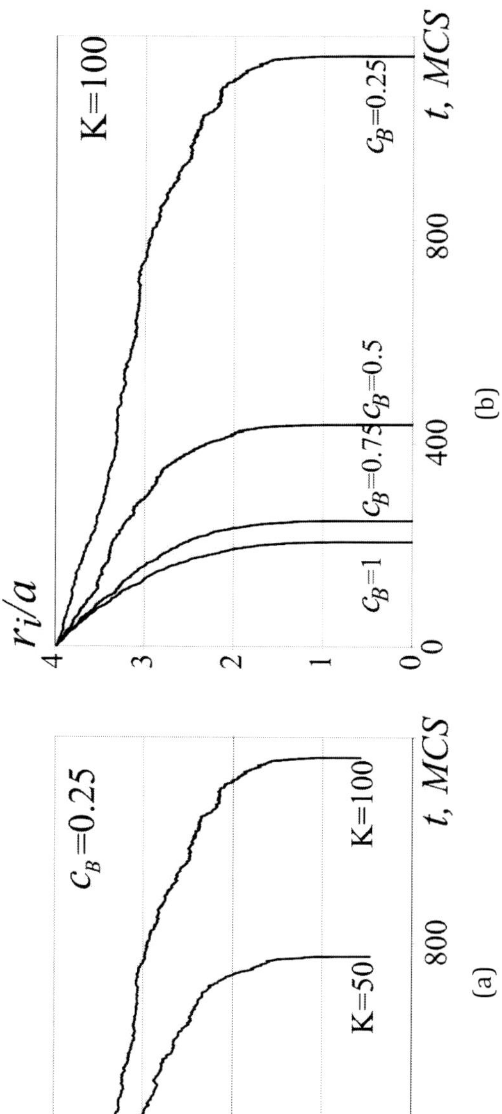

Figure 6.29 Collapse rate influenced by (a) ratio of frequencies and (b) volume fractions of components.

Acknowledgments

A large part of the presented results was obtained jointly with Prof. King-Ning Tu. Some of the results presented were discussed with late Prof. Ulrich Gösele. The work was supported by Fundamental Researches State Fund of Ukraine, Ministry of Education and Science of Ukraine.

References

1. A. D. Smigelskas, E. O. Kirkendall, *Trans. AIME*, **171**, 130 (1947).
2. Hideo Nakajima. *JOM*, **49**(6), 15–19 (1997).
3. Yin Yadong, Rioux Robert M, Erdonmez Can K, Hughes Steven, Somorjai Gabor A, Alivisatos A Paul, *Science*, **30430**, 711–714 (2004).
4. Yin Yadong, Erdonmez Can K, Cabot Andreu, Hughes Steven, Alivisatos A Paul, *Adv. Funct. Mater.*, **16**, 1389–99 (2006).
5. C. M. Wang, D. R. Baer, L. E. Thomas, J. E. Amonette, Jiji Antony, You Qiang, G. Duscher, *J. Appl. Phys.*, **98**, 094308 (2005).
6. R. Nakamura, D. Tokozakura, H. Nakajima, J. G. Lee, H. Mori, *J. Appl. Phys.*, **101**, 074303 (2007).
7. Fan Hong Jin, Knez Mato, Scholz Roland, Hesse Dietrich, Nielsch Kornelius, Zacharias Margit, Gösele Ulrich. *Nano Lett.*, 7, 993–997 (2007).
8. R. Nakamura, J.-G. Lee, H. Mori, H. Nakajima, *Phil. Mag.*, **88**, 257 (2008).
9. Nakamura R, Tokozakura D, Lee J-G, Mori H, Nakajima H, *Acta Mat.*, **56**, 5276–5284 (2008).
10. R. Nakamura, G. Matsubayashi, H. Tsuchiya, S. Fujimoto, H. Nakajima, *Acta Mat.*, **57**, 4261–4266 (2009).
11. K. N. Tu, U. Gösele, *Appl. Phys. Lett.*, **86**, 093111 (2005).
12. A. M. Gusak, T. V. Zaporozhets, K. N. Tu, U. Gösele, *Phil. Mag.*, **85**, 4445–4464 (2005).
13. A. M. Gusak, K. N. Tu, *Acta Mat.*, **57**, 3367–3373 (2009).
14. A. M. Gusak, T. V. Zaporozhets, *J. Phys.: Condens. Matter*, **21**, 415303 (2009).
15. F. Aldinger, *Acta. Metall.*, **22**, 923 (1974).
16. Y. E. Geguzin, *Diffusion Zone*, Nauka Publications, Moscow (1979).

17. Y. E. Geguzin, Yu I. Klinchuk, L. N. Paritskaya, *Fizika metallov i metallovedenie*, **43**, 602 (1977).
18. R. A. Masumura, B. B. Rath, C. S. Pande, *Acta Materialia*, **50**, 4535–4544 (2002).
19. I. M. Khalatnikov (Ed.) *Theory of Diffusive Decomposition of Solid Solutions*, Vol. 17 (1995).
20. K. N. Tu, J. W. Mayer, L. C. Feldman, *Electronic Thin Film Science*, MacMillan, Amsterdam (1992).
21. R. Kirchheim, *Acta Metall. Mater.*, **40**, 309 (1992).
22. M. A. Korhonen, P. Borgesen, K. N. Tu, C.-Y. Li, *J. Appl. Phys.*, **73**, 3790 (1993).
23. T. V. Zaporozhets, A. M. Gusak, K. N. Tu, S. G. Mhaisalkar, *J. Appl. Phys.*, **98,** 103508 (2005).
24. Annie T. Huang, A. M. Gusak, K. N. Tu, and Yi-Shao Lai, *Appl. Phys. Lett.*, **88**, 141911 (2006).
25. K. P. Gurov, A. M. Gusak, *Fizika metallov i metallovedenie*, **59**, 1062–1066 (1985).
26. A. M. Gusak, K. P. Gurov, *Proceedings of PTM-94, USA*, p. 1133 (1994).
27. A. M. Gusak, S. V. Kornienko, G. V. Lutsenko, Nonequilibrium vacancies in nanosystems. *Defects and Diffusion Forum*, **264**, 109–116, Trans Tech Publications, Switzerland (2007).
28. G. Brian Stephenson, *Acta Met.*, 10, 2663–2683 (1988).
29. M. J. H. van Dal, A. M. Gusak, C. Cserháti, A. A. Kodentsov, F. J. J. van Loo, *Phys. Rev. Lett.*, **86**, 3352 (2001).
30. F. H. vanDal, A. M. Gusak, C. Cserhati, A. A. Kodentsov, F. J. J. van Loo, *Phil. Mag. A*, **82**(5), 943–954 (2002).
31. A. A. Kodentsov, A. Paul, M. J. H. van Dal, C. Cserhati, A. M. Gusak, F. J. J. van Loo, Crit. *Rev. Sol. State*, **33** (2008), 210–233.
32. A. D. Marwick. *J. Phys. F*, **8**, 1849 (1978).
33. Marwick A. D. *J. Nucl. Mater.*, **135**, 68–76 (1985).
34. A. V. Nazarov, K. P. Gurov, *Fizika metallov i metallovedenie*, **37**, 496–500 (1973).
35. A. V. Evteev, E. V. Levchenko, I. V. Belova, G. E. Murch, **88**, 1525 (2008).
36. A. V. Evteev, E. V. Levchenko, I. V. Belova, G. E. Murch, *Phil. Mag.*, **87**, 3787 (2007).

37. A. V. Evteev, E. V. Levchenko, I. V. Belova, G. E. Murch, *Defect and Diffusion Forum*, Vol. 277, 21, Trans Tech Publications, Switzerland (2008).
38. A. M. Gusak, T. V. Zaporozhets, Yu O. Lyashenko, S. V. Kornienko, M. O. Pasichnyy, A. S. Shirinyan, *Diffusion-Controlled Solid State Reactions in Alloys, Thin-Films, and Nanosystems*, Wiley-VCH, Berlin (2010).
39. W. R. A. Johnson, P. J. White, *Phys. Rev. B*, **18**, 2939 (1978).
40. T. Pabisiak, A. Kiejna. *Solid State Commun.*, **144**, 324 (2007).
41. K. N. Tu, J. W. Mayer, L. C. Feldman, *Thin Film Science for Electrical Engineers and Materials Scientists*, Wiley, New York (1992).
42. J. R. Manning, *Diffusion Kinetics for Atoms in Crystals*, Van Nostrand, Princeton (1967).
43. A. M. Gusak, M. V. Yarmolenko, *J. Appl. Phys.*, **73**(10), 4881–4884 (1993).
44. O.M. Podolyan, T. V. Zaporozhets, *Ukr. J. Phys.*, **56**(9), 929–939 (2011).
45. G. Glodan, C. Cserhati, I. Beszeda, D. L. Beke, *Appl. Phys. Lett.*, **97**, 113109 (2010).
46. G. Glodan, C. Cserhati, I. Beszeda, D. L. Beke, *DIMAT 2011*, Dijon, France, Abstract O1.4.
47. Erdélyi Zoltán, Beke Dezső, *J. Mater. Sc.*, **46**(20), 6465–6483 (2011).
48. A. M. Gusak, F. Hodaj, T. V. Zaporozhets, *Phil. Mag.*, **91**(12), 741 (2011).

Chapter 7

Electrical Transport Properties of Doped Silicon Nanowires

Aya Seike and Iwao Ohdomari
Institute for Nanoscience and Nanotechnology, Waseda University, 1-6-1 Nishiwaseda, Shinjuku-ku, Tokyo 169-8050, Japan
a.seike@kurenai.waseda.jp

7.1 Introduction

Semiconductor technology contributes to sustained growth and success in the microelectronics and information technology industries. This requires continued significant improvements in ultra-high-performance transistors for aggressively scaled integrated circuits and systems on a chip. These are generally characterized as ultra-large scale integration (ULSI) at the circuit/system level, and nano–complementary metal oxide semiconductor (nano-CMOS) at the device level. In the past, such improvements have been achieved mainly by device shrinkage in a regime where iterative improvements were possible with decreases in length scales required. Now that device feature size has decreased deep into the nanometer regime, less than 10 nm, this iterative approach is no longer a valid approach

Silicon and Silicide Nanowires: Applications, Fabrication, and Properties
Edited by Yu Huang and King-Ning Tu

ISBN 978-981-4303-46-0 (Hardcover), 978-981-4303-47-7 (eBook)
www.panstanford.com

for performance improvement. For example, in aggressively scaled planar field effect devices such as metal-oxide-semiconductor field-effect transistors (MOSFETs), this results in high-leakage currents, short channel effects, threshold voltage (V_{th}) roll-off, and high resistance at source/drain contact.[1,2] To solve these problems, alternatives technologies qualitatively different than downsizing based on constant field scaling must be introduced (Fig. 7.1). This includes (1) Gate stack engineering including the introduction of (a) high-k alternative gate dielectric thin films to reduce gate leakage, and (b) metal gate electrodes to control V_{th}, (2) Channel engineering including strained-channels with Si/SiGe substrate,[3] stressor films as typified by SiC,[4,5] or III–V materials to enhance carrier mobility,[3] (3) Source-drain engineering including ultra-shallow junctions, and eventually, the introduction of metal source and drain materials. In 2007, several of those alternative technologies, such as Si/SiGe substrates, high-k alternative gate dielectric thin films, and metal gate electrodes were successfully implemented into high-performance ULSI products.[6] The introduction of three-dimensional (3-D) device architectures, that is, double gate, FINFETs, gate-all-around nanowire transistors, as an alternative to planar structures is being studied for the future generations of semiconductor technology enhancements. However, these technologies have only a limited and incomplete under-pinning by a validated science knowledge base. This is especially true for the 3-D device structures where there is not a scientific knowledge base that can be applied. Therefore the development of fundamental theories and knowledge for 3-D device structures is a significant and important challenge in moving forward into the future generations of ULSI and nano-CMOS technologies. The research in this chapter addresses a very important aspect of 3-D devices in general that is present in the nanowire MOSFETs that are discussed.

In the emerging 3-D transistor structures, the typical feature size is only a few tens of nanometers. In this regime, quantum effects such as quantum confinement and 1-D transport are significant. However, there is universal agreement that the 3-D transistors will differ from the planar bulk transistors in the following ways: (1) the transistor channel will be completely surrounded by semiconductor/dielectric interfaces, (2) the transistor channel has a better defined width and height dictated by performance goals. For this reason, at the beginning of this research program, we chose to investigate interfacial effects on dopant ions in 3-D configurations typified by nanowire MOSFETs. Our objective was to investigate the behavior of dopant ions inside confined channel regions, since their interface can act as sinks, and thereby induce segregation and/or pile-up of implanted dopant ions.[7–9] A number of studies have already revealed that strain at the Si/SiO_2 interface places a significant role in the extent of dopant atom pile-up and segregation.[10,11] Other important studies have also revealed that properly confined or channeled strain can enhance transconductance and carrier mobility.[12] These factors provided a rationale that the study on the stress in 3-D channel structures was both timely and significant. We have therefore initiated and completed a study on the effect of interfacial strain in carrier transport in protypical nanowire transistor structures.

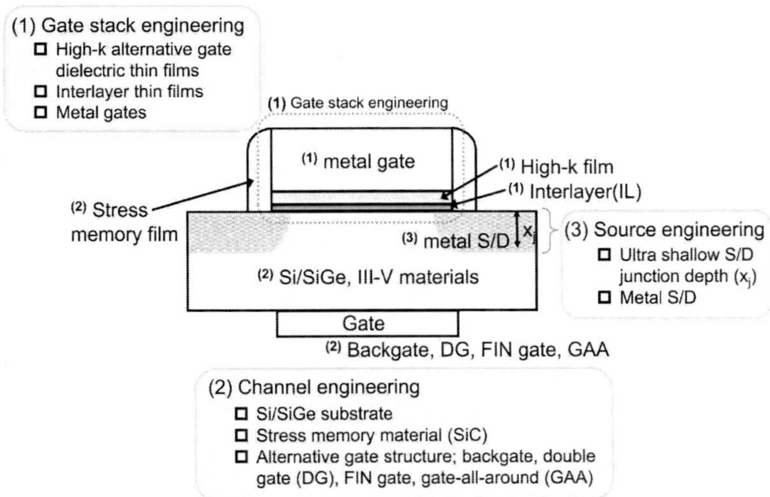

Figure 7.1 Schematic diagram of alternative technologies for high-performance transistors.

In the body of this chapter, we describe the results of comprehensive study directed toward gaining an increased understanding of carrier transport in *n*-type and *p*-type nanowire MOSFETs, and emphasizing correlations between controlled-stress introduced in the nanowire channel, and its effect on transistor performance.

7.2 Fabrication Processes and Electrical Measurements

7.2.1 Device Fabrication

To study effects of interfacial strain on carrier transport in 3-D nanowire geometries, Si nanowire FETs are fabricated using silicon-on-insulator (SOI) wafers as a device platform. The test devices are back gate control FETs with Si nanowires in a multi-channel arrangement. Figure 7.2 presents a schematic diagram of these Si nanowire FETs. P-type layers for nanowire fabrication were formed by separation process in oxygen implanted (SIMOX) wafers. The separated layers used in this study had a resistivity ranging from 7 to 14 Ωcm. The nanowire channels are in the <110> direction on (001) planar surface of the separated region. The initial thickness of buried oxide (BOX) is approximately 145 nm, and that of SOI layer is approximately 50 nm.

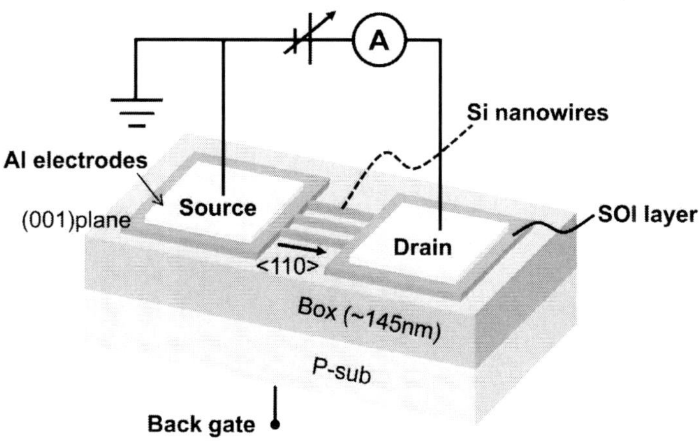

Figure 7.2 Schematic diagram of fabricated Si nanowire FETs.

Table 7.1 indicates the number of individual nanowires and their respective widths in each of the device wafers. The total projected widths of those multi-wire arrays are equal to eliminate geometry-dependent effects on the carrier conductivities within a given nanowire test device.

Table 7.1 Summary of targeted widths, number of nanowires, and the total projected widths of the nanowire FET arrays

Process targeted nanowire width (nm)	Number of nanowires	Total projected width (nm)
80	64	5000
155	32	5000
310	16	5000
2500	2	5000
5000	1	5000

The fabrication process steps for the Si nanowire FETs are presented in Fig. 7.3. Pads for source and drain electrodes are patterned by photolithography and dry etching using a $C_3F_8/O_2/Ar$ gas chemistry. Nanowires are then defined by electron beam (EB) lithography using the SAL601 SR-7 (1-methoxy-2-propyl acetate) negative EB resist (Rohm and Haas SAL601 SR-2). This is followed by dry etching using a $C_3F_8/O_2/Ar$ gas chemistry that we have developed in our laboratory. The Si nanowires are thermally oxidized at 850 °C and different times to control oxide-induced nanowire stress. Detailed oxidation conditions and thermal treatments after oxidation will be discussed in the next section of this chapter.

- Pad Lithography
- Pad Dry etching
- EB Lithography (Nanowires)
- Dry etching
- Oxidation (850°C, 3h)
- Channel Implantation (P for N-type, B for P-type 25keV, 5×10^{11}cm^{-2})

- Contact Lithography
- Contact Etching (w/ Buffered HF soln.)
- S/D Implantation (P for N-type, B for P-type 25keV, 5×10^{15}cm^{-2})
- Activation Annealing (950°C/10min, N_2 ambient)
- Metal Lithography
- Al deposition
- Al Lift-off
- IV measurement

Figure 7.3 Fabrication process steps for the Si nanowire FETs.

Phosphorus ions for the *n*-type nanowire FETs and boron ions for the *p*-type nanowire FETs are implanted into the nanowire channels with a dose of $5*10^{11}$ cm^{-2} at an energy of 25 keV to form n^{-}(minus) and p-(minus) transport channel regions. Contact holes in the bulk SOI at the ends of these nanowires are patterned by photolithography followed by dry etching. Phosphorus for the *n*-type, and boron for the *p*-type nanowire FETs are implanted into the source/drain contact pads with the dose of $5*10^{15}cm^{-2}$ for the formation, respectively, of n^+ and p^+ ohmic contacts to the nanowire channel regions. The implanted phosphorus and boron ions are thermally activated by furnace annealing at a temperature of 950 °C for 10 min in a N_2 ambient. Aluminum is then deposited and lifted off by acetone immersion and rinsing to define the contact geometries.

7.2.2 Methods of Electrical Characterization

I–V measurements are carried out for electrical characterization of the nanowire FETs. It is important to note that the fabricated nanowire FETs are turned on and off by a back gate voltage (V_{bg}) applied to the *p*-substrate. The transport mechanism is by majority carriers in accumulation mode, and this determines the device response properties. For example, transconductance (g_m) in the linear region is given by

$$g_m = \frac{\partial I_{ds}}{\partial V_{BG}} = \left(\frac{W}{L}\right)\mu C_i V_D, \tag{7.1}$$

where W is channel width, L is the channel length, μ is carrier mobility, C_i is the insulator capacitance, and V_D is the drain voltage; the source is grounded. The gate insulator capacitance is given by

$$C_i = \frac{\varepsilon_i}{t_{OX}}, \tag{7.2}$$

where ε_i is the insulator permittivity and t_{OX} is the physical thickness of gate dielectric film. Substituting Eq. 7.2 into Eq. 7.1 yields the following equation for transconductance:

$$g_m = \left(\frac{W}{L}\right)\left(\frac{\varepsilon_i}{t_{OX}}\right)\mu V_D. \tag{7.3}$$

Since W and t_{OX} in Eq. 7.3 are not constant, the value of g_m must be normalized for comparisons, and determination of stress effects. The following expression defines the channel width normalized transconductance:

$$g_{\mathrm{m}}\left(\frac{t_{\mathrm{OX}}}{W}\right)=\left(\frac{\varepsilon_{\mathrm{i}}}{L}\right)\mu V_{\mathrm{D}}. \tag{7.4}$$

In this study, the nanowires on a given SIMOX substrate wafer are connected in parallel. Therefore, Eq. 7.4 is written to included this device connection,

$$g_{\mathrm{m}}\left(\frac{t_{\mathrm{OX}}}{W_{\mathrm{m}}\times N}\right)=\left(\frac{\varepsilon_{\mathrm{i}}}{L}\right)\mu V_{\mathrm{D}}; \tag{7.5}$$

W_{m} is the measured wire width and N is the number of nanowires. The terms on the left-hand side of Eq. 7.5 are directly proportional to the carrier mobility (μ), and these terms, are hereafter treated as a measure of the channel mobility.

Raw data were then normalized by W_{m} and the gate oxide thickness (t_{OX}) as determined from cross-sectional TEM images (Fig. 7.4).

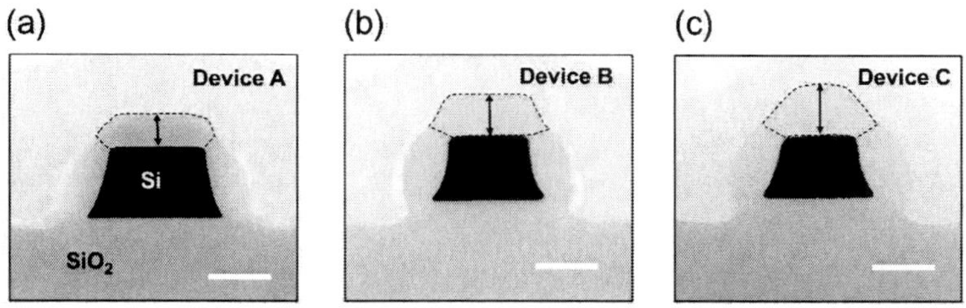

Figure 7.4 Cross-sectional TEM images of nanowires after the oxidation condition of (a) 850 °C/1 h (device A), (b) 850 °C/3 h (device B), (c) 850 °C/3 h + 1000 °C/30 min anneal (device C). The scale bar denotes 40 nm. The closed dashed lines denote the shapes, and the arrows denote the volumes of SiO_2 after oxidation and thermal treatment.

7.3 Introduction of Strain into Nanowire Channels by Oxidation, and Evaluation of Stress within Individual Nanowires

7.3.1 Stress Induced during Oxidation Using Pattern-Dependent Oxidation (PADOX) Theory

In the conventional fabrication processes for transistors, process temperatures are generally no higher than 1000 °C.[15] If the process temperature is above 960 °C, where viscoelastic relaxation takes place in SiO_2,[16] oxide films formed around the channel become viscous and completely relaxed, resulting minimal stress at the Si/SiO_2 interface. There is always a thermal component of the stress from differences in the expansion coefficients between Si and SiO_2.[17] This is generally smaller than oxide induced stress, so, there should be no significant strain in the channels processed, or exposed to temperatures >960 °C. However, when a process, especially, oxidation takes place below 960 °C, stress will accumulate at the Si/SiO_2 interface as the thermal oxidation proceeds. This is due in part to the difference in thermal expansion coefficient between Si and SiO_2,[16] and to intrinsic processes are well.[17] The strain at the interface is accumulated as oxidation proceeds, thus the oxidation induced strain is introduced inside the Si core. In this study, Si nanowires are oxidized at 850 °C to introduce strain inside Si nanowire channels, and different oxidation times and post-thermal treatments are performed to achieve, and thereby control stress levels for performance characterizations. The oxidation and post thermal conditions are summarized in Table 7.2. Oxidation conditions in dry O_2 at 1 atm are (1) at 850 °C for 1 h (device A), (2) at 850 °C for 3 h (device B), and (3) 850 °C for 3 h with a 30 min post-anneal at 1000 °C in an inert N_2 ambient (device C). The post annealing for device C are anticipated to relax stress significantly in the nanowires since the temperature is higher than 960 °C.

Table 7.2 Oxidation conditions for all devices

Device designation	Oxidation conditions (in dry O_2 at 1 atm)	Post-oxidation annealing conditions (in inert N_2)
Device A	850 °C for 1 h	N/A
Device B	850 °C for 3 h	N/A
Device C	850 °C for 3 h	1000 °C for 30 min

In this study, a pattern-dependent oxidation (PADOX)[13,14] has been used as a guideline to distinguish between controlled stresses under different processing conditions for nanowire FETs with a nanowire width of less than 80 nm. Since oxidation reduces W below the targeted patterning values in Table 7.1, nanowire FET devices will have smaller widths, hence the importance of PADOX as described below. PADOX, a characteristic phenomenon occurring with a patterned thin Si layers on SOI substrates, is associated with (i) a reduction of oxidation rates caused by volume expansion during oxidation and (ii) oxidation at the bottom of Si layer caused by the oxidant flux from the BOX layer. By recognizing the PADOX phenomena, effective values of transverse tensile stress can then be obtained for Si nanowire processing on SOI. In particular to evaluate the tensile strain profiles for nanowire FETs, three-dimensional molecular dynamics simulations have been used as a computer-aided design (CAD) tool.

Figures 7.4a,b,c display the cross-sectional TEM images of the fabricated nanowires for devices A, B, and C, respectively. As discussed above, the bottom channel widths of the nanowire (W_b) determined from the TEM images are changed from the patterned values in Table 7.1: (A) 85.4 ± 2.1 nm, (B) 70.5 ± 2.1 nm, and (C) 69.4 ± 2.1 nm, respectively. Note the following: (1) There is a relatively small volume expansion of SiO_2 at the top the Si nanowire in device A, and this is indicated by the closed dashed line (region A); (2) the volume expansion in region A and oxidation from the BOX layer resulting from PADOX is evident in devices B and C; and (3) the shape of the SiO_2 in device C is less well defined, and the vertical height of oxide in device C of region A is greater than in device B, as a result of the subsequent 1000 °C annealing in which viscoelastic relaxation takes place.[17,18] Both the shape of the SiO_2 and the increase in height of the oxide in device C indicate that the stress has been developed on top of the nanowire in devices B and C after oxidation, but it is significantly reduced in device C because of the viscoelastic relaxation during the high temperature anneal.

7.3.2 Three-Dimensional Molecular Dynamics Simulations of Stress Distributions in Nanowires

To evaluate the strain distribution inside a Si nanowire confined by thermally grown SiO_2 films, three-dimensional molecular dynamics simulations are performed.[19] The thermal oxidation is simulated by inserting O atoms in the midpoints of Si–Si bonds from the surface in layer-by-layer manner. In the simulation, the Si nanowire is directed along [1-10] direction, and the two dimensional periodic boundary conditions are introduced in [1-10] and [110] directions to allow a free expansion only along [001] direction. Figure 7.5 displays the horizontal stress tensor in the Si lattice obtained from these simulations. The region enclosed by the dashed line denotes compressive stress, and the region enclosed by the dark, solid line denotes tensile stress. The tensile stress predominantly occurs in the portion of the nanowire enclosed by the dark, solid line, where the oxide is allowed a free volume expansion in the [001] direction as defined in Fig. 7.5.[19] Compressive stress reduces the oxide viscosity and suppresses the oxidation.[14] As displayed in Fig. 7.5, compressive stress, which is localized at the bottom corners of the nanowire, causes oxidation retardation at the bottom Si edges. This result agrees well with the TEM images of the nanowires shown in Fig. 7.4a,b,c. In addition, the bottom Si edges are sharp as a result of oxidation retardation.

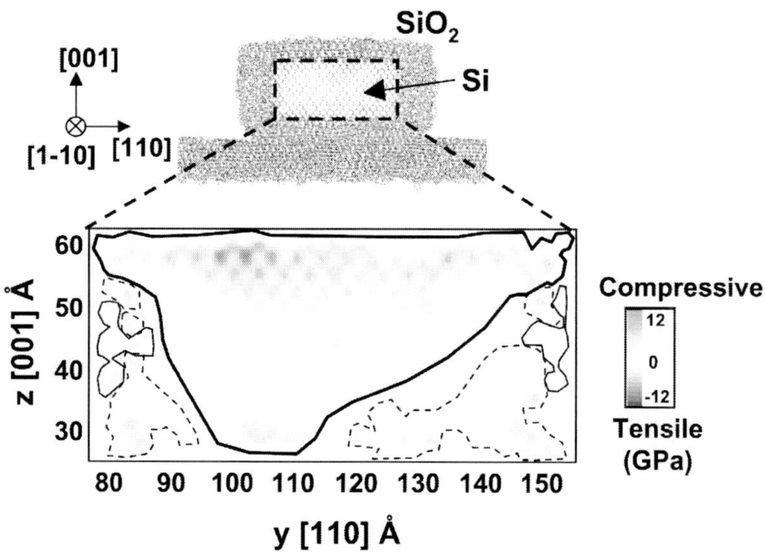

Figure 7.5 Molecular dynamics (MD) simulation results. Horizontal stress distribution in Si nanowires. The region enclosed by dashed line denotes compressive stress, and the region enclosed by solid line denotes tensile stress.

7.3.3 Evaluation of Induced Strain inside Si Nanowires by UV Raman Spectroscopy

Raman spectroscopy is a useful method for non-destructive strain characterization in Si structures, especially for evaluating the strain distribution in nanowires.[20] To qualitatively evaluate the strain distribution inside the fabricated nanowire devices and compare the results with MD simulations, Raman spectroscopy measurements are performed. An ultraviolet 354 nm laser is used in this study. Due to the laser mechanism and resonance effect, the sample measuring time can be reduced, and thus the sample heating effects as well.[21] The beam size was 100 μm*0.5 μm with an excitation energy of 150 mW at the laser tube output window; the penetration depth was 5 nm for these measurements. Figure 7.6 displays the Raman spectra for devices A, B, and C, as well as for a stress free Czochralski (Cz) silicon wafer, which serves as the supporting substrate of these three devices. Since the structure is the same in all the devices, all of the spectra have essentially the same Raman shifts caused by sample heating effects;[22] therefore, differences in the Raman response can be attributed to differences in the relative stress within the nanowires. The largest increase in the full-width half-maximum (FWHM) Raman spectrum and the largest Raman shift to the lower wavenumber are observed in device B, indicating that the largest amount of tensile stress exists in device B. Taking into account the penetration depth of the excitation source, this result agrees well with the tensile stress localized in the top part of the nanowire as obtained from the three-dimensional molecular dynamics simulation.[19]

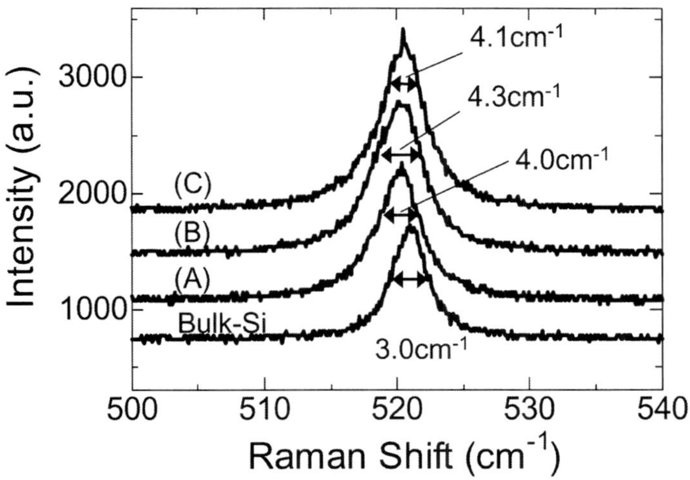

Figure 7.6 Intensity as a function of the Raman shift. Note the differences among the values of the FWHM Raman spectra of devices A, B, and C and bulk CZ-Si, which is the supported substrate of all three devices.

7.4 Electrical Characterization of Nanowire FETs

7.4.1 Effects of Stress on Carrier Transport in Nanowire FETs

The effects of stress on carrier transport have been reported in Ref. 23. The electron mobility, and hence the transconductance, is enhanced by either (i) transverse (in-plane) tensile stress, or (ii) the longitudinal compressive stress in Si (001)plane.

In Si, the conduction band edges exist along Δ; about 0.15 in from the boundary point χ at (2π/a)(100), and each conduction band edge has a spheroidal energy surface along the equivalent <100> direction in the Brillouin zone.[24,25] The spheroidal constant energy ellipsoids have an anisotropic mass, which is determined by cyclotron resonance measurements.[24] The cyclotron effective masses of these spheroidal constant energy surfaces are given by[25]

$$\frac{1}{m_C^*}=\sqrt{\frac{m_t(1-\gamma^2)+m_{.}\gamma^2}{m_t^2 m_{.}}}$$

where m_c^* is the cyclotron effective mass, γ is the directional cosine between the magnetic field and k_z (k = 0 at the conduction band edge), m_t is the transverse mass of the spheroids, and m_l is the longitudinal mass of the spheroid surfaces. For silicon, m_t and m_l are

$$m_t = 0.19m_0, \text{v}m_l = 0.98m_0,$$

where m_0 is the free electron rest mass in vacuum.

In the conduction band, the valleys are six-fold degenerate (Δ_6) with no applied strain. When a transverse tensile stress or the longitudinal compressive stress is applied in the Si(001) direction, the Δ_6 valleys are split into two-fold degenerate valleys (Δ_2) along the k_z direction, and four-fold degenerate valleys (Δ_4) in $k_x\, k_y$ plane of the Brillouin zone.[25] MOSFET operation also induces band splitting because carriers in the inversion layer are quantized in the MOS plane, and thereby form a two-dimensional electron gas.[25]

Due to the anisotropy of the effective mass of electrons in the conduction band, Δ_2 has the heavier effective mass of m_l perpendicular to the MOS interface and the lighter effective mass of m_t parallel to the MOS interface, and Δ_4 has m_t perpendicular to the MOS interface and m_t + m_l parallel to the MOS interface. Mobility along the MOS interface is therefore higher in the Δ_2 valleys than in the Δ_4 valleys due to this anisotropy.

Without strain, the electron population at room temperature is essentially the same in Δ_2 valleys and Δ_4 valleys. This is because the density of state in Δ_4 valleys is higher than in Δ_2 valleys and the ground state of subband in Δ_2 valleys (subband energy) is lower than Δ_4 valleys, resulting in comparable equal electron populations in both the Δ_2 and Δ_4 valleys. When a transverse tensile stress, or the longitudinal compressive stress is introduced in Si(001), (i) the sub-band energy for Δ_2 valleys is lowered, resulting in the preferential electron population in the Δ_2 valleys and additionally, (ii) intervalley scattering is suppressed due to the Δ_2 sub-band energy lowering when the sub-band energy difference between Δ_2 and Δ_4 is larger than the phonon energy contributing to the intervalley scattering. On the other hand, transverse compressive stress or longitudinal tensile stress in the Si(001) direction, produces an electron mobility degradation due to preferential electron population in Δ_4 valleys with a heavier transverse conductivity mass.

7.4.2 Electrical Characterization

7.4.2.1 Potential distribution inside the nanowire channels

It is important to specify where carrier transport takes place, since the fabricate nanowire devices are operated in the accumulation mode with a back gate bias. The MD simulation results show a small compressive stress can degradate the carrier mobility, at the back gate side. Thus, the potential distribution inside nanowire must be calculated by the Poisson and Continuity Equation Solver (PISCES)-II[26] to confirm that carrier transportation takes place in nanowire uniformly, and in the direction anticipated by the fabrication process and wafer alignment. Figure 7.7a is the schematic of the model used for this simulation, and Fig. 7.7b shows the electric potentials plotted against the depth with the nanowire device. W and t_{OX} for device B, as determined from the TEM image, were used for this calculation. The doping level of the Si nanowire is set equal to $5*10^{11}cm^{-2}$ of phosphorous, and V_{bg} is set at 40 V. As shown in the figure, the electrical potential is completely flat in the entire Si area (Fig. 7.7a), so that there is no significant potential variation inside the Si nanowire. Additionally, and equally important, carrier transport is not localized in a filament, or at the perimeter of the nanowire, but instead is uniform over the entire cross-sectional area.

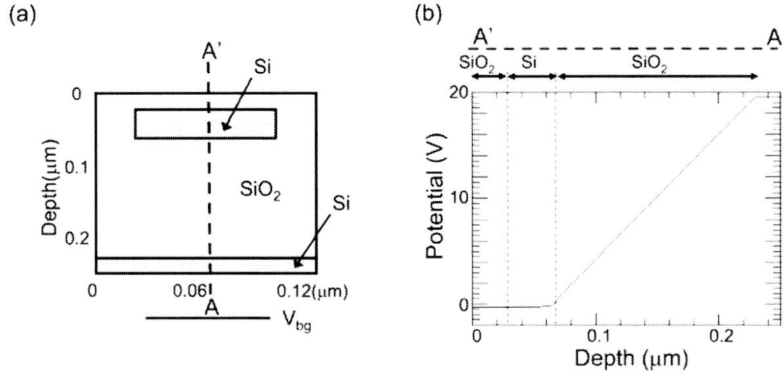

Figure 7.7 Simulation results. (a) Initial profile of the simulated nanowire device. The schematic diagram denotes the structure of nanowire device; *y* direction indicates the depth in μm, and *x* direction indicates the width in μm. The closed solid line denotes the Si nanowire (80 nm*20 nm), which is encapsulated with the SiO_2. (b) Electric potential against the depth of the nanowire device along the A'–A line.

7.4.2.2 I–V characteristics of nanowire FETs

I–V measurements were carried out to determine differences in the electrical properties that result from different levels of controlled strain in the three nwFETs that have characterized and discussed above. Figure 7.8 displays the I_{ds}–V_{bg} characteristics for both *n*-type nwFETs (N-nwFETs) and *p*-type nwFETs (P-nwFETs). It can be seen that the drain current (I_{ds}) for P-nwFETs increases as the back gate bias (V_{bg}) decreases below 0.0 V, and that the I_{ds} for N-nwFETs increases as V_{bg} increases above 0.0 V, indicating prototypical accumulation mode behavior. The values of transconductance (g_m) are normalized by W_b and t_{OX} as determined from the cross-sectional TEM images. The normalized values of g_m obtained from the analysis of these plots are (A) $1.06*10^{-7}$ S/μm, (B) $1.71*10^{-7}$ S/μm, (C) $7.47*10^{-8}$ S/μm for the P-nwFETs, and (A) $1.03*10^{-7}$ S/μm, (B) $4.65*10^{-7}$ S/μm (C) $1.84*10^{-7}$ S/μm for the N-nwFETs. The highest values of g_m are obtained in device B for both the N-nwFETs and P-nwFETs. The lower values of g_m in device C are attributed to the strain relaxation occurring during the post-oxidation annealing, as evident in the shape relaxation in region A shown in Fig. 7.4c.

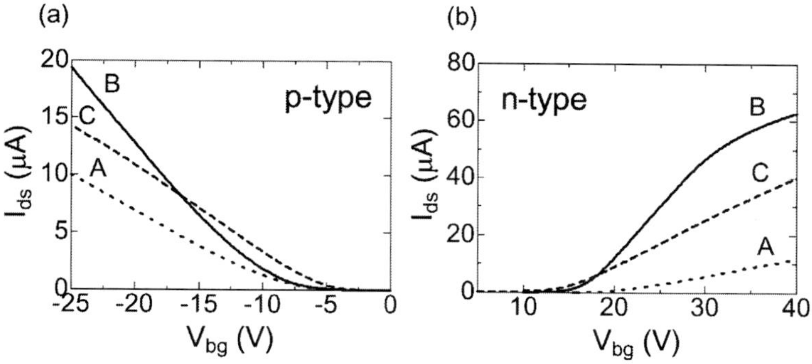

Figure 7.8 I_{ds}–V_{bg} characteristics of (a) *p*-type nanowire FETs and (b) *n*-type nanowire FETs. Nanowires are oxidized at (A) 850 °C/1 h (dotted lines), (B) 850 °C/3 h (solid lines), and (C) 850 °C/3 h + 1000 °C/30 min annealing (dashed lines).

Figure 7.9 shows normalized g_m values for each device. Open circles denote the corrected g_m values, and solid squares with a dashed line denote average values of the corrected g_m in each device. The average values of the normalized g_m in each device are (A) 1.07*10^{-7} S/μm, (B) 1.66*10^{-7} S/μm, (C) 6.29*10^{-8} S/μm for P-nwFETs, and (A) 1.46*10^{-7} S/μm, (B) 4.40*10^{-7} S/μm, (C) 1.83*10^{-7} S/μm for N-nwFETs. The values of g_m obtained in device B are larger, by a factor of 1.5 for P-nwFETs, and 3.0 for N-nwFETs, compared with the values obtained, respectively, in device A.

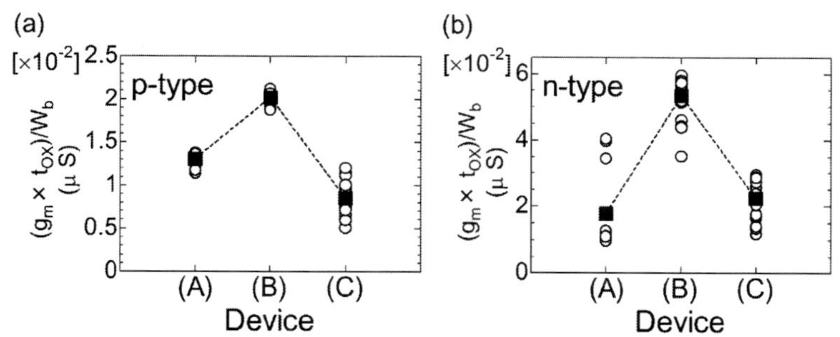

Figure 7.9 Values of g_m in each device for both (a) *p*-type and (b) *n*-type nanowire FETs. Open circles denote the corrected g_m values and solid squares denote the average values of g_m in each strain-controlled device.

Taking into account the tensile stress level differences in the nanowire channels as found in the Raman spectra, and determined from the MD simulation, the electric characterization results agree well with the g_m enhancements in both N-nwFETs and P-nwFETs, and the g_m enhancement in device B can be explained by band splitting and suppression of inter-valley scattering due to the transverse tensile stress in nanowires. The degradation of g_m in device C can be explained by the viscoelastic relaxation build-up of an increased compressive nanowire stress.

7.4.2.3 Size dependence of transconductance on nanowire size

The values of g_m in device B are plotted against the designed wire width (W_{mask}), as displayed in Fig. 7.10. Note that all the fabricated nanowire transistors has the same total wire width ($W_{total} = W_{m*}N$). Open circles denote the raw data of g_m and solid triangles denote the average values of g_m in each W_{mask}. The values of g_m for smaller W_{mask} are greater compared to the ones in larger W_{mask}. This corresponds to a relatively larger amount of tensile stress in the smaller W_{mask} nanowires, especially when the wire width is reduced to less than 150 nm. Note the values of g_m for W_{mask} = 80 nm are slightly overestimated because the processed wire width (W_m) is smaller than W_{mask}, resulting in a smaller W_{mask}. However, this effect is sufficiently small and can be neglected.

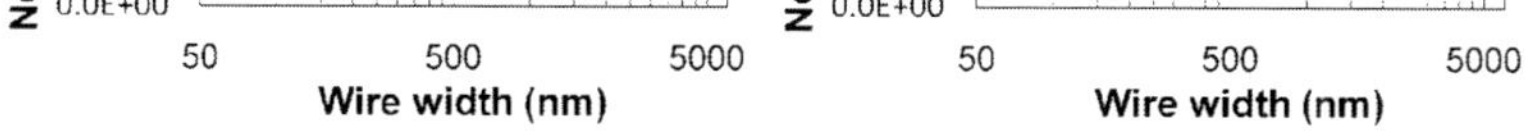

Figure 7.10 The value of g_m in sample B as a function of wire width; open circles denote the raw data of g_m and solid triangles denote the average values of g_m in each wire width.

7.5 Summary

This chapter has presented a comprehensive study directed toward gaining an increased understanding of electron and hole transport in

n-type and p-type strained nanowire MOSFETs that are targeted for applications in nano-CMOS circuits and systems for implementation as early as the 32 nm process node, and then extending into the 10 nm process node regime. In this chapter, we have treated 3-D systems for several tens to one-hundred nanometers, where quantum effects do not have to be addressed. In this nanoscale regime, stress plays the most important role in the carrier transport. Therefore, we studied the effect of stress on carrier transport, especially for channel transconductance (g_m). We now highlight the following results:

(1) Carrier transport in both n-type and p-type nanowire transistors can be enhanced by applying the appropriate stress direction and stress level inside the nanowire.
(2) Stress can be introduced and controlled by carefully selecting the oxidation and annealing conditions.
(3) Downsizing the diameter of the nanowire controls the stress gradient and ensures that a high level of stress is present across the entire nanowire cross-sectional area, thereby adding additional mobility and transconductance enhancement.

References

[1] J. R. Tucker, C. Wang, P. S. Carney, *Appl. Phys. Lett.*, **65**, 618 (1994).

[2] J. P. Snyder, C. R. Helms, Y. Nishi, *Appl. Phys. Lett.*, **67**, 1420 (1995).

[3] S. Takagi, T. Tezuka, T. Irisawa, S. Nakaharai, T. Numata, K. Usuda, *et al.*, *Solid State Electron.*, **51**, 526–536 (2007).

[4] T. Ghani, M. Armstrong, C. Auth, M. Bost, P. Charvat, G. Glass, *et al.*, *Tech. Dig.—Int. Electron Devices Meet.*, **2003**, 978 (2003).

[5] K-W. Ang, K-J. Chui, V. Bliznetsoc, C-H. Tung, A. Du, N. Balasunramanian, *et al.*, *Appl. Phys. Lett.*, 86, 093102 (2005).

[6] K. Mistry, *et al.*, "A 45 nm logic technology with High-k + metal gate transistors, strained silicon, 9 Cu interconnect layers, 193 nm dry patterning and 100% Pb-free packaging" IEDM Technical Digest, 2007, pp. 247–250.

[7] H.-H. Vuong, C. S. Rafferty, S.A. Eshraghi, J. Ning, J. R. McMacken, S. Chaudhry, et al., *J. Vac. Sci. Technol. B*, **18**(1), 428 (2000).

[8] P. B. Griffin, S. W. Crowder, J. M. Knight, *Appl. Phys. Lett.*, **67(4)**, 482 (1995).

[9] J. S. Johannessen, *et al.*, *J. Appl. Phys.*, **49**(8), 4453 (1978).

[10] T. Kimura, M. Hirose, Y. Osaka, *J. Appl. Phys.*, **56**, 932 (1984).

[11] N. R. Zangenberg, J. Fage-Pedersen, J. L. Hansen, A. N. Larsen, *J. Appl. Phys.*, **94**(6), 3883 (2003).

[12] S. Takagi, T. Maeda, T. Numata, T. Mizuno, K. Usuda, A. Tanabe, *et al.*, Proceedings IEEE 2004 of the International Conference on Microelectronic Test Structures, IEEE, New York, 2004, Vol. 17, p.133.

[13] M. Nagase, A. Fujiwara, K. Yamazaki, Y. Takahashi, K. Murase, K. Kurihara, *Microelectron. Eng.*, **41/42**, 527 (1998).

[14] M. Uematsu, H. Kageshima, K. Shiraishi, M. Nagase, S. Horiguchi, Y. Takahashi, *Solid-State Electron.*, **48**, 1073 (2004).

[15] S. M. Sze, *Semiconductor Devices*, 2nd ed., John Wiley & Sons, New Jersey (2002).

[16] E. P. EerNisse, *Appl. Phys. Lett.*, **30**, 290 (1977).

[17] J. T. Fitch, C. H. Bjorkman, G. Lucovsky, F. H. Pollak, X. Yin, *J. Vac. Sci. Technol. B*, **7**(4), 775 (1989).

[18] J. T. Fitch, G. Lucovsky, *J. Vac. Sci. Technol. B*, **7**(2), 153 (1989).

[19] H. Ohta, T. Watanabe, I. Ohdomari, *Jpn. J. Appl. Phys.*, Part 1, **46**, 3277 (2007).

[20] S. Piscanec, M. Cantoro, A. C. Ferrari, J. A. Zapien, Y. Lifshitz, S. T. Lee, S. Hofmann, J. Robertson, *Phys. Rev. B*, **68**, 241312 (R) (2003).

[21] A. Ogura, K. Yamasaki, D. Kosemura, S. Tanaka, I. Chiba, R. Shimidzu, *Jpn. J. Appl. Phys.*, Part 1, **45**, 3007 (2006).

[22] I. D. Wolf, *J. Raman Spectrosc.*, **30**, 877 (1999).

[23] T. Vogelsang, K. R. Hofmann, *Appl. Phys. Lett.*, **63**, 186 (1993).

[24] C. Kittel, *Introduction to Solid State Physics*, 8th ed., John Wiley and Sons, New York (2005).

[25] T. Hamaguchi, *Handoutai Buturi*, Asakura Shoten, Tokyo, p. 257 (2001).

[26] M. R. Pinto, C. S. Rafferty, R. W. Dutton, "PISCES-II: Poisson and Continuity Equation Solver," Stanford University Technical Report, 1984.

Chapter 8

Silicon Nanowires and Related Nanostructures as Lithium-Ion Battery Anodes

Liangbing Hu, Lifeng Cui, Seung Sae Hong, James McDonough, and Yi Cui
Department of Materials Science and Engineering, Stanford University, Stanford, CA 94305, USA.
yicui@stanford.edu

8.1 Lithium-Ion Batteries and Different Types of Anodes

Energy is used ubiquitously in current society and energy storage is unquestionably important. Batteries represent a key type of energy storage technology which is widely used. It is an old technology and many types of batteries have been developed historically. Figure 8.1a provides a comparison of energy storage devices in terms of gravimetric and volumetric energy densities, which are two key parameters in most applications in addition to cycle and calendar life, cost and safety.[1] In general, batteries can store much more energy per volume or weight when compared to other forms of energy

Silicon and Silicide Nanowires: Applications, Fabrication, and Properties
Edited by Yu Huang and King-Ning Tu

ISBN 978-981-4303-46-0 (Hardcover), 978-981-4303-47-7 (eBook)
www.panstanford.com

storage devices such as capacitors and supercapacitors.[2] There are two types of batteries: primary batteries and secondary batteries. Secondary batteries are rechargeable, while primary batteries are not. For example, lead acid batteries, widely used in vehicles, can be cycled for thousands of cycles. As shown in Fig. 8.1a, Li-ion batteries have the highest energy densities of the group, making them the most promising candidates for next generation, advanced energy storage applications. They can operate within a higher voltage stability window and have a much lower self-discharge rate than other types of rechargeable batteries.

Li-ion batteries can have different shapes: cylindrical, coin-shaped or flat. They have a multilayered device structure including encapsulation materials, current collectors (typically Al ~15 μm for the positive electrode and Cu ~15 μm for the negative electrode), a positive electrode ~100 μm, a negative electrode ~100 μm, and a separator ~20 μm (Fig. 8.1b). Compared with other types of batteries, Li-ion batteries are more complex to manufacture. In industry, the electrodes materials are generally mixed with conductive additives for enhanced electrical conductivity and made into viscous slurries. The electrode slurries are then roll-to-roll coated onto metal substrates. Then, the electrodes and separators are rolled together into a cylindrical shape in a dry room, followed by electrolyte injection and encapsulation.[3] Due to the complex manufacturing process and the relatively expensive materials, Li-ion batteries are more expensive than other secondary batteries.

The operation of Li-ion batteries is shown in Fig. 8.1c. Li ions are shuttled back and forth between negative and positive electrodes during charge-discharge cycles. During the charging process, Li-ions are extracted from positive electrode materials, normally Li-metal oxides, and inserted into negative electrode materials, normally graphite. The process is reversed during the discharging process. The voltage of the cell is determined by the Gibbs free energy difference. The electrode potentials for various electrode materials can be found in handbooks. The ideal battery voltage is the potential difference between the two electrodes when the cell is charged or discharged infinitely slowly. However, the voltage deviates from the theoretical value when the cells are operating quickly and away from equilibrium.[4] Ideally, the discharge voltage is the same as the charge voltage. The difference in the real, non-ideal case is reflected in a voltage loss, which is typically due to the ohmic resistance

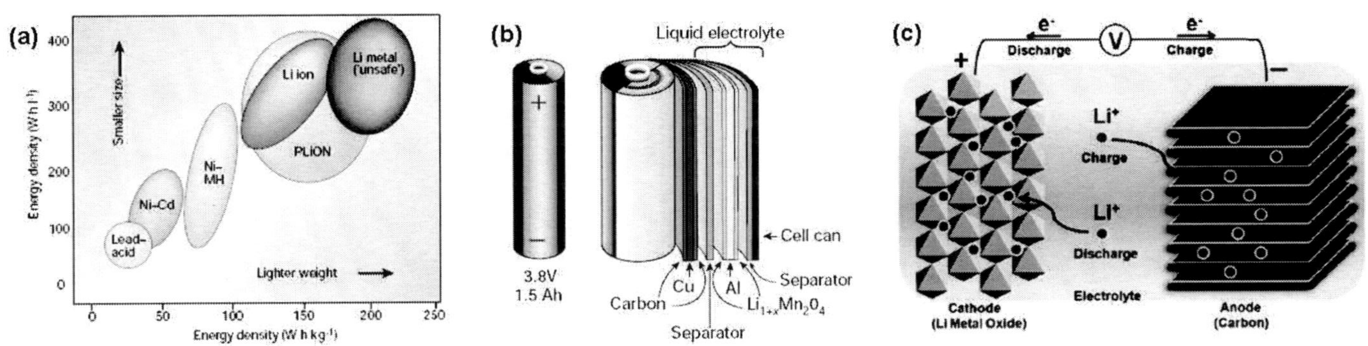

Figure 8.1 (a) Comparison of the types of batteries in terms of gravimetric and volumetric energy densities with units of Wh/kg and Wh/L, respectively. (b) Schematic drawing of the shape and the components of a cylindrical Li-ion battery. (c) Schematic diagram of a conventional Li-ion battery showing the typical cathode and anode structure. During the charging process, Li ions are extracted from the cathode and inserted into anode.

and kinetic limitations with regards to electron transport and ion diffusion among other factors. The separators should be efficient ionic conductors and electrical insulators and typically have a trilayer structure with thicknesses ~20 μm.[5] The liquid electrolyte consists of a solid lithium salt such as $LiPF_6$, dissolved in a organic solvents such as a mixture of carbonates, amides and imides.[6]

For the new generations of rechargeable Li-ion batteries, the advances in electrode materials are crucial. Figure 8.2 lists the various electrodes materials in terms of potential *versus* Li^+/Li^0 and their theoretical capacities. The figure can be used to calculate the theoretical voltage and energy density for various combinations of positive electrodes and negative electrodes. Engineering the shapes of the electrodes, especially into nanoscale materials, could benefit Li-ion battery applications. Nanosized electrodes comprised of materials from Fig. 8.2 would have the following benefits: better strain release and shorter diffusion lengths for ions for high power applications, among the others. In terms of mechanisms, there are three unique types that electrodes may exhibit: an intercalation mechanism in electrodes such as graphite and lithium titanate; an alloy mechanism in electrodes such as Si, Sn, and Al, and a conversion mechanism in electrodes such as CoO, NiO, and Fe_3O_4. Current industrial applications tend to utilize graphite as the anode, and $LiCoO_2$, $LiMn_2O_4$, $LiFePO_4$, or other Li metal oxides as the cathode, due to their low costs and appropriate voltage stability windows

8.2 Advantages and Challenges of Silicon Anodes

Due to its highest known theoretical capacity (~4200 mAh/g *vs.* ~370 mAh/g for graphite) and extreme natural abundance on earth, Si holds great excitement as an anode material for Li-ion batteries. Si has been investigated as a potential candidate for Li-ion batteries since the 1970s, and it has grabbed more attention in the past few years, due to the applications of nanostructured Si for Li-ion batteries.[7] A few studies have shown large capacities, ~4000 mAh/g, which is close to the theoretical capacity.

Coulometric titration studies have been carried out for Li–Si system. The Si/Li phase diagram is shown in Fig. 8.3a. There are multiple phases when T < 100 °C, which apply for most real Li-ion

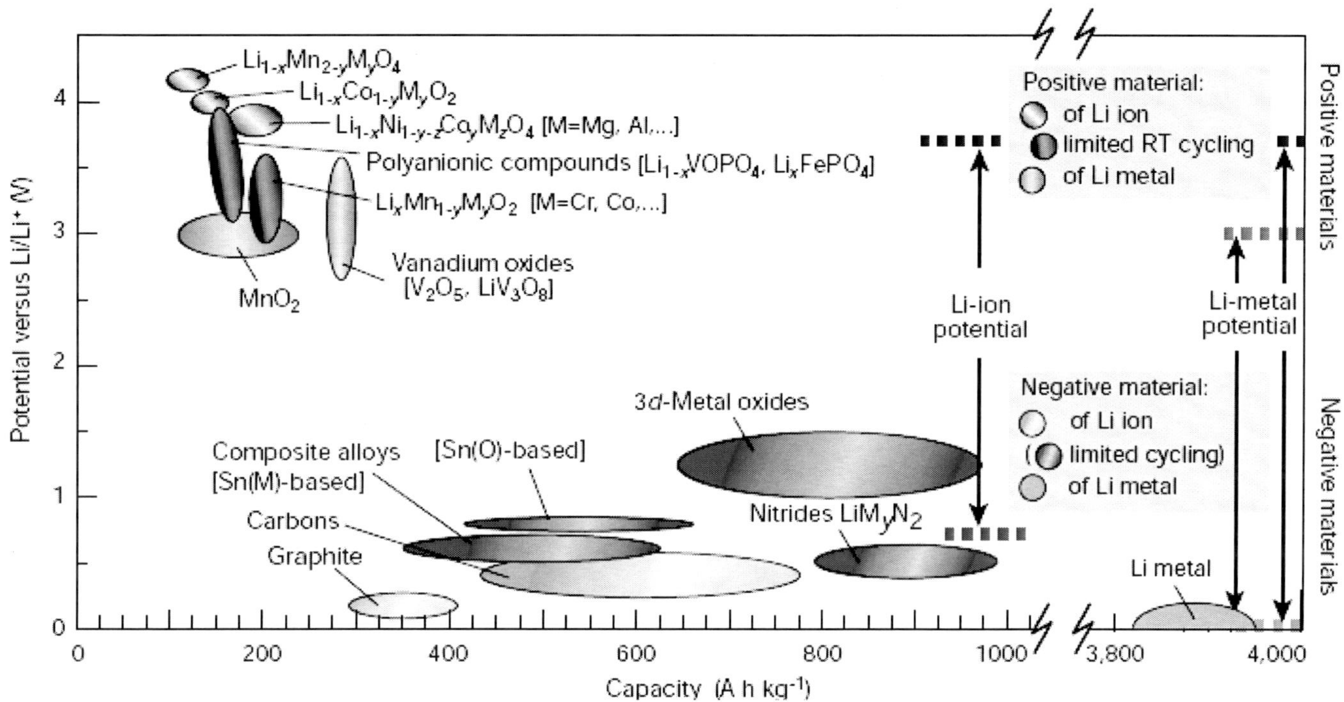

Figure 8.2 Plot of various types of electrode materials in terms of potential *versus* Li^+/Li and capacity. The battery voltage is the potential difference between the positive electrode and the negative electrode.

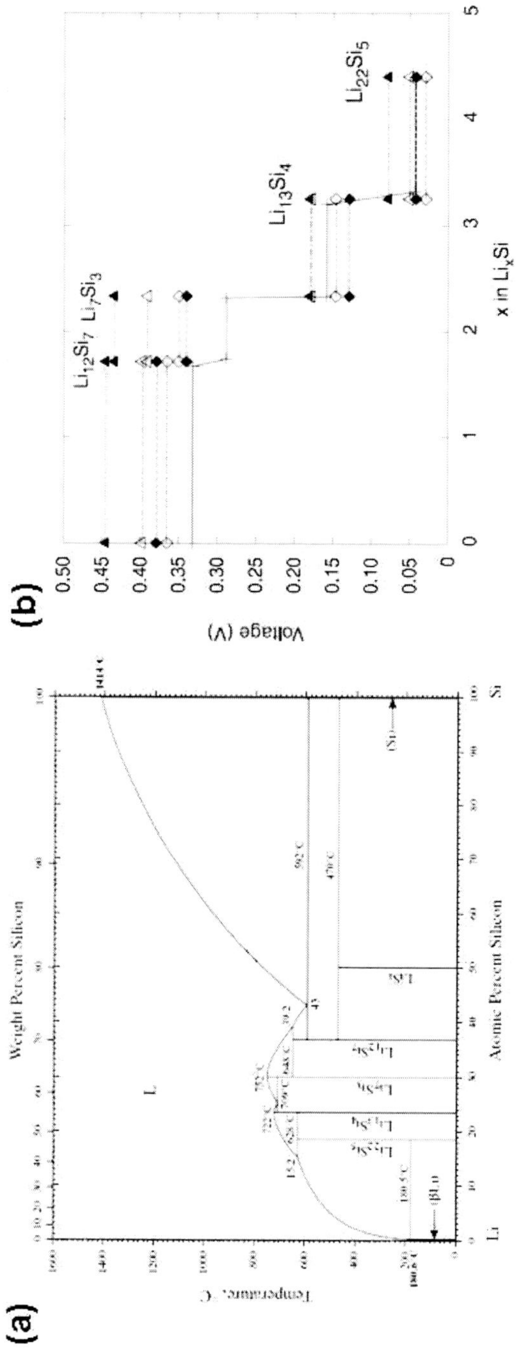

Figure 8.3 (a) Li–Si equilibrium phase diagram. (b) Voltage profile of the reaction of Li with Si.

batteries. Various phases form during the charging and discharging of the Si anode with Li-ions. The potentials of different phases *versus* Li^+/Li^0 are plotted in Fig. 8.3b. The capacity can be easily obtained with the molar mass. For $Li_{22}Si_5$, a theoretical capacity of 4200 mAh/g is achieved.

One major challenge for Si anodes is its high volume expansion, which results in several problems for its use as an anode: (1) disconnection of active electrode materials with the current collector which leads to fast decay of the capacity; (2) unstable solid–electrolyte interphase (SEI) which causes continuous loss of the electrolyte and limits the Columbic efficiency (CE) and the cycling performance. These phenomena have been predicted in theory and widely observed in experiments. Structural change will accompany Li-ion insertion into Si anodes. Dahn *et al.* have carried out a study about the volume change of Si upon Li-ion insertion (Fig. 8.4).[8] The volume changes are almost linearly correlated to the number of Li ions inserted. The results are useful for rational design of Si anodes to accommodate the volume expansion for improving the cycling performance.

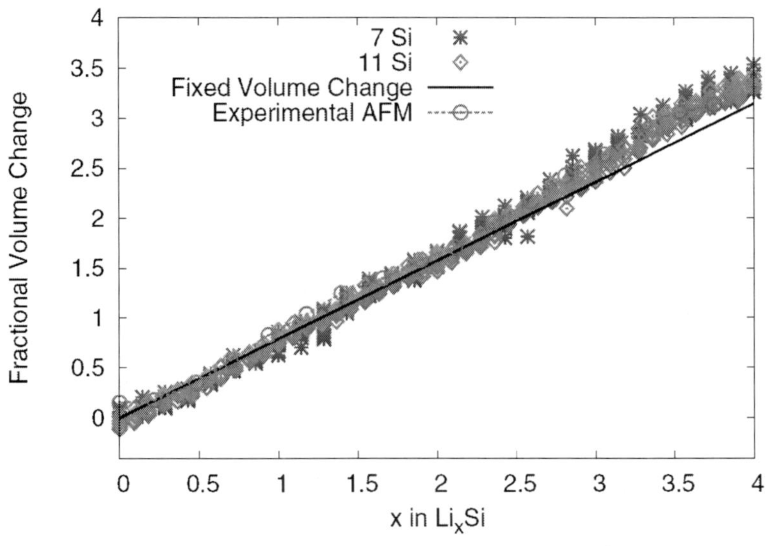

Figure 8.4 Fractional volume changes obtained after full structural optimization of 7 and 11 Si cells compared to experimental *in situ* AFM studies.

8.3 Thin Film Silicon Anodes and Microsized Particles

Si thin films on Cu foil provide an excellent tool for studying the volumetric expansion of a Si anode during lithiation/delithiation cycling. Generally, Si is sputtered on Cu or Ni films with thicknesses of 50–500 nm. A ~1 cm^2 disk is punched after film deposition and the Si mass is calculated, which is used to calculate the charge/discharge rate. The rate, C/n, gives the time needed to charge or discharge the cell in n hours. The half-cells are fabricated in a glove box using a punched Si anode and a piece of shiny lithium metal as the counter electrode. A microporous separator us used between the Li metal and the Si anode. The electrolyte used for testing is normally 1 M $LiPF_6$ in an equal volume mixture of ethylene carbonate/dimethyl carbonate. Throughout the experiment, the charge/discharge voltage profile, cycling performance, and the CE are monitored. The film morphologies after cycling are examined by SEM, AFM, and TEM.

Figure 8.5a shows the surface morphology of a Si film on a metal substrate after 30 charge/discharge cycles. It is clear that the continuous Si film was broken into island structures.[9] It is believed that large strains or interfacial fracture energies were formed in the Si thin film during the lithiation process. The analytical formula for the interfacial energy was derived by Chow et al. to be

$$U = \frac{h}{2}\frac{E_f}{1-v_f^2}\varepsilon_{cx}^{*2}$$

where ε_{cx}^{*} is the critical longitudinal strain applied to the system to induce debonding, h is the thickness of the film, and E_f and υ_f are the modulus and Poisson's ratio of the film, respectively.[10] Fig. 8.5b shows the cycling performance comparison between 250 nm and 1 µm Si. The 250 nm Si shows a higher capacity and longer cycling lifetime than 1 µm Si. The interfacial energy is proportional to the Si film thickness. With a simplified model, Kumta *et al.* obtained a range of stresses between 640 Mpa < σ < 3.0 Gpa that develop in the film. The shear strength of copper at room temperature is 3.7 Gpa. Plastic deformation is probable because the stress developed during the first lithiation is of the order of the shear strength of copper. Kumta *et al.* also observed excellent capacity retention, even though the Si island structure is formed and the gaps are widened between the 1st and 30th cycles. This may be due to some electrochemical volume expansion/contraction-induced incremental plastic strain for the Cu substrate. As cycling proceeded further, the Si thin film delaminated from the substrate. For delamination to occur in a compressively stressed film on a substrate, an initial decohesion from an interface imperfection needs to be present. A diffusion barrier interfacial layer that has a very low modulus may improve the cycle life of the a-Si thin film on a polycrystalline Cu substrate anode system.

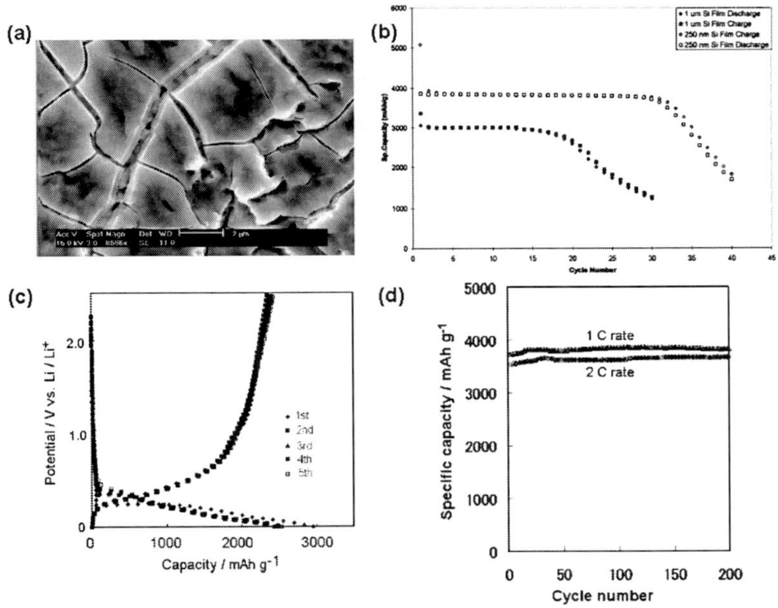

Figure 8.5 (a) SEM of a 250-nm Si film after 30 cycles. (b) Comparison of galvanostatic cell cycling of 250 nm and 1.0 μm Si film on Cu foil at C/2.5. (c) Constant current charge/discharge curves of Li in propylene carbonate (PC) containing 1 M $LiClO_4$ of a 150 nm Si deposited on a Ni film at 1 C. (d) Cycling performance of 50 nm Si on Ni film.

There has been much effort for improving performance of Si thin films as anodes. Si thin films on porous Cu or Ni substrates have been tested by Park *et al.*[11] In a separate study, Takamura *et al.* investigated a thin, n-type doped Si on Ni film and they achieved very large specific capacity and long cycle life.[12] A 500 Å thick Si film gave a charge capacity over 3500 mAh g^{-1} being maintained during 200 cycles under a 2 C charge/discharge rate, while a 1500 Å film revealed around 2200 mAh g^{-1} during 200 cycles under a 1 C rate. Fig. 8.5c and (d) show the voltage profile and cycling of an n-type doped Si on Ni film. Excellent cycling and power have been achieved with little capacity decay after 200 cycles. However, the thickness is too low for many applications and the initial capacity loss needs to be minimized more. These are extremely challenging for Si thin films because the large stresses and volume expansion are proportional to the film thickness. In our opinion, Si thin films on metal substrates provide an excellent platform to investigate the fundamentals. It may find applications in thin-film batteries where the total energy need is small.

Another rational approach for Si anodes is to use the standard slurry method already widely used in the Li-ion battery industry In this method, the electrode materials ~70–85% by weight, conductive additive ~10–20% by weight, and polymer binder ~5–10% are mixed well. Solvent is added and the dispersion is stirred for 24 hours. The final dispersion, often called slurry, is blade coated onto metal substrates with dry thicknesses of ~50–100 μm. The electrodes on metal substrates are then vacuum-dried at 80 °C 120 °C for a few hours , followed by Li-ion battery device assembly in a glove box. The same approach has been applied for microsized Si particle anodes. However, Si experiences huge volume changes while other anode materials, such as $Li_4Ti_5O_{12}$ (LTO) and graphite, do not. The traditional slurry method can leads to outstanding cycling performance for graphite or LTO, without capacity decay for >1000 cycles. However, for microsized Si, the anode performance is typically poor, with lower CE and capacity retention. One possible explanation is the elasticity of the polymer binder. The polymer used for LTO or graphite is polyvinylidene fluoride (PVDF). During the delithiation process, Si contracts substantially, while PVDF and the conductivity additive, cannot contract, which leads to the detachment of Si with the conductive pathway. This loss of Si inside polymer-based structures can be minimized by adding more conductive additive, up to 30% by weight, but the performance improvement is still not good enough to allow microsized silicon for practical applications.[13]

The capacity of silicon must be limited during cycling for it to be of practical use. Based on the same slurry method, Krause *et al.* did a nice study about crystalline, microsized Si particles as the anode in 2007 with the constant current/constant voltage (CCCV) method.[14] Commercial Li-ion batteries are typically charged at a constant current to a fixed voltage and then held by the charger at constant voltage until the current decreases to a certain value. In the conditioning cycling method, crystalline Si is first partially cycled to amorphous. After their conditioning cyclings, the Si anode can be cycled normally with good CE and little capacity loss for the first cycle. As shown in Fig. 8.6a, crystalline Si particles are first lithiated

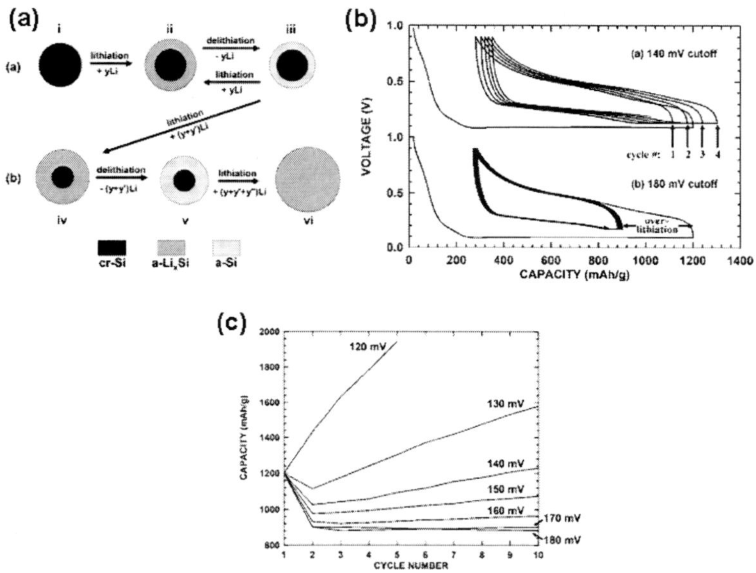

Figure 8.6 (a) An illustration of how a single Si particle changes its structure during cycling. First, a crystalline silicon particle is partially lithiated, which results in a two-phase microstructure of partially lithiated silicon and unlithiated crystalline silicon (ii). Because the diffusion of lithium is from the surface, the lithiated silicon is believed to form a shell around a core made up of the unlithiated silicon. If the silicon is delithiated, the particle has an amorphous silicon shell surrounding a crystalline silicon core (iii). If the same amount of lithium is added and removed during lithiation and delithiation in each cycle and the voltage is maintained above 170 mV, the silicon is reversibly converted between structures (iii) and (ii). If Si is exposed to voltages below 170 mV, more and more lithium can be added to the silicon particle, which causes the lithiation of more crystalline silicon (iv) and then eventually leads to the lithiation of all the crystalline silicon (vi). (b) Voltage curves of two of the cells. After first lithiating to 1200 mAh/g, one cell was cycled with a cutoff voltage of 140 mV and the other with a 180 mV cutoff. (c) Capacity *versus* cycle number for Cr–Si cells with different voltage cutoffs measured at room temperature.

partially, which results in a two-phase structure, with amorphous Si on the shell and crystalline Si in the core, which is due to the low diffusion constant of Si. For the first delithiation, the two-phase structure is maintained between (ii) and (iii). Ideally, good cycling performance will be achieved if the Si anode is being cycled between (ii) and (iii). If the cell is exposed to a voltage less than 0.17 V, more and more Si will participate in the lithiation/delithiation process, and eventually all of the crystalline Si becomes amorphous. Therefore, an accurate cell voltage monitoring during the lithiation process is important to maintain the two phase structure for good cycling performance. Krause *et al.* designed a excellent method to find the lithiation potential for crystalline Si for the two phase structure. In Fig. 8.6b, the voltage profiles are different for cutoffs of 140 mV and 180 mV. The capacity increases from cycle 1 to 4 for a cutoff of 140 mV, which indicates more and more Si particles are participating in the lithiation/delithiation process. For a cutoff of 180 mV, the profiles are the same after the first cycle, which indicates the processes are between (ii) and (iii). Figure 8.6c shows the cycling performance with different lower cutoffs. With the CCCV strategy, they achieved excellent cycling performance for microsized particles.

8.4 Vapor–Liquid–Solid (VLS)-Grown SiNWs as High-Capacity Anode

Silicon is an attractive alloy-type anode material because of its highest known capacity (4200 mAh/g). However lithium insertion into and extraction from silicon are accompanied by a huge volume change (up to 400%), which induces a strong stress on the silicon particles and causes pulverization and rapid capacity fading.[15] To overcome this issue, several approaches have been suggested, which include the preparation of nanosized active materials,[16,17] active/inactive composite materials[18] and Si-based carbon composites.[19] These approaches have resulted in improvements of the electrochemical performance of Si-based anodes but only to a limited extent.

Figure 8.7a illustrates the pulverization issues of Si films or particles after repeated Li^+ intercalation. We have designed a Si anode using a novel nanowire (NW) approach,[20] where silicon nanowire (SiNW) active material was directly grown onto stainless metal current collectors via a gold catalyzed VLS method as indicated in Figure 8.7b In this design, the nanowire geometry is used to exploit the nanometer size in the diameter to avoid pulverization, while simultaneously ensuring good electronic contact with the current collector. A schematic of the Si nanowire anode architecture is shown in Fig. 8.7b, where the nanowires are grown directly on the metallic current collector.

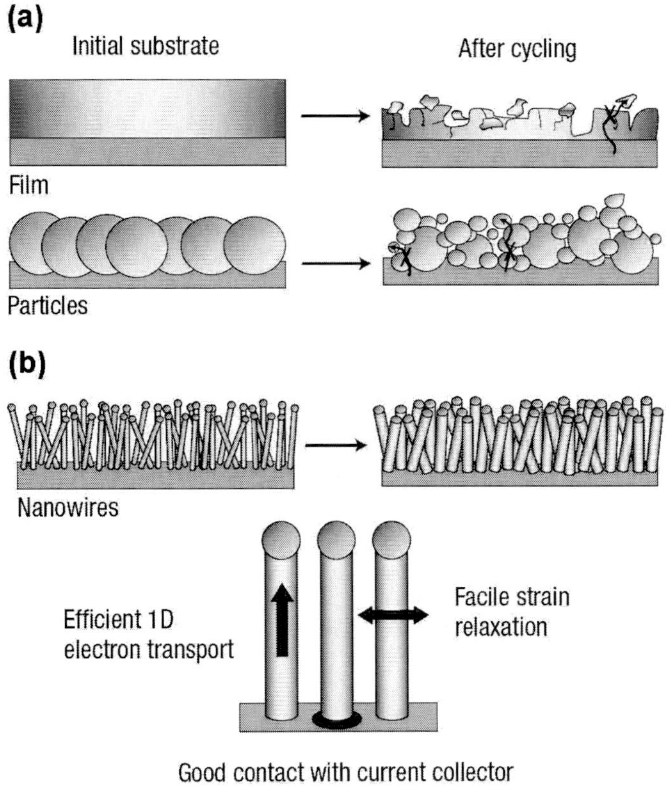

Figure 8.7 Schematic of morphological changes that occur in Si during electrochemical cycling. (a) The volume of silicon anodes changes by about 400% during cycling. As a result, Si films and particles tend to pulverize during cycling. Much of the material loses contact with the current collector, resulting in poor transport of electrons, as indicated by the arrow. (b) NWs grown directly on the current collector do not pulverize or break into smaller particles after cycling. Rather, facile strain relaxation in the NWs allows them to increase in diameter and length without breaking. This NW anode design has each NW connecting with the current collector, allowing for efficient 1-D electron transport down the length of every NW.

This geometry has several advantages. First, the small nanowire diameters can relieve the strain and allow for better accommodation of the large volume changes without the initiation of fracture. Second, each nanowire is electrically connected to the current collector such that all of the active material (all nanowires) can participate in the charge storage and contribute to the capacity. Third, the Si nanowires have direct 1-D electronic pathways allowing for efficient charge transport. In electrode microstructures based on particles, electronic charge carriers must move through small inter-particle contact areas. In addition, as every nanowire is connected to the current collector, the need for binders or conducting additives is eliminated.

Figures 8.8a,b show the typical scanning electron microscope images of VLS-grown SiNWs. The SiNWs have an average diameter of ~89 nm and lengths up to several tens of microns. The cross-section SEM image (Fig. 8.8b) shows that the SiNWs grow vertically off of the substrate. Figure 8.8c shows the voltage profile of the SiNWs in a pouch type half-cell test, where the average charge (lithium alloying) potential is 0.15 V and the average discharge (lithium de-alloying) potential is 0.4 V. The low discharge voltage of 0.4 V *versus* lithium indicates that Si is an appropriate anode material for Li-ion batteries. Figure 8.8d shows the cycling performance of a SiNW anode in a half-cell test at rate of C/5. The SiNWs demonstrate a super high capacity of 2800 mAh/g, eight times that of the current commercial graphite value of 340 mAh/g. This cell also shows good capacity retention over 30 battery cycles and a high CE of >98%. Even at a high rate of 1 C, the SiNW still demonstrate a high capacity of ~2100 mAh/g, which is six times larger than that of graphite.

The structural morphology changes during Li insertion were studied to understand the high capacity and good cyclability of our SiNW electrodes. Pristine, unreacted SiNWs are crystalline with smooth sidewalls and have an average diameter of ~89 nm. Cross-sectional scanning electron microscopy (SEM) images show that the Si NWs grow directly off the substrate and have good contact with the stainless steel current collector. After charging with Li, the SiNWs have roughly textured sidewalls, and the average diameter increased to ~141 nm. Despite the large volume change, the Si NWs remain intact and do not break into smaller particles. They also appear to remain in contact with the current collector, suggesting minimal capacity fade due to electrically disconnected material during cycling.

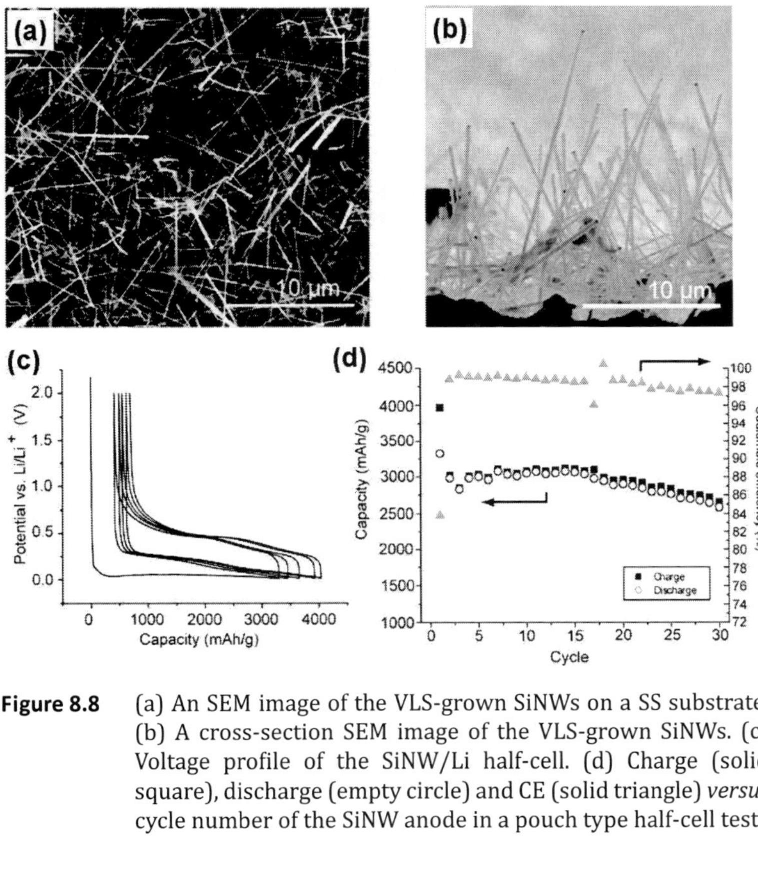

Figure 8.8 (a) An SEM image of the VLS-grown SiNWs on a SS substrate. (b) A cross-section SEM image of the VLS-grown SiNWs. (c) Voltage profile of the SiNW/Li half-cell. (d) Charge (solid square), discharge (empty circle) and CE (solid triangle) *versus* cycle number of the SiNW anode in a pouch type half-cell test.

8.5 Surface Characterization and Electrochemical Analysis of the Solid–Electrolyte Interphase (SEI) on Silicon Nanowires

The operation potential of the negative electrode is low and close to that of metallic lithium. In such a low potential range (0.1–0.5 V *vs.* Li/Li+), the organic electrolyte can be decomposed or experience electrochemical reduction so that it creates a surface layer on the surface of the negative electrode. This layer consists of a by-product of the electrolyte salt and solvent, which is insoluble to the electrolyte. This layer, called the solid–electrolyte interphase (SEI) is an important feature of the negative electrode material. For example, the property of the SEI determines the CE of initial cycles, as well as the long term stability of the electrode material.[21–23] For the silicon anode system, there have been several studies focused on the SEI formation of amorphous Si thin films and particles.[24] In the particular case of the silicon nanowire (SiNW) anode, C. Chan and coworkers studied the SEI formation using X-ray photoelectron spectroscopy (XPS), scanning electron microscopy (SEM), and electrochemical impedance spectroscopy (EIS).[25] Because of their large surface areas, the SiNWs are expected to have a higher reactivity with the electrolyte, which implies that the SEI formation can affect the long term cycling performance of the cell significantly.

Ex-situ SEM of the SiNW anode shows dramatic changes of the surface morphology due to the SEI formation (Figs. 8.9a–h). The morphology changes when the potential is near 0.1 V, indicating that that SEI formation is not limited to the intermediate potential region (0.3–0.7 V) as previously thought but rather occurs in parallel with Li insertion at lower potentials. As the potential gets lower than the initial plateau near 0.1 V and approaches to the lithiation limit (10 mV), the nanowires and the SEI form a composite film, made of thick films and large particles (Fig. 8.9b). During the discharge, a large portion of the SEI dissolves into the electrolyte and the composite film comes back to the SiNW shape (Fig. 8.9c) resembling the original morphology (Fig. 8.9a) except for some disconnected particles (Fig. 8.9d). These observations are the same for the second cycle (Figs. 8.9e–h), indicating that the substantial part of SEI is reversibly deposited and dissolved during the cell operation.

XPS gives similar information based on the elemental and bonding analysis on the SiNW surface. Figures 8.9i,j show the composition and the molecular species of the surface at different potentials, respectively. During the first lithiation cycle, the SEI grows so thick that the silicon signal disappears. The composition varies during cycling; however, two molecular species—lithium fluoride (LiF) and lithium carbonate (Li_2CO_3) are the major components during complete charge–discharge cycles. Although this inorganic SEI is stable and insoluble regardless of the potential, the organic components of the SEI, such as PEO-containing oligomers, are rather unstable and soluble depending on the potential range.

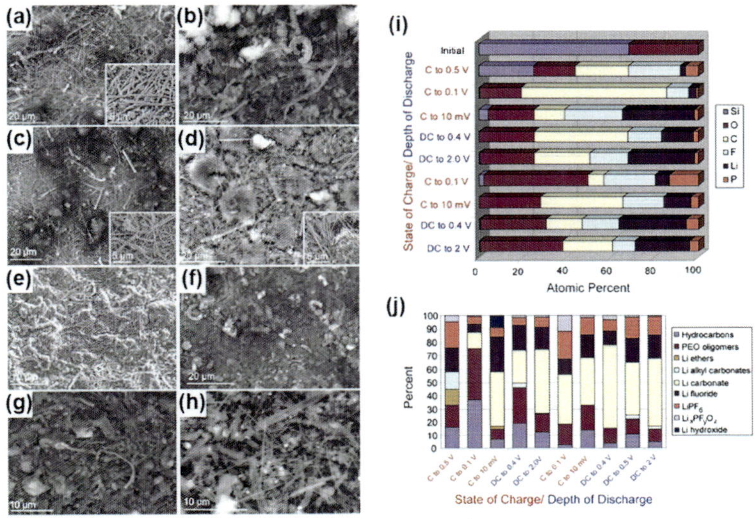

Figure 8.9 (a–h) SEM images of SiNWs and SEI layer at different potentials: the first charge to (a) 0.1 V and (b) 10 mV; the first discharge to (c) 0.4 V and (d) 2.0 V; the second charge to (e) 0.1 V and (f) 10 mV; the second discharge to (g) 0.4 V and (h) 2.0 V. (i), (j): Surface composition (i) and molecular species (j) of the SiNW samples at different potentials, as derived from the XPS scan. 1 M LiPF6 in 1:1 ethylene carbonate:diethyl carbonate was used as the electrolyte.[25]

Ruffo *et al.* studied the SEI formation on the SiNW anode by impedance spectroscopy (IS), which is a useful tool to study the electrochemical behavior of materials.[26] In this experiment, a small sinusoidal stimulus is applied to an electrochemical cell and its response is measured. The ac response of an electrochemical system at different frequencies can reveal the surface properties of the SiNW anode, especially with regards to the SEI formation. The impedance spectra are analyzed by using Nyquist plots—the imaginary part of the impedance is plotted as a function of the real part over a wide range of frequencies. A typical Nyquiest plot of the SiNW half-cell is given in Fig. 8.10a. The size of the half-circle in the intermediate frequency domain (10 Hz $< f <$ 20 kHz) represents the surface contribution of impedance due to the SEI layer. The surface resistance shows a minimum for intermediate states of charge (SOC) around 1000 mAh/g (Fig. 8.10b). The slight increase at a high SOC (>3000 mAh/g) is attributed to the growth in thickness of the

SEI when the potential becomes more negative.[24] XPS analysis[25] also suggests that the inorganic components of the SEI contribute substantially to the growth. The surface resistance increase at low SOC (<300 mAh/g) is regarded as the porous and inhomogeneous formation of the SEI surface at low lithium composition.

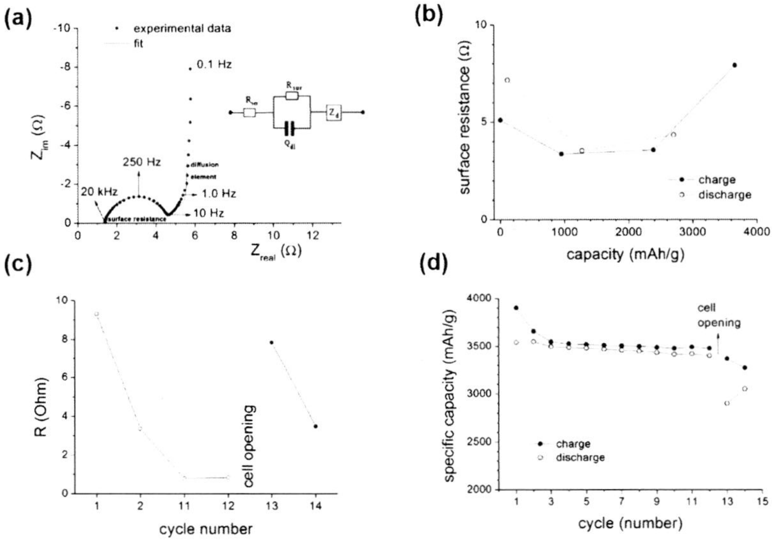

Figure 8.10 (a) Typical Nyquist plot showing the impedance of the SiNW anode at intermediate SOC (3-electrode SiNW/Li/Li cell). Inset indicates the model circuit diagram comprising the ohmic resistance (R_{ser}), surface resistance (R_{sur}) and capacitance (Q_{dl}), and diffusion component (Z_d). (b) Surface resistance *versus* SOC for the second cycle. (c) Surface resistance *versus* cycle number for the charge state of intermediate SOC (900 mAh/g SOC). The cell was opened after 12 cycles and reassembled. (d) Specific capacity of charge/discharge cycling *versus* cycle numbers for the same cell as (c).

During the first several cycles, the surface resistance decreases substantially and it becomes an order smaller than the original magnitude, which may indicate the stabilization of the SEI layer (Fig. 8.10c). However, if the cell is opened and re-assembled, the resistance went back to the first cycle value even though the cell was kept in electrolyte. More interestingly, the CE of the cycle after the cell opening is very close to that of the first cycle. This observation indicates that the majority of the interface resistance may come from the external organic SEI/liquid electrolyte interface, rather than the charge transfer across the SiNW/SEI interface. Because the inner SEI mainly consists of the inorganic compounds, as the XPS report showed, the stabilization of the outermost organic SEI would be more important to improve the cell performance such as the CE and the long term stability.[25]

8.6 Si Core–Shell Structures for Anodes

Following our VLS grown single crystal SiNW anode study, we have demonstrated a Si crystalline-amorphous (c-a) core–shell NW design resulting in significant improvements in the power rate and cycling life.[27] Also, these Si core–shell NWs can be catalytically synthesized using the surface of ordinary type 304 stainless steel such that the expensive gold catalyst is no longer needed. Previous studies, mostly involving thin films, have suggested that a-Si has a superior cycling performance compared to c-Si[28,29]. One explanation for this is that the volume expansion of a-Si on Li insertion is homogeneous and causes less pulverization compared to the crystalline material case.[28] Another interesting behavior of a-Si is that it reacts with lithium at slightly higher potentials than c-Si does,[20,28] which leads to our idea of using c-a core–shell Si NWs as an anode material. When limiting the charging potential, it should be possible to utilize only the amorphous shell material for Li^+ storage while preserving the crystalline core as mechanical support and efficient electron transport pathways, as indicated by Fig. 8.11a.

Core–shell NWs for solar cells and transistors have been shown previously but not for batteries.[30–32] Lieber and coworkers have grown Au-catalyzed c-Si NWs in a chemical vapor–deposition (CVD) furnace at 435 °C and coated a-Si shells at an elevated temperature of 450 °C.[33] Here, we developed a simple method to grow c-a core–shell Si NWs without changing the growth condition (Fig. 8.11a). Our growth was conducted directly on a type 304 stainless steel (SS) current collector without Au as catalyst in a SiH_4 CVD furnace. We found that large flow rates, high pressure and high temperatures promote the yield of c-a core–shell Si NWs. After the growth, no metal catalyst particles were found at the ends of Si NWs. Nickel in SS 304 is believed to catalyze the growth most likely by a direct vapor–solid

(VS) mechanism since it has previously been observed to catalyze such growth.[34–36]

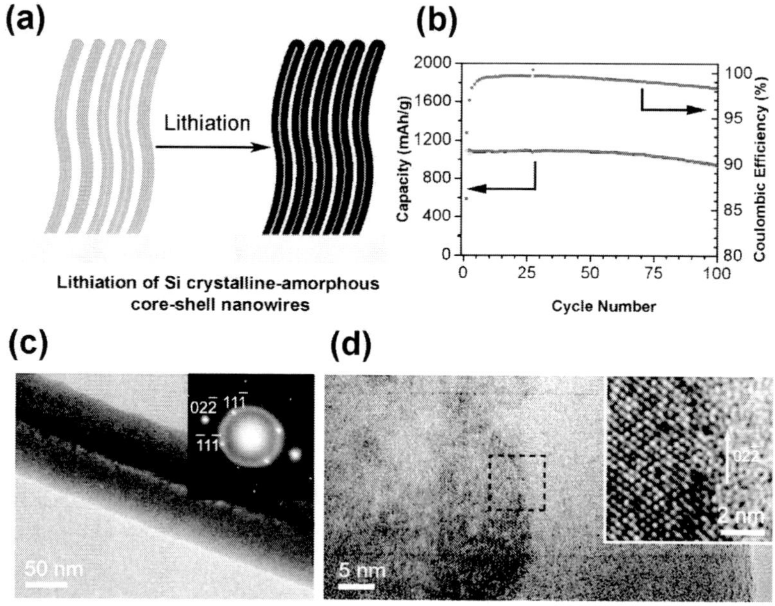

Figure 8.11 (a) Schematic illustration of the lithiation of the Si c-a core–shell NWs grown on a SS substrate. (b) Cycling performance of the Si core–shell NWs at a C/5 rate. (c) A TEM image of a core–shell NW. (d) Zoomed-in TEM image of the core–shell NW.

Previous studies have suggested that crystalline Si does not react with Li^+ above 0.12 V *versus* Li^+/Li while a-Si can largely react with lithium well above this voltage. When limiting the charge (lithium alloying) potential to above 0.15 V, the crystalline core will not react with Li^+ ion and will be preserved throughout the battery cycling. This way the intact core helps maintain the NWs' 1-D structure without severe breaking or damage and provide an electron conduction pathway. This clever design has enabled us to obtain significantly better cycle life. Figure 8.11b shows the cycling performance of these core–shell NWs at a rate of C/5 when limiting the charge voltage above 0.15 V. The core–shell NWs with a 0.15 V cutoff and the same current density show significantly improved cycle life compared to those of pure crystalline SiNWs. The core–shell NWs showed a capacity of ~1060 mAh/g with 85% discharge capacity retention after 100 cycles. Despite the fact that the discharge capacity with a 0.15 V cutoff is lower than that with a 10 mV cutoff (2800 mAh/g), the value is three times that of carbon and is already high enough to be technologically significant. What is more important is to have good capacity retention over a long cycle life. In addition, the CE (86.2% for the first cycle, 99–99.7% for cycles 5–70, 98.4–99% for cycle 70–100) at a 150 mV cutoff is excellent, which provides strong evidence of little fatigue during cycling. Figures 8.11c,d show TEM images of a Si c-a core–shell NW. The core–shell structure can be clearly seen by contrast in Fig. 8.11c. Figure 8.11d shows a zoomed-in image of the NW, where the c-a interface can be clearly identified. The inset image in Fig. 8.11d is the zoomed-in image of the dashed square at the core–shell interface, where the lattice structure of the crystalline Si can be clearly seen.

To confirm that the c-a core–shell structure still remained intact for robust mechanical and electrical support after cycling, we have investigated the morphology and structure of cycled c-a NWs (delithiated at the end of cycling) with scanning electron microscopy (SEM) and TEM. Two samples cycled with 150 and 10 mV cutoffs are compared in Fig. 8.12. Figure 8.12a shows the SEM image of the c-a core–shell NWs after 15 cycles using a 150 mV cutoff, where the NWs apparently still maintain distinct wire shapes. Figure 8.12b shows the TEM image of a corresponding single NW, which still clearly shows a core–shell structure by contrast. A SAED pattern [Fig. 8.12b (inset)] confirms that the core is still crystalline Si. The bright and dark contrast of the crystalline Si core might be caused by the straining effect of the expansion of amorphous shell during lithium insertion. Figure 8.12c shows the SEM image of the c-a core–shell NWs after three cycles using a cutoff of 10 mV. Figure 8.12d is the TEM image and SAED pattern (inset) of corresponding NWs, indicating a completely amorphous structure without a crystalline core. This observation was expected because the crystalline core would react with lithium at 10 mV and become totally amorphous upon cycling, which was also indicated by previous studies.[10,28]

Following our Si c-a core–shell design, we introduced a new design of carbon–silicon (C–Si) core–shell nanowires for high loading densities and cheap production costs.[37] We have synthesized C–Si core–shell NWs by CVD of amorphous Si onto carbon nanofibers (CNFs) as illustrated in Fig. 8.13a. The SEM images of bare CNFs and a-Si coated CNFs are shown in Fig. 8.13b,c. In a similar fashion to the crystalline Si core in the c-a core–shell Si NWs, carbon cores function as efficient electron transport pathways and stable mechanical supports. However, the difference is that the carbon core, due to its small capacity, has little structural or volumetric changes with charging potentials down to 10 mV *versus* lithium metal. Charging to this low potential allows a much higher usage (>2000 mAh/g) of the specific charge capacity of a-Si.

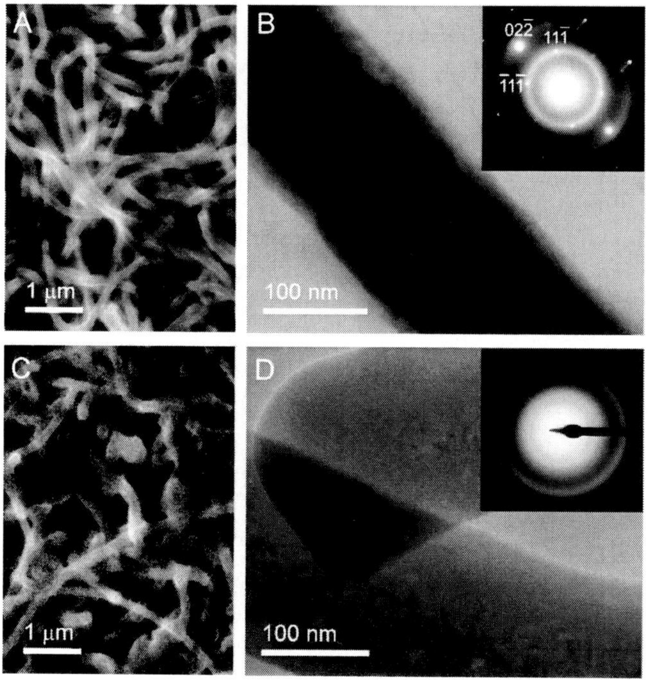

Figure 8.12 SEM, TEM, and SAED images of the c-a core–shell NWs after cycling at 0.2 C (0.85 A/g) and different low-voltage cutoffs. (A) SEM image of NWs after 15 cycles with a cutoff of 150 mV. (B) TEM and SAED (inset) images of a NW after 15 cycles with a cutoff of 150 mV. (C) SEM image of NWs after three cycles with a cutoff of 10 mV. (D) TEM and SAED (inset) images of NWs after three cycles with a cutoff of 10 mV.

CNFs were purchased commercially from Sigma-Aldrich and used as obtained. The silicon coating on the CNFs was done using SiH_4 CVD. After the Si coating, the SS substrates can function directly as current collectors and the C–Si core–shell NWs are the active material, for which no binding or conductive additives are needed.

Figure 8.13d shows the cycling performance of C–Si core–shell NW anode at a rate of C/5. These C–Si NWs show very little capacity fading after 55 cycles. Because CNFs are commercially available in large quantities, the mass production of C–Si core–shell NWs is expected to be easily achievable. We have successfully demonstrated high mass loadings of these core–shell NWs and achieved a per-area capacity of ~4 mAh/cm^2, which is comparable to commercial values.

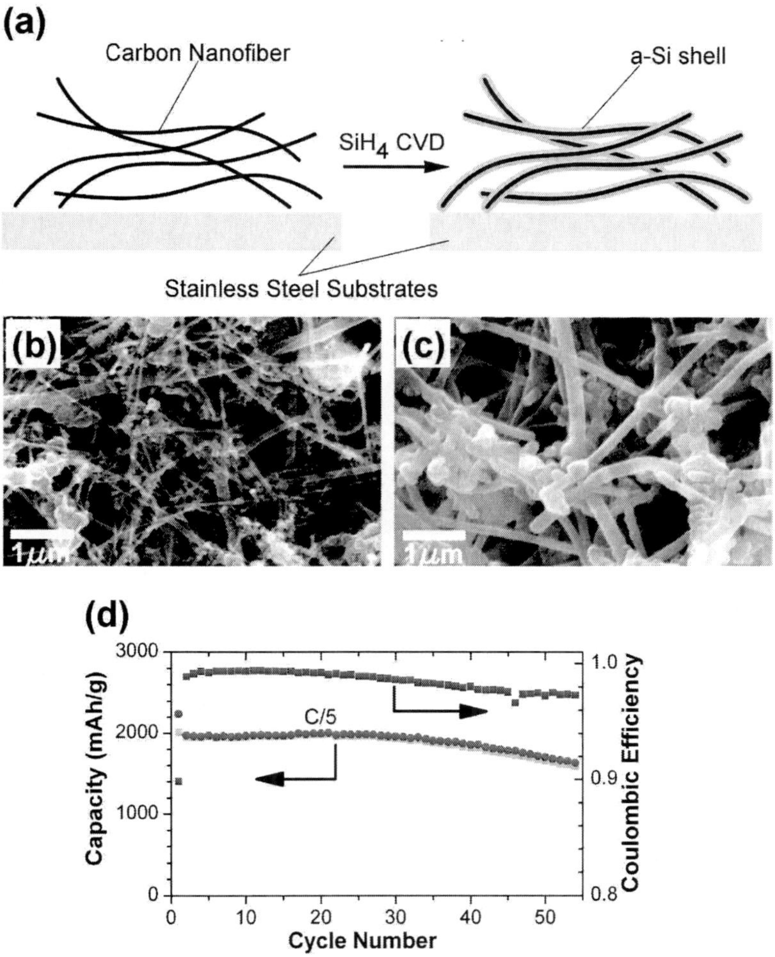

Figure 8.13 (a) Schematic illustration of Si coating onto CNFs. (b) SEM image of CNFs before Si coating. (c) SEM image of a-Si coated CNFs. (d) Half-cell cycling performance of the C–Si core–shell NW anode.

A full cell made of a$LiCoO_2$ cathode and C–Si NW anode was also investigated. The mass loading density of cathode is 10 mg/cm^2 and C–Si NW anode about 1 mg/cm^2. Figure 8.14a shows the cycling performance of the C–Si NW anode in the full cell. The cycling rate used is C*/5 (1 C* = 140 mA/g) according to the specific capacity of $LiCoO_2$ cathode. In this cell, the specific capacity according to the weight of C–Si NWs is ~1400 mAh/g and the corresponding $LiCoO_2$ capacity is ~140 mAh/g. The first-cycle CE is 86% and around 98.5% for subsequent cycles. The full cell has a capacity retention of 80% after 30 cycles. We believe the capacity loss during the first 30 cycles of the full cell test is mainly due to the decay of the $LiCoO_2$ (purchased from Sigma-Aldrich) cathode because C–Si NWs showed little fading during the first 30 cycles in half-cell tests (see Fig. 8.14a). For a comparative study, we have made a cell using $LiCoO_2$ (10 mg/cm^2) as the cathode and Li foil as the counter electrode, which

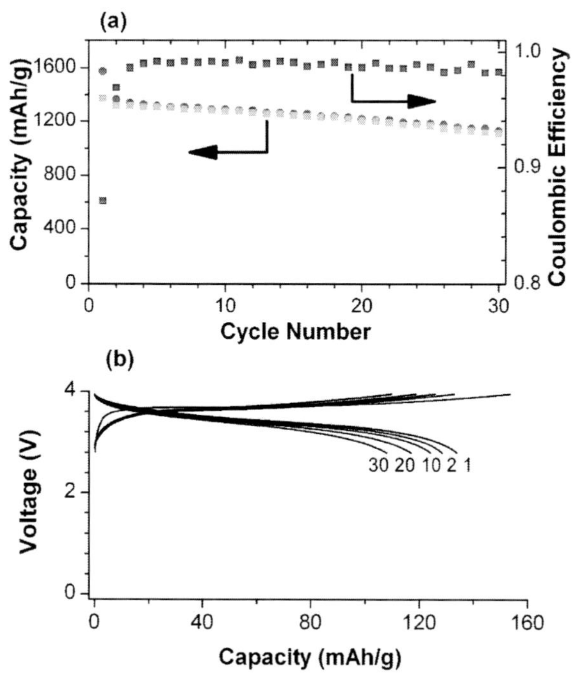

Figure 8.14 (a) Cycling performance of C–Si core–shell NWs in a $LiCoO_2$/NW full cell, cycled between 3.5–4 V. (b) Voltage profile of the full cell in (a), where the specific capacity is calculated according to the weight of $LiCoO_2$.

also has a capacity retention of ~80% after 30 cycles. The voltage profile of the $LiCoO_2$/C–Si NW full cell is shown in Fig. 8.14b. The full cell discharges at an average voltage of ~3.3 V, which is about 0.3 V lower than a commercial $LiCoO_2$/graphite cell (~3.6 V) but still applicable. This lower voltage than $LiCoO_2$/graphite system is due to the higher discharge (lithium de-alloying) potential of Si anode compared to graphite. No plateaus were observed because C–Si NWs have a sloping voltage profile (except the 1st charge).

8.7 Other Si Nanostructures

Besides nanowire and core shell structures, there are several other types of nanostructures based on Si. The key design rule is that the Si material remains in electrical contact with the current collector during the charging and discharging processes, and the electrode has enough space for Si expansion. Figure 8.15 summarizes the types of nanostructures which have been actively investigated in the past few years. All these structures have shown certain promise for practical applications and provide excellent platforms for investigating the fundamentals. Note that most of the Si nanoelectrodes are binder-free, that is, there is no polymer binder used as in the traditional slurry-based electrode. The reason is that the polymer binder is not elastic enough to follow the mechanical expansion and contraction, which disrupts the conducting channel between the conductive additive and Si anode materials. The method used to provide electrical conduction between the Si materials in the nanoscale is crucial. Normally, a metal coating or a carbon coating of carbonized polymers is used.

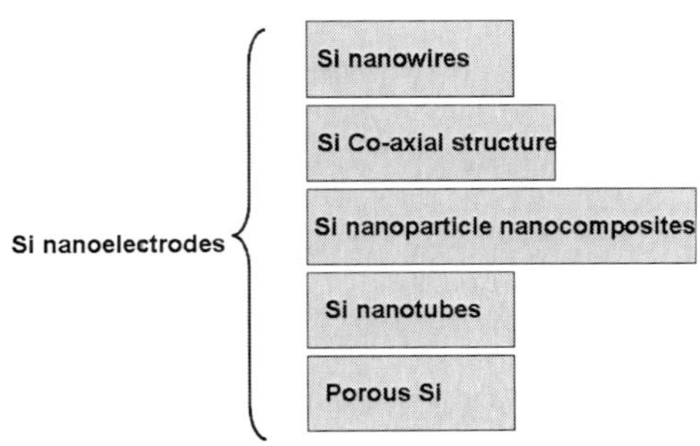

Figure 8.15 Outline of various Si nanostructures for efficient Li-ion battery anode.

As discussed in the section of Si thin films and microsized particles, Si will tend to break during the charge–discharge process. The huge stress built in Si will cause the electrode to break into small islands, and the breaking continues with cycling, which causes fast capacity decay in the batteries. The mechanical stresses are less for smaller particles. Therefore, there should be a critical size of thin films or particles. It has been observed that the Si thin films can pass 200 cycles with little capacity decay. The same phenomenon has been observed for Si particles, as the mechanical strain induced by volume changes can be reduced by employing nanoscale particles. Cho *et al.* found that the fracture toughnesses of lithiated silicon nanoparticles smaller than 20 nm are greatly enhanced.[38] In their study, they reported a synthetic method using reverse micelles at high pressures and temperatures in a bomb that produces Si nanoparticles with different sizes. There are several other reports about the synthesis of Si nanoparticles at high temperatures. Room temperature syntheses have also been demonstrated. However, the obtained Si are not fully crystallized and often coated with SiO_x. In the study by Cho *et al.*, 10 nm sized Si nanoparticles showed higher capacities, up to 4210 mAh/g and 3380 mAh/g, compared to 20 nm sized particles. However, due to the large surface-to-volume ratio resulting from the decreased formation of the SEI, 10 nm sized particles lead to lower CEs. After adding a carbon overcoating, both the CE and capacity retention were greatly improved to 89% and 96%, respectively. Due to the excellent performance, Si nanoparticles are believed to have promising potential for real applications. Incorporating Si nanoparticles into electrospun nanofibers, or mixing Si nanoparticles with nanoscale conductive additives such as carbon nanotubes (CNTs) and graphene have caught much research attention recently.

Recently, Yushin *et al.* designed a Si nanoparticle–Carbon hierarchical structure with a bottom up approach.[39] The nanoscale rigid and robust spheres with irregular channels have large internal porosities which can accommodate large volume expansions. Figure 8.16a shows a schematic of the Si–C composite granule formed through a bottom up approach. Si nanoparticles are electrically contacted to one another via the carbon backbone. The deposition of Si nanoparticles was conducted by means of low pressure silane decomposition at 550 °C. They found that Si nanoparticles densely coated the carbon surface. Then, a wet granulation process is used, where the primary particles are self-assembled into larger spheres by a combination of viscous and capillary forces. Figure 8.16b shows the microsized spherical granules. Although the surfaces of Si nanoparticles are coated with carbon, the small nanoparticles are still visible on the surface (Fig. 8.16c). The diameters of the particles are in the range of ~15–35 μm. Coin cells based on such microsized granules show excellent performance, giving ~3670 mAh/g at C/20, the highest value for Si particle–based structures. They also demonstrated high rate, 8 C, performance. Such a strategy could be extended to cathode nanoparticles.

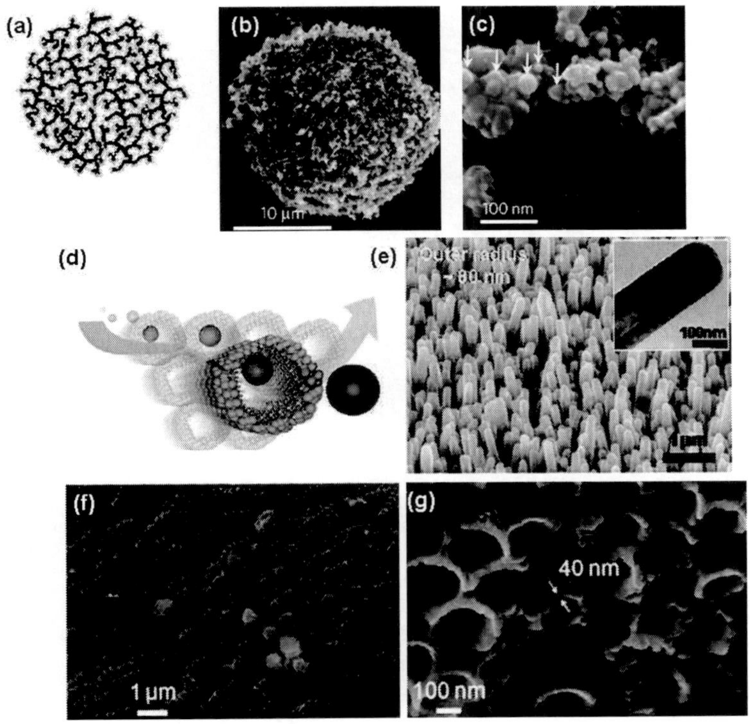

Figure 8.16 Si nanosctructure examples: (a) Schematic of Si–C nanocomposite granule formation through a hierarchical bottom-up assembly, where the Si particles are electrically connected through internal channels. (b) and (c) SEM images of C–Si nanocomposite spherical granules at different magnifications. (d) Schematic diagram of Li-ion pathway in Si nanotubes. (e) SEM image of a vertically aligned Si nanotube array with R_{out} ~80 nm. (f) and (g) SEM images of the 3-D porous C–Si particles.

Another nanoscale design of a Si anode involves Si nanotubes. The advantages of Si nanotubes are mainly two-fold: (1) Si nanotubes provide large free spaces for volume expansion. (2) The large surface areas of Si nanotubes provide large, accessible surfaces for electrolyte, which will enable high rate operation (Fig. 8.16d). There are a few reports about Si nanotube structures for Li-ion anode applications. In 2009, Cho and Cui *et al.* demonstrated the Si anode with an alumina template and etching method.[40] The silicon nanotubes have outer diameters of ~200–250 nm and wall thicknesses of ~40 nm. The Si nanotubes exhibit impressive performance, 3247 mAh/g and 89% CE. They attribute the high CE of the first cycle to the thin carbon layer, which minimizes the direct contact between Si and electrolyte and promotes the formation of a stable SEI. The morphologies of Si nanotubes remain after 200 cycles and no pulverization occurred even though the thickness of the 40 nm wall expands to ~300 nm. In a separate study by Song *et al.*, arrays of sealed Si nanotubes were fabricated.[41] The motivation is also to use the free space inside the Si nanotubes to benefit the mechanics without creating additional surfaces for accelerated SEI formation. They used a sacrificial template method for the Si nanotube fabrication. The tubes have dimensions of R_{out} ~60–80 nm, h ~3–5 μm, shell thicknesses of ~30 nm and R_{in} ~30 nm. Figure 8.16e shows the SEM image of Si nanotube arrays on the substrate. Lithiation cycles show anisotropic behavior, where the lengths change by ~35% and the diameters change by ~120%. After the first and second cycles, the Si nanotubes showed reversible behavior, with only small increases in the diameters.

Three-dimensional porous Si particles have been applied for Li-ion batteries. A thin pore-wall size of ~40 nm can accommodate large strains without pulverization. Figures 8.16f,g show the SEM images of 3-D porous Si particles.[42] The sizes of Si particles are ~20 μm. The discharge capacities are 2820, 2668, 2471, and 2158 mAh/g at rates of 0.2, 1, 2 and 3 C, respectively. Excellent performance at high rates is due to the interconnected 3-D porous structures that provide fast lithium mobility. After cycling for 100 times, the 3-D porous Si particles maintain their porous morphologies. HRTEM study shows that the crystallite size in the cycles sample decreased from 100 nm to 2–5 nm.

8.8 Solution-Processed Si Nanostructures

Given that the reduced prices of batteries constitute a major goal, solution-processed nanostructures are preferred for practical applications. Therefore, traditional slurry methods have been attempted. As we discussed above, one major problem for slurry-based methods is the loss of electrical contact between Si and the percolating conductive additives. Elastic polymers are preferable to be used as the binders. Cui *et al.* investigated a composite structure using traditional slurry methods, where carboxymethyl cellulose (CMC) is used as the binder and multiwalled CNTs are used as the conductive additives.[43] Note that PVDF and conductive carbon are usually used in the traditional slurry method. Figure 8.17a shows the SEM of the composite electrode, where CNTs and Si nanowires are seen entangled together. In their comparison study, the carbon black does not form the percolating network, while CNTs instead form the percolative network. CNTs have much lower percolation thresholds due to the large aspect ratios. The CNTs in the composite structures not only form a highly conductive pathway but also provide extra volume for the Si expansion. Figure 8.17b shows the device performance comparison, where the carbon coated Si NWs with multiwalled CNTs shows much better performance than carbon black–based electrodes. In their study, they also studied the final overcoating of the composite structure with a thin layer of CNTs. No additional advantages were observed, as shown in Figure 8.17b. The Si NW and CNT composite structures display much better performance than other reports in the literature that used ball milling of Si and CNTs or direct CVD growth of CNTs on Si.

In a separate study, Cui *et al.* investigated a binder-free Si nanoparticle and CNT nanostructure based on solution processing.[44] Commercial Si nanoparticles and CNTs are dispersed in water with the assistance of a surfactant. Freestanding thin films of Si NPs and CNTs are formed after coating and delamination of the films from stainless steel substrates. To provide better electrical contact, a short CVD growth of Si is performed to overcoat and electrically glue Si NPs and CNTs together. The deposited a-Si passivates the surface of CNTs, and thus, the irreversible reaction with Li^+ is greatly reduced. Figure 8.17c shows the SEM of the composite electrodes after CVD processing. In such free standing films, CNTs provide sufficient electrical conductivities such that they themselves can function as the current collectors. The half-cell was tested at a rate of C/10. With the incorporation of Si NPs, the cell demonstrates good CE and has a capacity retention of 75% after 50 cycles. Figure 8.17d shows the voltage profile. The first charge has a short plateau at ~0.18V due to the lithation potential of pure a-Si deposited from SiH_4 CVD. The second plateau at ~0.09 V is the lithiation potential of pure C–Si from Si NPs. The free standing films are bifunctional, that is, they function as both the electrode and current collector. Such solution-processed structures provide an excellent new type of electrode structure for Si anode applications.

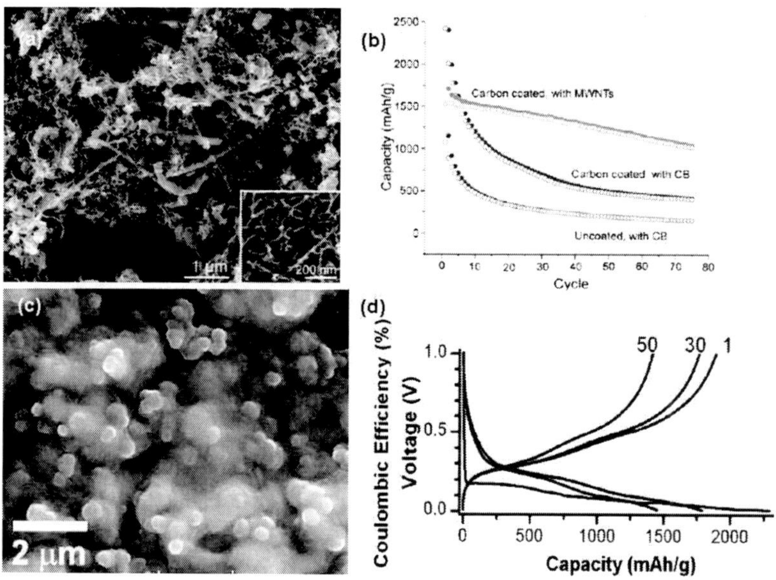

Figure 8.17 (a) SEM image of Si NW composite electrodes with 78% Si NWs, 12% CMC binder and 10% multiwalled CNTs. (b) Capacity *versus* cycling data for Si NWs uncoated with carbon black, carbon-coated and mixed with carbon black, and carbon-coated and mixed with multiwalled CNTs with an additional nanotube overcoat layer. (c) SEM of CNT-Si NP composite film after SiH_4 treatment. (d) Charge and discharge voltage profile using a CNT-Si NP film as working electrode between 0.01–1 V.

8.9 Some Fundamental Aspects

This section is devoted to the discussion of some fundamental aspects of Si nanowires and other nanostructures for Li-ion batteries, which include the mechanical modeling results, electrode design criteria along with the fate of Si anodes after multiple charge-discharge cycles such as softening and porous structure formation and high rate performance. Dempsey *et al.* have studied the design criteria by treating Si as nanospheres embedded in an inert matrix (Figure 8.18a).[45] In particular, their study predicts that small Si sizes are more stable. In order to determine the size of Si that will lead to stable performance, a stability index, κ, was defined which is respect to the crack radius ρ, where $\kappa = b^*G^*\, dG/d\rho$. The more negative the κ value is, the more stable of Si structure (Fig. 8.18b) is. Based on Griffith's theory, a crack will continue growing as long as the energy released G during propagation is greater for the energy G_c required to create a new crack surface. From Fig. 8.18c, the crack radius at

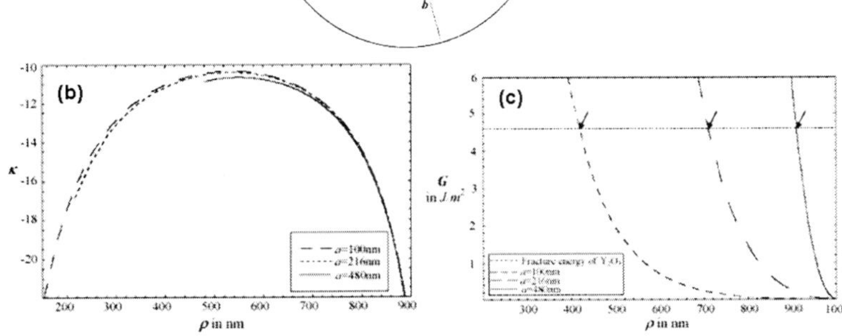

Figure 8.18 (a) Electrode configuration for modeling: Spherical active sites [shaded area, radius of (a)] and the inert matrix [blank, radius of (b)]. Δ is the expansion area and the ρ is the crack size. (b) Stability plots for systems with different active Si site volume fractions. (c) Griffith's criterion for an electrode that comprises of Si active sites and Y_2O_3 matrix. Arrows indicate the distances at which crack growth stops.

which the two energies intersect can be predict, at which point the crack growth will stop. Note the horizontal line is the fracture energy of the material, which is a material constant. The active site is Si and the matrix is Y_2O_3 in this case. Note that these numbers are assumed, but the patterns will remain the same if the actual numbers are used.

Verbrugge *et al.* did the first modeling of the Si nanowire system, especially for diffusion-induced stress (DIS).[46] They developed a mathematical model by combining the surface energy with DIS in a Si nanowire electrode. They modeled the tensile stresses in both the tangential and axial directions. For nanoscale materials, the surface energy can reduce the total tensile stresses which serve to make the nanomaterials more stable. The surface strain energy is significant in magnitude compared to the bulk strain energy. In their modeling, the bulk of the nanowire will be assumed to be an isotropic, linearly elastic solid. For the work by Cui *et al.*, nanowire electrodes are long wires grown on a substrate and one of the ends of the wire is always attached to the substrate while the other end is free to expand. Also, mechanical equilibrium is always achieved due to the fact that atomic diffusion in solids is a much slower process than elastic deformation. Figure 8.19a shows the radial stress inside Si at different radial locations at different times, where R is the radius of the nanowire and r is the location. One can tell that there is a significant decrease in radial stresses for nanoscale materials. Also, they plotted the radial and tangential stresses at $T = 0.076$ for cylindrical electrodes with different dimensions (Fig. 8.19b). It is clear that the effect of surface tension becomes much more significant as the radius decreases. These results make it clear that when the influence of surface tension is incorporated, smaller (nanoscale) wires see a more compressive stress environment. The compressive stress environment makes Si nanowires less likely to propagate cracks.

Theoretical and experimental data have confirmed that the volume increases linearly with Li insertion into Si anodes. *A* > 400% volume increase is possible when nanoscale Si structures are used for anodes in Li-ion batteries. For rational design of electrodes, the simple volume consideration needs to be included. One example is the Si nanotube array by Song *et al.*[41] As shown in Fig. 8.20a, the Si nanotubes expand dramatically upon lithiation and retain the structure expansion to a certain level even after the delithiation. They did excellent work on Si nanotubes and came out with a design

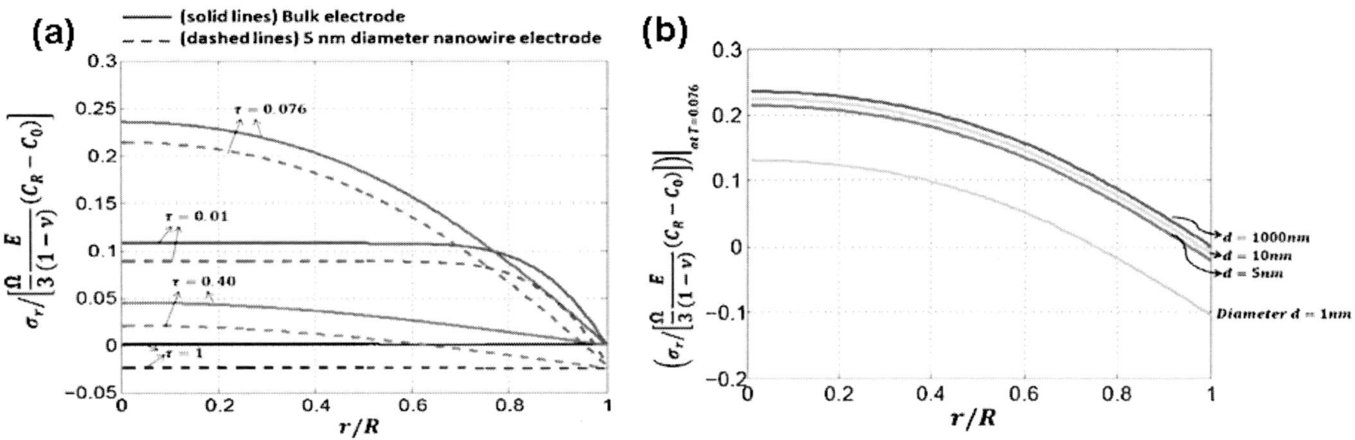

Figure 8.19 (a) Radial stresses inside an electrode at different radial locations at different times, where solid lines represents stresses in large electrodes, that is, no surface effects are considered. Dashed lines represent stresses in 5 nm diameter nanowire electrodes, that is, surface effects are significant. (b) Radial stresses at a dimensionless time T = 0.076 for electrodes with different diameters. The results are for plane strain and generalized plane strain conditions.

rule for Si nanotubes. One array structure is illustrated in Fig. 8.20b, where Si nanotubes have certain inner and outer diameters. For a give area fraction, the optimized radii are plotted in Fig. 8.20c. The area fraction of Si is $f = \dfrac{\pi(R_{\text{out}}^2 - R_{\text{in}}^2)}{d^2}$. The outer diameter increases to αR_{outer}, where α = 160% and it corresponds to 410% volume expansion. Therefore, the general design rule is $R_{\text{in}} = R_{\text{out}}\sqrt{1 - \dfrac{4f\alpha^2}{\pi}}$. For each area fraction, there exist optimal outer and inner diameters. The design criteria can provide excellent guidelines. However, this solely geometrical consideration does not include the flow of $SiLi_x$ during the lithiation process, or so-called viscous flow.

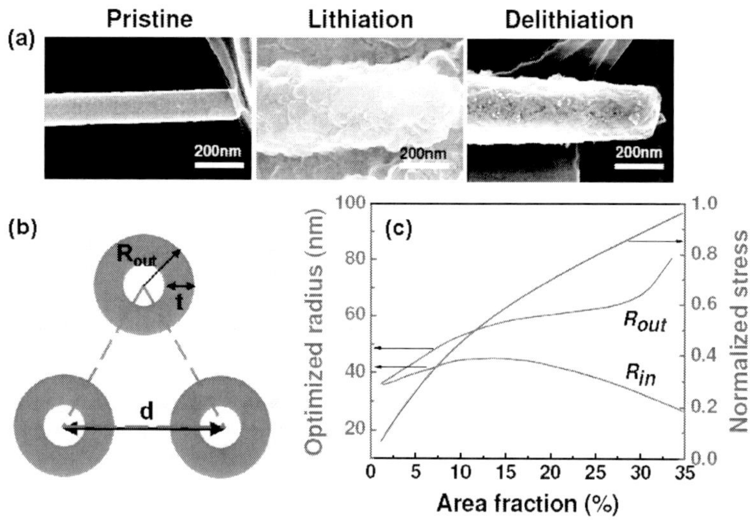

Figure 8.20 (a) Morphological changes of the surfaces of Si NTs before and after Li cycling. The left, center, and right panels are SEM images of Si NTs before the reaction and after full lithiation and the first cycle (delithiation), respectively. (b) Schematic illustration of a triangular array of Si NTs. (c) Optimized Si NT inner and outer radii (R_{in} and R_{out}, respectively), as well as normalized maximum stress *versus* area fraction .

One interesting phenomenon for Si anodes is its increasing softening with increasing Li concentration. Qi *et al.* did a first-principle study recently.[47] They developed a functional form of the relationship between averaged elastic moduli and the Li concentration in the Li-Si alloy. They used *ab initio* molecular dynamics (MD) simulation. They found that the moduli of the Li–Si alloys decrease almost linearly with the Li concentration. When it is fully lithiated, the moduli drops to about an order of magnitude of their original values of pure Si. Figure 8.21a shows the bulk modulus B *versus* Li fraction, where the Li fraction is given by $x/(1 + x)$ in Li_xSi. Softening happens for both crystalline and amorphous Si. They also found the formation energy, $E_f = E(Li_xSi) - [xE(Li) + E(Si)]$. E denotes the total energy. Figure 8.21b shows the Young's modulus *versus* Li fraction, which shows a strong dependence on the Li fraction. The softening of Si upon lithiation is important for the mechanical study and Si anode design. Because their method is based entirely on Density functional theory (DFT) formalism, it should be more accurate than others. Also, the mechanical values such as the elastic properties of Li–Si from experiments are lacking, so the study by Qi *et al.* could be insightful for further understanding.

The other important phenomenon for Si nanostructure anodes, particularly Si nanowires, is the porous structure formation. In the charging process (lithium insertion), Si can take up about 4.4 Li per Si, and its volume upon the full lithiation is even larger, up to 400% of the original. As Li is extracted in the discharging process, the expanded active materials shrink, but the final volume at the end of the discharging is still larger than the original, which indicates formation of void space inside the active materials. Thus, the pore formation depends directly on the volumetric expansion of the active materials in the charging process and the atomic diffusion of the active materials in the discharging process. When the atomic diffusion of active materials is slow during the discharging process, the active materials tend to form pores more efficiently because the slow diffusion prohibits the active materials from returning to their original structures. Instead, thermodynamically metastable structures with pores generated inside are preferred. Cui *et al.* studied the stepwise pore evolution of Si nanowires.[48] During the lithiation-delithiation processing, Si nanowires become more and more porous, as illustrated in Fig. 8.22a. The pore properties were well characterized by BET, TEM and SEM. In the alloying–dealloying processing, they found that the pore sizes are tunable by the number of cycles. As shown in Fig. 8.22b, the nanopore sizes increase with cycling, indicated by BET (Brunauer–Emmett–Teller) measurements. This observation is reflected in the pore size increases in the pore size distribution. During the lithiation–delithiation processing, the

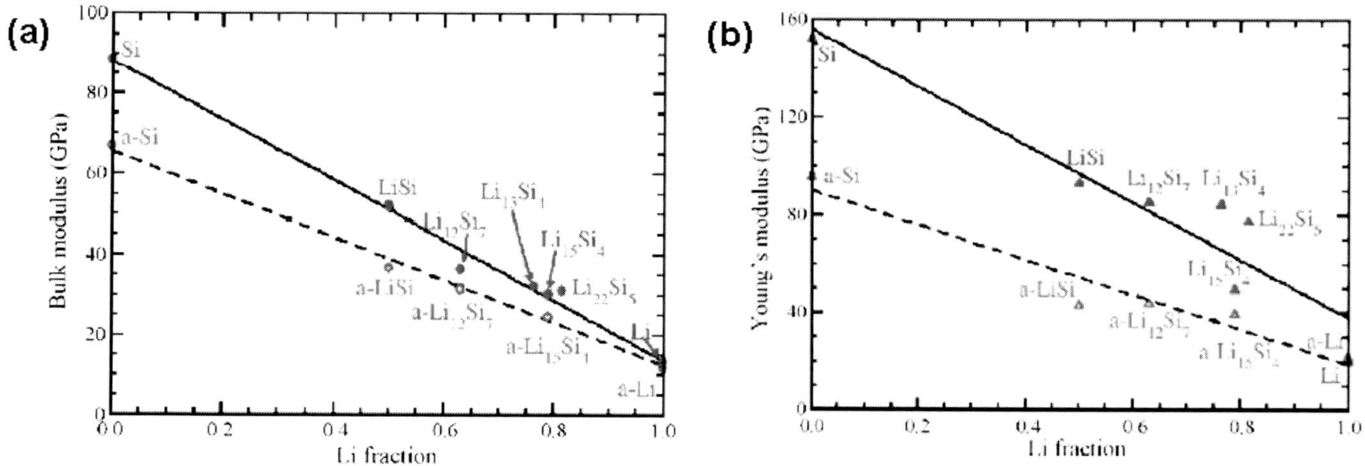

Figure 8.21 (a) Bulk modulus and (c) Young's modulus E of Li–Si alloys plotted as a function of Li fraction in crystalline (solid symbols) and in amorphous (open symbols) phases for the alloy Li_xSi. The Li fraction in the alloy is given by $x/(1 + x)$. Solid and broken lines show linear fits for the crystalline and the amorphous systems, respectively.

Si nanoparticles are likely agglomerate in order to minimize surface energy. Figures 8.22c,d show the SEM images of Si nanowires after 5 and 10 cycles. The smooth Si nanowires become completely porous with large diameter increases. The images are for Si nanowires under the discharged state. It is likely that some Si material comes off from Si nanowires. The porous structure of such Si nanowires has been proposed and tested for double layer supercapacitor applications.

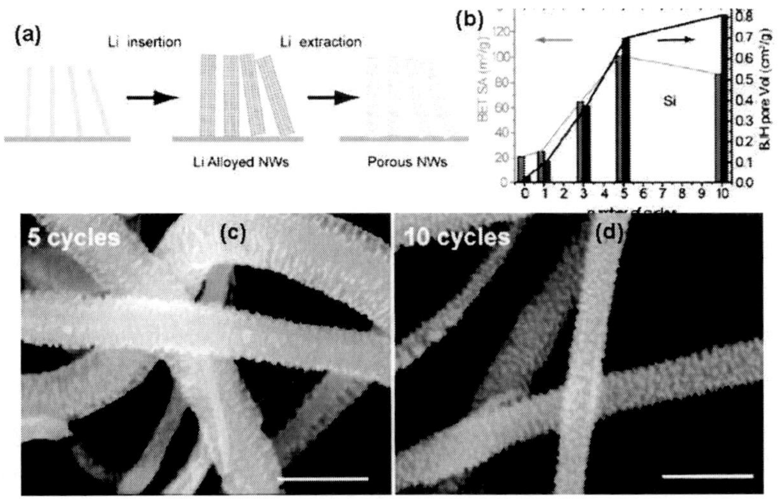

Figure 8.22 (a) Schematic illustration of pore generation by the alloying/dealloying process. As-grown silicon nanowires expand upon electrochemical Li-ion insertion. As Li ions are extracted from these 1-D nanomaterials, the nanostructures shrink but to volumes larger than the original. The increased volumes suggest pore formation inside the nanostructures. (b) The BET surface area and pore volume changes for different numbers of Li-battery cycles. The increase in pore volume is more abrupt than the increase in BET surface area. Very small pore volumes at 0 Li-battery cycles indicate that interwire spacing is negligible in our measurement range (<300 nm) for both materials. (c) and (d) SEM images of SiNWs after 5 and 10 Li-battery cycles. All the scale bars are 200 nm.

Rate dependence studies of Si nanowires or other related nanostructures are also important. High rates, that is, high power, are important for electrical vehicle and other applications. In a Li-ion battery, the rate is limited by the Li-diffusion in Si and the electrolyte diffusion through the separator and into the electrodes. Takamura *et al.* have tested the Li diffusion constant and it is 10^{-9} cm^2/s for vacuum-deposited Si and 10^{-11} cm^2/s for a silicon single crystal wafer.[49] The larger value for the vacuum-deposited film is attributed to the loose structure of the deposited films. Therefore, for high rate performance, vacuum-deposited Si is preferred over single crystal Si. Si nanostructures could lead to much better performance than bulk Si due to the smaller distance for Li-ion to travel in the Si electrode material. Si nanostructures typically have sizes of ~100 nm in one or more dimensions. The diffusion time for Li in 100 nm Si is ~0.1 s for vacuum-deposited Si and 10 s for crystalline Si. Therefore, for nanostructured Si, the limiting factor for the rate dependence is not due to the Li in the Si anode but rather to other factors such as Li^+ transport in the electrolyte. A typical diffusion constant is 10^{-6} cm^2/s for Li^+ in the electrolyte, which corresponds to ~100–2500 s for L ~100 μm. Figure 8.23a shows the rate dependent performance of 3-D porous crystalline Si particles.[42] There is a clear decrease of capacity as the rate increases from 0.2 C to 3 C. The capacity retention is 72% at a 3 C rate. The 28% loss compared with a 0.2 C rate is possibly due to the mass transfer of Li ions. Note that most of the devices are tested with the constant voltage cutoff. As the current increases, the overpotential increases and the actual cell voltage will deviate from the ideal cell voltage. Figure 8.23b shows the rate dependent performance for Si nanotubes.[40] As the rate increases from 0.2 C to 5 C, the capacity retention is excellent. One should keep in mind that the electrode thickness should be kept in mind when comparing the rate dependence performance, which will greatly affect the transport for charges and mass and, in return, affect the rate-dependence performances.

8.10 Remaining Challenges and Commercialization

There is no question that Si anodes will continue to be the most promising anode for Li-ion batteries. Although Si and its alloys have been explored for effective anode applications, nanostructured Si is still in its infancy and much potential remains to be explored. Due to its ~10-fold theoretical capacity increase compared to current state of art graphite technology, it surely holds the greatest potential for industry applications. On the other hand, due to its huge volume

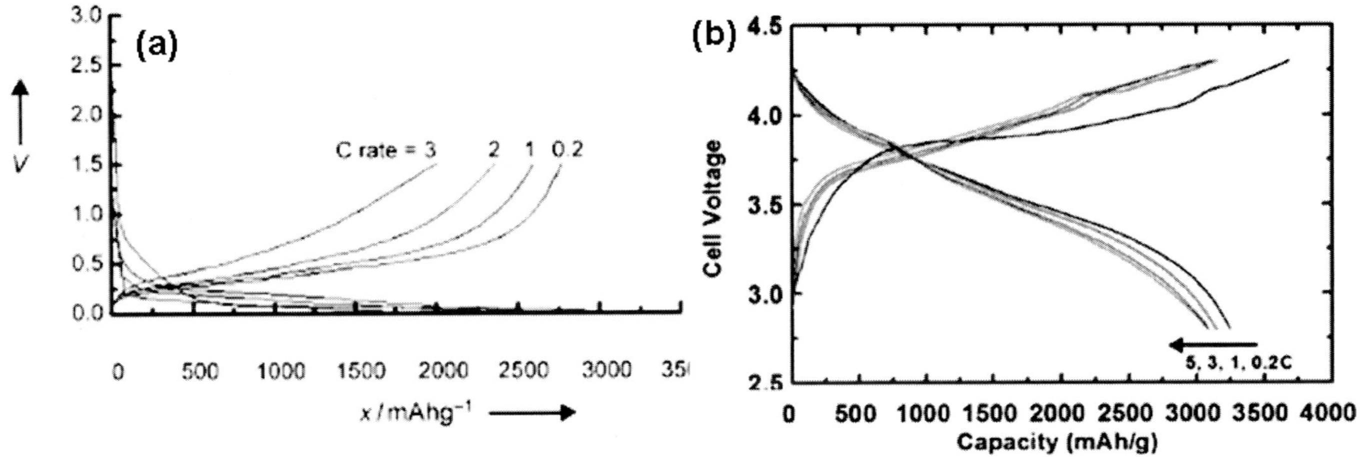

Figure 8.23 (a) Voltage profiles of the 3-D porous C–Si particles at different C rates between 0 and 1.5 V in coin-type cells. (b) Rate capability performance of the Si nanotubes in pouch-type Li-ion cells (cathode was $LiCoO_2$) between 2.75 and 4.3 V for 200 cycles. The rate was increased from 0.2 C to 5 C.

changes in the battery cell, there is much fundamental research to do. For example, the rate-dependence performance for Si nanowires has been studied. However, its full potential for high rates or high-power applications still needs to be explored. Si nanowire diameters and lengths need to be optimized for this application. Detailed modeling of charge transfer and mass transfer is needed. Impedance studies will provide much insight about the limiting factors for high rate applications.

As we discussed, Si nanowires become soft and porous, and the pores show stepwise increases in their sizes with cycling. Eventually, the porous Si nanowires tend to break which will cause them to lose electrical contact with the substrate, and thus cause capacity fading. Critical size studies need to be done to guide us in designing optimized wire diameters and lengths. Si nanowires or other nanostructures have larger surface areas per mass compared to bulk Si, which can release the mechanical strain more effectively and tend to be more stable under charge/discharge cycles. However, the larger surface area will lead to a large SEI which is not desirable for the CE. Therefore, an optimized Si nanowire diameter is needed. On top of it, uniform nanowire diameters are important for achieving the desired performance, which involves tuning the growth conditions. The SEI for Si nanowires has been discussed. However, a more complete understanding of the SEI and its effects for Li-ion batteries is still needed. Forming a stable SEI for a Si anode as in a graphite anode is critical for a long cycling lifetime. This will involve the electrode design, optimized conditioning cycling strategy, choice of the right electrolyte and many other factors. For example, Si nanotubes have been found to expand inward. Such electrodes show great potential for forming stable SEIs. Using the right cycling strategies, such as the CCCV method discussed above, Si nanowire structures with maintained crystalline Si cores to avoid breaking can be achieved. $LiPF_6$ in EC:DEC has been used widely for Si anodes. However, other electrolytes or adding other additives in $LiPF_6$ could modify the SEI greatly, with the potential to achieve a stable SEI. Currently, the CE for half-cell is typically ~98.0–99.0% and lower for full cells for Si anodes. A high CE, especially for full cells, close to 99.99%, is needed for practical electrical vehicle applications, where >200 cycles are needed. Surface coatings such as PEO on Si nanostructures, especially with elastic polymers, will modify the SEI as well. To pair up with cathodes materials such as MnO_2 which do have Li, prelithiated Si anodes are needed. Delamination of Si nanowires or other nanostructures from current collectors might be an issue. Methods to improve the binding between the substrate and electrode need to be investigated. Volume changes at the full cell level should be considered. To compensate for the large volume increases for the anode during the charging process, an appropriate pairing with a well-chosen cathode will be required, where the cathode will contract during the charging. Safety tests of Si anodes are still needed for applications. At elevated temperatures, many side reactions will happen which will release heat. Temperature dependent cycling from −30 C to 60 C is needed.

Si has been explored for commercial products already. Often, Si is used as a component in Si/graphite composites. ~20% Si will not affect the CE and cycling of the anode but will still increase the capacity. Several large companies have been actively involved with the commercialization of Si anode technology, such as Samsung, 3M, Mitsui, GM and LG Chem. Small startups such as Amprius, Envia, Nexeon, Inorganic Specialist, and Enovix have also get involved in the commercialization of such technology. Amprius and Nexeon are mainly focusing on Si nanowire based systems, while 3M is focusing on Si alloys. Roll-to-roll continuous CVD for Si nanowire deposition or other lower cost processes for Si nanowires are needed for practical commercialization. At the end, the requirements for Si anodes should be the following: high CE ~99.99%, long cycling ~500 cycles, scalable and low-cost processing and safe-cell operation.

References

1. Tarascon, J. M., Armand, M., Issues and challenges facing rechargeable lithium batteries. *Nature*, 2001, **414**(6861), 359–367.
2. Chu, A. Braatz, P., Comparison of commercial supercapacitors and high-power lithium-ion batteries for power-assist applications in hybrid electric vehicles I. Initial characterization. *Journal of Power Sources*, 2002, **112**(1), 236–246.
3. Reddy, T. B., Linden, D. *Handbook of Batteries*, 2001.
4. Huggins, R. A., *Advanced Batteries: Materials Science Aspects*, 2008.
5. Arora, P., Zhang, Z. M., Battery separators. *Chemical Reviews*, 2004, 104(10), 4419–4462.
6. Xu, K., Nonaqueous liquid electrolytes for lithium-based rechargeable batteries. *Chemical Reviews*, 2004, **104**(10), 4303–4417.

7. Chan, C. K., Peng, H. L., Liu, G., McIlwrath, K., Zhang, X. F., Huggins, R. A., *et al.* High-performance lithium battery anodes using silicon nanowires. *Nature Nanotechnology*, 2008, **3**(1), 31–35.
8. Chevrier, V. L., Dahn, J. R., First principles model of amorphous silicon lithiation. *Journal of the Electrochemical Society*, 2009, **156**(6), A454–A458.
9. Maranchi, J. P., Hepp, A. F., Evans, A. G., Nuhfer, N. T., Kumta, P. N. Interfacial properties of the a-Si/Cu : Active–inactive thin-film anode system for lithium-ion batteries. *Journal of the Electrochemical Society*, 2006, **153**(6), A1246–A1253.
10. Chow, T. S., Liu, C. A., Penwell, R. C. Direct determination of interfacial energy between brittle and polymeric films. *Journal of Polymer Science Part B-Polymer Physics*, 1976, **14**(7), 1305–1310.
11. Lee, K. L., Jung, J. Y., Lee, S. W., Moon, H. S., Park, J. W., Electrochemical characteristics of a-Si thin film anode for Li-ion rechargeable batteries. *Journal of Power Sources*, 2004, **129**(2), 270–274.
12. Ohara, S., Suzuki, J., Sekine, K., Takamura, T., A thin film silicon anode for Li-ion batteries having a very large specific capacity and long cycle life. *Journal of Power Sources*, 2004, **136**(2), 303–306.
13. Liu, W. R., Guo, Z. Z., Young, W. S., Shieh, D. T., Wu, H. C., Yang, M. H., Wu, N. L., Effect of electrode structure on performance of Si anode in Li-ion batteries: Si particle size and conductive additive. *Journal of Power Sources*, 2005, **140**(1), 139–144.
14. Obrovac, M. N., Krause, L. J., Reversible cycling of crystalline silicon powder. *Journal of the Electrochemical Society*, 2007, **154**(2), A103–A108.
15. Obrovac, M. N., Christensen, L., Structural changes in silicon anodes during lithium insertion/extraction. *Electrochemical and Solid State Letters*, 2004, **7**(5), A93–A96.
16. Kasavajjula, U., Wang, C. S., Appleby, A. J., Nano- and bulk-silicon-based insertion anodes for lithium-ion secondary cells. *Journal of Power Sources*, 2007, **163**(2), 1003–1039.
17. Li, H., Huang, X. J., Chen, L. Q., Zhou, G. W., Zhang, Z., Yu, D. P., *et al.* The crystal structural evolution of nano-Si anode caused by lithium insertion and extraction at room temperature. *Solid State Ionics*, 2000, **135**(1–4), 181–191.
18. Mao, O., Turner, R. L., Courtney, I. A., Fredericksen, B. D., Buckett, M. I., Krause, L. J., Dahn, J. R., Active/inactive nanocomposites as anodes for Li-ion batteries. *Electrochemical and Solid State Letters*, 1999, **2**(1), 3–5.

19. Yang, J., Wang, B. F., Wang, K., Liu, Y., Xie, J. Y., Wen, Z. S., Si/C composites for high capacity lithium storage materials. *Electrochemical and Solid State Letters*, 2003, **6**(8), A154–A156.

20. Chan, C. K., Peng, H. L., Liu, G., McIlwrath, K., Zhang, X. F., Huggins, R. A., Cui, Y., High-performance lithium battery anodes using silicon nanowires. *Nature Nanotechnology*, 2008, **3**, 31–35.

21. Peled, E., The electrochemical-behavior of alkali and alkaline-earth metals in non-aqueous battery systems—The solid electrolyte interphase model. *Journal of the Electrochemical Society*, 1979, **126**(12), 2047–2051.

22. Aurbach, D., Markovsky, B., Levi, M. D., Levi, E., Schechter, A., Moshkovich, M., Cohen, Y., New insights into the interactions between electrode materials and electrolyte solutions for advanced nonaqueous batteries. *Journal of Power Sources*, 1999, **81**, 95–111.

23. Gnanaraj, J. S., Thompson, R. W., DiCarlo, J. F., Abraham, K. M., The role of carbonate solvents on lithium intercalation into graphite. *Journal of the Electrochemical Society*, 2007, **154**(3), A185–A191.

24. Lee, Y. M., Lee, J. Y., Shim, H. T., Lee, J. K., Park, J. K., SEI layer formation on amorphous Si thin electrode during precycling. *Journal of the Electrochemical Society*, 2007, **154**(6), A515–A519.

25. Chan, C. K., Ruffo, R., Hong, S. S., Cui, Y., Surface chemistry and morphology of the solid electrolyte interphase on silicon nanowire lithium-ion battery anodes. *Journal of Power Sources*, 2009, **189**(2), 1132–1140.

26. Ruffo, R., Hong, S. S., Chan, C. K., Huggins, R. A., Cui, Y., Impedance analysis of silicon nanowire lithium ion battery anodes. *Journal of Physical Chemistry C*, 2009, **113**(26), 11390–11398.

27. Cui, L. F., Ruffo, R., Chan, C. K., Peng, H. L., Cui, Y., Crystalline-amorphous core-shell silicon nanowires for high-capacity and high-current battery electrodes. *Nano Letters*, 2009, **9**(1), 491–495.

28. Maranchi, J. P., Hepp, A. F., Kumta, P. N., High capacity, reversible silicon thin-film anodes for lithium-ion batteries. *Electrochemical and Solid State Letters*, 2003, **6**(9), A198–A201.

29. Ohara, S., Suzuki, J. J., Sekine, K., Takamura, T., Attainment of high rate capability of Si film as the anode of Li-ion batteries. *Electrochemistry*, 2003, **71**(12), 1126–1128.

30. Tian, B. Z., Zheng, X. L., Kempa, T. J., Fang, Y., Yu, N. F., Yu, G. H., Huang, J. L., Lieber, C. M., Coaxial silicon nanowires as solar cells and nanoelectronic power sources. *Nature*, 2007, **449**, U885–U888.

31. Goldberger, J., Hochbaum, A. I., Fan, R., Yang, P. D., Silicon vertically integrated nanowire field effect transistors. *Nano Letters*, 2006, **6**(5), 973–977.

32. Bryllert, T., Wernersson, L. E., Froberg, L. E., Samuelson, L., Vertical high-mobility wrap-gated InAs nanowire transistor. *IEEE Electron Device Letters*, 2006, **27**(5), 323–325.

33. Dong, Y. J., Yu, G. H., McAlpine, M. C., Lu, W., Lieber, C. M., Si/a-Si core/shell nanowires as nonvolatile crossbar switches. *Nano Letters*, 2008, **8**, 386–391.

34. Liu, Z. Q., Xie, S. S., Zhou, W. Y., Sun, L. F., Li, Y. B., Tang, D. S., et al., Catalytic synthesis of straight silicon nanowires over Fe containing silica gel substrates by chemical vapor deposition. *Journal of Crystal Growth*, 2001, **224**(3–4), 230–234.

35. Li, C., Gu, C., Liu, Z. T., Mi, J. X., Yang, Y., Iron-catalytic growth of prism-shaped single-crystal silicon nanowires by chemical vapor deposition of silane. *Chemical Physics Letters*, 2005, **411**(1–3), 198–202.

36. Chen, X. H., Xing, Y. J., Xu, J., Xiang, J., Yu, D. P., Rational growth of highly oriented amorphous silicon nanowire films. *Chemical Physics Letters*, 2003, **374**(5–6), 626–630.

37. Cui, L. F., Yang, Y., Hsu, C. M., Cui, Y., Carbon-silicon core-shell nanowires as high capacity electrode for lithium ion batteries. *Nano Letters*, 2009, **9**(9), 3370–3374.

38. Kim, H., Seo, M., Park, M. H., Cho, J., A critical size of silicon nano-anodes for lithium rechargeable batteries. *Angewandte Chemie-International Edition*, **49**(12), 2146–2149.

39. Magasinski, A., Dixon, P., Hertzberg, B., Kvit, A., Ayala, J., Yushin, G., High-performance lithium-ion anodes using a hierarchical bottom-up approach. *Nature Materials*, **9**(4), 353–358.

40. Park, M. H., Kim, M. G., Joo, J., Kim, K., Kim, J., Ahn, S., *et al.* Silicon nanotube battery anodes. *Nano Letters*, 2009, **9**(11), 3844–3847.

41. Song, T., Xia, J. L., Lee, J. H., Lee, D. H., Kwon, M. S., Choi, J. M., *et al.*, Arrays of sealed silicon nanotubes as anodes for lithium-ion batteries. *Nano Letters*, **10**(5), 1710–1716.

42. Kim, H., Han, B., Choo, J., Cho, J., Three-dimensional porous silicon particles for use in high-performance lithium secondary batteries. *Angewandte Chemie-International Edition*, 2008, **47**(52), 10151–10154.

43. Chan, C. K., Patel, R. N., O'Connell, M. J., Korgel, B. A., Cui, Y., Solution-grown silicon nanowires for lithium-ion battery anodes. *ACS Nano*, **4**(3), 1443–1450.

44. Cui, L. F., Hu., L., Cui, Yi, Light-weight free-standing carbon nanotube-silicon films for anodes of lithium ion batteries. *ACS Nano*, 2010, **4**(7), 3671–3678.

45. Aifantis, K. E., Hackney, S. A., Dempsey, J. P., Design criteria for nanostructured Li-ion batteries. *Journal of Power Sources*, 2007, **165**(2), 874–879.

46. Deshpande, R., Cheng, Y. T., Verbrugge, M. W., Modeling diffusion-induced stress in nanowire electrode structures. *Journal of Power Sources*, **195**(15), 5081–5088.

47. Shenoy, V. B., Johari, P., Qi, Y., Elastic softening of amorphous and crystalline Li–Si phases with increasing Li concentration: A first-principles study. *Journal of Power Sources*, **195**(19), 6825–6830.

48. Choi, J. W., McDonough, J., Jeong, S., Yoo, J. S., Chan, C. K., Cui, Y., Stepwise nanopore evolution in one-dimensional nanostructures. *Nano Letters* **10**(4), 1409–1413.

49. Yoshimura, K., Suzuki, J., Sekine, K., Takamura, T., Measurement of the diffusion rate of Li in silicon by the use of bipolar cells. *Journal of Power Sources*, 2007, **174**(2), 653–657.

Chapter 9

Porous Silicon Nanowires

Yongquan Qu[a,b] **and Xiangfeng Duan**[a,c]
[a]*Department of Chemistry and Biochemistry, University of California, Los Angeles, CA 90095, USA*
[b]*Center for Applied Chemical Research, Frontier Institute of Science and Technology, Xi'an Jiaotong University, Xi'an, 710049, China*
[c]*California Nanosystems Institute, University of California, Los Angeles, CA 90095, USA*
yongquan@mail.xjtu.edu.cn, xduan@chem.ucla.edu

9.1 Introduction

One-dimensional nanostructure has attracted significant attention due to its unique chemical and physical properties associated with the geometric feature of one-dimension. Among various one dimensional nanomaterials, silicon nanowires have been widely explored for many applications including nanoscale electronics,[1,2] flexible large area electronics,[3–6] thermoelectrics,[7] solar energy conversion,[8–10] battery electrodes,[11] and sensors.[12–14] However, silicon nanowires are rarely investigated as an optically active material because of the indirect band gap of silicon. On the other hand, silicon nanostructures with critical dimension below 10 nm, such as

Silicon and Silicide Nanowires: Applications, Fabrication, and Properties
Edited by Yu Huang and King-Ning Tu

ISBN 978-981-4303-46-0 (Hardcover), 978-981-4303-47-7 (eBook)
www.panstanford.com

silicon quantum dots and nanoporous silicon, show strong photo- and electro-luminescence in visible light range and thus have the potential applications in the fields of silicon-based optoelectronics, bioimaging and biosensors.[15–19] Recently, single crystalline porous silicon nanowires have been reported with both electrical and optical activities.[20,21] The visible photoluminescence properties of porous silicon nanowires are attributed to the deep quantum confinement effect and/or complex surface states.[20,21] Combining other fantastic properties of porous silicon nanowire, including large surface area, rich surface chemistry, high chemical reactivity, biodegradability and biocompatibility, it has been considered as an interesting material for potential applications in a wide range of fields, such as optics, electronics, optoelectronics, chemical sensors, energy production, and biology. In this chapter, we summarize the recent progress on the synthesis, characterization, pore formation mechanism, and properties of porous silicon nanowires. The applications of porous silicon nanowires as the novel photocatalysts and drug delivery and anode materials for lithium-ion battery are also discussed.

9.2 Synthesis of Porous Silicon Nanowires

Metal-assisted electroless chemical etching method is employed to synthesize porous silicon nanowires. Originally, this method was developed to produce porous silicon with excellent and repeatable optical properties.[22,23] Lately, it was also employed to synthesize vertical silicon nanowire arrays in an etchant solution composed of HF and a selected oxidant such as H_2O_2, $AgNO_3$, $KAuCl_4$, and K_2PtCl_6.[24–32] This method has been widely studied for the synthesis of silicon nanowires starting from various types of silicon wafers with different doping levels and orientations.[24–32] Recent investigations demonstrate it can be adopted to synthesize porous silicon nanowires by finely adjusting reaction conditions including the doping level of starting silicon wafers, concentration of oxidant and etching duration.[20,21,33] In general, the metal-assisted chemical etching reactions can be classified into two types: one-step reaction and two-step reaction. Both approaches are found to effectively produce porous silicon nanowires.

9.2.1 One-Step Chemical Etching

One-step chemical etching involves the immersion of clean p-type silicon pieces into an etchant solution containing 0.02 M $AgNO_3$ and 5 M HF.[20] The surface morphology of as-synthesized silicon nanowires is determined by resistivity of starting silicon wafer as shown in the transmission electron microscopy (TEM) images (Fig. 9.1). Etching produced silicon nanowires with rough surfaces when lightly doped silicon wafer (10 Ω·cm) was used. Porous silicon nanowires were obtained from silicon wafer with a resistivity of less than 0.005 Ω·cm. As-grown porous silicon nanowires were vertically oriented on silicon substrate as shown scanning electron microscopy (SEM) image in Fig. 9.2a. The selected-area electron diffraction (SAED) (Fig. 9.2b) and high-resolution transmission electron microscopy (HRTEM) (Fig. 9.2d–e) indicate that porous silicon nanowires maintain the single crystalline structure as same as that of the starting silicon wafer. The observed increase in surface roughness and porosity with decreasing resistivity can be a result of low resistivity or high dopant concentration of silicon wafers. Crystal defects and impurities (around dopants) at the silicon surface are thought to serve as nucleation sites for pore formation. As a result, higher dopant concentrations may create a lower energy barrier for charge injection across the silicon surface and a larger thermodynamic driving force for pore formation near dopants.

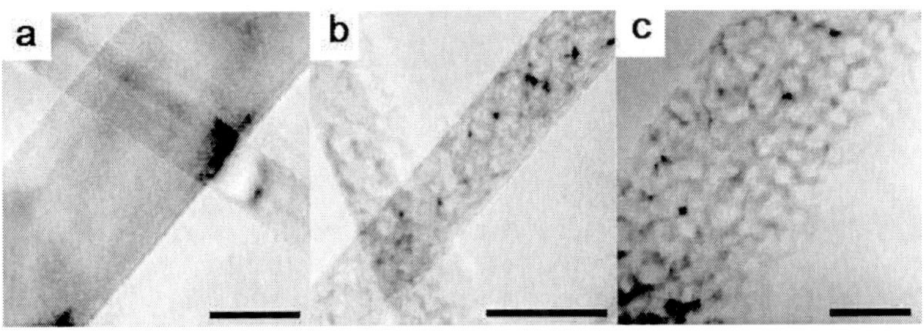

Figure 9.1 TEM images of silicon nanowires synthesized through one-step etching from p-type silicon wafers with various resistivities. (a) 10 Ω·cm; (b) 0.01 Ω·cm and (c) 0.005 Ω·cm. Scale bar is 100 nm for (a) and (b), and 50 nm for (c). Reprinted with permission from Ref. 20. Copyright (2009) American Chemical Society.

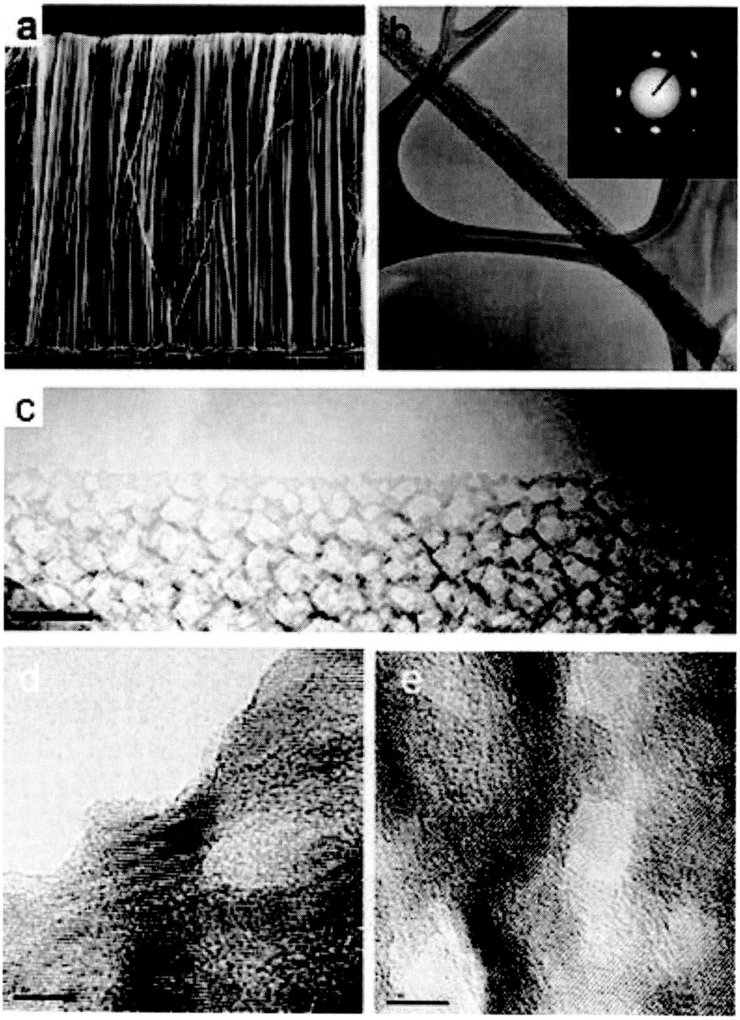

Figure 9.2 Structural characterization of porous silicon nanowires synthesized from p-Si (100) wafer with a resistivity less than 0.005 Ω·cm. (a) Cross-sectional SEM image of the porous silicon nanowire array. (b, c) TEM images of porous silicon nanowires. SAED is shown in inset b. (c, d) Phase contrast TEM images of the silicon lattice surrounding pores near the surface and in the interior of the nanowire, respectively. Scale bars are 10 μm, 200 nm, 50 nm, 5 nm and 5 nm for a, b, c, d, and e, respectively. Reprinted with permission from Ref. 20. Copyright (2009) American Chemical Society.

9.2.2 Two-Step Chemical Etching

Synthesis of silicon nanowire arrays starting from n-Si wafers with various resistivities has also been studied through one-step reaction.[21] The commercially available n-type Si(100) wafers with resistivities of 0.008–0.02, 0.3–0.8, and 1–5 Ω·cm were used. For all synthesis, the silicon pieces were soaked in an etching solution composed of 0.02 M $AgNO_3$ and 5 M HF for same time. The chemical etching on the lightly doped silicon substrate resulted in the well-aligned vertical silicon nanowire arrays as shown in Fig. 9.3a (1–5 Ω·cm silicon wafer) and Fig. 9.3c (0.3–0.8 Ω·cm silicon wafer). Solid single crystalline structures of as-synthesized silicon nanowires were observed as shown in TEM images of Figs. 9.3b and d for n-type Si (100) 1–5 and 0.3–0.8 Ω·cm wafers, respectively. These observations are consistent with previous studies.[25] In contrast, no vertical nanowires (Fig. 9.3e) can be observed when highly doped n-type Si(100) wafers with a resistivity 0.008–0.02 Ω·cm were used under the same conditions. Such growth phenomena can be explained by the dopant concentration of starting silicon wafers. In one-step reaction, the silver ions play dual roles: oxidants and ion sources for the formation of Ag nanoparticles as the etching catalysts. The silver nanoparticles preferentially nucleate near defective sites (e.g., dopants) where the energy barrier for redox reaction is lower. Therefore, a heavily doped silicon wafer with a high concentration of dopants results in a high density of nucleated Ag nanoparticles. When the highly doped silicon wafer (0.008–0.02 Ω·cm) was used, nucleated Ag nanoparticles with a high density are in close proximity of each other under the same reaction conditions for Ag deposition with short time. Hence, each nanoparticle may not be effectively confined in its own pit to ensure vertical etching. Even if the vertical etch does happen at the initial stage, the high concentration of defects of formed silicon nanowires and high concentration of Ag^+ ions in etchant solution will facilitate the nucleation of Ag nanoparticles on the side wall of silicon nanowires. Lateral etching pathways become possible, which can prevent the formation of vertical nanowire arrays. According to the above analysis, it is possible to produce the vertical nanowire array on highly doped n-Si(100) wafer if the available amount of Ag^+ could be reduced in the solution.

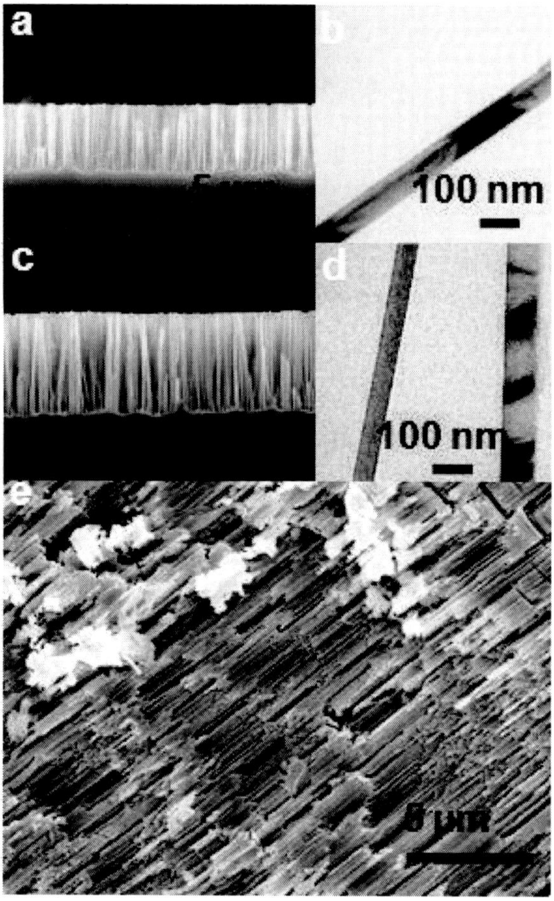

Figure 9.3 (a) Cross-sectional SEM image of the silicon nanowire array obtained by one-step etching from n-Si(100) wafer (1–5 Ω·cm) in a solution containing 5 M HF and 0.02 M $AgNO_3$ for 2 hours; (b) TEM image of the silicon nanowire in (a); (c) Cross-sectional SEM image of the silicon nanowire array obtained by one-step etching from n-Si(100) wafer (0.3–0.8 Ω·cm) in a solution containing 5 M HF and 0.02 M $AgNO_3$ for 2 hours; (d) TEM image of the silicon nanowire in (c); (e) SEM image of n-Si(100) wafer (0.008–0.02 Ω·cm) after etching in a solution containing 5 M HF and 0.02 M $AgNO_3$ for 2 hours. Reprinted with permission from Ref. 21. Copyright (2009) American Chemical Society.

However, too low concentration of $AgNO_3$ results in a slow etching rate. Herein, a two-step metal-assisted electroless chemical etching method was adopted to grow silicon nanowire arrays from heavily doped silicon wafer.[21,33] In this method, the etching catalysts, Ag nanoparticles, are pre-deposited on the surface of highly doped n-Si(100) wafer with a resistivity of 0.008–0.02 Ω·cm. The followed electroless chemical etching process is carried out in an etchant solution containing various concentrations of H_2O_2 and 4.8 M HF. Therefore, it is possible to produce silicon nanowires since pre-deposited Ag nanoparticles only function as the etching catalysts and no extra etching catalysts could be deposited on the sidewall of nanowires from etching solution. Extensive studies show that the concentration of H_2O_2 and doping level of silicon wafer are main factors to control morphology of as-synthesized silicon nanowires.

9.2.2.1 Effect of [H_2O_2]

The investigation of the concentration of H_2O_2 was carried out on silicon wafer with a resistivity of 0.008–0.02 Ω·cm.[21] The morphology evolution of the as-grown silicon nanowires from the highly doped n-Si (100) wafers with different concentrations of H_2O_2 is summarized in Fig. 9.4. The well-aligned silicon nanowire arrays were obtained at all concentrations of H_2O_2. The higher concentration of H_2O_2 was, the longer length of as-synthesized silicon nanowires was. When the concentration of H_2O_2 was 0.1 M, nonporous nanowires were obtained as shown in TEM images (Fig. 9.4b,d) for 30 and 60 min etching, respectively. The smooth surface state of the silicon nanowires with 30 min etching has been observed (Fig. 9.4b), while longer etching duration (60 min) led to the rough surfaces of the porous silicon nanowires. With increasing the etching time at the concentration of 0.15 M H_2O_2, the silicon nanowires started to evolve from solid nanowires with rough surface (Fig. 9.4f) to porous shells surrounding the solid cores (Fig. 9.4h). Increasing the concentration of H_2O_2 over 0.2 M eventually produced entirely porous nanowires as shown in Fig. 9.4. The TEM studies clearly demonstrate that the increase portion of the nanowires becomes porous with the increase of the H_2O_2 concentration and reaction time.

HRTEM and SAED (inset) patterns from as-grown silicon nanowires with 0.1 M H_2O_2 for 30 min, 0.2 M H_2O_2 for 30 min, and 0.3 M H_2O_2 for 60 min are shown in Fig. 9.5. Similar to that of porous silicon nanowires synthesized from p-type Si(100) wafers, the continuous lattice fringes and the single crystal-like diffraction pattern are observed in all samples, demonstrating that the etching process did not destroy the crystalline integrity of the porous silicon nanowires.

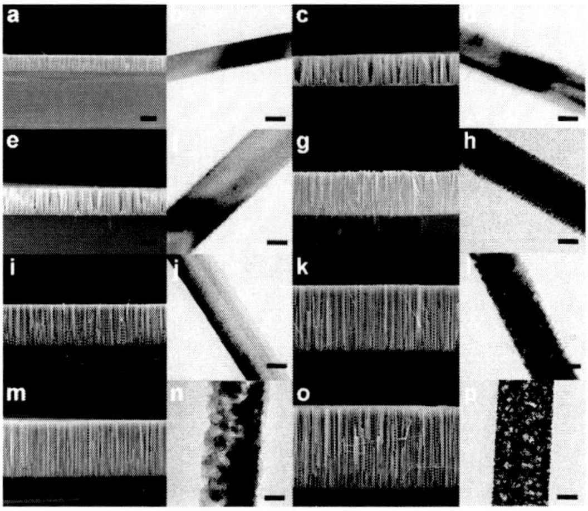

Figure 9.4 SEM and TEM images show the evolution of silicon nanowire morphology from n-Si(100) with 0.008–0.02 Ω·cm resistivity in etchant solutions composed of 4.8 M HF and variable concentrations of H_2O_2 through a two-step reaction. (a, b) 0.1 M H_2O_2 for 30 min; (c, d) 0.1 M H_2O_2 for 60 min; (e, f) 0.15 M H_2O_2 for 30 min; (g, h) 0.15 M H_2O_2 for 60 min; (i, j) 0.2 M H_2O_2 for 30 min; (k, l) 0.2 M H_2O_2 for 60 min; (m, n) 0.3 M H_2O_2 for 30 min; and (o, p) 0.3 M H_2O_2 for 60 min. The scale bars for all SEM and TEM images is 10 μm and 60 nm, respectively. Reprinted with permission from Ref. 21. Copyright (2009) American Chemical Society.

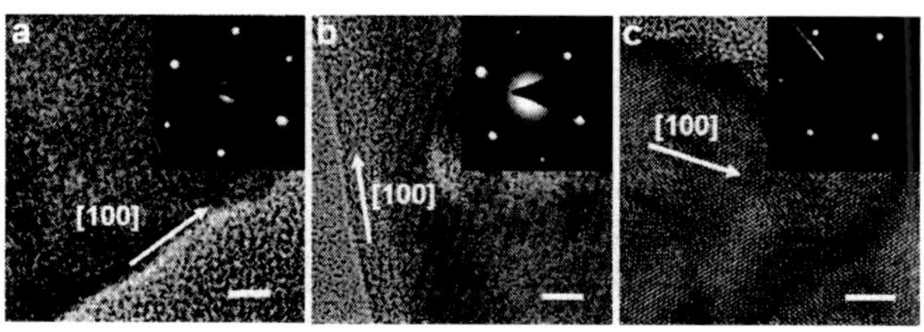

Figure 9.5 High-resolution TEM images with SAED patterns of nonporous and porous silicon nanowires. (a) Nanowires obtained from the reaction in 0.1 M H_2O_2 and 4.8 M HF for 30 min. (b) Nanowires obtained from the reaction in 0.2 M H_2O_2 and 4.8 M HF for 30 min. (c) Nanowires obtained from the reaction in 0.3 M H_2O_2 and 4.8 M HF for 60 min. All scale bars are 5 nm. Reprinted with permission from Ref. 21. Copyright (2009) American Chemical Society.

9.2.3 Effect of Doping Levels of Silicon Wafers

Four different n-type silicon (100) wafers with increasing doping levels (resistivity: 1–5 Ω·cm, 0.3–0.8 Ω·cm, 0.008–0.016 Ω·cm and 0.001–0.002 Ω·cm) were used to study the effect of doping levels of starting silicon wafers. For all four silicon wafers, the uniform vertical silicon nanowire arrays were obtained under 0.1 M, 0.3 M, and 0.6 M H_2O_2 concentrations. With increasing the concentration of H_2O_2, the length of as-synthesized silicon nanowires from each silicon wafer became longer. When lightly doped silicon wafers with resistivities of 1–5 Ω·cm and 0.3–0.8 Ω·cm were used, only solid silicon nanowires were obtained. The surface of nanowires was either smooth or rough, which was determined by the concentration of H_2O_2. When the resistivity of starting silicon wafer was lowered to 0.008–0.016 Ω·cm, the porous silicon nanowires were observed at H_2O_2 concentration of 0.3 M. Only solid nanowires with smooth or rough surfaces were observed with a H_2O_2 concentration below 0.3 M. For silicon wafer with a resistivity 0.001–0.002 Ω·cm, pore structures were already observed even at a low H_2O_2 concentration of 0.1 M. The results clearly demonstrate that heavily doped n-Si(100) wafer is favorable to form porous silicon nanowires at the same etching conditions.

9.2.4 Mechanism of Formation of Porous Silicon Nanowires

It is well accepted that the formation of silicon nanowire array is *via* a nanoelectrochemical reaction process.[24–32] In the two-step chemical etching, silicon is oxidized into silicon oxide with the assistance of the Ag nanoparticle at the interface of Ag/Si. At the same time, electrons are transferred to Ag nanoparticles to reduce H_2O_2 on the surface of metal particles. The removal of silicon oxide by HF makes the etching continue to happen. The possible chemical reactions are listed as the following:

$$\text{Cathode:}\quad H_2O_2 + 2H^+ + 2e^- \rightarrow 2H_2O \qquad E^0 = 1.78\ \text{V} \tag{9.1}$$

$$\text{Anode:}\quad Si + 2H_2O \rightarrow SiO_2 + 4H^+ + 4e^- \qquad E^0 = 0.91\ \text{V} \tag{9.2}$$

$$SiO_2 + 6HF \rightarrow [SiF_6]^{2-} + 2H_2O + 2H^+ \tag{9.3}$$

$$\text{The total reaction is:}\quad Si + 2H_2O_2 + 6F^- + 4H^+ \rightarrow [SiF_6]^{2-} + 4H_2O \tag{9.4}$$

The standard potential for the total reaction is 2.69 V, indicating that the etching reaction can happen spontaneously. According to the Nernst equation of the total reaction: $\Delta E = \Delta E_0 - \frac{0.059}{4}\left\{\log[SiF_6^{2-}]/[H_2O_2]^2 - \log[H^+]^4[F^-]^6\right\}$, increasing the H_2O_2 concentration results in a higher potential for the etching process. Hence, the driving force for the etching process becomes larger and thus increases the etching rate. Since H_2O_2 itself cannot etch the silicon off, above reactions cannot be used to explain the formation of porous structure. One possibility for the formation of porous structures is the nucleation of metal nanoparticle catalysts on the side wall of the silicon nanowires during the etching process. Previous studies suggested that the silver nanoparticles trapped at the roots of silicon nanowires can be converted into Ag^+ ion by H_2O_2 as follows:[25]

$$\text{Ag dissolving: } 2Ag+H_2O_2+2H^+ \rightarrow 2Ag^+ +2H_2O \tag{9.5}$$

$$\text{Ag}^+ \text{ reduction: } Si+4Ag^+ +6F^- \rightarrow 4Ag+SiF_6^{2-} \tag{9.6}$$

Considering the standard potentials of 2.57 V for Eqs. 9.1 and 9.5, the dissolving process of Ag metal is also thermodynamically favored in the presence of H_2O_2. At low concentration of H_2O_2, the small amount of Ag^+ ions could be generated at the interface of Ag/Si. They can quickly withdraw electrons from silicon near the Ag/Si interface and be recovered into original Ag metal again as shown in Eq. 9.6 at the interface of Ag/Si also increases. Then, the generated Ag^+ ions may not be 100% recovered into the original Ag nanoparticles and have more chance to diffuse out. When the concentration of H_2O_2 in the etching solution reach a certain threshold, out-diffused Ag^+ ions can re-nucleate on the side wall of silicon nanowires by extracting electrons from the silicon nanowires and create new etching sites for new etching pathway.

Doping level of starting silicon wafer is another important factor for the formation of porous silicon nanowires. Since only few available defective sites within lightly doped silicon wafer, there is less chance for out-diffused Ag^+ ions to re-nucleate on the side wall of silicon nanowires. The cross-sectional SEM images of as-produced silicon nanowire arrays without any further treatment show a clean surface of nanowires from 1–5 Ω·cm and 0.3–0.8 Ω·cm silicon wafers (Fig. 9.6a,b). In contrast, nanowires from two heavily doped wafers (Fig. 9.6c,d) exhibit rough surfaces with large numbers of specks. HRTEM and energy-dispersive X-ray (EDX) spectrum (Fig. 9.7a–d) analysis indicate those specks are Ag nanoparticles. Such an observation confirms that dopants as active sites for nucleation of silver nanoparticles on the side wall of silicon nanowires. With continuous Ag nucleation on the side wall of the highly doped nanowires, the size of the Ag at the root of the highly doped silicon nanowires will become smaller than original size. The direct evidence of the shrinking Ag nanoparticles is the size evolution of diameter of as-synthesized silicon nanowires from top to bottom. The trapezoid structure was observed in nanowires starting from highly doped silicon wafers (Fig. 9.7e1–e3). The SEM images clearly show that diameter near the tip is smaller than that near the roots, indicating the size of the Ag nanoparticles reduces during the etching process. On the other hand, the diameter of nanowires from lightly doped wafers (Fig. 9.7e4–e6) is mostly unchanged, indicating that the size of the Ag does not significantly vary during the growth of silicon nanowires. According to the above analysis and observation, the concentration of oxidant H_2O_2 and doping level of silicon wafer are main factors for the formation of porous silicon nanowires for two-step etching.

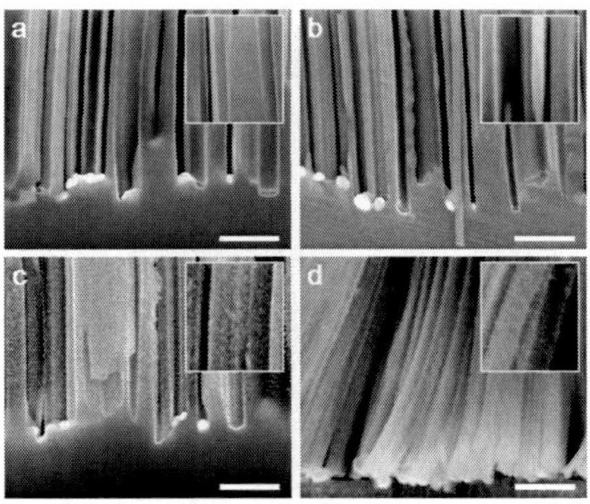

Figure 9.6 Cross-sectional SEM image of nanowires evolved from four types wafers with reaction time of 15 min in the etchant solutions containing 0.3 M H_2O_2 without immersing in the nitric acid solution. The insets are enlarged SEM images for each type of the nanowires. (a) Silicon nanowires from 1–5 Ω·cm wafer. (b) Silicon nanowires from 0.3–0.8 Ω·cm wafer. (c) Silicon nanowires from 0.008–0.016 Ω·cm wafer. (d) Silicon nanowires from 0.001–0.002 Ω·cm wafer. Reprinted with permission from Ref. 33. Copyright (2011) American Chemical Society.

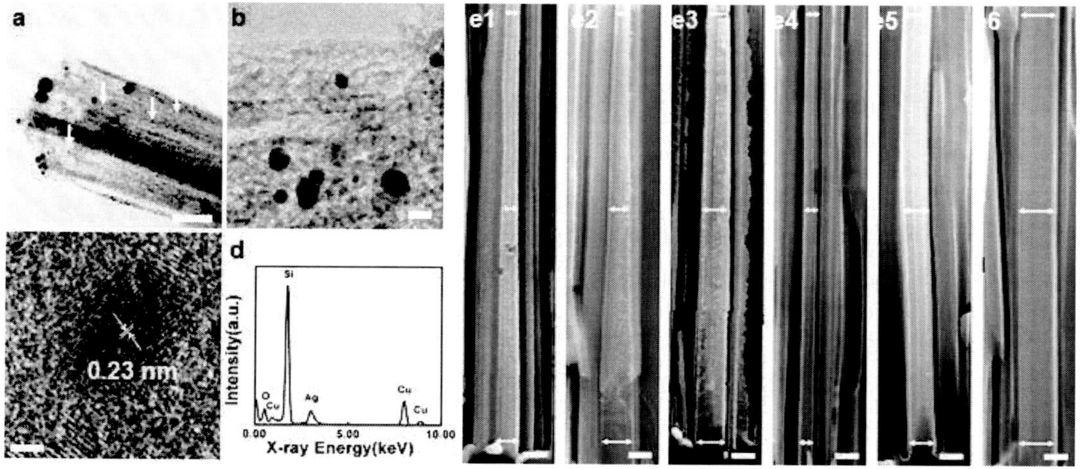

Figure 9.7 (a, b) TEM images of a porous silicon nanowire decorated by Ag nanoparticles on the sidewall. The silicon nanowire was made from 0.008–0.016 Ω·cm silicon wafer in 0.3 M H_2O_2 for 15 min without the treatment in the concentrated nitric acid. Small Ag nanoparticles are indicated by the white arrows. The scale bars are 80 nm and 20 nm for (a) and (b), respectively. (c) HRTEM of Ag nanoparticles on the sidewall of the porous silicon nanowire. The lattice spacing 0.23 nm corresponds to Ag (111) plane. The scale bar is 2 nm. (d) An EDX spectrum recorded from one nanowire. (e1–e3) Cross-sectional SEM images of nanowires from 0.008–0.016 Ω·cm silicon wafer showing the trapezoid structure. The diameter of the upper part is smaller than the lower part, indicating the shrinking size of Ag as the reaction proceeds. (e4–e6) Cross-sectional SEM images of nanowires from 0.3–0.8 Ω·cm silicon wafer showing nearly constant wire width/diameter. The scale bar is 200 nm for e1–e6. Reprinted with permission from Ref. 33. Copyright (2011) American Chemical Society.

The formation of porous silicon nanowires from lightly doped n-Si(100) wafer have also been reported using two-step reaction.[34,35] The reactions required either higher oxidizer concentration or high temperature. Although the detail of pore formation mechanism is not clear, high concentration of oxidizer and high reaction temperature may play important roles for pore formation.

Another work reports the formation of porous silicon nanowires regardless of silicon resistivity using two-step reactions.[36] The main difference of this work to our studies[21,33] is the existence of surfactant ethanol in etching solution. Concentrations of H_2O_2 and ethanol were critical for the formation of porous structures. In this work, the metal ions are proposed to be responsible for the pore formation since no metal nanoparticles are detected within the pores of silicon nanowires.

9.3 Properties of Porous Silicon Nanowire

Silicon quantum dots and porous silicon have been extensively studied for the applications of bioimaging, optoelectronics, and solar energy harvesting. Given the one-dimensional nature of porous nanowire and the small size of the crystalline silicon nanowall of porous nanowires, the interesting optical and electrical properties should be expected for this material.

9.3.1 Optical Properties

Both n-type and p-type porous silicon nanowires show a broad visible emission near infrared range.[20,21] Figure 9.8 shows photoluminescence spectrum of bulk porous silicon nanowire and confocal photoluminescence image of individual nanowire synthesized from 0.008–0.02 Ω·cm silicon wafer. The visible emission centered around 650 nm was observed as shown in Fig. 9.8a. It may be generated from the deep quantum confinement in the porous structure. The broad emission may be attributed to broad size distribution of silicon nanowalls in the porous structure. The overlapping between the fluorescence image (Fig. 9.8b) and the optical microscopy image (Fig. 9.8c) indicates the origination of photoluminescence from silicon nanowires themselves. Further surface treatments can be used to tune the photoluminescence properties of porous silicon nanowires. For example, HF-treatment on porous silicon nanowires results in a red-shift of photoluminscence spectrum due to the surface Si–H bonds.[37] The blue-shift of photoluminscence spectrum is observed for porous silicon nanowires after selenization treatment at high temperature.[35]

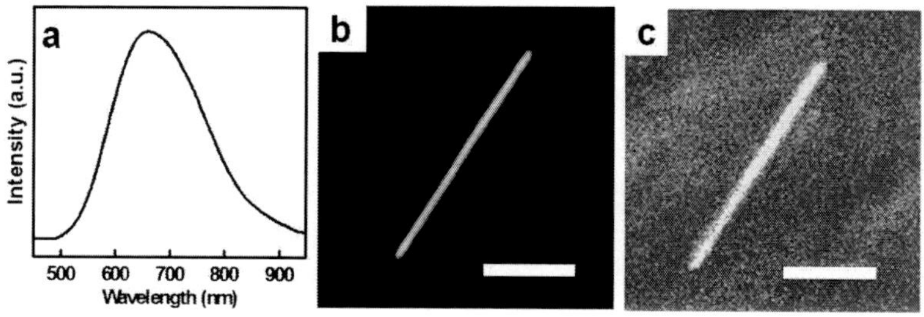

Figure 9.8 (a) Photoluminescence spectrum of porous silicon nanowires obtained with 60 min etching in a solution with 0.30 M H_2O_2. (b) Optical micrograph of a single porous silicon nanowire. (c) Confocal photoluminescence image of the same single porous silicon nanowire. The scale bar in (b) and (c) is 3 μm. Reprinted with permission from Ref. 21. Copyright (2009) American Chemical Society.

9.3.2 Electrical Properties

The investigation of the electrical properties was carried out on porous silicon nanowires from n-Si(100) wafer with a resistivity of 0.008–0.02 Ω·cm. Electrical transport studies on individual nanowire used a simple back-gated device configuration on a silicon substrate, where an underlying silicon was used as the back gate, 100 nm thick silicon nitride (SiN_x) as the gate dielectric, and Ti/Au (100 nm/50 nm) film deposited as the source and drain electrodes. A typical fabricated device is shown in Fig. 9.9a. A set of drain current versus drain bias voltage (I_d-V_d) relations (Fig. 9.9b) at various gate voltages (V_g) in steps of 5 V suggest an n-channel transistor behavior for a porous silicon nanowire device. This electrical characterization of porous silicon nanowires is consistent with the doping type of the starting silicon wafer. However, the overall resistance is rather large compared to that of starting silicon wafer. It can be attributed to: the porous structure which reduces the current conducting volume and increases large surface depletion and band gap due to quantum confinement effect; chemical etching which removes the dopants; and non-ideal source drain contact.

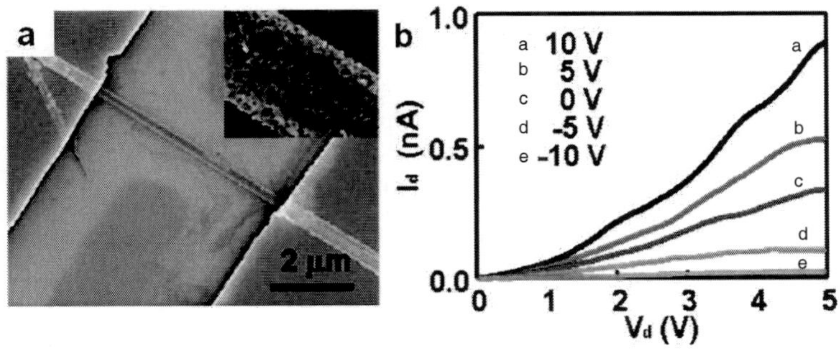

Figure 9.9 (a) SEM image of a porous silicon nanowire device. The inset shows a higher resolution SEM image highlighting the porous structure. The scale bar in inset is 100 nm. (b) Drain current (I_d) versus drain voltage (V_d) at variable gate voltages. Reprinted with permission from Ref. 21. Copyright (2009) American Chemical Society.

9.3.3 Porosity of the Porous Silicon Nanowires

A recent work has systematically summarized the porosity and specific surface area of the porous and non-porous nanowires obtained from silicon wafers with various resistivities or under different etching conditions.[33] The study indicates that the surface area of as-synthesized silicon nanowires can be readily tuned by controlling the resistivity of starting silicon wafer and reaction duration. The results are summarized in Fig. 9.10. For a reaction time of 30 min in an etching solution of 0.3 M H_2O_2 and 4.8 M HF, the surface areas are 30, 40, 240, and 370 $m^2 \cdot g^{-1}$ for silicon nanowires synthesized from silicon wafers with resistivities of 1–5, 0.3–0.8, 0.008–0.016, 0.001–0.002 Ω·cm, respectively. The results are consistent to the characterization of nonporous silicon nanowires from lightly doped silicon wafers (1–5 and 0.3–0.8 Ω·cm) and of porous structure from highly doped silicon wafers (0.008–0.016 and 0.001–0.002 Ω·cm). For silicon nanowires starting from same silicon wafer (0.008–0.016 Ω·cm) in an etching solution of 0.3 M H_2O_2 and 4.8 M HF, the surface areas also increase from 160 $m^2 \cdot g^{-1}$, 240 $m^2 \cdot g^{-1}$ to 337 $m^2 \cdot g^{-1}$ for 15, 30, and 60 min etching, respectively. The pore distribution of silicon nanowires synthesized from lightly doped wafers is flat over all the range, indicating the lack of pores. On the other hand, the porous nanowires synthesized from highly doped silicon wafers have a broad pore size distribution. The average pore sizes are 10.0 nm and 9.7 nm for nanowires synthesized from the silicon wafers with resistivity of 0.001–0.002 Ω·cm and 0.008–0.016 Ω·cm, respectively.

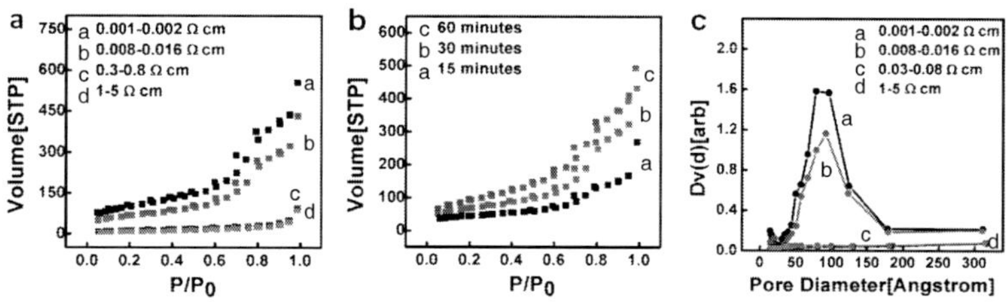

Figure 9.10 (a) Nitrogen adsorption/desorption isotherms of porous and nonporous silicon nanowires with various dopant concentrations. (b) Nitrogen adsorption/desorption isotherms of 0.08–0.016 Ω·cm porous nanowires with various reaction time. (c) Corresponding BJH pore size distributions for samples in (a). Reprinted with permission from Ref. 33. Copyright (2011) American Chemical Society.

9.4 Applications of Porous Silicon Nanowire

Porous materials have attracted considerable interest due to the potential applications in broad areas including catalysis, integrated optics, energy harvesting and biotechnology.[38–43] For example, porous materials are widely used as supports for catalysts because of the abundant pores, large surface areas and the ease of recycling compared to nanopowders. Porous silicon nanowires possess large surface area and exhibit optical and electrical activities. Combining the features of biocompatibility and biodegradability, it has been investigated in the fields of photocatalysis, lithium battery, and drug delivery.

9.4.1 Photocatalysis

Compared to TiO_2, the most widely used photocatalyst absorbing UV light only (5% overall sunlight[45,46]), the porous silicon nanowires show a much broader absorption (Fig. 9.11a) spanning over the entire spectral range from UV to visible and near IR (Fig. 9.11a).

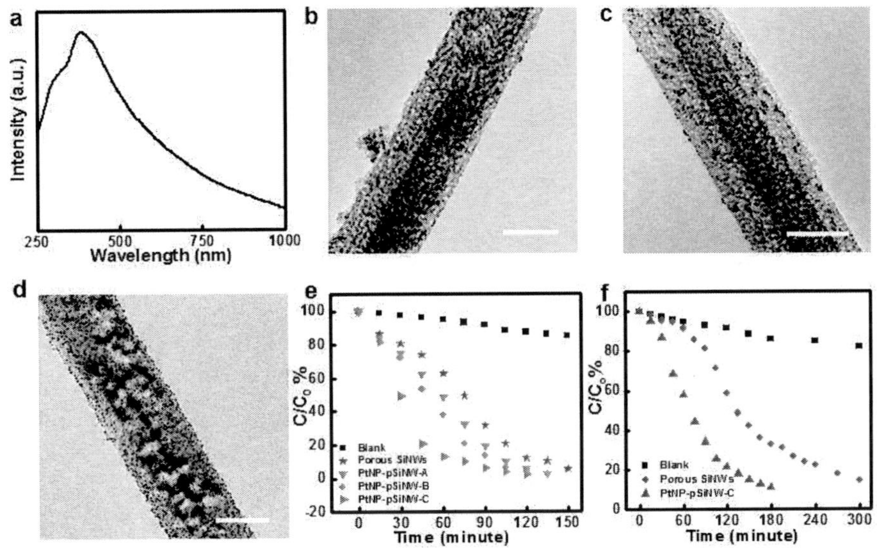

Figure 9.11 Characterization and photoactivity of the porous SiNWs and PtNP loaded porous SiNWs. (a) Absorption spectrum of the porous SiNWs. The concentration of porous SiNWs was set at 0.1 mg/ml. (b) TEM image of the PtNP-pSiNW-A. (c) TEM image of the PtNP-pSiNW-B. (d) TEM image of the PtNP-pSiNW-C. (e) IC degradation catalyzed by the porous SiNWs and Pt loaded porous SiNWs. Squares represent the IC under the light irradiation without photocatalysts. Stars, down-triangles, circles, and right-triangles represent the catalytic behavior of the porous SiNWs, PtNP-pSiNW-A, PtNP-pSiNW-B and PtNP-pSiNW-C, respectively. (f) 4NP degradation catalyzed by the porous SiNWs and PtNP loaded porous SiNWs. Squares represent the 4NP without photocatalysts. Diamonds and up-triangles represent the catalytic behavior of the porous SiNWs and the PtNP-pSiNW-C, respectively. The scale bar is 100 nm. [Qu, Y. Q., Zhong, X., Li, Y. J., Liao, L., Huang, Y., Duan, X. F., *J. Mater. Chem.*, 2010, **20**, 3590–3594.][44] Reproduced by permission of the Royal Society of Chemistry.

The visible emission near red infrared (IR) indicates that excitons generated within the porous silicon nanowires could be energetic enough to drive photoelectrochemical reactions. Therefore, porous silicon nanowires can be the one of potential candidates for solar energy conversion.

Photocatalytic activity of porous silicon nanowires synthesized from highly doped n-Si(100) wafers (0.01–0.02 Ω·cm) has been investigated.[44] In order to enhance the photocatalytic activity, porous silicon nanowires were modified with Pt nanoparticles (PtNPs). The difference in their Fermi levels can introduce a Schottky barrier, which facilitates the separation of photogenerated electron-hole pairs. The as-synthesized PVP-capped PtNPs[47] typically have negative surface charges and can be readily attached onto APTMS modified porous silicon nanowires through electrostatic force. Controlled densities of PtNPs on porous silicon nanowires were achieved by adding various amounts of PtNPs into a fixed volume of porous SiNW suspension. Three catalysts with different densities of PtNPs were prepared and named as PtNP-pSiNW-A (Fig. 9.11b), PtNP-pSiNW-B (Fig. 9.11c) and PtNP-pSiNW-C (Fig. 9.11d), respectively.

Photodegradation of indigo carmine (IC) and 4-nitrophenol (4NP) were carried out to evaluate the photoactivity of porous silicon nanowires. The photocatalytic activity of the four catalysts on the degradation of IC is compared in Fig. 9.11e. After 60 min of irradiation, the percentages of degraded IC are 4.7%, 37.2%, 51.1%, 62.2%, and 86.9% for control experiment without catalysts, porous SiNWs, PtNP-pSiNW-A, PtNP-pSiNW-B, and PtNP-pSiNW-C, respectively. The results clearly indicate that porous silicon nanowires can function as effective photocatalysts and Pt-loaded porous silicon nanowires exhibit much more efficiency than the porous silicon nanowires only. The porous SiNWs and PtNP-pSiNW-C were also used to degrade 4NP (Fig. 9.11f). After 90 min of light irradiation for the porous SiNWs and the PtNP-pSiNW-C, 18.8% and 66.0%, respectively, of the 4NP was degraded. The results further demonstrate the photocatalytic ability of the porous SiNWs and improved photoactivity of porous silicon nanowires by introducing PtNPs.

9.4.2 Platform for Drug Delivery

Combining the attractive properties of porous silicon nanowires including high surface area, intrinsic luminescence, and biocompatiblity, porous silicon nanowires exhibit a promising application for biological systems.[48] Recently, we reported the tailored porous silicon nanowires with 200 ~ 400 nm length served as a platform for drug delivery in biological systems. The delivery system is established with a pH-responsive nanovalve-system consisting of an aromatic amino group and a nanovalve (cyclodextrin) as illustrated in Fig. 9.12a. The nanovalve can be tightly closed at the physiological pH of 7.4 to seal cargo within the pores and be released under the acidic conditions (pH<6.0) to open the pores.

The fluorescent biological staining dye Hoeschst 33342 was chosen as a model cargo to verify the functional operation of the drug delivery system based on porous silicon nanowires. Figure 9.12a shows the release profiles of cargo dye Hoeschst 33342 in water by monitoring the concentration of cargo dye as a function of time. When pH of water was 7.0, the release profile gave a flat baseline (Fig. 9.12b), indicating the nanovalves on porous silicon nanowires remained tightly closed. With lowering the pH, the nanovalves detached from porous silicon nanowires and released the cargo. In the control experiment without nanovalves, the release profile of porous silicon nanowires displayed a flat one for the cargo was washed away during the sample preparation. The similar release profile was also observed in DMEM cell culture medium (Fig. 9.12c).

9.4.3 Lithium-Ion Battery

Silicon has the highest known theoretical Li storage capacity of 4200 mAh g^{-1}, which is much higher than that of the commercially available graphites (372 mAh g^{-1}) nowadays. Nanoporous structure can improve the battery performance due to the following benefits: (1) flexibility to accommodate large swing in volume; (2) short length for Li diffusion and (3) acceleration of charge transfer. Therefore, porous silicon nanostructures are the potential anode materials for lithium-ion battery.

A recent study indicates that the capacities of lithium-ion batteries using porous silicon nanowires as anodes are much higher than those using nonporous silicon nanoparticles.[49] With the same content of carbon black, the initial capacities of porous silicon nanowires (2172 mAh g^{-1}) is five times higher than those of 56 nm silicon nanoparticles (477 mAh g^{-1}). After 20 cycles, porous silicon nanowires still exhibit a capacity of 815 mAh g^{-1}, which is

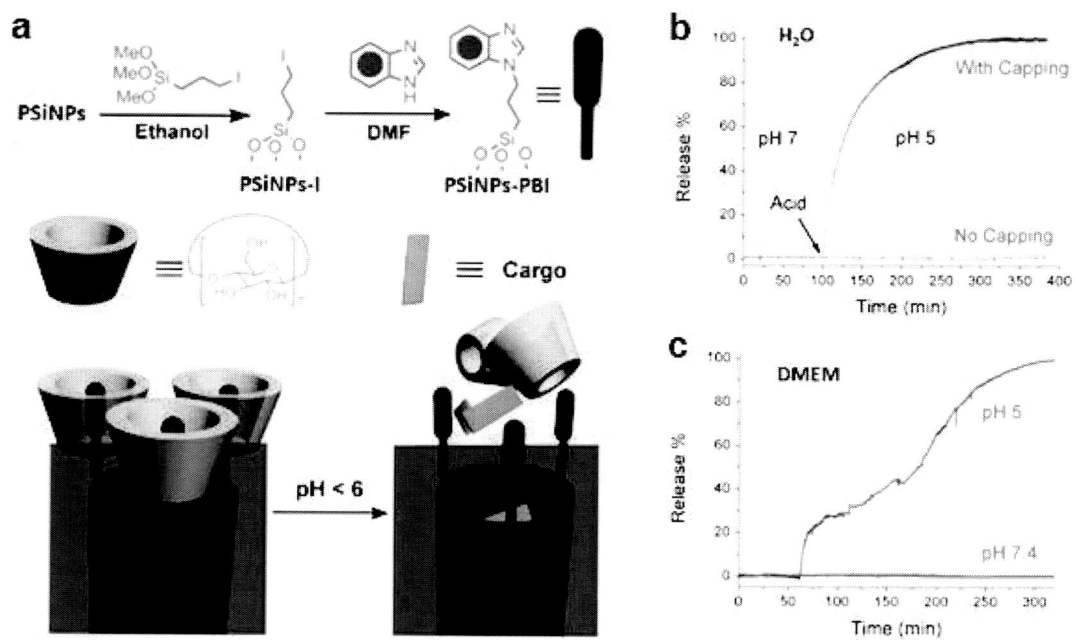

Figure 9.12 (a) Schematic illustration of pH-operated mechanism of the nanovalves. (b) Plot of the release of Hoechst 33342 from PSiNPs in aqueous solution. The straight, horizontal line shows the release profile from the control experiment in which no cyclodextrin was used for capping. (c) Release profile generated in DMEM tissue culture medium. The irregular, curved line shows the functioning of the nanovalve when acid was added at 60 min. The red line shows the absence of leakage when the pH was unchanged. Reprinted with permission from Ref. 48. Copyright (2011) American Chemical Society.

3.36 times higher than that of silicon nanoparticles. The highly accessible surfaces and short charge and Li diffusion lengths are used to explain the higher capacity of lithium-ion batteries based on porous silicon nanowires. Kinetic study also indicates that porous silicon nanostructures could enhance the electrochemical activity and results in a low energy barrier for alloying process.

9.5 Conclusion

Porous silicon nanowires starting from n or p type silicon wafers have been synthesized through Ag-assisted electroless chemical etching method. Current reports demonstrate that large scale n-type Si (100) porous/non-porous single crystalline silicon nanowire arrays can be synthesized in a highly controlled manner by systematically tuning the dopant concentration of the silicon wafers and the concentration of oxidants. Electrical transport measurements show these nanowires are conductive and optical studies indicate that they can exhibit visible luminescence. The combination of electrical and optical properties in such a porous silicon nanowire may open new opportunities for nanoscale optoelectronic devices, solar energy harvesting and conversion, and sensors. Porous silicon nanowires have been demonstrated as effective photocatalysts and drug delivery and anodes materials for lithium-ion battery.

Acknowledgment

We acknowledge support by the NIH Director's New Innovator Award Program, part of the NIH Roadmap for Medical Research, through Grant 1DP2OD004342-01. We acknowledge Electron Imaging Center for Nanomachines (EICN) at UCLA for support in TEM, and Nanoelectronics Research Facility at UCLA for support in etching process and device fabrication.

References

(1) Cui, Y., Duan, X., Hu, J., Lieber, C. M., *J. Phys. Chem. B*, 2000, **104**, 5213–5216.

(2) Huang, Y., Duan, X. F., Cui, Y., Lauhon, L. J., Kim, K. H., Lieber, C. M., *Science*, 2001, **294**, 1313–1317.

(3) McAlpine, M. C., Friedman, R. S., Jin, S., Lin, K. H., Wang, W. U., Lieber, C. M., *Nano Lett.*, 2003, **3**, 1531–1535.

(4) Duan, X., *MRS Bull.*, 2007, **32**, 134–141.

(5) Duan, X., *IEEE Trans. Electron Devices* 2008, *55*, 3056–3062.

(6) Javey, A., Nam, S., Friedman, R. S., Yan, H., Lieber, C. M., *Nano Lett.*, 2007, **7**, 773–777.

(7) Hochbaum, A. I., Chen, R., Delgado, R. D., Liang, W., Garnett, E. C., Najarian, M., Majumdar, A., Yang, D., *Nature*, 2008, **451**, 163–168.

(8) Tian, B., Zheng, X., Kempa, T. J., Fang, Y., Yu, N., Huang, J., Lieber, C. M., *Nature* 2007, **449**, 885–890.

(9) Hwang, Y., Bukai, A., Yang, P., *Nano. Lett.*, 2009, **9**, 410–415.

(10) Garnett, E. C., Yang, P., *J. Am. Chem. Soc.*, 2008, **130**, 9224–9225.

(11) Chan, C. K., Peng, H., Liu, G., McIlwrath, K, Zhang, X., Huggins, R. A., Cui, Y., *Nat. Nanotechnol.*, 2008, **3**, 31–35.

(12) Zheng, G., Patolsky, F., Cui, Y., Wang, W. U., Lieber, C. M., *Nat. Biotechnol.*, 2005, **23**, 1294–1301.

(13) Patolsky, F., Timko, B. P., Yu, G., Fang, Y., Greytak, A. B., Zheng, G., Lieber, C. M., *Science*, 2006, **313**, 1100–1104.

(14) Yang, C., Barrelet, C. J., Capasso, F., Lieber, C. M., *Nano Lett.*, 2006, **6**, 2929–2934.

(15) Lin, V. S. Y., Motesharei, K., Dancil, K. P. S., Sailor M. J., Ghadiri, M. R., *Science*, 1997, **278**, 840–843.

(16) Tasciotti, E., Liu, X. W., Bhavane, R., Plant, K., Leonard, A. D., Price, B. K., *et al.*, *Nat. Nanotechnol.*, 2008, **3**, 151–157.

(17) Canham, L. T., *Appl. Phys. Lett.*, 1990, **57**, 1046–1048.

(18) Cullis, A. G., Canham, L. T., Calcott, P. D. J., *J. Appl. Phys.*, 1997, **82**, 909–965.

(19) Lin, V. S. Y., Motesharei, K., Dancil, K. P. S., Sailor, M. J., Ghadiri, M. R., *Science*, 1997, **278**, 840–843.

(20) Hochbaum, A. I., Gargas, D., Hwang, Y. J., Yang, P. D., *Nano Lett.*, 2009, **9**(10), 3550–3554

(21) Qu, Y. Q., Liao, L., Li, Y. J., Zhang, H., Huang, Y., Duan, X. F., *Nano Lett.*, 2009, **9**(12), 4539–4543.

(22) Archer, R. J., *J. Phys. Chem. Solids*, 1960, **14**, 104–110.

(23) Li, X., Bohn, P. W., *Appl. Phys. Lett.*, 2000, **77**, 2572–2574.

(24) Peng, K. Q., Yan, Y. J., Gao, S. P., Zhu, J., *Adv. Mater.*, 2002, **14**, 1164–1167.

(25) Zhang, M. L., Peng, K. Q., Fan, X., Jie, J. S., Zhang, R. Q., Lee, S. T., Wong, N. B., *J. Phys. Chem. C*, 2008, **112**, 4444–4450.

(26) Peng, K. Q., Xu, Y., Wu, Y., Yan, Y. J., Lee, S. T., Zhu, J., *Small*, 2005, **1**, 1062–1067.

(27) Peng, K. Q., Zhu, J. J., *Electroanal. Chem.*, 2003, **558**, 35–39.

(28) Peng, K. Q., Hu, J. J., Yan, Y. J., Wu, Y., Fang, H., Xu, Y., Lee, S. T., Zhu, J., *Adv. Funct. Mater.*, 2006, **16**, 387–394.

(29) Peng, K. Q., Fang, H., Hu, J. J., Wu, Y., Zhu, J, Yan, Y. J., Lee, S. T., *Chem.Eur. J.*, 2006, **12**, 7942–7947.

(30) Huang, Z. P., Fang, H., Zhu, J., *Adv. Mater.*, 2007, **19**, 744–748.

(31) Huang, Z. P., Shimizu, T., Senz, S., Zhang, Z., Zhang, X. X., Lee, W., Geyer, N., Goüsele, U., *Nano. Lett.*, 2009, **9**, 2519–2525.

(32) Peng, K. Q., Zhang, M. L., Lu, A. J., Wong, N. B., Zhang, R. Q., Lee, S. T., *Appl. Phys. Lett.*, 2007, **90**, 163123.

(33) Zhong, X., Qu, Y. Q., Lin, Y. C., Liao, L., Duan, X. F., *ACS Appl. Mater. Interfaces.*, 2011, **3**, 261–270.

(34) Chen, H. H., Zou, R. J., Wang, N., Sun, Y. G., Tian, Q. W., Wu, J. H., *et al.*, *J. Mater. Chem.*, 2011, **21**, 801–805.

(35) Lin, L. H., Sun, X. Z., Tao, R., Feng J. Y., Zhang, Z. J., *Nanotechnology*, 2011, **22**, 075203.

(36) Chiappini, C., Liu, X. W., Fakhoury, J. R., Ferrari, M., *Adv. Funct. Mater.*, 2010, **20**, 2231–2239.

(37) Lin, L. H., Guo, S. P., Sun, X. Z., Feng J. Y., Zhang, Z. J., *Nanoscale Res. Lett.*, 2011, **5**, 1822–1828.

(38) Davis, M. E., *Nature*, 2002, **417**(6891), 813–821.

(39) Sanchez, C., Lebeau, B., Chaput, F., Boilot, J. P., *Adv. Mater.*, 2003, **15**, 1969–1994.

(40) Bartl, M. H., Boettcher, S. W., Frindell, K. L., Stucky, G. D., *Acc. Chem. Res.*, 2005, **38**, 263–271.

(41) Seayad, A. M., Antonelli, D. M., *Adv. Mater.*, 2004, **16**, 765–777.

(42) Taguchi, A., Schuth, F., *Micropor. Mesopor. Mater.*, 2005, **77**, 1–45.

(43) Slowing, I. I, Vivero-Escoto, J. L., Wu, C. W., Lin, V. S. Y., *Adv. Drug Deliv. Rev.*, 2008, **60**, 1278–1288.

(44) Qu, Y. Q., Zhong, X., Li, Y. J., Liao, L., Huang, Y., Duan, X. F., *J. Mater. Chem.*, 2010, **20**, 3590–3594.

(45) Yoo, K. H., Kang, K. S., Chen, Y., Han, K. J., Kim, *J. Nanotechnol.*, 2008, **19**, 505202.

(46) Zhang, H. J., Chen, G. H., Bahnemann, D. W., *J. Mater. Chem.*, 2009, **19**(29), 5089–5121.

(47) Tsung, C. K., Kuhn, J. N., Huang, W. Y., Aliaga, C., Hung, L. I., Somorjai, G. A., Yang, P. D., *J. Am. Chem. Soc.*, 2009, **131**, 5816–5822.

(48) Xue, M., Zhong, X., Shaposhnik, Z., Qu, Y. Q., Tamanoi, F., Duan X. F., Zink, J. I., *J. Am. Chem. Soc.*, 2011, **133**, 8798–8801.

(49) Wang X. L., Han, X. Q., *ACS Appl. Mater. Interfaces*, 2010, **2**, 3709–3713.

Chapter 10

Nanoscale Contact Engineering for Si Nanowire Devices

Yung-Chen Lin and Yu Huang
Department of Materials Science and Engineering, University of California, Los Angeles, CA 90095, USA.
yhuang@seas.ucla.edu.

10.1 Scope of the Chapter

One-dimensional nanostructures, such as nanotubes and nanowires (NWs), are attractive building blocks for nanoelectronics due to their unique morphology, size, and electronic properties. In particular, substantial efforts have been made for the development of nanoscale transistors based on Si NWs due to their potential to replace conventional planar metal-oxide-semiconductor field-effect transistors (MOSFETs) in integrated circuits, to open new opportunities in flexible macroelectronics and highly sensitive biosensors. Making reliable electrical contact to individual NW devices is one of the key factors that determine the device performance and reliability. This chapter reviews the common synthetic approaches, growth kinetics, electronic properties, and device applications of nanoscale silicides formed through solid-state reaction.

Silicon and Silicide Nanowires: Applications, Fabrication, and Properties
Edited by Yu Huang and King-Ning Tu

ISBN 978-981-4303-46-0 (Hardcover), 978-981-4303-47-7 (eBook)
www.panstanford.com

10.2 Introduction

10.2.1 The Challenges of Modern Transistor for Contact Engineering

As addressed by the scaling rule, the device dimensions of complementary metal-oxide-semiconductor (CMOS) transistors have been shrunk to improve the performances for ultra-large-scale integrated circuits (ULSI). The shrinkage of the contact feature size or carrier channel size has caused the increase of resistances in both source/drain regions, leading to the degradation of electrical characteristics.[1,2] This degradation becomes unavoidable in the advanced short-channel devices and thus design strategies need to be imposed to current nanotechnology. In 1925, a field-effect transistor (FET) device structure with only a conducting channel (no junctions) was proposed. The FET was only a resistor that could be depleted of charge carriers by gate modulation. To turn the device completely off, the device needs to fully deplete its body of carriers. Therefore, it required a very thin nanoscale channel for which the technology did not exist at the time. Recent explorations of NW or thin-film transistor architectures have raised practical interest in these simple gated resistors. Especially for the NW transistors, the thin body structure displaying better gate control, allowing researchers to explore diverse device structures and to investigate new operation functions. However, for this type of thin body transistor or NW FET, the contact or sheet resistance is still high. The silicided NW FET atomically flate Si-silicide interfaces in place of the polycrystalline Si source/drain contacts and has attracted broad research interest. By combining the NW structure and silicide technology, it is possible to solve the short-channel problem and achieve high device performance for ultrashort-channel device in the future.

10.2.2 NW Transistor and Silicided NW Transistor

Adopting the NW FET structure can bypass the short-channel effect issue since the thin body of NW excludes large depletion region. Hence, degradation of gated control and source/drain leakage can be avoided. The charge carriers in the NW can be fully depleted and thus the NW FET can be function as junctionless FET or metal heterojunction FET. When proper silicides are used as the contact materials the contact resistance can be effectively reduced making silicided NW FET promising candidate to achieve high-performance transistor operation.[3]

Si NWs are often grown by vapor–liquid–solid (VLS) method, during which the diameter of the NWs can be controlled by the size of the metal catalyst. In addition, the Si NW doping can be done by using the *in situ* doping. As the NW transistor usually functions as junctionless FET (no source/drain implantation), to achieve high-performance, the doping concentration of NWs is usually kept at the level of $\sim 10^{17} \sim 10^{18}$ cm^{-3}. In these doping ranges, the source/drain contact resistance will not be too large to limit the current. Therefore, in this operation mode, the device is always in the on-state and the accumulation-mode operation is applied to reach higher on-current and lower subthreshold voltage (V_t). The same charge voltage is applied to deplete the NW to reach the off-state. The metal heterojunction FET can also be fulfilled in the NW structure. With the silicide technology, nickel and platinum silicide has been grown in the Si NW and form the silicide/Si/silicide heterojunction structure. With various doping concentration of Si, the device can function in the normal-on or normal-off mode, and in both modes the NW transistor can achieve high performance, due to the better gated control in the NW structure.

10.2.3 The Properties and Applications of Metal Silicides

Silicide materials with diversified properties (Table 10.1) have been widely used in electronics such as device contacts (NiSi, $TiSi_2$, $CoSi_2$),[4–10] optical light-emitting diodes (β-$FeSi_2$, $CrSi_2$),[11,12] thermoelectrics ($MnSi_{1.7}$),[13] photovoltaic devices ($BaSi_2$)[14–16] and spintronics devices ($Fe_xCo_{1-x}Si$, Fe_3Si, Co_2MnSi)[17–19] shown in as Fig. 10.1. Silicides with low contact resistivity, good interface property and good thermal stability are candidates for device contact applications. Some metallic silicides are also magnetic which draw considerable interest for its potential application in spintronics due to its compatibility with current Si technology.

Table 10.1 Electronic and optical properties of silicide materials

Materials	Properties	Resistivity (μΩ cm)	Band gap (eV)
δ-Ni_2Si	Metallic [22]	22	
NiSi	Metallic [22]	10–18	
$NiSi_2$	Metallic [23]	50	
C54-$TiSi_2$	Metallic [22]	14	
Co_2Si	Metallic (Ferromagnet) [24]	40	
CoSi	Metallic (diamagnetic semimetal) [22]	120	
$CoSi_2$	Metallic [22]	24	
Pd_2Si	Metallic [22]	30	
$MoSi_2$	Metallic [22]	100	
WSi_2	Metallic [22]	70	
MnSi	Metallic (Helical magnet) [25]	220	
$MnSi_{1.7}$	Semiconducting [25]	4100	0.45–0.47 Indirect 0.78–0.83 Direct
PtSi	Metallic [22]	28	
Pt_2Si	Metallic [22]	6.5	
$ErSi_2$	Metallic [26]	30	
$CrSi_2$	Semiconducting [22]	600	0.35–0.67 Indirect 0.5–0.9 Direct
β-$FeSi_2$	Semiconducting [12]	>1000	0.83–0.89 Indirect 0.5–0.9 Direct
FeSi	Semiconducting [27,28]	300	0.05
Fe_3Si	Metallic (Ferromagnet) [29]	70	

Source: Adapted from Refs. 22–29.

10.3 Synthetic Approaches to Nanoscale Silicides

Various approaches have been developed to grow nanoscale silicide structures, which can be summarized to three categories:

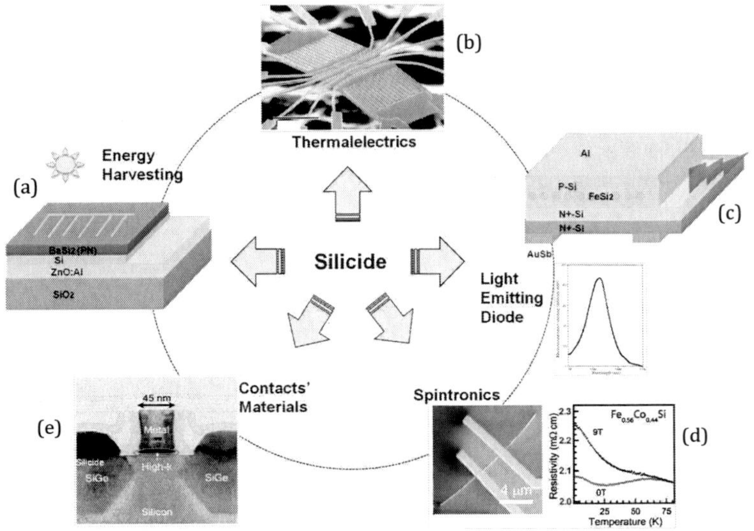

Figure 10.1 Applications of silicide materials: (a) Photovoltaic device[14–16]; (b) Thermalelectrics[20] [Reprinted by permission from Macmillan Publishers Ltd. Nature, copyright (2008)]; (c) Light-emitting diode[12]; (d) Spintronics[21] [Reprinted by permission from American Chemical Society, copyright (2008)]; (e) Contacts materials [Reprinted with permission from Intel Corp.].

(1) endotaxy, (2) chemical vapor deposition (CVD), and (3) solid-state reaction (Fig. 10.2). For the endotaxial growth, transition metals, rare-earth (RE) metals or noble metals were deposited onto a heated Si substrate and subsequently reacted with the substrate to form self-assembled silicide nanostructures. In order to have the lowest strain energy against the substrate, silicide materials adjusted their shape into nanostructures with small anisotropic lattice mismatch (~0%) in the larger (longitudinal) direction and larger mismatch (~ >3%) in the smaller (axial) direction. Figure 10.2a,b showed $CoSi_2$ NW (NWs) islands grown on Si(100) substrate at 750 °C. Cobalt was deposited by sublimation onto a heated substrate and reacted with Si to form $CoSi_2$.[30] $CoSi_2$ would self-assemble into two distinct shapes, rectangular and NW (NWs) shape in Fig. 10.2b. For rectangular shape, $CoSi_2$ adopts A-type epitaxial relation, Si(111)//$CoSi_2$(111), with the Si substrate; for NW shape, B-type epitaxial relation appeared on one side of interface, Si(111)//$CoSi_2$(-11-1) and Si(1-51)//$CoSi_2$(1-11) epitaxial relation shows on the other side. These interface configurations correspond to the coherent interface

relations between $CoSi_2$ nanostructures and Si(100) substrate with lower interfacial energy. A variety of metals such as Ti, Mn, Fe, Co, Ni, and Pt have been reported to form silicide nanostructures on the Si substrate through endotaxy growth.[31-34] However, no effective control methods are available so far over the growth length, growth position of the silicide NWs or NW diameter with this approach.

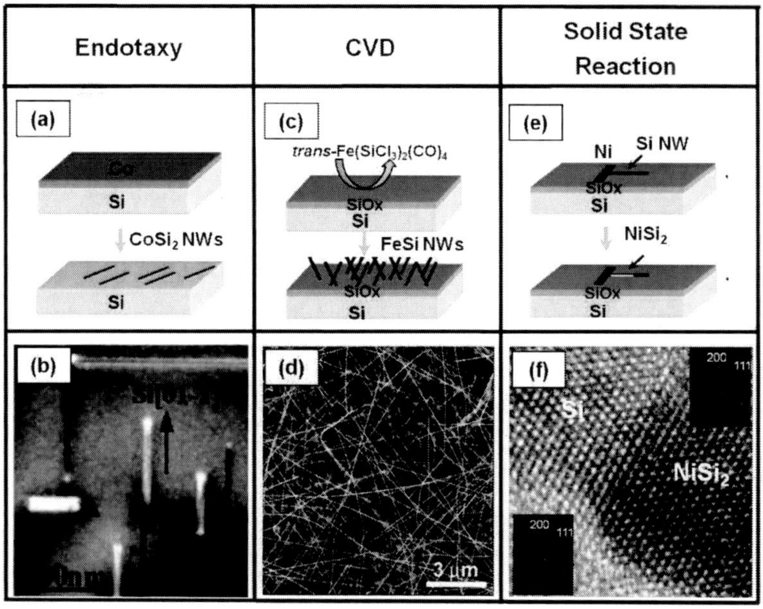

Figure 10.2 Growth technologies of silicide material endotaxy: (a) Formation of NW islands on a heating substrate. (b) Plane-view TEM showing silicide nanostructures with rectangular and NW shapes. Reprinted from Ref. 31, Copyright (2011), with permission from Elsevier. (c) Schematic of CVD method: FeSi NW was grown from single-source precursor *trans*-$Fe(SiCl_3)_2(CO)_4$ by CVD. (d) Representative SEM images of FeSi NWs. Reprinted with permission from American Chemical Society, Copyright (2006). (e) Formation of Ni silicide NWs by solid-state reaction method. (f) HRTEM image of $NiSi_2$/Si epitaxial interface. Inset images show the fast Fourier transform (FFT) of Si[0-11] zone axis and $NiSi_2$[0-11] zone axis. Reprinted with permission from American Chemical Society, Copyright (2010).

CVD is a popular synthetic route to form silicide nanostructures which is implemented through the reactions between metal and/

or Si precursors. CVD method could be conducted with versatile precursors, reactions between Si vapor precursor (SiH_4) and metal thin film (e.g., nickel), metal vapor precursors (e.g., metal halogen precursors such as metal chlorine or metal iodine vaporized at high temperature) with Si substrate, reaction with both Si and metal vapor precursors or reactions from a single source of precursor containing both Si and metal. Figure 10.2c and d demonstrate the use of a single vapor precursor (*trans*-$Fe(SiCl_3)_2(CO)_4$)[35] as the iron and Si source to grow FeSi NWs in a CVD setting. The precursor decomposed at elevated temperatures to form a NW structure on the SiO_2/Si substrate. In this manner, a wider range of single-crystal silicide NWs can be synthesized and used to explore the properties of various silicide materials. In a CVD setting, vapor precursor can be supplied continuously to synthesize abundant free standing silicide NWs which may be scaled up to mass production. In addition, the silicide growth is not limited to the growth substrates. (e.g., Si, sapphire, SiO_2 substrates) since the metal and Si source can be both supplied through vapor precursors. But as the precursors may under different decomposition routes at different temperature zones and/or growth pressures, mixed product of Si, silicide thin film, different silicide nanostructures, and sometimes metal film, metal particles can be found in the same reaction chamber.[37–40] Therefore the CVD reaction conditions need to be strictly controlled in order to obtain desired silicide products.

Solid-state reaction is another approach to silicide nanostructures and, because of its reaction nature, to form silicide/Si interface which can be explored as contacts for various device applications based on Si NWs. This approach has been used extensively in microelectronic industry to form metallic silicide contacts to Si through the interface reaction of Si and the metal. In short, metal is deposited as thin film on top of Si NWs and then at elevated temperatures metal atoms diffuse into and react with Si to form silicides, which normally adopt an epitaxial relation with the Si lattice. A variety of silicide contacts with excellent performance by solid-state reaction method have been demonstrated such as $NiSi_2$/Si, NiSi/Si, Ni_2Si/Si, $CoSi_2$/Si, PtSi/Si and MnSi/Si for nanoelectronics.[36, 41–46] Figure 10.2e shows the schematic of a typical solid-state reaction set up. Ni contacts were deposited on top of Si NWs, which were then subject to annealing at elevated temperatures. Ni atoms would diffuse into Si NW template and transform Si into Ni silicides. $NiSi_2$ was found to form at

550 °C,[43] which maintains a sharp clean epitaxial interface with Si lattice (Si[1-10]//$NiSi_2$[1-10] and Si(111)//$NiSi_2$(111)) as show in the high-resolution transmission electronic microscopic (HRTEM) image in Fig. 10.2f. The smooth and high-quality epitaxial interface between silicides and Si achieved in solid-state reactions makes this approach desirable for device fabrication at nanoscale, which is also compatible with current microelectronic processing techniques.

10.4 Contact Formation through Solid-State Reaction

10.4.1 Introduction of Silicide/Si Heterostructure by Solid-State Reaction

While Si NW devices have caught increasing attention for future device applications, it has been difficult to achieve low resistance, high stability, or less defective contacts at this scale due to the small contact area, surface traps, or non-conformal coverage between electrodes and NW structure.[47,48] For solid-state reactions confined in Si NW structures, the silicides form directly on the silicide/Si interface resulting in a fresh and few-defect metal/Si interface that may contribute to high device performance. With lithographically patterned metal pads, silicide/Si/silicide structures can be achieved with two silicide regions serving as source and drain electrodes, and the Si as the transistor channel. It has also been demonstrated that silicides can grow nearly up to lengths in micrometers into the Si NW from the metal contacts, which has been explored for device scaling to achieve short-channel devices.[44,45,49] Studies indicate that the junction leakage is small (~60 nA at the 0.5 V) in the Ni silicide/Si/Ni silicide heterostructure even with an ultrashort Si channel of ~5 nm, which may be attributed to the perfect Ni silicide/Si interface and the low concentration of Ni atoms existing in the Si channel.[50] The current-voltage (*I–V*) measurement across the ultrashort-channel silicide/Si heterostructure device shows a rectified curve which indicates the existence of Schottky barrier in the Ni silicide/Si/Ni silicide hetero-junction. This suggests that the semiconducting properties of Si are preserved and the flat interface does not enhance thermal emission or current tunneling which are normally observed at a rough or spike-like interface, even though the semiconducting channel is less than 10 nm. The impressive results imply that the Ni silicide/Si/Ni silicide heterostructure can be applied for ultrashort-channel devices.

10.4.2 The Growth of Silicide NWs by Solid-State Reaction

Wu *et al.*[49] first demonstrated the idea to implement the silicide technology at nanoscale for making silicide NWs by solid-state reaction. As illustrated in Fig. 10.3a, Si NWs were grown by VLS method and then Ni films were deposited with comparable thickness to the average NW diameter. The Ni-coated Si NWs were then annealed at 550 °C to form NiSi. The excess Ni was etched away by acid and followed by the post-annealing at 600 °C at forming gas (N_2:H_2 = 90:10). This process will render fully silicided NWs that are single crystal in nature as shown in the Fig. 10.3b,c. The different NiSi NW growth directions, NiSi[1-11] (Fig. 10.3b) and NiSi[001] (Fig. 10.3c) may be explained by different growth directions of the starting Si NWs.

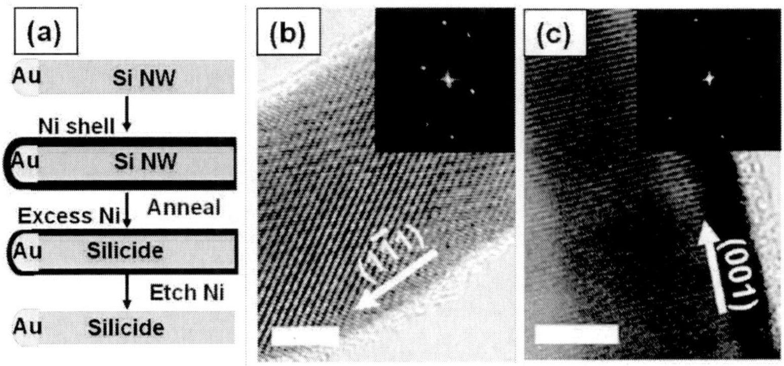

Figure 10.3 Schematic diagram of Ni silicide NW growth by solid-state reaction. (a) Growth of fully silicided NWs: deposit Ni film to cover the Si NW and then annealing at 550 °C to transform Si NW into silicide NW. Then remove the excess Ni metal. (b) TEM image of a 20-nm NiSi NW. Inset, Fourier transform of the image showing the [10-1] zone axis of NiSi.[49] (c) TEM image of a 32 nm NiSi NW. Inset, Fourier transform of the image indicating the [2-10] zone axis of NiSi.[49] The scale bars in (b) and (c) are 5 nm. Reprinted by permission from Macmillan Publishers Ltd., Nature, copyright (2004).

10.4.3 Forming Silicide/Si NW Heterostructure by Solid-State Reaction

To achieve the growth of silicide/Si heterostructure, it is crucial that metal elements possess higher diffusivity in Si. Due to the fast diffusion of metal atom into the Si NWs, silicides can grow into the NWs, forming silicide/Si heterojunction. Several silicide/Si heterostructures have been investigated with the metallic contacts such as Ni, Pd, Pt, Co and Mn to the Si NWs.[36,41,43–45,51] Silicide/Si heterostructures mostly adopt the lateral diffusion model of the metal atoms from the metal contacts into Si NWs and its subsequent transformation to silicide materials as shown in Fig. 10.4a.[44] The heterostructures are generally formed by patterning the Si NW template with metal pads followed by annealing at elevated temperatures to perform the silicidation (Fig. 10.4a). The silicide/Si/silicide heterostructure or fully silicided NW can be controlled through the annealing time (Fig. 10.4a,b–d). There are various material systems that have demonstrated such heterostructures such as NiSi/Si, $CoSi_2$/Si, $NiSi_2$/Si, PtSi/Si and MnSi/Si.[36,44–46,49] In these systems, epitaxial relations are all observed between silicide and Si with atomically sharp interface and few interface defects, which is desirable for high-performance nanoelectronics. Taken PtSi and MnSi as the examples, both silicide systems can form the silicide/Si heterostructure. Surprisingly, although accompanied with large lattice mismatch (~8.8% in PtSi/Si, ~24.5% in MnSi/Si), both silicides mange to keep the epitaxial relation and a flat interface with Si when advancing into the Si NWs (Fig. 10.4e–h).[36,44] Even though the Si region is below 10 nm (Fig. 10.4e), the silicide regions will not merge and a clear interface is maintained. The excellent interface properties and junction properties allow silicide heterojunction to potentially work as ultrashort channel devices.

10.5 Silicide Growth Mechanism

10.5.1 The Growth Phases of Nickel Silicides in the NW Structure

The growth mechanism of silicides through solid-state reaction is well established by studies in thin film or bulk growth model, where

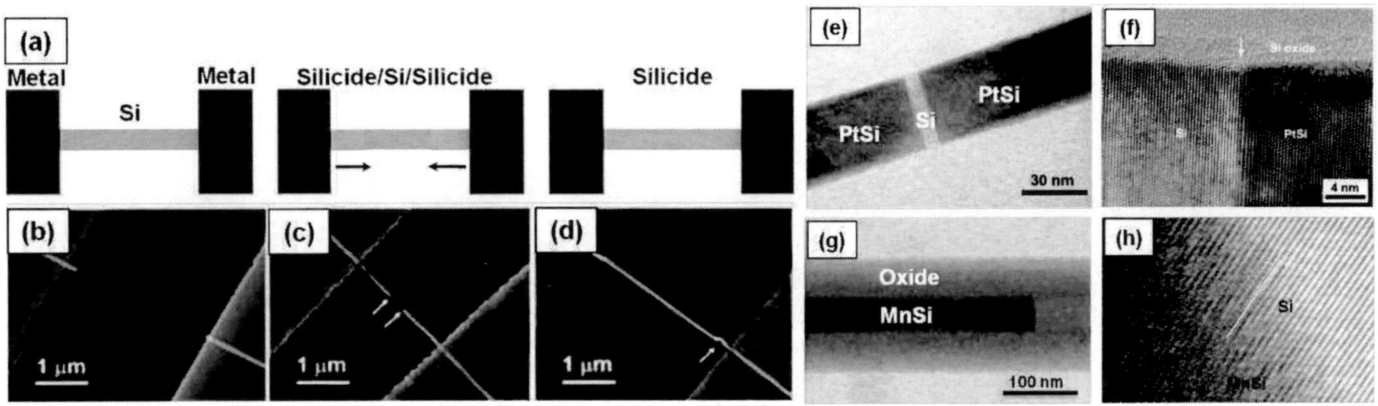

Figure 10.4 TEM analysis for silicide/Si/silicide heterostructure.[36,44] (a) A schematic shows to form a silicide/Si heterostructure and silicide NWs. (b–d) show SEM images of a patterned Si NW, partially silicided NW and fully silicided NW, respectively. (e) A TEM image of a PtSi/Si/PtSi heterostructure with only ~8 nm Si region. (f) A higher magnification image of a PtSi/Si interface. (g) TEM image of MnSi NW and MnSi/Si interface wrapped in the oxide shell. (h) HRTEM image of MnSi/Si interface. Reprinted with permission from American Chemical Society, Copyright (2008 and 2010).

Ni_2Si normally appear as the first growth phase in the Ni silicide system and other phases start to appear at fixed sequence (e.g., metal-rich phases (Ni_3Si_2), and then NiSi below 800 °C or Ni_3Si_2, NiSi and then $NiSi_2$ beyond 800 °C).[52,53] The growth rate or length of the silicide phase depends on Ni atom supply at the silicide/Si interface. Owing to the highest diffusion rate of Ni in Ni_2Si phase (D_{Ni,Ni_2Si} $\sim10^{-9}\sim10^{-10}$ cm²/S, $D_{Ni,NiSi}\sim10^{-11}$ cm²/S, and $D_{Ni,NiSi_2}\sim10^{-12}\sim10^{-14}$ cm²/S),[54–56] in the thin film or bulk growth systems, Ni_2Si usually emerges as the first growth phase winning the growth competition with other Ni silicide phases.

After Ni_2Si grows beyond its critical length or thickness, the dominance of the subsequent growth phase is attributed to the diffusion of Ni or Si atoms which determines the supply for the formation and hence the composition of next silicides. For example, at 600 °C, Ni is the dominant diffusing species ($D_{Si}/D_{Ni}\sim1/100$–$1/10$ and $D_{Ni,Si}\sim3.5\times10^{-8}$ cm²/S)[57] and Ni_2Si grows first. When Ni_2Si grows beyond the critical length, the supply of Ni to the Ni_2Si/Si interfacial layer decreases and the next more Si-rich phase (e.g., Ni_3Si_2 or NiSi) will tend to form. A change in supply of atoms to the interface determines when the growth of the next phase will initiate. This sequence observation in the thin film or bulk system is well accepted and investigated. For silicide growth at nanoscale, however, a systematic investigation and observation is still lacking. Wu *et al.*[49] observed NiSi epitaxial growth in Si[110], Si[111], Si[121] NWs by two-step annealing at 550 °C. Weber *et al.* have grown $NiSi_2$ as the first (epitaxial) phase in Si[110] NWs at 480 °C. Lu *et al.*[45,58] have achieved NiSi or $NiSi_2$ as the first phase (epitaxial phase) in Si NWs through point contact reaction at 500~650 °C. These observed silicide phases and their appearing sequence at nanoscale deviate from the thin film system, wherein δ-Ni_2Si usually appears as the first phase in these temperature ranges, and thus suggesting different growth kinetic may exist at this scale.

A few mechanisms have been proposed to explain the different observed first silicide phases in Si NWs. Dellas *et al.*[59,60] has proposed an orientation dependent growth for Ni silicides in Si NWs and observed θ-Ni_2Si epi-growth in Si[121] NWs and $NiSi_2$ epi-growth in Si[111] NWs. They suggested that the first appearing silicide phase was determined by the relative interfacial energies between the epitaxial silicide and the Si lattice, adopting the silicide phase with smaller lattice mismatch with Si. For example, due to the facet growth of Si[121] NWs, surrounding planes of a Si[121] NW nearly form a rectangular shape with the two long sides bound by (111) planes with larger area and the other two by a combination of (110) and (311) type planes. The facets surrounding the Si NW are important to initial silicide phase growth. As the first silicide phase appears, it tends to adopt the smallest lattice mismatch with surrounding Si facet planes. θ-Ni_2Si is nearly perfectly matched to the Si (111) facet surface with θ-Ni_2Si[001]//Si[111] and Ni_2Si(100)//Si(112) epitaxial relations. The in-plane distance between Ni_2Si(100) and Si(112) is with small lattice mismatch $\sim$ –5.1% and 0.1% in the two directions, leading to θ-Ni_2Si growth as the first phase in Si[121] NWs. Similarly, $NiSi_2$ appears as the first phase in Si[111] NWs adopting $NiSi_2$[1-10]//Si[1-10] and $NiSi_2$[111]//Si[111] epitaxial relations for lower the interfacial energy with small interfacial mismatch ($\sim$0.9%).[59] Utilizing these facet surfaces, the first silicide phase growth could be controlled and explained. While it is a possible explanation for the first observed phase in Si NWs, it does not explain the other observed first Ni silicide phase reported, such as NiSi, indicating contributions from mechanisms other than the interface lattice match. In order to solve the discrepancy of first formation phases, kinetic factors, for example, have also proposed to be important in nanosilicide formation processes. However, more convincing mechanism is still lacking and it needs further investigation.

10.5.2 Growth-Limiting Steps in the Nickel Silicide System at Nanoscale

The kinetics of silicide formation at nanoscale may be different from bulk and should be investigated in order set up the diffusion model for the growth rate and growth length control of silicide materials at this scale. Lu *et al.*[58] and Yaish *et al.*[61] have proposed a silicide growth mechanism at nanoscale that is determined by three processes: dissolution of Ni atoms into the Si NWs (with or without oxide barrier), diffusion of Ni atoms at the surface (Si NW/native oxide interface) or through bulk NWs, and interface reaction between metal and Si at the silicide/Si interface (in Fig. 10.5). The relative speed of these three steps will lead to different silicide growth behavior observed in Si NW. With the thin oxide barrier (native oxide) between Si NW and metal contact, silicide growth would be determined by Ni diffusion along the NW (parabolic growth behavior) or the interface reaction (linear growth behavior). However, with thick oxide barrier between the metal and the Si NW, the dissolution of Ni atoms into Si NW may become the limiting step because of the difficulty for Ni atoms to penetrate the thick oxide barrier. In this case, the silicide growth will be determined by the supply of Ni and not the Ni diffusion in the NW.

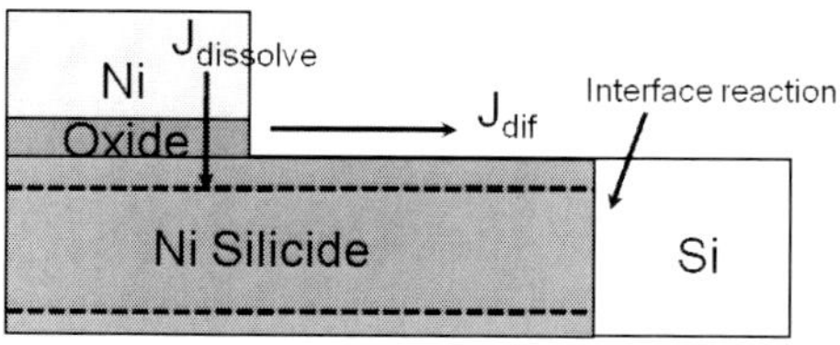

Figure 10.5 Schematic diagram of Ni diffusion in one-dimensional NW. Thin oxide barrier exists between Ni pad and Si NW. The dash line indicates the proposed diffusion thickness of surface diffusion. Reprinted with permission from Ref. 61. Copyright [2011], American Institute of Physics.

According to Yaish *et al.*'s calculation and experiment data, the parabolic silicide growth curve (silicide length versus time) can be obtained with the thin oxide barrier (shorter exposure time to ambient before metal deposition) and linear curve (length versus time) is observed for the thick oxide barrier (longer exposure time to ambient before metal deposition). Furthermore, with the steady state approach, their calculation model can fit the experimental data (growth length versus time) well by only considering the surface diffusion in the NW and dissolution of Ni atoms around the metal/NW interface. Thus they come to the conclusions that (a) poor contact between metal and Si NW or increasing the exposure time to air before Ni deposition will change the Ni silicide growth behavior from parabolic growth to linear growth due to slow dissolution of Ni atoms through the oxide barrier. (b) Ni atoms mainly diffuse at the interface between Si NW and native oxide. The argument is that Ni surface diffusivity may be over one hundred time larger than bulk diffusivity ($D_{\mathrm{Ni,surface}} \sim 10^{-10}$–$10^{-11}$ cm^2/s, $D_{\mathrm{Ni,bulk}} \sim 10^{-12}$–$10^{-15}$ cm^2/s).[61–63] Based on the equation, $D_{\mathrm{app}} = D_{\mathrm{b}} + g^*D_{\mathrm{s}}$, the apparent diffusivity through the NW is determined by adding bulk diffusivity (D_{b}) to g^*D_{s}, where g is the cross-section area ratio between surface and bulk, and D_{s} is the surface diffusivity for Ni. Therefore, in NW geometry, with large g value (large surface to NW cross-section ratio), surface diffusion can take over the bulk diffusion through the NW. However, a flat coherent epitaxial interface between silicide/Si has been constantly observed in Si NWs, for which the surface diffusion model cannot offer a sound explanation. As the of diffusivity values used in this work are quoted from the bulk systems, it is possible that at nanoscale, the diffusivities of Ni at the surface or in the bulk of silicide have been modified from its bulk counter parts such that Ni surface diffusivity may not dominate over the bulk diffusivity in NW structures. To gauge that, the silicide growth with different Si NW diameters and an *in situ* study are necessary in order to calculate the Ni diffusivity in the silicide NW and observe the nucleation in the parabolic growth behavior.[64–66] On the other hand, the contribution from interface reaction control was completely excluded in Yaish *et al.*'s model, which had been reported to govern $NiSi_2$ growth in Si NW.[51] These mechanisms are the possible growth-limiting steps in the NW structure.

10.5.3 Nucleation-Controlled Growth or Interfacial-Limited Growth of Nickel Silicide

Chou and Lu *et al.* have studied the growth of NiSi and $NiSi_2$ in Si NWs, in which interface reaction appear to be the rate-control step.[41,42] In the NiSi study (Fig. 10.6a inset schematic), Si NW and Ni NW assemble into a crossed structure and form a point contact, indicating a poor contact (with native oxide barrier and small contact) between Si NW and Ni NW as in the Fig. 10.6a. In the $NiSi_2$ study (Fig. 10.6c inset schematic), NWs are cast onto a TEM grid with pre-deposited Ni dots, wherein Ni dots and Si NW appears to form a better contact (without or only thin oxide) compared to NiSi case. In their observations, NiSi grows from far end of the Si NWs and the silicide/Si interface keeps sharp and flat with single-crystal structure. They proposed that the reaction is assisted by Ni interstitial diffusion through the Si NW and the growth rate is limited by the dissolution of Ni atoms through the thin oxide barrier into Si NW. The linear growth behavior is observed of the NiSi NW in the Si NW of 20 nm in diameter over the temperatures ranging from 500 to 650 °C. Homogeneous nucleation of NiSi is observed at the interface initiating from the middle of the NW. The homogenous nucleation is defined by nucleation of a single atomic layer of silicide at Si/silicide interface, that is, without a giant step (Fig. 10.6a–d). The activation energy of the epitaxial growth is determined to be 1.25 eV/atom. Compared to the activation energy of interstitial diffusion of Ni in Si ~0.47 eV/atom, it indicates that the growth may be controlled by the interface reaction at the silicide/Si interface, although slow dissolution of Ni atoms through thin oxide may also contribute.[45,58]

Furthermore, repeating homogeneous and heterogeneous nucleation growth behavior is also observed by Chou *et al.* They reported the observation of both homogeneous (Fig. 10.6c,d) and heterogeneous nucleation (Fig. 10.6e–g) of epitaxial silicide of $NiSi_2$ in Si [110] NWs, by *in situ* HRTEM. Owing to the excellent lattice match between Si and $NiSi_2$, a giant epitaxial step of 2–8 nm wide forms at the interface between Si and $NiSi_2$ during the growth of the latter. The step formation results in two silicide/Si interfaces parallel to each other in TEM observations (Fig. 10.6e–g). This step (Fig. 10.6g) indicates heterogeneous nucleation of $NiSi_2$ occurs at the intersection where the step meets the interfaces. The overall growth rate of $NiSi_2$ in the case of heterogeneous nucleation is faster than that of the homogeneous nucleation, which is explained by the much shorter incubation time of heterogeneous nucleation of $NiSi_2$ (Fig. 10.6g) when compared to homogeneous one (Fig. 10.6d). Comparing the growth rates of homogeneous ~0.0415 nm/s and heterogeneous nucleation ~0.133 nm/s, it is possible to be a reaction control rather than dissolution control, considering similar Ni supply in both cases.

Chou and Lu *et al.*'s studies suggest that with a point contact the silicide growth mechanism can be Ni dissolution-limited or interface reaction–controlled. Both cases assume that Ni diffusion in the NW is the relatively fast step and hence constant Ni supply through oxide (dissolution) or constant interface reaction rate will both lead to the linear growth behavior of silicide NWs. Additionally, with better Ni and Si NW contact with thin oxide or no oxide barrier, the first growth phase is switched to $NiSi_2$, with faster growth rate owing to the co-existence of both homo- and heterogeneous nucleation at the interface. Linear growth behavior is also observed indicating interfacial-limited growth and the fast diffusion of Ni atoms in the silicide NW.

10.5.4 Stress-Limited Growth of Nickel Silicide Phases

Through understanding of silicides formation at nanoscale, control of the kinetic process may become available. At nanoscale, the diameter of NW become small and is easily influenced by the

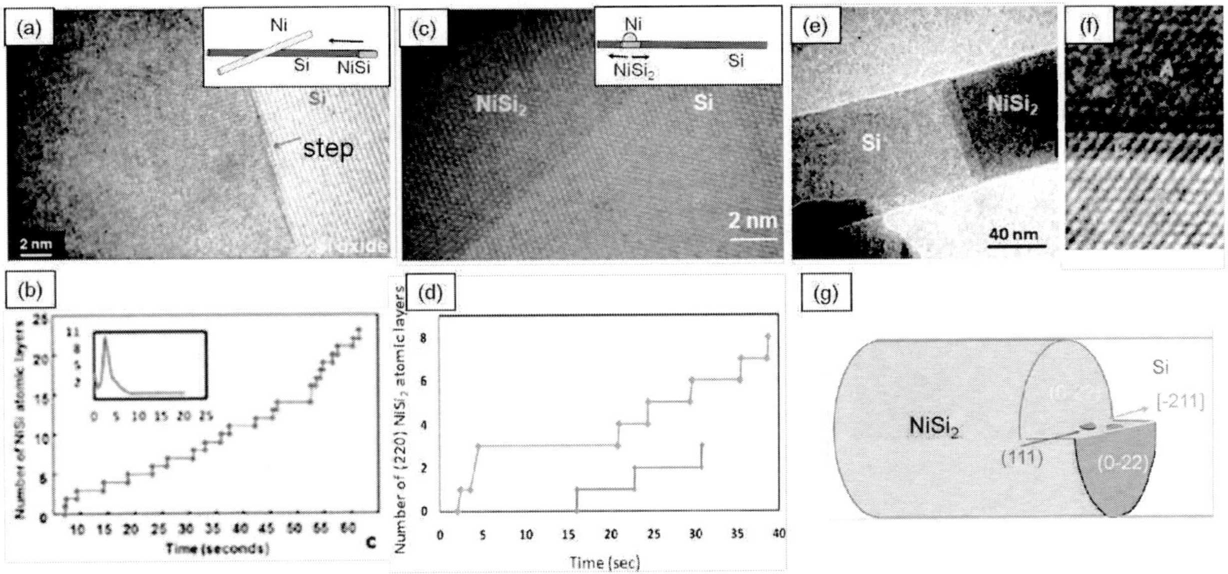

Figure 10.6 *In situ* TEM images of silicide/Si interfaces.[41,42] (a) High-resolution TEM image of NiSi/Si interface. Inset shows a point contact. (b) Stepwise growth curve of NiSi epitaxial growth with Si by homogeneous nucleation. Inset shows a statistic distribution of incubation time. (c) High-resolution image of single interface of $NiSi_2$ and Si. It shows one atomic layer growth and indicates a homogeneous nucleation. Inset shows a schematic of Ni dot contacting with a Si NW. (d) Stepwise growth curve of $NiSi_2$ epitaxial growth with Si by homogeneous nucleation. The upper and lower curves are from two different sets of data. It shows incubation time during growth. (e) and (f) *In situ* TEM images with different . Reprinted with permission from American Chemical Society, Copyright (2008 and 2009).

external force, volume expansion or thermal conductivity difference. For example, residual stress can readily build up in nanostructures which may have strong influence on the nucleation of silicide or diffusion of metal atoms. In order to investigate the stress effect on the silicide phase transformation in Si NWs, Lin *et al.*[43] utilize Si/SiO*x* core–shell NWs with various oxide shell thicknesses to examine the silicide growth kinetics, as shown in Fig. 10.7. In their studies, $NiSi_2$

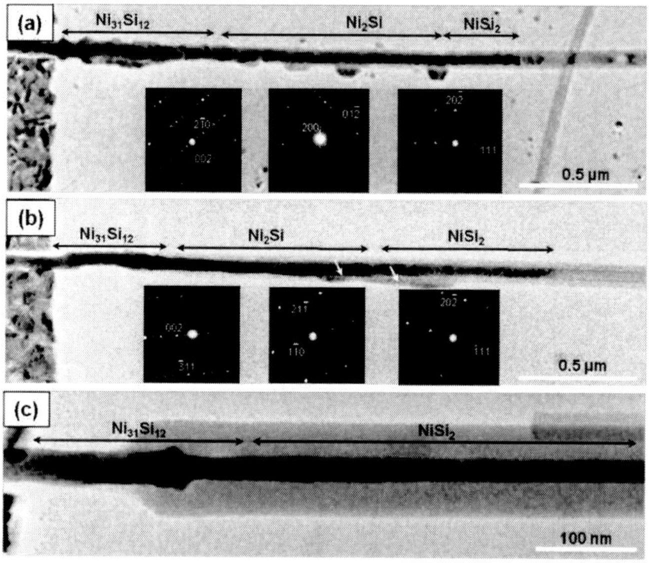

Figure 10.7 Ni silicide growth in Si/SiO*x* core–shell NWs with various oxide shell thickness.[43] (a) Silicide growth in a Si/SiO*x* with 5~15 nm oxide shell at 550 °C for 8 min. Inset diffraction patterns represent $Ni_{31}Si_{12}$[120], Ni_2Si[021] and $NiSi_2$[-12-1] zone axis, respectively, showing the three phases are $Ni_{31}Si_{12}$ close to the pad, Ni_2Si in the middle and $NiSi_2$ at interface. Extrusion structures are frequently observed with the same crystal structure to its surrounding silicide phase. (b) Silicide growth in the Si/SiOx NW with ~25 nm oxide shell thickness at 550 °C for 8 min. The inset electron diffraction patterns show that the phases are identified as $Ni_{31}Si_{12}$[130] close to the pad, Ni_2Si[113] in the middle and $NiSi_2$[-12-1] at interface, respectively. (c) Silicide growth in a Si/SiO*x* core–shell NW with ~50 nm oxide shell at 550 °C for 8 min. The silicide phases identified by electron diffraction are $Ni_{31}Si_{12}$[130] and $NiSi_2$[-12-1]. The growth of the $Ni_{31}Si_{12}$ in Si/SiO*x* is limited to a short length, and Ni_2Si is not present. Reprinted with permission from American Chemical Society, Copyright (2010).

is found to be the first phase, followed by Ni_2Si and then $Ni_{31}Si_{12}$. It is observed that with increasing oxide shell the growth rate of all silicides will be increasingly retarded in the Si/SiO*x* core–shell NWs, which is attributed to the slower Ni diffusion rate due to the compressive stress from the oxide shell. It is also proposed that the oxide shell limits the volume expansion accompanied with silicide transformation from Si and hence limits the growth of particular silicide phases with larger volume expansion. For example, the growth of Ni-rich phases, $Ni_{31}Si_{12}$ and Ni_2Si, are limited in the oxide shell due to the relatively large volume expansion. The atomic volume of Si is 20.01 $Å^3$ per Si atom, and those of $Ni_{31}Si_{12}$ and Ni_2Si are 39.46 $Å^3$ and 32.15 $Å^3$, respectively. Therefore upon transformation to the metal-rich silicides, the volume of the Si NW will expand several times, which can lead to significant stress pile-up in the confined NW core. Therefore, it is believed that the strong confinement of the thick oxide shell ~50 nm seriously retards the diffusion rate of nickel atoms to an extent that it completely annihilates the formation of Ni_2Si and significantly limits the formation of $Ni_{31}Si_{12}$ into the oxide shell region (in Fig. 10.7c). This work demonstrates the possibility to modulate the growth of different silicide phases in Si NW in a controllable fashion. Coupled with deeper understanding of silicide growth kinetics at nanoscale, nanosilicide formation may be engineered specifically towards desirable device functions.

10.5.5 Diffusion-Limited Growth of Nickel Silicide

Ogata *et al.*[67] demonstrated the kinetic studies of nickel silicide growth by *in situ* TEM. Fig. 10.8 shows nickel silicide growth length versus annealing time for as-grown NWs (with a native oxide layer) and for thermally oxidized NWs. It clearly shows that length of nickel silicide, L_{NixSiy}, follows a square root dependence with annealing time. Moreover, L_{NixSiy} is significantly shorter for oxidized NWs at any given time. By exploiting different nickel pad thickness, the metal source–limiting issue can be eliminated due to the same parabolic trend observed. Therefore, this parabolic growth is a clear evidence for a diffusion-limited growth. Utilizing the Deal–Grove-like model, the whole system behaved as the interface-limited growth mechanism with a linear growth rate ($L \sim K_t$). As the growth time increases, the diffusion-limited growth will start to take over as the main growth mechanism ($L \sim \sqrt{D_t}$). By fitting the parabolic curves,

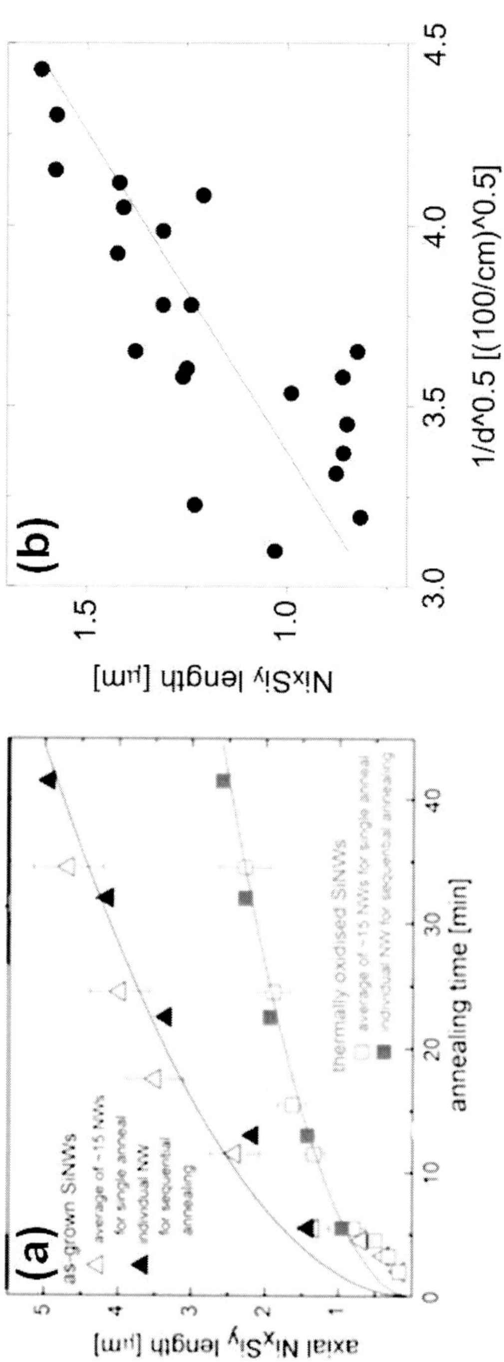

Figure 10.8 Ni silicide growth length versus time: (a) Silicide growth length for bared and oxidized SiNWs at ~440 °C. (b) Silicidation rate versus diameter for [111] orientated SiNWs. The fitted line represents an inverse square root dependence. Reprinted with permission from Ref. 67. Copyright [2011], Institute of Physics.

the extrapolation diffusion coefficients for as-grown and oxidized NWs are $D_{\mathrm{Ni}_{\text{as-grown}}} = 3.6 \times 10^{-12}\,\mathrm{cm^2/s}$ and $D_{\mathrm{Ni}_{\text{oxided}}} = 9.6 \times 10^{-12}\,\mathrm{cm^2/s}$. These results also indicated that oxide shell exerted a compressive stress on the silicide NWs and retarded the nickel diffusion. This oxide confined growth was also previously observed by Lin *et al.* Compared the L_{NixSiy} for these [111] orientated Si NW with different diameters (Fig. 10.7b), the growth approximately follow the tendency, which can be fitted by an inverse square root dependence. The diameter dependence can be modeled by combining the grain boundary diffusion with bulk diffusion for nickel ($D_{\mathrm{Ni}_{\mathrm{eff}}} = (1-g)D_{\mathrm{B}} + gD_{\mathrm{s}}$). Here, the grain boundary diffusivity is replaced by the surface diffusivity. However, if the surface diffusion has the contribution to the system, the concave or wedge-shape silicide at interface should be observed. However, for Ogata *et al.*'s observation, nickel silicide always shows step growth. Thus, the validity of this model (the domination of surface diffusion) still needs to be verified.

10.6 New Technical Approaches or Structures for Low-Contact Resistance FET and Short-Channel Device

10.6.1 The Challenging for the Low-Device Junction Resistance

The conventional FET can be classified as the junction FET. The opposite type of doping for the source/drain and channel is necessary for this device structure. As the device scales down, the main issue is the short-channel effect due to the depletion region across the channel. Other issues such as degradation of sub-threshold slopes, drain-induced barrier–lowering (DIBL) and source/drain leakage all come from the similar reasons related to the source/drain junction. Industry has several approaches to achieve the shallow junction which relieves the short-channel effect. However, the *p-n* junction around source and drain still has higher sheet resistance, which may become a current limit for ultrashort channel device. Therefore, major architectural changes have been proposed to facilitate downsizing toward the perceived limit of ~30 nm. Recently, several groups have suggested that short-channel effects can be suppressed by imposing a junctionless structure or a metal heterojunction barrier at the source/drain in place of heavily doped junctions.

10.6.2 Comparison of Junction FET, Junctionless FET, and Metal Heterojunction FET

For the junctionless FET (Fig. 10.9a), the conduction in the junctionless gated transistor is based on the transport of majority carriers in the bulk of the device, through a body channel from source to drain. Therefore, drain current in the on-state is proportional to the channel doping and the carrier mobility and not dependant on the gate oxide capacitance (dielectric constant or thickness). Speed (or "intrinsic delay") is also independent of the gate oxide thickness. The extreme efforts currently underway to reduce gate oxide thicknesses can therefore be avoided. This junctionless FET needs a thinner body or one-dimensional wire structure to provide the better gated control. The complete depletion is necessary for the junctionless FET to behave as the off-state. Therefore, the junctionless FET can only be fulfilled for the SOI technology or in a NW structure.

Metal heterojunction FET is another new structure to alleviate the short-channel effect (without source/drain doping) and reduce the source/drain sheet resistance. A heterolayer, silicides, grows at the source/drain in place of a heavily doped homojunction (Fig. 10.9b). The Schottky barrier which forms naturally between a metal silicide source/drain and the Si channel offers a number of advantages such as alleviating shallow source/drain extensions, junction capacitance, parasitic bipolar and latch-up phenomena, and fluctuations due to finite number of randomly placed dopants in the channel. A conventional self-aligned silicide process is employed to create the Schottky barrier MOSFET (SBMOSFET) structure. Gate-induced electric fields then render the source barrier nearly transparent to tunneling in the on-state, $V_g > V_t$, as in Fig. 10.9c, while carriers remain to be confined by the Schottky barrier in the off state, $V_g < V_t$ as in Fig. 10.9c. A low barrier height of ~0.2–0.3 eV (Table 10.2) is needed at the source in order to induce large tunnel currents in the on-state at practical gate voltage or electric field. Conversely, a relatively high barrier is needed to suppress "reverse" tunneling by opposite type of carriers at the drain. Because the Fowler–Nordheim transmission probability for internal field emission is exponentially

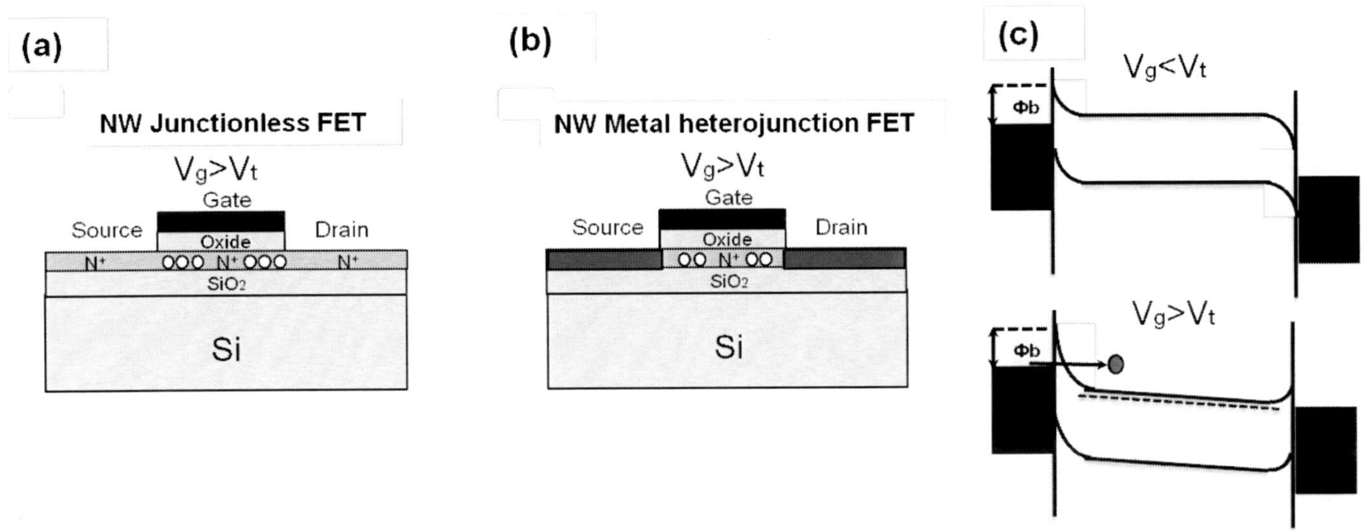

Figure 10.9 Junctionless FET and metal heterojunction FET based on SOI or NW structures. (a) The junctionless FET (b) Metal heterojunction FET: the source and drain regions are replaced by the metal contact (silicides) ErSi2 is used for *n*-channel contact. (c) Band diagrams of *n*-channel metal heterojunction device. With low gate bias ($V_g < V_t$), source barrier width is too wide for tunneling. With high gate bias ($V_g > V_t$), large electric fields induce tunneling from the source.

dependent upon barrier height as $\Phi b^{3/2}$, very large on/off ratio can be achieved at low temperatures in metal heterojunction MOSFETs.

According to the literatures, PtSi and $ErSi_2$[44,68] have successfully experimented as the *p*- and *n*-type metal heterojunction FET on bulk Si substrate. PtSi source/drain *p*-MOSFETs show excellent scaling to ~25 nm channel length with greatly simplified processing and reduced contact, sheet and parasitic resistance. Junction capacitance is effectively eliminated, and much smaller. Complementary *n*-type devices have also been demonstrated recently based on the low ~0.39 eV *n*-barrier of $ErSi_2$. Due to the high doping concentration in the channel, the off-current is still the limitation for this device structure. However, for the NW structure, the fully depleted characteristic may suppress the strong source/drain leakage. Therefore, it may be more practical for NWs to achieve the metal heterojunciton FET.

Table 10.2 Work function of silicide materials to the *n*-type Si

Materials	Schottky barrier height (eV)	Materials	Schottky barrier height (eV)
TiSi2	0.60	HfSi	0.53
VSi2	0.64	TaSi2	0.59
CrSi2	0.57	WSi2	0.65
MnSi	0.65	ReSi2	0.77
MnSi1.7	0.67	OsSi1.8	0.85
FeSi2	0.68	IrSi	0.93
CoSi	0.68	Ir2Si 3	0.85
CoSi2	0.64	IrSi3	0.94
Ni2Si	0.66	Pt2Si	0.85
NiSi	0.65	PtSi	0.88
NiSi 2	0.66	SiAu	0.80
YSi1.7	0.39	GdSi2	0.37
ZrSi2	0.55	DySi 2	0.37
NbSi2	0.63	HoSi2	0.37
MoSi2	0.65	ErSi2	0.39
Ru2Si3	0.72	ErSi1.7	0.40
RhSi	0.75		
Pd2Si	0.71		

Source: Adapted from Ref. 22

10.7 Electronic Properties of Silicide NWs and Silicide/Si/Silicide Heterostructures

10.7.1 The Resistivity of Silicide Materials

The intrinsic resistance of various silicide NWs have been investigated for its potential application as device contact in the silicide/Si/silicide heterostructures.[44,46,49] For instance, using four-terminal measurement (Fig. 10.10a,b), it has been shown that the resistivity of NiSi NW is around 9.5 μΩ·cm (Fig. 10.10a), close to the value of NiSi single crystal. The resistivity of PtSi silicide NW is found to be ~28.6 μΩ·cm, which compares favorably to the previously reported values in PtSi thin film materials (45–68 μΩ·cm).[69] Different low-resistance silicide NWs have been synthesized and studied such as Ni_2Si (20 μΩ·cm),[70] $Ni_{31}Si_{12}$ (51 μΩ·cm),[71] $NiSi_2$ (30 μΩ·cm),[72] Ni_3Si (72 μΩ·cm),[73] Co_2Si (213 μΩ·cm)[38] and CoSi (510 μΩ·cm).[74] In addition, these silicides usually show a high breakdown current $>10^8$ A/cm^2,[44,49] which makes them suitable to serve as the contact materials. In order to achieve the high device performance, strain Si is implemented to enhance the carrier mobility. Habicht *et al.*[75] shows that the contact resistivity of NiSi/Si will not change much in the SSOI (stressed Si on insulator) substrate (Fig. 10.10c). These results also indicated that nickel silicide can keep the low contact resistance and applied to the high-performance devices.

10.7.2 Low-Resistivity Contacts: Ohmic Contacts

Silicide/Si/Silicide heterostructures have been explored for the applications of FETs. Weber *et al.*[46] have constructed an intrinsic Si NW FET based on $NiSi_2$/intrinsic-Si/$NiSi_2$ heterostructure and demonstrated a high on/off ratio ~10^7. Wu *et al.*[49] explored the Ni silicide/*p*-Si heterostructure and obtained high on-current and good on-off ratio ~10^6. However, without high doping region around the contact area in Si NWs, both saturation current and transconductance are limited due to the large Schottky barrier between metal or silicide contacts to Si NWs.

On the other hand, silicide with suitable work function can reduce the contact Schottky barrier height and make a good Ohmic contact to the Si material (Table 10.2). Lin *et al.*[44] reported that a normal-off *p*-channel device can be fabricated using intrinsic Si NW

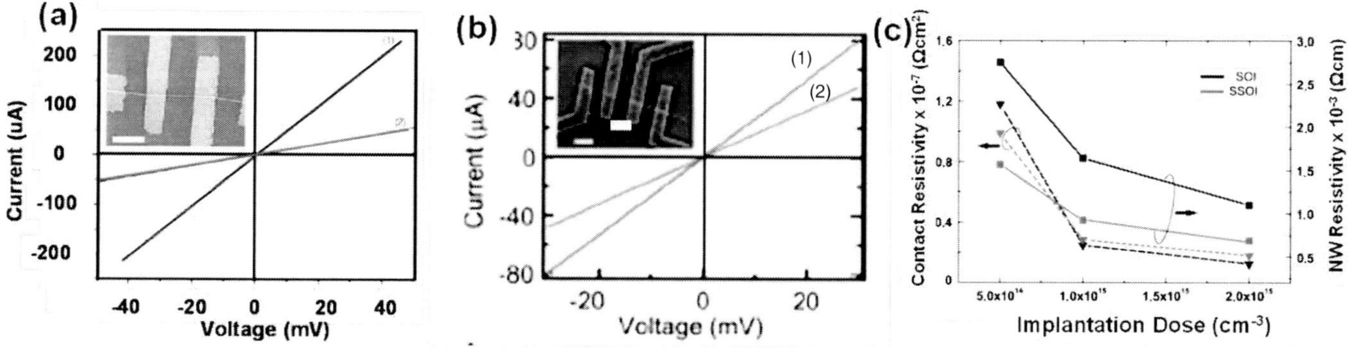

Figure 10.10 Resistance measurements on individual single-crystal silicide NWs.[44,49,75] (a) I–V of a 29 nm NiSi NW, with (1) and (2) lines corresponding to four- and two-terminal measurements, respectively. The scale bar in the image is 1 μm. (b) I–V of a 50 nm diameter PtSi NW device. The scale bar is 3 μm). (c) Resistivity of SOI and SSOI NWs (Stressed Si: 0.8% elastic strain) and the NiSi/Si contact resistivity *as* a function of As+ ion implantation dose. The width of the NWs was 80 nm. Reprinted with permission from American Chemical Society, Copyright (2008).

without the high doping concentration exploiting PtSi as the contact material. PtSi is used to make a *p*-channel device because of its small barrier height ~0.2 eV for hole transport in intrinsic Si, as shown in the schematic in Fig. 10.11a. Additionally, the formation of atomically clean PtSi/Si interface prevents Fermi level pinning and helps to maintain a low Schottky barrier determined by the work function rather than by interface states. This has enabled them to fabricate high-performance NW transistors from intrinsic Si NWs using PtSi as the source and drain contacts. To make the device (Fig. 10.11b), a 7 nm thick layer of HfO_2 was deposited on the surface of PtSi/ Si/ PtSi heterostructure device using atomic layer deposition process to serve as gate dielectrics, and a top Cr/Au electrode was defined with e-beam lithography and deposited using e-beam evaporation.

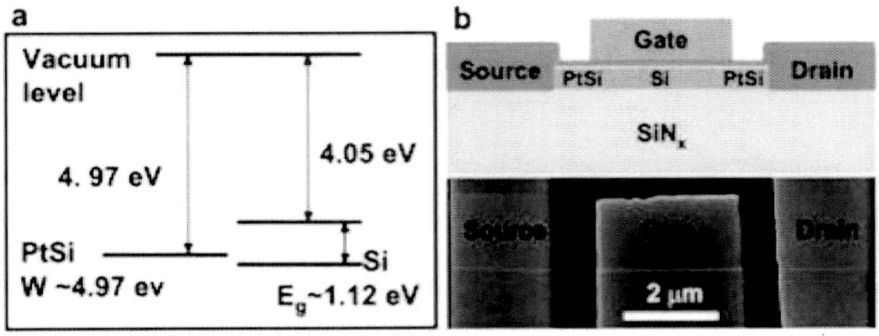

Figure 10.11 PtSi/i-Si/PtSi NW heterostructures as high-performance *p*-channel enhancement mode transistors.[44] (a) Relative energy band alignment between PtSi and Si. (b) (top) Schematic of PtSi/i-Si/PtSi NW heterostructure device with HfO_2 ~7 nm as gate dielectrics. (bottom) SEM images of a device (scale bar, 2 μm). Reprinted with permission from American Chemical Society, Copyright (2008).

Electrical measurement shows nearly perfect transistor characteristics. Figure 10.12a shows good Ohmic contact properties in the lower V_d biasing region and large saturation current. In the I_d–V_g curve (Fig. 10.12b), the device shows the ambipolar transport and that it is a *p*-type enhanced mode transistor with a high on-off ratio ~10^7. These results clearly demonstrate that a high-performance *p*-channel enhancement mode transistor has been obtained from a PtSi/Si/PtSi heterostructure using intrinsic Si NW. A normalized transconductance of ~0.3 mS/μm and an impressive 110 mV/decade subthreshold swing at the current minimum has been reported, representing the best-performed intrinsic-Si NW FET at the time. In current NW technology, highly doped NWs are usually used to reduce the Schottky barrier width to achieve Ohmic contact. However the dopant distribution in NWs so far is difficult to control and a large reverse gate voltage is normally necessary to turn off these heavily doped devices. Thus, the PtSi contact approach is an alternative approach to make Ohmic contact and achieve the normal-off and high on/off ratio silicide heterostructure FETs.

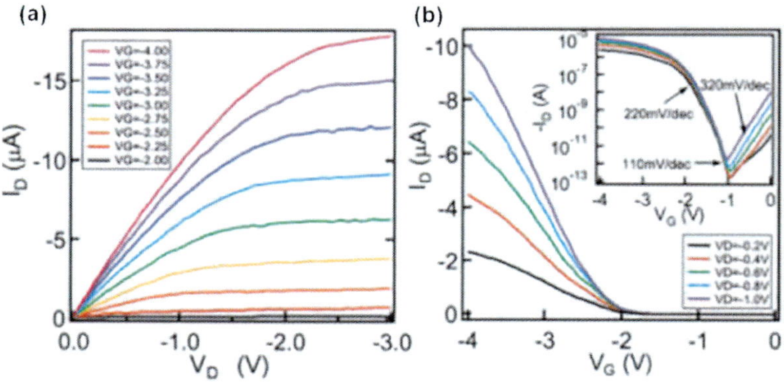

Figure 10.12 PtSi/i-Si/PtSi NW heterostructures as high-performance *p*-channel enhancement mode transistors.[44] (a) I_d vs. V_d increase with negative gate voltages (V_{gs}) (b) I_d vs. V_g. The inset shows $-I_d$ vs. V_g in the exponential scale, highlighting the on/off ratio >10^7, and showing ambipolar transport properties. Reprinted with permission from American Chemical Society, Copyright (2008).

10.7.3 Conductive Contacts and Beyond: Magnetic Contacts and Schottky Contacts

The formation of silicide/Si/silicide NW heterostructure discussed above is an interesting approach to make nearly perfect contact to Si NWs due to the single crystalline and the few-defect interface properties. These unique properties and structures may lead to less transport carrier scattering compared to current thin film structure, which can be explored for coherent spin transport in spintronic applications. Si-based spintronics attracts much interest for its excellent properties such as long spin relaxation time, large diffusion length, as well as its compatibility with industrial process. Exploiting ferromagnetic silicide, such as manganese monosilicide (MnSi), the nanosilicide technique can be used to create single-crystal MnSi/Si/MnSi NW heterostructure with nearly perfect ferromagnetic-semiconductor interface and variable Si channel length, and therefore allow the spin transport studies in short-channel Si NW devices with Schottky tunneling junction.

Theoretical studies have suggested that up to 45% spin polarization can be achieved in MnSi ferromagnetic state, suggesting that this material may function as an effective ferromagnetic contact for spin injection in nanoscale Si devices.[77] Thus, MnSi attracts much more interests for spintronics in recent years. Higgins *et al.*[76] investigated the transport properties of CVD MnSi NW (the diameter ~252 nm). In their studies, they found the magnetic ordering temperature (paramagnetic to helicalmagnetic) is around 35 K, which was reported in the bulk or thin film systems (Fig. 10.13a,b). Additionally, the transition field from helical magnetic state to ferromagnetic state is also observed at around 0.6 T. The signature of helical ordering and field transition further indicated that MnSi NW ~252 nm has the similar properties compared with bulk MnSi. Interestingly, Lin *et al.*[36] studied the MnSi nanowrie (~50 nm), formed by solid-state reaction method (Fig. 10.13c,d). In their results, the MnSi NW shows a magnetic ordering at ~30 K and negative magnetoresistance. However, the maximum magnetoresistance happens at lower temperature and saturates below 10 K, which is different from the bulk systems and Higgins *et al.*'s results (~30 K). A small magnetoresistance hysteresis is also observed under the sweeping magnetic field. In Lin's studies, MnSi seems to behave as the ferromagnet under the low temperature which also verifies that MnSi has the spin polarization

To explore the spin injection and detection in the Si NWs, Lin *et al.* fabricated a MnSi/*p*-Si/MnSi heterostructure device and investigated the spin transport behavior through this structure. The equivalent circuit of a MnSi/*p*-Si/MnSi NW heterostructure comprises of two Schottky diodes connected back to back (Fig. 10.14a). The work function difference suggests a Schottky barrier ~0.45 eV between MnSi and *p*-type Si (Fig. 10.14a). When the bias voltage is across the whole junction, the device displays typical charge transport characteristics across a barrier and the transport behavior do not vary much with temperature range (Fig. 10.14b).

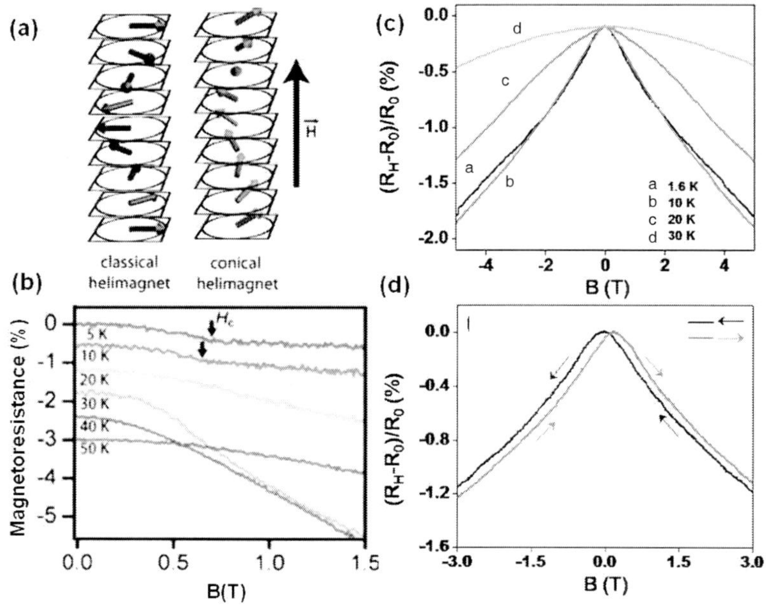

Figure 10.13 Helimagnets and magnetotransport of MnSi NW devices.[36,76] (a) helimagnetic ordering (left) and a conical helimagnet (right) produced by an applied magnetic field. (b) Magnetoresistance as a function of magnetic field at various temperatures above and below the transition temperature (offset for clarity). The diameter of MnSi NW is around 252 nm. (c) Magnetoresistance of 50 nm MnSi NW, forming by solid-state reaction, at different temperatures. (d) Magnetoresistance hysteresis of 50 nm MnSi NW at 1.6 K. The arrows indicate the sweeping direction of magnetic field. Reprinted with permission from American Chemical Society, Copyright (2010).

A positive bias applied to one MnSi contact can cause hole injection through the MnSi/Si Schottky diode in the reverse bias regime (Fig. 10.14c,d). When the biasing voltage is large enough (>0.5 V), the I_{rev} shows little dependence on temperature, indicating that the thermal emission is no longer dominant carrier transport mechanism (Fig. 10.14c,d). Instead, the tunneling mechanism takes over in charge transport, where the coherent spin transport can happen.

Indeed, when the bias is higher ~ +1.0 V (tunneling regime), a region with higher resistance is observed at low magnetic field (Fig. 10.15a,b). This behavior is similar to the giant magnetoresistance (GMR) effect reported in Ni/Si/Ni Schottky tunneling junction due to the spin carriers scattering.[78] If the spin orientation of the injection and detection contact is the same, spin carriers will flow through the device without being scattered, leading to a lower resistance. Otherwise, if the spin carriers in the Si NW encounter a different spin orientation at the detection ferromagnetic contact, they will be scattered at the detection contact, leading to an abrupt increase in resistance. According to these studies, the hysteresis and spin scattering (GMR) signal clearly demonstrate electrical spin injection and detection in Si NW devices via a Schottky tunneling barrier with the ferromagnetic MnSi contact. This research demonstrates the versatile opportunities for the silicide/Si-based NW structure, indicating more device functions achievable with the nanosilicide technology.

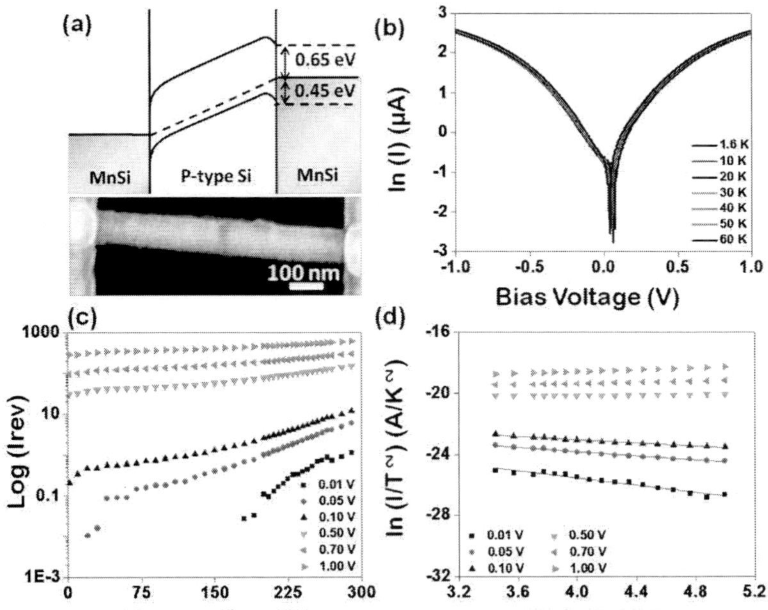

Figure 10.14 Transport properties of MnSi/p-Si/MnSi heterojunction.[36] (a) Band diagram of MnSi/p-Si/MnSi heterostructure. The positive bias is applied at the left end of MnSi. A Schottky barrier ~0.45 eV is predicted for hole injection into p-Si. The SEM images show a MnSi/p-Si/MnSi heterojunction. The Si gap is ~20 nm. (b) I–V characteristics of heterostructure at various temperatures. (c) Current versus temperature dependence at different reverse biasing voltages. (d) Arrhenius plot for heterostructure at various biasing voltages between 200 K and 290 K. Reprinted with permission from American Chemical Society, Copyright (2010).

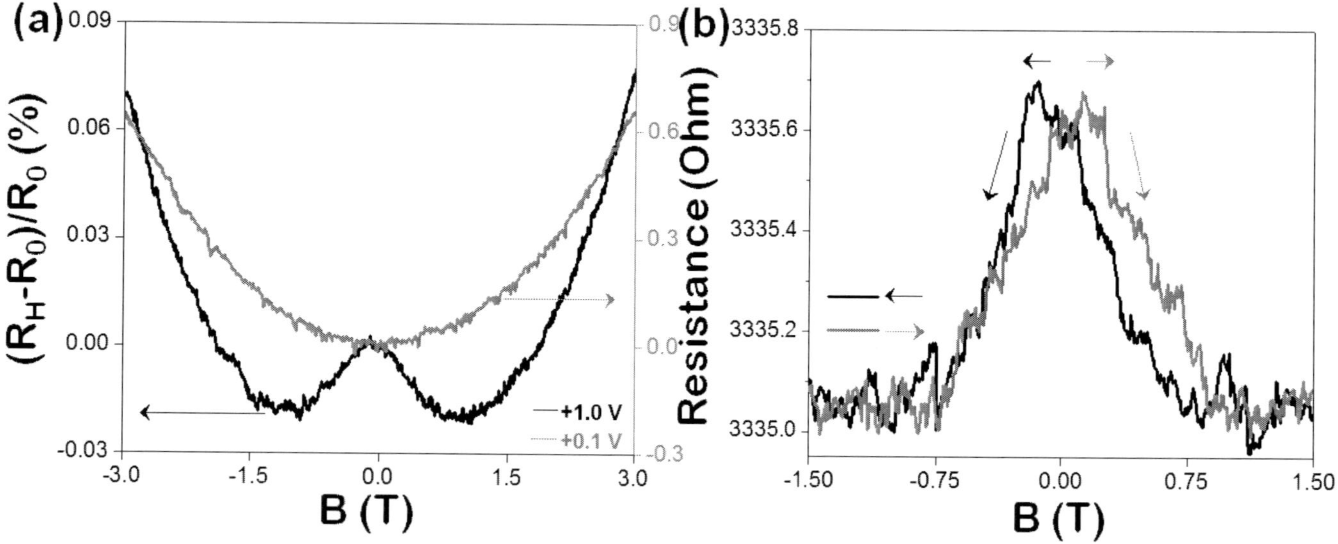

Figure 10.15 Transport properties of MnSi/*p*-Si/MnSi heterojunction.[36] (a) Magnetoresistance of heterostructure versus magnetic field. R_H is the resistance of the heterostructure device at the certain magnetic field. R_0 represents the device resistance at zero field. (b) Resistance change of heterostructure as a function of magnetic field. Reprinted with permission from American Chemical Society, Copyright (2010).

10.7.4 High-Mobility Field-Effect Transistor and Short-Channel Device

With advances to the high-performance FET, silicide/Si heterostructure is adopted as a channel scaling technology. Ni silicide can grow and advance into the Si NW to form a silicide/Si/silicide heterostructure where Si channel region can be as small as ~2 nm,[58] while keeping sharp silicide/Si interface. Hu *et al.*[79] utilized the intrinsic Ge/Si core–shell NW structure (with high hole-carrier mobility) and Ni silicide technology (for channel scaling) to achieve the high-performance transistor. In the intrinsic Ge/Si core–shell NW, high-mobility free hole gas can be confined around the Ge/Si interface sand serving as the conducting channel. The Ni electrodes are put down and annealed to form germanide/silicide structure into the core–shell NW. Similar to the formation of silicide/Si/silicide, $NiGe_xSi_y$ (Ni germanide silicide) can grow and advance into the Ge/Si core–shell NWs from the Ni contact pads, and scale the Ge/Si channel to deep nanometer region (Fig. 10.16a). 70 and 40 nm channel length devices are both achieved and studied (Fig. 16b–e). A large normalized transconductance ~6.2 μS/μm for 40-nm device is achieved adopting this structure. This is the highest value reported for the Si-based device with the comparable channel length. The I_{on} and g_m of 40 nm device are superior to 70 nm device. By calculating the intrinsic delay, the 40 nm device can operate with 2 THz intrinsic speed that represents the highest frequency among NWFETs and *p*-MOSFETs. This work demonstrates yet another example of the great promises of nanosilicide technology in high-performance nanodevice applications.

10.8 Conclusion

Nanoscale metal silicides have garnered significant interest during the past few years. The fact that silicides can form a clean, flat and sharp single crystal epitaxial interface with Si NWs through solid-state reaction mechanism especially gained them fame for their superior function as the perfect contact material for Si NW based devices. Within this framework, various silicide/Si/silicide heterostructures have been fabricated through lateral diffusion of metal atoms from the metal contacts into Si NWs and the subsequence transformation

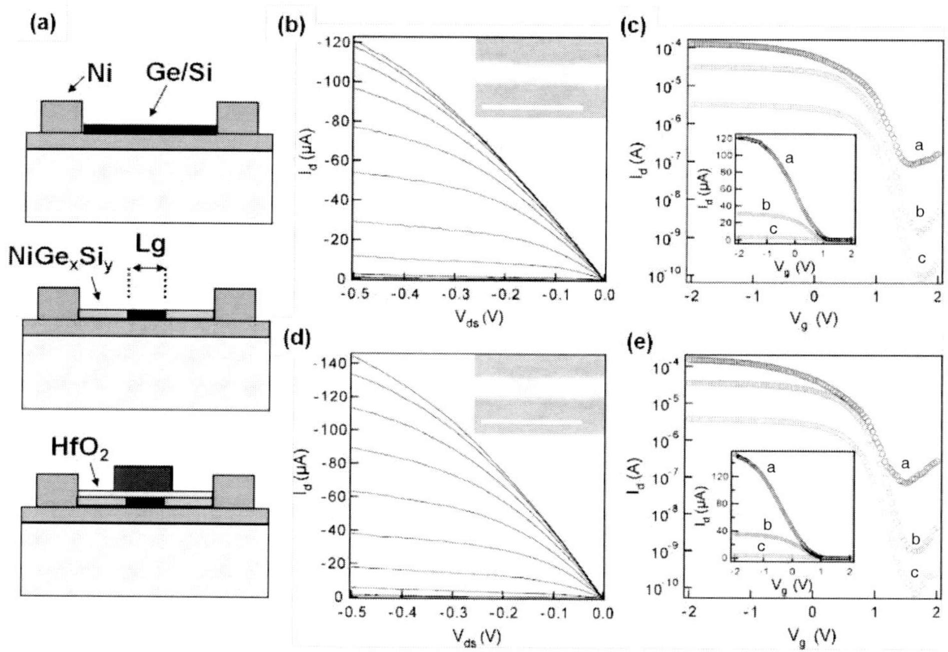

Figure 10.16 Transport properties of Ge/Si core–shell NW FET.[79] (a) Fabrication of sub-100 nm short-channel device. (b–d) Transport characteristics of sub-100 nm channel length Ge/Si NWFETs. (b,d) I_d– V_{ds} data for L ~70 and 40 nm devices, respectively; V_g varies from −2 to 2 V in increments of 0.4 V from top to bottom. The insets show respective SEM images of the 70 and 40 nm devices; scale bars are 100 nm in both images. (c,e) I_d–V_g data for L ~70 and 40 nm devices, respectively. Insets show data on linear scale. The blue, red, and green data points correspond to V_{ds} = −10, −100, and −500 mV, respectively. Reprinted with permission from American Chemical Society, Copyright (2008).

of Si to silicides. Exploring this structure, high-performance *p*-MOSFET has been demonstrated with intrinsic Si NWs with high work function PtSi contact, spin transport has been detected in the MnSi/*p*-Si/MnSi heterostructures, and a high-performance high-speed *p*-MOSFET has been achieved with NiGe*x*Si*y*/(Ge/Si)/NiGe*x*Si*y* structure with sub-100 nm Ge/Si channel. These works demonstrate the bright future lying ahead for nanosilicide technology, as well as raise the need for better understanding and control of the silicide formation at nanoscale. To date, studies on the diffusion and phase transformation processes of silicides in Si NWs have indicated possible deviations from the well-established silicide formation techniques in bulk and the thin film system. Although a few mechanisms have been proposed, a systematic understanding is not yet available to explain all observed experimental results. While advances in Si NW device applications is much desired, systematic study and rational control of the grow kinetics of nanosilicides is equally important.

References

1. J. P. Colinge, I. Ferain, A. Kranti, C. W. Lee, N. D. Akhavan, P. Razavi, R. Yan, and R. Yu, *Sci. Adv. Mater.*, 2011**, 3**, 477–482.
2. J. P. Colinge, C. W. Lee, A. Afzalian, N. D. Akhavan, R. Yan, I. Ferain, *et al.*, *Nat. Nanotechnol.*, 2010, **5**, 225–229.
3. A. M. Ionescu, *Nat. Nanotechnol.*, 2010, **5**, 178–179.
4. F. Deng, R. A. Johnson, P. M. Asbeck, S. S. Lau, W. B. Dubbelday, T. Hsiao, and J. Woo, *J. Appl. Phys.*, 1997, **81**, 8047–8051.
5. H. Iwai, T. Ohguro, and S.-i. Ohmi, *Microelectron. Eng*, 2002, **60**, 157–169.
6. H. Jiang, C. M. Osburn, Z. G. Xiao, G. McGuire, and G. A. Rozgonyi, *J. Electrochem. Soc.*, 1992, **139**, 211–218.
7. J. M. Larson and J. P. Snyder, *IEEE Trans. Electron. Devices*, 2006, **53**, 1048–1058.
8. P. G. McCafferty,A. Sellai, P. Dawson and H. Elabd, *Solid State Electron.*, 1996, **39**, 583–592.
9. T. Morimoto, *IEEE Trans. Electron. Devices*, 1995, **42**, 915–922.
10. K. N. Tu, W. N. Hammer, and J. O. Olowolafe, *J. Appl. Phys.*, 1980, **51**, 1663–1668.

11. V. Bellani, G. Guizzetti, F. Marabelli, A. Piaggi, A. Borghesi, F. Nava, V. N. Antonov, O. Jepsen, O. K. Andersen, and V. V. Nemoshkalenko, *Phys. Rev. B*, 1992, **46**, 9380–9389.
12. D. Leong, M. Harry, K. J. Reeson, and K. P. Homewood, *Nature*, 1997, **387**, 686–688.
13. Q. R. Hou, Z. M. Wang, and Y. J. He, *Appl. Phys. a-Mater.*, 2005, **80**, 1807–1811.
14. Y. Ichikawa, M. Kobayashi, M. Sasase, and T. Suemasu, *Appl. Surf. Sci.*, 2008, **254**, 7963–7967.
15. T. Saito, Y. Matsumoto, M. Suzuno, M. Takeishi, R. Sasaki, T. Suemasu, and N. Usami, *Appl. Phys. Express*, 2010, **3**, 068001.
16. T. Suemasu, K. Morita, and M. Kobayashi, *J. Cryst. Growth*, 2007, **301**, 680–683.
17. J. Herfort, H. P. Schonherr, and K. H. Ploog, *Appl. Phys. Lett.*, 2003, **83**, 3912–3914.
18. Y. Sakuraba, M. Hattori, M. Oogane, Y. Ando, H. Kato, A. Sakuma, T. Miyazaki, and H. Kubota, *Appl. Phys. Lett.*, 2006, **88**, 192508.
19. J. P. DeGrave, A. L. Schmitt, R. S. Selinsky, J. M. Higgins, D. J. Keavney, and S. Jin, *Nano Lett.*, 2011, **11**, 4431–4437.
20. A. I. Boukai, Y. Bunimovich, J. Tahir-Kheli, J. K. Yu, W. A. Goddard, and J. R. Heath, *Nature*, 2008, **451**, 168–171.
21. A. L. Schmitt, J. M. Higgins, and S. Jin, *Nano Lett.*, 2008, **8**, 810–815.
22. E. Bucher, S. Schulz, M. C. Luxsteiner, P. Munz, U. Gubler, and F. Greuter, *Appl. Phys. a-Mater.*, 1986, **40**, 71–77.
23. S. P. Murarka, *J. Vac. Sci. Technol.*, 1980, **17**, 775–792.
24. Q. Huang, A. S. Tan, J. M. Tan, I. S. Goh, Z. Z. Dong, C. K. On, and T. Osipowicz, *Appl. Phys. a-Mater.*, 2003, **76**, 439–443.
25. M. Deisenberg, and K. N. Tu, *J. Appl. Phys.*, 1982, **53**, 6885–6890.
26. M. Q. Huda, and K. Sakamoto, *Mat. Sci. Eng. B-Solid*, 2002, **89**, 378–381.
27. S. J. Oh, J. W. Allen, and J. M. Lawrence, *Phys. Rev B*, 1987, **35**, 2267–2272.
28. K. Firmest, J. F. Disuse, E. Bucher, and G. Apply, *Ann. Phys-Leipzig*, 1996, **5**, 175–183.
29. J. Thomas, J. Schumann, H. Vineberg, E. Arushanov, R. Engelhard, O. G. Schmidt, and T. Gemming, *Nanotechnology*, 2009, **20**, 235604.
30. Z. A. He, D. J. Smith, and P. A. Bennett, *Phys. Rev. Lett.*, 2004, **93**, 256102.

31. P. A. Bennett, Z. He, D. J. Smith, and F. M. Ross, *Thin Solid Films*, 2011, **519**, 8434–8440.

32. H. C. Hsu, W. W. Wu, H. F. Hsu, and L. J. Chen, *Nano Lett*, 2007, **7**, 885–889.

33. Y. Ohira, T. Tanji, M. Yoshimura, and K. Ueda, *Jpn. J. Appl. Phys.*, 2008, **47**, 6138–6141.

34. R. T. Tung, *Mater. Chem. Phys.*, 1992, **32**, 107–133.

35. A. L. Schmitt, M. J. Bierman, D. Schmeisser, F. J. Himpsel, and S. Jin, *Nano Lett.*, 2006, **6**, 1617–1621.

36. Y. C. Lin, Y. Chen, A. Shaios, and Y. Huang, *Nano Lett.*, 2010, **10**, 2281–2287.

37. K. Kang, S. K. Kim, C. J. Kim, and M. H. Jo, *Nano Lett.*, 2008, **8**, 431–436.

38. Y. Q. Qu, J. W. Bai, L. Liao, R. Cheng, Y. C. Lin, Y. Huang, T. Guo, and X. F. Duan, *Chem. Commun.*, 2011, **47**, 1255–1257.

39. K. Seo, S. Lee, H. Yoon, J. In, K. S. K. Varadwaj, Y. Jo, M. H. Jung, J. Kim, and B. Kim, *ACS Nano.*, 2009, **3**, 1145–1150.

40. K. Seo, H. Yoon, S. W. Ryu, S. Lee, Y. Jo, M. H. Jung, J. Kim, Y. K. Choi, and B. Kim, *ACS Nano.*, 2010, **4**, 2569–2576.

41. Y. C. Chou, W. W. Wu, L. J. Chen, and K. N. Tu, *Nano Lett.*, 2009, **9**, 2337–2342.

42. Y. C. Chou, W. W. Wu, S. L. Cheng, B. Y. Yoo, N. Myung, L. J. Chen, and K. N. Tu, *Nano Lett.*, 2008, **8**, 2194–2199.

43. Y. C. Lin, Y. Chen, D. Xu, and Y. Huang, *Nano Lett.*, 2010, **10**, 4721–4726.

44. Y. C. Lin, K. C. Lu, W. W. Wu, J. W. Bai, L. J. Chen, K. N. Tu, and Y. Huang, *Nano Lett.*, 2008, **8**, 913–918.

45. K. C. Lu, K. N. Tu, W. W. Wu, L. J. Chen, B. Y. Yoo, and N. V. Myung, *Appl. Phys. Lett.*, 2007, **90**, 253111.

46. W. M. Weber, L. Geelhaar, A. P. Graham, E. Unger, G. S. Duesberg, M. Liebau, W. Pamler, C. Cheze, H. Riechert, P. Lugli, and F. Kreupl, *Nano Lett.*, 2006, **6**, 2660–2666.

47. J. Appenzeller, J. Knoch, M. I. Bjork, H. Riel, H. Schmid, and W. Riess, *IEEE T Electron. Dev.*, 2008, **55**, 2827–2845.

48. M. Zhang, J. Knoch, J. Appenzeller, and S. Mantl, *IEEE Electron. Device L*, 2007, **28**, 223–225.

49. Y. Wu, J. Xiang, C. Yang, W. Lu, and C. M. Lieber, *Nature*, 2004, **430**, 61–65.

50. W. W. Wu, K. C. Lu, K. N. Chen, P. H. Yeh, C. W. Wang, Y. C. Lin, and Y. Huang, *Appl. Phys. Lett.*, 2010, **97**, 203110.

51. Y. C. Chou, W. W. Wu, C. Y. Lee, C. Y. Liu, L. J. Chen, and K. N. Tu, *J. Phys. Chem. C*, 2011, **115**, 397–401.

52. L. R. Zheng, L. S. Hung, and J. W. Mayer, *J. Vac. Sci. Technol. A*, 1983, **1**, 758–761.

53. C. Lavoie, F. M. d'Heurle, C. Detavernier, and C. Cabral, *Microelectron. Eng.*, 2003, **70**, 144–157.

54. P. Gas, *Appl. Surf. Sci.*, 1989, **38**, 178–184.

55. J. H. Gulpen, A. A. Kodentsov, and F. J. J. Vanloo, *Zeitschrift Fur Metallkunde*, 1995, **86**, 530–539.

56. C. Hayzelden, and J. L. Batstone, *J. Appl. Phys.*, 1993, **73**, 8279–8289.

57. F. H. M. Spit, D. Gupta, and K. N. Tu, *Phys. Rev. B*, 1989, **39**, 1255–1260.

58. K. C. Lu, W. W. Wu, H. W. Wu, C. M. Tanner, J. P. Chang, L. J. Chen, and K. N. Tu, *Nano Lett.*, 2007, **7**, 2389–2394.

59. N. S. Dallas, B. Z. Liu, S. M. Richfield, C. M. Richfield, T. S. Mayer, and S. E. Money, *J. Appl. Phys.*, 2009, **105**, 094309.

60. S. M. Woodruff, N. S. Dellas, B. Z. Liu, S. M. Eichfeld, T. S. Mayer, J. M. Redwing and S. E. Mohney, *J. Vac. Sci. Technol. B*, 2008, **26**, 1592–1596.

61. Y. E. Yaish, A. Katsman, G. M. Cohen, and M. Beregovsky, *J. Appl. Phys.*, 2011, **109**, 094303.

62. O. V. Aleksandrov and V. V. Kozlovski, *Semiconductors*, 2009, **43**, 885–891.

63. A. Katsman, Y. Yaish, E. Rabkin, and M. Beregovsky, *J. Electron. Mater.*, 2010, **39**, 365–370.

64. L. J. Chen, H. C. Hsu, W. W. Wu, and H. F. Hsu, *Nano Lett.*, 2007, **7**, 885–889.

65. L. J. Chen, and W. W. Wu, *Mat. Sci. Eng. R*, 2010, **70**, 303–319.

66. S. Hofmann, R. Sharma, C. T. Wirth, F. Cervantes-Sodi, C. Ducati, T. Kasama, R. E. Dunin-Borkowski, J. Drucker, P. Bennett, and J. Robertson, *Nat. Mater.*, 2008, **7**, 372–375.

67. K. Ogata,E. Sutter, X. Zhu, and S. Hofmann, *Nanotechnology*, 2011, **22**, 365305.

68. J. Kedzierski, P. Xuan, E. H. Anderson, J. Bokor, K. Tsu-Jae, and H. Chenming. In: *Complementary Silicide Source/Drain Thin-Body MOSFETs for the 20 nm Gate Length Regime*, Electron Devices Meeting, 2000. IEDM Technical Digest. International, 2000, 57–60.

69. D. Xu, J. P. McCaffrey, S. R. Das, G. C. Aers, and L. E. Erickson, *Appl. Phys. Lett.*, 1996, **68**, 3588–3590.

70. Y. P. Song, A. L. Schmitt, and S. Jin, *Nano Lett.*, 2007, **7**, 965–969.

71. C. Y. Lee, M. P. Lu, K. F. Liao, W. W. Wu, and L. J. Chen, *Appl. Phys. Lett.*, 2008, **93**, 113109.

72. C. Y. Lee, M. P. Lu, K. F. Liao, W. F. Lee, C. T. Huang, S. Y. Chen, and L. J. Chen, *J. Phys. Chem. C*, 2009, **113**, 2286–2289.

73. Y. P. Song and S. Jin, *Appl. Phys. Lett.*, 2007, **90**, 113109.

74. A. L. Schmitt, L. Zhu, D. Schmeisser, F. J. Himpsel, and S. Jin, *J. Phys. Chem. B*, 2006, **110**, 18142–18146.

75. S. Habicht, Q. T. Zhao, S. F. Feste, L. Knoll, S. Trellenkamp, B. Ghyselen, and S. Mantl, *Nanotechnology*, 2010, **21**, 105701.

76. J. M. Higgins, R. H. Ding, J. P. DeGrave, and S. Jin, *Nano Lett.*, 2010, **10**, 1605–1610.

77. H. Wu, M. Hortamani, P. Kratzer, and M. Scheffler, *Phys Rev. Lett.*, 2004, **92**, 237202.

78. Y. Q. Jia, R. C. Shi, and S. Y. Chou, *IEEE Trans. Magn.*, 1996, **32**, 4707–4709.

79. Y. J. Hu, J. Xiang, G. C. Liang, H. Yan, and C. M. Lieber, *Nano Lett.*, 2008, **8**, 925–930.

Index